New Trends in Diabetes, Hypertension and Cardiovascular Diseases

New Trends in Diabetes, Hypertension and Cardiovascular Diseases

Editors

Yutang Wang
Dianna Magliano

Basel • Beijing • Wuhan • Barcelona • Belgrade • Novi Sad • Cluj • Manchester

Editors
Yutang Wang
Federation University
Australia
Ballarat
Australia

Dianna Magliano
Monash University
Melbourne
Australia

Editorial Office
MDPI
St. Alban-Anlage 66
4052 Basel, Switzerland

This is a reprint of articles from the Special Issue published online in the open access journal *International Journal of Molecular Sciences* (ISSN 1422-0067) (available at: https://www.mdpi.com/journal/ijms/special_issues/9S2VWC1OY8).

For citation purposes, cite each article independently as indicated on the article page online and as indicated below:

Lastname, A.A.; Lastname, B.B. Article Title. *Journal Name* **Year**, *Volume Number*, Page Range.

ISBN 978-3-7258-0537-2 (Hbk)
ISBN 978-3-7258-0538-9 (PDF)
doi.org/10.3390/books978-3-7258-0538-9

Contents

Yutang Wang and Dianna J. Magliano
Special Issue: "New Trends in Diabetes, Hypertension, and Cardiovascular Diseases"
Reprinted from: *Int. J. Mol. Sci.* **2024**, 25, 2711, doi:10.3390/ijms25052711 1

Racha El Hage, Nada Al-Arawe and Irene Hinterseher
The Role of the Gut Microbiome and Trimethylamine Oxide in Atherosclerosis and Age-Related
Disease
Reprinted from: *Int. J. Mol. Sci.* **2023**, 24, 2399, doi:10.3390/ijms24032399 4

**Iwona Świątkiewicz, Marcin Wróblewski, Jarosław Nuszkiewicz, Paweł Sutkowy,
Joanna Wróblewska and Alina Woźniak**
The Role of Oxidative Stress Enhanced by Adiposity in Cardiometabolic Diseases
Reprinted from: *Int. J. Mol. Sci.* **2023**, 24, 6382, doi:10.3390/ijms24076382 28

**Daniela Maria Tanase, Emilia Valasciuc, Evelina Maria Gosav, Anca Ouatu,
Oana Nicoleta Buliga-Finis, Mariana Floria, et al.**
Portrayal of NLRP3 Inflammasome in Atherosclerosis: Current Knowledge and Therapeutic
Targets
Reprinted from: *Int. J. Mol. Sci.* **2023**, 24, 8162, doi:10.3390/ijms24098162 52

**Stanley T. Lewis, Frank Greenway, Tori R. Tucker, Michael Alexander, Levonika K. Jackson,
Scott A. Hepford, et al.**
A Receptor Story: Insulin Resistance Pathophysiology and Physiologic Insulin Resensitization's
Role as a Treatment Modality
Reprinted from: *Int. J. Mol. Sci.* **2023**, 24, 10927, doi:10.3390/ijms241310927 81

Yunwen Hu, Zhaohua Cai and Ben He
Smooth Muscle Heterogeneity and Plasticity in Health and Aortic Aneurysmal Disease
Reprinted from: *Int. J. Mol. Sci.* **2023**, 24, 11701, doi:10.3390/ijms241411701 92

**Nikolaos Koumallos, Evangelia Sigala, Theodoros Milas, Nikolaos G. Baikoussis,
Dimitrios Aragiannis, Skevos Sideris and Konstantinos Tsioufis**
Angiotensin Regulation of Vascular Homeostasis: Exploring the Role of ROS and RAS Blockers
Reprinted from: *Int. J. Mol. Sci.* **2023**, 24, 12111, doi:10.3390/ijms241512111 114

Roman Roy, Joshua Wilcox, Andrew J. Webb and Kevin O'Gallagher
Dysfunctional and Dysregulated Nitric Oxide Synthases in Cardiovascular Disease:
Mechanisms and Therapeutic Potential
Reprinted from: *Int. J. Mol. Sci.* **2023**, 24, 15200, doi:10.3390/ijms242015200 127

**Janette van der Linden, Lianne Trap, Caroline V. Scherer, Anton J. M. Roks, A. H. Jan Danser,
Ingrid van der Pluijm and Caroline Cheng**
Model Systems to Study the Mechanism of Vascular Aging
Reprinted from: *Int. J. Mol. Sci.* **2023**, 24, 15379, doi:10.3390/ijms242015379 148

**Tatiana Ivanova, Maria Churnosova, Maria Abramova, Denis Plotnikov, Irina Ponomarenko,
Evgeny Reshetnikov, et al.**
Sex-Specific Features of the Correlation between GWAS-Noticeable Polymorphisms and
Hypertension in Europeans of Russia
Reprinted from: *Int. J. Mol. Sci.* **2023**, 24, 7799, doi:10.3390/ijms24097799 168

Irina Shilenok, Ksenia Kobzeva, Tatiana Stetskaya, Maxim Freidin, Maria Soldatova, Alexey Deykin, et al.
SERPINE1 mRNA Binding Protein 1 Is Associated with Ischemic Stroke Risk: A Comprehensive Molecular–Genetic and Bioinformatics Analysis of *SERBP1* SNPs
Reprinted from: *Int. J. Mol. Sci.* **2023**, *24*, 8716, doi:10.3390/ijms24108716 . . . . **195**

Barbara Toffoli, Federica Tonon, Fabiola Giudici, Tommaso Ferretti, Elena Ghirigato, Matilde Contessa, et al.
Preliminary Study on the Effect of a Night Shift on Blood Pressure and Clock Gene Expression
Reprinted from: *Int. J. Mol. Sci.* **2023**, *24*, 9309, doi:10.3390/ijms24119309 . . . . **216**

Dmitry Yu. Oshchepkov, Yulia V. Makovka, Mikhail P. Ponomarenko, Olga E. Redina and Arcady L. Markel
Age-Dependent Changes in the Relationships between Traits Associated with the Pathogenesis of Stress-Sensitive Hypertension in ISIAH Rats
Reprinted from: *Int. J. Mol. Sci.* **2023**, *24*, 10984, doi:10.3390/ijms241310984 . . . . **230**

Mohamed Abo-Aly, Elica Shokri, Lakshman Chelvarajan, Wadea M. Tarhuni, Himi Tripathi and Ahmed Abdel-Latif
Prognostic Significance of Activated Monocytes in Patients with ST-Elevation Myocardial Infarction
Reprinted from: *Int. J. Mol. Sci.* **2023**, *24*, 11342, doi:10.3390/ijms241411342 . . . . **248**

Mohammed F. Gholam, Niharika Bala, Yunus E. Dogan and Abdel A. Alli
Augmentation of Cathepsin Isoforms in Diabetic db/db Mouse Kidneys Is Associated with an Increase in Renal MARCKS Expression and Proteolysis
Reprinted from: *Int. J. Mol. Sci.* **2023**, *24*, 12484, doi:10.3390/ijms241512484 . . . . **260**

Laris Achlaug, Lama Awwad, Irina Langier Goncalves, Tomer Goldenberg and Ami Aronheim
Tumor Growth Ameliorates Cardiac Dysfunction and Suppresses Fibrosis in a Mouse Model for Duchenne Muscular Dystrophy
Reprinted from: *Int. J. Mol. Sci.* **2023**, *24*, 12595, doi:10.3390/ijms241612595 . . . . **277**

Yutang Wang, Owen Sargisson, Dinh Tam Nguyen, Ketura Parker, Stephan J. R. Pyke, Ahmed Alramahi, et al.
Effect of Hydralazine on Angiotensin II-Induced Abdominal Aortic Aneurysm in Apolipoprotein E-Deficient Mice
Reprinted from: *Int. J. Mol. Sci.* **2023**, *24*, 15955, doi:10.3390/ijms242115955 . . . . **289**

International Journal of
Molecular Sciences

Special Issue: "New Trends in Diabetes, Hypertension, and Cardiovascular Diseases"

Yutang Wang [1,*] and Dianna J. Magliano [2]

[1] Discipline of Life Science, Institute of Innovation, Science and Sustainability, Federation University Australia, Ballarat, VIC 3350, Australia
[2] Diabetes and Population Health, Baker Heart and Diabetes Institute, Melbourne, VIC 3004, Australia
* Correspondence: yutang.wang@federation.edu.au

Cardiovascular disease (CVD) encompasses a range of disorders affecting the heart and blood vessels, including coronary heart disease and cerebrovascular disease [1] and conditions such as aortic aneurysms and lower-extremity peripheral artery disease [2]. It remains the leading cause of death worldwide, claiming 17.9 million lives annually [3], despite the availability of numerous therapeutic drugs [4]. Thus, further research is warranted to enhance CVD treatment and reduce its morbidity and mortality [5].

Diabetes is a leading cause of blindness, kidney failure, heart attacks, stroke, and lower-limb amputation, causing an estimated 2 million deaths annually [6]. In 2021, 529 million individuals (prevalence, 6.1%) were afflicted with diabetes, a figure that is projected to soar to 1.21 billion by 2050 [7]. Diabetes represents a significant public health concern worldwide [8]. Despite remarkable strides in prevention and treatment, further research is required. This is particularly important given the recent introduction of new efficacious medications and the developments in AI technology, the latter being useful in the development of precision medicine algorithms to treat people with diabetes [9].

Hypertension impacts 1.28 billion individuals aged 30–79 worldwide [10], serving as a critical risk factor for ischemic heart disease, stroke, chronic kidney disease, and dementia, with approximately 10 million deaths attributed to it annually [11]. The number of people suffering from hypertension doubled from 1990 to 2019 [12], with only 22.5% of hypertensive adults having their condition under control [13]. Hence, improved detection, treatment, and control strategies are imperative.

Diabetes and hypertension are conditions commonly found in aging populations; they share common pathogenic pathways [14] and are both strong independent risk factors for CVD. Reducing blood glucose and blood pressure levels can significantly contribute to mitigating CVD-related morbidity and mortality.

This Special Issue aims to present the latest perspectives and research findings on diabetes, hypertension, and CVD, comprising eight review articles and eight original research articles.

The first review, "The Role of the Gut Microbiome and Trimethylamine Oxide in Atherosclerosis and Age-Related Disease", explores the contribution of the gut microbiome and trimethylamine oxide to CVD, emphasizing the potential utility of reducing trimethylamine oxide as a novel therapeutic strategy.

The second review, "The Role of Oxidative Stress Enhanced by Adiposity in Cardiometabolic Diseases", delves into the contribution of adiposity-associated oxidative stress to cardiometabolic diseases, offering insights into the mechanisms underlying this interplay.

The third review, "Portrayal of NLRP3 Inflammasome in Atherosclerosis: Current Knowledge and Therapeutic Targets", examines the role of the NLRP3 inflammasome in atherosclerosis, highlighting its potential as a therapeutic target.

It is followed by "A Receptor Story: Insulin Resistance Pathophysiology and Physiological Insulin Resensitization's Role as a Treatment Modality", which discusses the

Citation: Wang, Y.; Magliano, D.J. Special Issue: "New Trends in Diabetes, Hypertension, and Cardiovascular Diseases". *Int. J. Mol. Sci.* **2024**, *25*, 2711. https://doi.org/10.3390/ijms25052711

Received: 6 February 2024
Accepted: 22 February 2024
Published: 27 February 2024

pathophysiology of insulin resistance, emphasizing the therapeutic potential of dynamic exogenous insulin administration.

Next, "Smooth Muscle Heterogeneity and Plasticity in Health and Aortic Aneurysmal Disease" explores vascular smooth muscle cell phenotypes in aortic aneurysms, calling for further research into their relationship with aneurysm formation.

The sixth review, "Angiotensin Regulation of Vascular Homeostasis: Exploring the Role of ROS and RAS Blockers", highlights the role of the renin–angiotensin system (RAS) in CVD pathogenesis and the potential of RAS blockers to reduce vascular oxidative stress.

Next, "Dysfunctional and Dysregulated Nitric Oxide Synthases in Cardiovascular Disease: Mechanisms and Therapeutic Potential" examines the contribution of nitric oxide synthases to cardiovascular health and their potential as therapeutic targets.

Finally, "Model Systems to Study the Mechanism of Vascular Aging" explores various models of vascular aging, providing insights into the aging process.

The original research articles delve into various aspects of CVD, shedding light on sex-specific features, genetic associations, and the effects of lifestyle factors, such as night shifts, on disease risk.

The first, "Sex-Specific Features of the Correlation between GWAS-Noticeable Polymorphisms and Hypertension in Europeans in Russia", examines the sex-specific features of genome-wide-association-studies-identified single-nucleotide polymorphisms (SNPs) in relation to hypertension. This study revealed that hypertension-associated SNPs had a more pronounced effect on susceptibility in women than in men.

The second research paper, "SERPINE1 mRNA Binding Protein 1 Is Associated with Ischemic Stroke Risk: A Comprehensive Molecular–Genetic and Bioinformatics Analysis of SERBP1 SNPs", investigated the association between genetic variants of serpin family E member 1 (SERBP1) and ischemic stroke risk. The findings suggest that SERBP1 SNPs may serve as novel genetic markers for stroke.

The third, "Preliminary Study on the Effect of a Night Shift on Blood Pressure and Clock Gene Expression", examined the effect of night shifts on ambulatory blood pressure and clock gene expression. The study revealed that night shifts negatively impacted ambulatory blood pressure, leading to a non-dipping blood pressure status, which is a risk factor for CVD. Moreover, night shifts were associated with an increase in clock gene expression.

The fourth research paper, "Age-Dependent Changes in the Relationships between Traits Associated with the Pathogenesis of Stress-Sensitive Hypertension in ISIAH Rats", investigated the impact of age on the development of stress-sensitive hypertension. The study identified age as a factor affecting hypertension manifestation.

The fifth research study, "Prognostic Significance of Activated Monocytes in Patients with ST-Elevation Myocardial Infarction", examined the prognostic efficacy of monocyte subsets in patients with ST-elevation myocardial infarction. The study found that elevated levels of nonclassical ($CD14^+CD16^{++}$) monocytes and their subsets predicted worse clinical outcomes in these patients.

Next, "Augmentation of Cathepsin Isoforms in Diabetic db/db Mouse Kidneys Is Associated with an Increase in Renal MARCKS Expression and Proteolysis" investigated the expression and proteolysis of key proteins in the kidneys of diabetic db/db mice. The study revealed a significant cleavage of the myristoylated alanine-rich C-kinase substrate (MARCKS) family of proteins, important in maintaining normal renal function, in diabetic mice. This increase in MARCKS cleavage was associated with an increase in the protein expression of cathepsin isoforms.

Following this, "Tumor Growth Ameliorates Cardiac Dysfunction and Suppresses Fibrosis in a Mouse Model for Duchenne Muscular Dystrophy" explored the association between heart failure and cancer in mice. The study found that tumour growth ameliorated cardiac dysfunction and reduced overall fibrosis, suggesting potential novel strategies to improve cardiac function and fibrosis-associated diseases by understanding tumour paradigms.

Finally, "Effect of Hydralazine on Angiotensin II-Induced Abdominal Aortic Aneurysm in Apolipoprotein E-Deficient Mice" investigated the impact of hydralazine on abdominal aortic aneurysms in a mouse model. The study concluded that hydralazine inhibited the formation and rupture of abdominal aortic aneurysms, attributed to its anti-inflammatory and anti-apoptotic properties.

In conclusion, this Special Issue provides a comprehensive overview of recent developments in research on hypertension, insulin resistance/diabetes, and CVD, including atherosclerosis, aortic aneurysms, and myocardial infarction [15]. These articles offer insights into CVD pathogenesis and facilitate the development of new therapeutic approaches.

Author Contributions: Y.W. prepared the manuscript. Y.W. and D.J.M. revised the manuscript. All authors have read and agreed to the published version of the manuscript.

Conflicts of Interest: The authors declare no conflicts of interest.

References

1. Wang, Y.; Emeto, T.I.; Lee, J.; Marshman, L.; Moran, C.; Seto, S.W.; Golledge, J. Mouse models of intracranial aneurysm. *Brain Pathol.* **2015**, *25*, 237–247. [CrossRef]
2. Firnhaber, J.M.; Powell, C.S. Lower Extremity Peripheral Artery Disease: Diagnosis and Treatment. *Am. Fam. Physician* **2019**, *99*, 362–369. [PubMed]
3. World Health Organization, Cardiovascular Diseases. 2024. Available online: https://www.who.int/health-topics/cardiovascular-diseases#tab=tab_1 (accessed on 4 January 2024).
4. Hong, C.C. The grand challenge of discovering new cardiovascular drugs. *Front. Drug Discov.* **2022**, *2*, 1027401. [CrossRef] [PubMed]
5. Roth, G.A.; Mensah, G.A.; Johnson, C.O.; Addolorato, G.; Ammirati, E.; Baddour, L.M.; Barengo, N.C.; Beaton, A.Z.; Benjamin, E.J.; Benziger, C.P.; et al. Global Burden of Cardiovascular Diseases and Risk Factors, 1990–2019: Update From the GBD 2019 Study. *J. Am. Coll. Cardiol.* **2020**, *76*, 2982–3021. [CrossRef] [PubMed]
6. World Health Organization, Key Facts-Diabetes. 2023. Available online: https://www.who.int/news-room/fact-sheets/detail/diabetes (accessed on 18 October 2023).
7. GBD 2021 Diabetes Collaborators. Global, regional, and national burden of diabetes from 1990 to 2021, with projections of prevalence to 2050: A systematic analysis for the Global Burden of Disease Study 2021. *Lancet* **2023**, *402*, 203–234. [CrossRef] [PubMed]
8. American Diabetes Association. Introduction: Standards of Medical Care in Diabetes—2022. *Diabetes Care* **2021**, *45* (Suppl. S1), S1–S2.
9. Tobias, D.K.; Merino, J.; Ahmad, A.; Aiken, C.; Benham, J.L.; Bodhini, D.; Clark, A.L.; Colclough, K.; Corcoy, R.; Cromer, S.J.; et al. Second international consensus report on gaps and opportunities for the clinical translation of precision diabetes medicine. *Nat. Med.* **2023**, *29*, 2438–2457. [CrossRef]
10. World Health Organization. Hypertension Key Facts. 2023. Available online: https://www.who.int/news-room/fact-sheets/detail/hypertension (accessed on 20 November 2023).
11. World Heart Federation. World Hypertension Day: Taking Action against the Silent Epidemic of High Blood Pressure. 2022. Available online: https://world-heart-federation.org/news/world-hypertension-day-taking-action-against-the-silent-epidemic-of-high-blood-pressure/ (accessed on 1 February 2024).
12. NCD Risk Factor Collaboration. Worldwide Trends in hypertension prevalence and progress in treatment and control from 1990 to 2019: A pooled analysis of 1201 population-representative studies with 104 million participants. *Lancet* **2021**, *398*, 957–980. [CrossRef]
13. Centers for Disease Control and Prevention, Facts about Hypertension. 2023. Available online: https://www.cdc.gov/bloodpressure/facts.htm#:~:text=About%201%20in%204%20adults,22.5%25,%2027.0%20million). (accessed on 2 February 2024).
14. Sunkara, N.; Ahsan, C.H. Hypertension in diabetes and the risk of cardiovascular disease. *Cardiovasc. Endocrinol.* **2017**, *6*, 33–38. [CrossRef]
15. Wang, Y.; Sargisson, O.; Nguyen, D.T.; Parker, K.; Pyke, S.J.R.; Alramahi, A.; Thihlum, L.; Fang, Y.; Wallace, M.E.; Berzins, S.P.; et al. Effect of Hydralazine on Angiotensin II-Induced Abdominal Aortic Aneurysm in Apolipoprotein E-Deficient Mice. *Int. J. Mol. Sci.* **2023**, *24*, 15955. [CrossRef] [PubMed]

International Journal of
Molecular Sciences

Review

The Role of the Gut Microbiome and Trimethylamine Oxide in Atherosclerosis and Age-Related Disease

Racha El Hage [1], Nada Al-Arawe [1,2] and Irene Hinterseher [1,2,3,*]

[1] Department of Vascular Surgery, Universitätsklinikum Ruppin-Brandenburg, Medizinische Hochschule Brandenburg, Fehrbelliner Str. 38, 16816 Neuruppin, Germany

[2] Vascular Surgery Clinic, Charité—Universitätsmedizin Berlin, Corporate Member of Freie Universität Berlin and Humboldt-Universität zu Berlin, and Berlin Institute of Health (BIH), 10117 Berlin, Germany

[3] Fakultät für Gesundheitswissenschaften Brandenburg, Gemeinsame Fakultät der Universität Potsdam, der Medizinischen Hochschule Brandenburg Theodor Fontane und der Brandenburgischen Technischen Universität Cottbus—Senftenberg, Karl-Liebknecht-Str. 24–25, 14476 Potsdam, Germany

* Correspondence: irene.hinterseher@charite.de; Tel.: +49-3391-39-47110

Abstract: The gut microbiome plays a major role in human health, and gut microbial imbalance or dysbiosis is associated with disease development. Modulation in the gut microbiome can be used to treat or prevent different diseases. Gut dysbiosis increases with aging, and it has been associated with the impairment of gut barrier function leading to the leakage of harmful metabolites such as trimethylamine (TMA). TMA is a gut metabolite resulting from dietary amines that originate from animal-based foods. TMA enters the portal circulation and is oxidized by the hepatic enzyme into trimethylamine oxide (TMAO). Increased TMAO levels have been reported in elderly people. High TMAO levels are linked to peripheral artery disease (PAD), endothelial senescence, and vascular aging. Emerging evidence showed the beneficial role of probiotics and prebiotics in the management of several atherogenic risk factors through the remodeling of the gut microbiota, thus leading to a reduction in TMAO levels and atherosclerotic lesions. Despite the promising outcomes in different studies, the definite mechanisms of gut dysbiosis and microbiota-derived TMAO involved in atherosclerosis remain not fully understood. More studies are still required to focus on the molecular mechanisms and precise treatments targeting gut microbiota and leading to atheroprotective effects.

Keywords: gut microbiome; atherosclerosis; TMAO; aging; gut dysbiosis; probiotics; short chain fatty acids

Citation: El Hage, R.; Al-Arawe, N.; Hinterseher, I. The Role of the Gut Microbiome and Trimethylamine Oxide in Atherosclerosis and Age-Related Disease. *Int. J. Mol. Sci.* **2023**, *24*, 2399. https://doi.org/10.3390/ijms24032399

Academic Editors: Yutang Wang and Dianna Magliano

Received: 6 January 2023
Revised: 20 January 2023
Accepted: 21 January 2023
Published: 25 January 2023

1. Introduction

Atherosclerosis is a chronic disease that affects medium and large arteries of the body through a major increase in the lipoproteins of their intimal layer [1]. Risk factors associated with atherosclerosis include diabetes mellitus, hypertension, dyslipidemia, obesity, and smoking [2]. Diabetes mellitus is a major public health problem that has been a leading cause for mortalities worldwide. Diabetes is characterized by elevated levels of blood glucose, which leads over time to serious damage to the heart, blood vessels, eyes, kidneys, and nerves [3]. As a result, diabetes mellitus is associated with accelerated atherosclerosis, leading to vascular lesions that include cardiovascular disease (CVD), coronary artery disease, cerebrovascular disease, and peripheral arterial disease (PAD), with CVD being the major cause of premature death in diabetes. Another consequence of diabetes mellitus is microangiopathy that occurs in the colon, and which has been reported to be more common in diabetics than non-diabetics [4]. Microangiopathy can also occur in the retina, skin (specifically foot skin/diabetic foot ulcer), nerve, kidney, muscle, and heart of diabetic patients and is associated with the thickening of the capillary basement membrane [5]. Other mediators of diabetes that cause vascular complications include dyslipidemia, chronic hyperglycemia, and insulin resistance. Dyslipidemia and chronic inflammation are among the major causes of the development of atherosclerosis, which causes chronic accumulation of lipid-rich plaque in the arteries in diabetic patients [6,7]. Therefore, the regulation of these chronic diseases is crucial.

The human gut microbiome is a huge microbial community that plays a vital role in human health. With the development in research, the influence of intestinal flora on human diseases has been gradually revealed. Dysbiosis in the gut microbiota (GM) has been reported to have adverse health effects on the human body that lead to a variety of chronic diseases. Regulation of the GM can provide a potential target for the prevention and treatment of disease. The fermentation products of the gut microbiota are by far the most well studied, and they have been described to have a key role in the maintenance of the gut microbial ecology and the modulation of host immunity and metabolic disease [8–12]. The major fermentation products of the GM that result from dietary fibers are short chain fatty acids (SCFAs), with the most abundant metabolites being acetate, propionate, and butyrate. [12,13]. SCFAs can function as a macronutrient energy source and hormone-like signaling molecules that enter the portal circulation to signal through specific host receptor systems in order to regulate the innate immunity and host metabolism. Most studies linking the GM to disease designate SCFAs as potential disease-moderating or prevention factors in metabolic disease, intestinal immunity, cancer, and liver disease [12,13]. Recent reports have indicated that dysbiosis is increased with aging, and that the GM of elderly people is enriched with pro-inflammatory commensals and fewer beneficial microbes [14]. Dysbiosis is presumed to be the primary cause of age-associated morbidities, and, consequently, the premature death of elderly people [14]. Gut dysbiosis leads to a disruption of the microbial metabolites, impaired function of the gastrointestinal tract, and increased leakage of the gut [14]. These events enhance systemic inflammation which is associated with aging, termed inflammaging, and they, consequently, result in aging-associated pathologies [14].

Trimethylamine (TMA) is a byproduct generated from the gut microbial metabolism of dietary amines such as choline, betaine, and carnitine that originate from animal-based foods [15,16]. TMA is absorbed into the portal circulation and is oxidized by the liver into trimethylamine-N-oxide (TMAO) using the flavin monooxygenase enzyme (Figure 1) [16,17]. More attention has been directed upon circulating TMAO due to its pro-inflammatory, pro-atherogenic, and pro-thrombotic properties [18–22]. Several factors, such as diet, gut microbial flora, drug administration, and liver flavin monooxygenase activity, influence the plasma TMAO level [16]. TMAO has been described as vital for lipid balance and for the increase in scavenger receptors, such as CD36 and scavenger receptor class A type 1 (SR-A1), that contribute to the surge of fat accumulation in foam cells, which in turn play a major part in atherosclerotic plaque progress [16,18,23]. In addition, the hepatic enzyme flavin-containing monooxygenase 3 (FMO3) is considered the most active in converting TMA into TMAO, leading to higher plasma TMAO levels. High levels of plasma TMAO have been linked to an alteration of reverse cholesterol transport [16,24], hyperlipidemia and hyperglycemia [16,24,25], and the overexpression of inflammatory markers including tumor necrosis factor alpha (TNF-α), interleukin-6 (IL-6), c-reactive protein (CRP) [26,27], and insulin resistance [16,25], which all lead to the promotion of atherosclerosis [16,19,25,28,29]. It has also been reported that high plasma levels of TMAO metabolite are related to the prognosis of 5-year all-cause mortality in stable patients diagnosed with peripheral artery disease [28,30]. Furthermore, Brunt et al. confirmed that circulating TMAO was high in older compared with younger adults, and that elevated TMAO was correlated with a higher carotid–femoral pulse wave velocity (PWV). Their findings in humans represented the first link between the age-related increase in circulating TMAO with higher aortic stiffness and blood pressure (BP). Nevertheless, as both aortic stiffness and BP are aging risk factors, it might be that both outcomes are in fact not causally associated with TMAO [31].

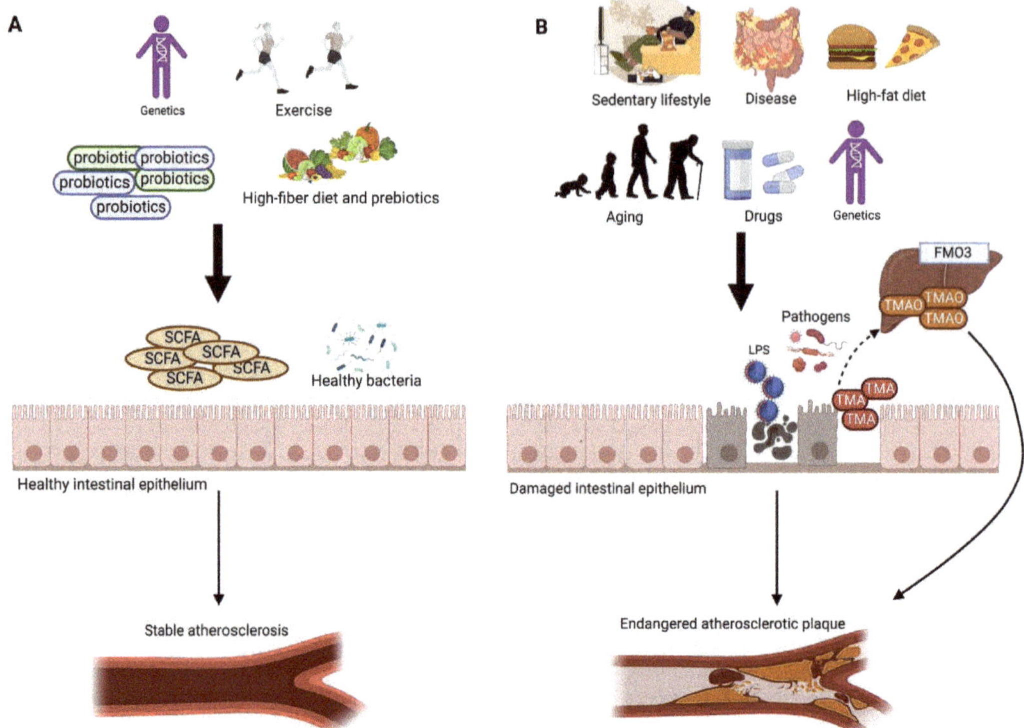

Figure 1. Different factors, such as the gut microbiome, diet, lifestyle, and genetics, play a major role in the development of atherosclerosis. (**A**) High-fiber diet, active lifestyle, and the intake of probiotics increase the abundance of beneficial gut bacteria and the amount of SCFAs, which support the tight junctions and protect the intestinal epithelium, thus preventing harmful metabolites from entering the circulation. This contributes to a stable atherosclerotic plaque. (**B**) High-fat diet, sedentary lifestyle, intestinal disease (e.g., Crohn's disease, irritable bowel disease), the intake of different drugs (e.g., antibiotics), and aging lead to a disruption in the gut microbial profile, resulting in higher abundance of pathogenic bacteria and lower amounts of SCFAs. These effects lead to damage in the gut epithelium; thus, a leaky gut allows the translocation of LPS, TMA, and other damaging metabolites into the circulation. High amounts of plasma LPS and TMAO result in an endangered atherosclerotic plaque.

2. TMAO, TMA-Producing Gut Bacteria, and Atherosclerosis

Previous studies described that many human gut colonizing bacteria are capable of producing TMA, which leads to an increase of TMAO levels in plasma. These gut bacteria include *Streptococcus sanguinis, Desulfovibrio alaskensis, Desulfovibrio desulfuricans, Acinetobacter, Serratia, Escherichia coli, Citrobacter, Klebsiella pneumoniae, Providencia, Shigella, Achiomobacter,* and *Sporosorcine,* which belong to the Firmicutes and Actinobacteria phyla. [16,23]. On the other hand, bacteria belonging to phylum Bacteroidetes are not able to produce TMA [16,32]. Several previous studies have investigated the impact of host factors, such as diet and dietary compounds, on TMAO plasma levels, and it was reported that higher plasma TMAO levels have been linked to an animal-based diet [33,34] compared to vegetarians (Figure 1) [19,35]. Mainly, TMA is generated by the enzymes produced by the gut microbiota, and its levels are dependent on the amount of precursors available and the abundance and activity of bacteria catalyzing TMA formation [36], which compete with the host for these precursors that are usually absorbed as essential nutrients [36]. As the quantification of TMA producers in the gut is limited, it is more difficult to understand the distribution of these bacteria in the gut and specify their niches, which is important for designing effective and sustainable treatment strategies to minimize TMAO plasma concentrations [36]. In order to discover the abundance and diversity of TMA-forming gut bacteria, Rath et al. developed assays that targeted key genes encoding enzymes responsible for TMA formation from choline (choline-TMA lyase, *CutC*), carnitine (carnitine monooxygenase, *CntA*), and betaine (subunit B of betaine reductase, *grdH*) [17]. In addition to the association of TMAO with PAD and atherosclerosis, advancing age has also been strongly linked to TMAO levels. As age increases, the host's physiology and function are altered. For example, the epithelial integrity of the colon, which is needed to promote the influx of bacterial metabolites, including TMAO, might be reduced [37,38]. Rath et al. reported that there was an association between carotid intima-media thickness (IMT) and TMAO only in individuals above 65 years of age, which indicates that aging people are principally affected by this metabolite [38]. Age-related associations between TMAO plasma levels and health parameters have not yet been reported in patients with PAD. Preclinical studies in both animal and in vitro models on human-derived material have emphasized the contribution of TMAO to endothelial senescence and vascular aging [39,40]; however, age-specific effects of TMAO have still not been fully studied. Many studies have investigated the link between diet and TMAO levels, reporting different results; however, an analysis of gut microbiota has only recently been included [38]. Rath et al. were able to provide important information regarding the formation of TMAO in the general population, and they have elaborated on the functional role of the gut microbiota and specific foods, clarifying the increased levels of TMAO with increasing age [38].

3. Gut Dysbiosis, Aging, and TMAO Levels

Gut dysbiosis is the disruption in the gut microbiome that is associated with different diseases. Dysbiosis disturbs the gut barrier function, leading to the leakage of harmful metabolites, such as lipopolysaccharides (LPS), and other bacterial components, such as peptidoglycans, into the circulation, which triggers an inflammatory response leading to atherosclerosis (Figure 1) [41–43]. LPS can stimulate the uptake of modified low-density lipoprotein (LDL) and reduce the efflux of cholesterol from foam cells, promoting monocyte recruitment and macrophage foam cell formation [41,43,44]. LPS can induce vascular inflammation directly or by producing pro-inflammatory factors from immune cells [44]. The increased production of pro-inflammatory cytokines promotes oxidative stress and oxidized LDL (oxLDL), increasing the risk of hypertension via nitric oxide synthase inhibition. This effect reduces vasodilator nitric oxide levels and increases levels of vasoconstrictor endothelin-1 [41].

Aging, Gut Dysbiosis, and TMAO

Aging leads to several changes in cells, tissues, and organs [45] and is influenced by an individual's genetics, lifestyle, and environment [46]. The term "immunoscence" first appeared a few decades ago to refer to impaired or faulty immune responses leading to a decrease in the ability to trigger the immune response and effectively produce antibodies against different pathogens [47,48]. The gut microbiome undergoes dynamic changes through time, and gut dysbiosis is an age-related complication caused by host senescence, changes in nutritional behavior, drug use, and the lifestyle of aged people [48]. The changes in the gut microbiome include shifts in bacterial composition and metabolic function [49]. In humans, age-related gut dysbiosis is characterized by increased inter-individual variation and decreased species diversity; specifically, a loss of Clostridiales and Bifidobacterium, an enrichment of Proteobacteria, Lactobacilli, and an overrepresentation of pathobionts such as Enterobacteriaceae [49–52]. However, the major gut microbiota aging feature is the decreased ratio of Firmicutes/Bacteroidetes [53]. Schneeberger et al. reported that aged mice showed a decrease in beneficial gut bacteria, such as in Clostridium members of cluster IV that produce SCFAs and *Akkermansia muciniphila*, and an increase in pro-inflammatory microbes [14,54]. Overall, the decrease in intestinal commensal microbes diversity is associated with increased susceptibility to pathogen infection accompanied by disturbance of the gut mucosal barrier and enrichment in pro-inflammatory cytokines; all these events have a detrimental consequence in aging [14,55]. Recent studies suggested that gut dysbiosis is associated with the development of several chronic diseases including cardiovascular disease and other metabolic disorders [56].

Many studies have reported a close relationship between TMAO levels, aging, and age-related diseases. Several animal models have been used to identify mechanisms that underlie TMAO's role in senescence [57]. Cell senescence involves many processes including increased production of reactive oxygen species (ROS), mitochondrial dysfunction, and senescence-associated secretory phenotype (SASP) [57]. Ke et al. reported that senescence-accelerated prone mouse strain 8 (SAMP8) and senescence-accelerated mouse resistant 1 (SAMR1) were treated with 1.5% (w/v) TMAO for 16 weeks to induce vascular aging and advanced vascular aging processes, respectively [40]. Many potential mechanisms underlie TMAO's role in aging, including the inhibition of sirtuin 1 (SIRT1) expression, which increases oxidative stress and results in the activation of the p53/p21/Rb pathway. Increased P53 and P21 acetylation and reduced CDK2, cyclinE1, and Rb phosphorylation are followed by enhanced endothelial cell senescence and vascular aging [58]. In addition, TMAO increases the accumulation of ROS, matrix metalloproteinase 2 (MMP2), and matrix metalloproteinase 9 (MMP9) in vivo and in vitro, which are associated to oxidative stress in cells [59]. Furthermore, high TMAO levels are linked to increased expression of pro-inflammatory cytokines, such as TNF-α and IL-1β, as well as decreased production of anti-inflammatory cytokines such as IL-10 [26].

4. SCFAs and Their Function in Atherosclerosis

Short chain fatty acids (SCFAs) are the primary fermentation products of dietary fibers and non-digestible carbohydrates (NDCs) [60,61]. NDCs are an important fraction of dietary fibers, and SCFAs are the main products of the favorable saccharolytic fermentation of NDC in the gut [62]. In elderly people, the level of carbohydrate-derived SCFAs is decreased, while the metabolites resulting from protein fermentation, such as phenols, ammonia, and branched fatty acids, are increased. This indicates a shift from saccharolytic fermentation to unfavorable proteolytic fermentation [14,50]. A shift in SCFA production occurs progressively during the aging process, and it is accelerated upon the usage of antibiotics and changes in diet [50].

Several bacterial families, including anaerobic Bacteroides, Bifidobacterium, Eubacterium, Streptococcus, Lactobacillus, clostridial clusters IV and XIVa of Firmicutes, including species of Eubacterium, Roseburia, Faecalibacterium, and Coprococcus, are involved in the production of SCFAs [61,63]. SCFAs mediate the interaction between the gut, diet, and microbiota, highlighting their essential role in intestinal health [61,64–66]. The major SCFAs are acetate, propionate, and butyrate, which account for approximately 90% of the total SCFAs formed in the human colon by colonic microorganisms [67–69]. They help in the regulation of host metabolic processes to achieve host homeostasis. SCFAs have a high abundance in the gastrointestinal tract (GIT) and are utilized by intestinal epithelial cells (IECs) [70]; they are able to modify several crucial cellular processes including gene expression, proliferation, differentiation, and chemokines production [71].

The function of SCFAs is mediated by the activation of six G protein-coupled receptors (GPCR) encoded by the human genome: GPR41 (free fatty acid receptor 3; FFAR3), GPR42, GPR43 (FFAR2), GPR109a (HCAR2), GPR164 (OR51E1), and OR51E2. GPCR41 and GPCR43 are expressed in adipose tissue, intestines, and immune cells [71].

The epithelial cells are in direct contact with high SCFA concentrations, so SCFAs are uptaken into the IEC cytosol via monocarboxylate transporters, such as monocarboxylate transporter-1 (Slc16a1) and the sodium-dependent monocarboxylate transporter-1 (Slc5a8) [72]. Several studies showed that SCFAs improve immune defenses; for instance, butyrate increases the expression of many antimicrobial peptides including LL-37 and

CAP-18 [73]. SCFAs are believed to have a universal anti-inflammatory effect by upregulating anti-inflammatory cytokines and downregulating pro-inflammatory cytokines [7,74]. Kim et al. showed that SCFAs enhance the production of cytokines and chemokines, including TNF-α, IL-6, CXCL1, and CXCL10, in ICE in vitro [75]. Butyrate has been demonstrated to reduce atherosclerotic development in animal models via the reduction of pro-inflammatory factors [76]. Aguilar et al. showed in an atherosclerotic mice model with atherosclerosis-prone apolipoprotein E-deleted (ApoE$-$/$-$) that consumed a diet containing 1% butyrate for 10 weeks, atherosclerotic lesions in the aorta were reduced by 50%, suggesting a more stable fibrous cap. Moreover, the mice showed a lower macrophage infiltration and elevated collagen deposition; this phenomenon was linked to a decreased CD36 expression in both endothelial cells and macrophages, nuclear factor-κB (NF-κB) activation, and pro-inflammatory cytokines production [76].

The inhibition of histone deacetylases (HDACs), stimulation of histone acetyltransferase, and stabilization of hypoxia-inducible factor (HIF) activity are the second major mechanisms involved in SCFAs' action [70,77–79]. This results in the regulation of gene expression and the inhibition of a vast array of downstream consequences. However, the understanding of SCFA-mediated inhibition of HDACs is still unclear [80]. In this context, SCFAs increase IECs' oxygen consumption, causing a decrease in oxygen tension and stabilization of HIF [70,81].

Propionate has been reported to act as an anti-atherosclerotic agent via its positive effects on immunity and immune system components, and its ability to reduce plasma lipid levels [82]. Haghikia et al. reported that propionate is able to control cholesterol hemostasis and reduce the aortic atherosclerotic lesion area in ApoE$-$/$-$ mice fed a high-fat diet (HFD) [83]. In addition, they stated that propionate increases T regulatory (Treg) cell numbers, thus elevating the IL-10 levels in the intestinal wall. IL-10 suppresses the expression of transmembrane transporter Niemann–Pick C1-like 1 (NPC1L1), which is responsible for intestinal cholesterol absorption. Their results were further translated into humans, in which propionate supplementation was able to reduce LDL and TC in hypercholesterolaemic patients [83]. This indicates the potential therapeutic effect of propionate, which modulates the intestinal immune system, thus improving cardiovascular health and preventing atherosclerotic cardiovascular disease [83]. Similarly, Bartolomaeus at al. reported the immunomodulatory effect of propionate, its ability to reduce atherosclerotic lesion in ApoE$-$/$-$ mice, and the susceptibility to cardiac ventricular arrhythmias of angiotensin II–infused wild-type NMRI mice [84].

The metabolic pathways of both acetate and propionate participate in the regulation of lipid biosynthesis [85]. The ratio of propionate to acetate is a crucial factor in lipid metabolism, in which acetate contributes to lipid synthesis, while propionate reduces fat deposition in the liver and visceral organs [85,86].

5. Butyrate-Producing Bacteria and Atherosclerosis

Butyrate has been reported to have an important role in inflammatory diseases in addition to its significant lipid-lowering, anti-oxidant, and insulin resistance-improving effects [7,87,88]. Butyrate also serves as a primary fuel for colonocytes [89]. It acts as a histone deacetylase inhibitor and ligand to GPCRs, affecting cellular signaling in target cells such as enteroendocrine cells [89]. Metagenomics studies reported that diabetes is associated with an altered gut microbiota. An altered gut microbiota leads to a shift in SCFA production. For instance, a lower abundance of butyrate-producing bacteria has been detected in patients with type 2 diabetes [89]. As a result, treatment strategies for diabetes have been developed to increase intestinal levels of butyrate. These strategies involve supplementation with butyrate-producing bacteria, together with dietary fiber, or via fecal microbial transplant from healthy subjects [89].

Moreover, studies have provided evidence that butyrate is able to regulate the antioxidant effect of NF-*k*B in endothelial cells and macrophage-mediated lipid metabolism [7]. It is also able to alleviate the production of pro-inflammatory cytokines, such as IL-1β,

TNF-α, IL-6 [90], IL-12, and interferon-γ (IF-γ), and it is able to upregulate the production of anti-inflammatory IL-10 by monocytes in vitro [91,92]. Furthermore, butyrate is able to attenuate the release of vascular cell adhesion molecule-1 (VCAM1) and chemotaxis protein-1 (MCP1/CCL2) in vitro, thus reducing the migration and adhesion of monocytes in the lesion area [93]. Increasing evidence shows that butyrate can regulate the occurrence and development of atherosclerosis [7,94]. As such, butyrate is considered a collateral in the prevention and treatment of atherosclerosis [90].

Considering the treatment strategies that involve increasing intestinal butyrate levels, next generation probiotics and butyrate-producing bacteria together with prebiotics could be a promising approach. Different next generation probiotics have shown to have a positive effect in diabetes and atherosclerosis.

To start with, *A. muciniphila*, a mucin-degrading gut bacterium, has been inversely linked to diabetes, inflammation, and metabolic disorders [95]. This bacterium has probiotic properties and has been reported to be more abundant in healthy subjects than in patients with diabetes or other metabolic disorders [95–97]. The probiotic properties of *A. muciniphila* could be associated with its ability to modulate mucus thickness and gut barrier integrity [95]. Studies using mice have shown that supplementation of *A. muciniphila* resulted in the restoration of mucus thickness that was disrupted in obese and type 2 diabetic mice because of HFD. In addition, this treatment with *A. muciniphila* was able to improve the metabolic profile and reduce the level of serum lipopolysaccharides (LPSs) [95,98]. High levels of serum LPSs have been linked to gut permeability, and, thus, the disruption of intestinal mucus [95,99]. Li et al. reported that *A. muciniphila* reduces atherosclerotic lesions by improving metabolic endotoxemia-induced inflammation through the restoration of the gut barrier [100]. Studies indicate that, despite using mucin as a source of nutrients, *A. muciniphila* is positively linked to mucus thickness and intestinal barrier integrity in both humans and animals [95,98,101]. Moreover, human and animal studies have also shown that *A. muciniphila* is able to improve insulin sensitivity and glucose homeostasis, in addition to modulating obesity by regulating metabolism and energy homeostasis [98,102]. Bodogai et al. reported that the series of inflammatory events that manifest in insulin resistance occur in aged mice as a result of the decrease in abundance of *A. muciniphila* [103]. This has been linked to the reduction of the mucin layer in the colon, which in turn leads to the loss of butyrate-producing commensal bacteria, such as *Intestinimonas butyriciproducens (I. butyriciproducens)*, *Faecalibacterium prausnitzii (F. prausnitzii)*, *Roseburia faecis (R. faecis)*, and *Anaerostipes butyraticus (A. butyraticus)*, thus leading to an SCFA reduction, specifically of butyrate both in the gut lumen and in the circulation. The reduction in butyrate promotes dysbiosis and leakage from the gut in aged mice, which sustains inflammaging [103].

Another butyrate-producing gut bacterium that contributes to different diseases, such as type 2 diabetes, atherosclerosis, antiphospholipid syndrome, and inflammatory bowel disease, is *Roseburia intestinalis*. *R. intestinalis* has been shown to reduce intestinal inflammation by enhancing the proliferation of Tregs and stimulating the secretion of anti-inflammatory cytokines IL-10, TGF-b, and thymic stromal lymphopoietin (TSLP) [61]. These results proposed a significant immunomodulating and anti-inflammatory effect of the butyrate produced by *R. intestinalis* in the gut [104]. Kasahara et al. reported that *R. intestinalis* was able to reduce endotoxemia, inflammatory markers in plasma and aorta, and the extent of atherosclerotic lesions. These positive effects were attained after the interaction of *R. intestinalis* with dietary plant polysaccharides [105].

F. prausnitzii, a butyrate-producing gut bacterium, is another next generation probiotic that is dominant in healthy adults and is described for its anti-inflammatory properties and its potential therapeutic effect in patients in Crohn's disease [13,106,107]. In addition to its beneficial role in bowel disease, *F. prausnitzii* could have a positive effect in obese and diabetic patients due to its ability to produce butyrate. Butyrate has been reported to activate GPCR, thus facilitating the downstream control of gut alterations during obesity and diabetes [108,109].

6. Prebiotics in Atherosclerosis

Prebiotics are non-digestible dietary products that can be fermented by the gut microflora and stimulate the growth of beneficial bacteria that colonize the gut. Prebiotics and probiotics are both able to modulate the gut microbiome, resulting in favorable effects for the host. A study by Chen et al. discussed the positive effect of resveratrol (RSV), which is a natural polyphenol with prebiotic benefits, on gut health and, specifically, its ability to cause a reduction in TMAO levels in vivo [110]. RSV naturally occurs in grapes, berries, and other dietary constituents, and is described to be beneficial in the treatment of many metabolic diseases, including atherosclerosis [111]; however, its bioavailability is not high. Evidence elucidated that phenolic phytochemicals with poor bioavailability can act through remodeling the gut microbiota. It has been reported that a polyphenol-rich cranberry extract and metformin were able to reduce diet-induced metabolic syndrome in mice by altering the gut microbiota [112,113]. For instance, studies found that consumption of RSV can modulate the growth of specific gut microbiota in vivo; this included an increase in the Bacteroidetes-to-Firmicutes ratio, and the growth of *Bacteroides, Lactobacillus,* and *Bifidobacterium* [114–118]. As such, RSV was suggested as a potential prebiotic that could promote the growth of beneficial bacteria that confer health benefits to the host. Considering the association between TMAO levels, the gut microbiota, bile acid (BA) metabolism, and atherosclerosis, Chen et al. examined the effect of RSV on TMAO-induced atherosclerosis and the other mentioned factors in C57BL/6J and ApoE/mice. They were able to prove that RSV attenuated TMAO-induced atherosclerosis by decreasing TMAO levels and increasing hepatic BA neosynthesis through a remodeling of the gut microbiota. Moreover, they also showed that RSV-induced BA neosynthesis was partially mediated via downregulation of the enterohepatic farnesoid X receptor–fibroblast growth factor 15 (FXR/FGF15) axis [110].

7. Probiotics, TMAO, and Atherosclerotic Lesions

Probiotics are defined as "live strains of strictly selected microorganisms which, when administered in adequate amounts, confer a health benefit on the host" [119,120]. To date, conventional probiotics include lactic acid bacteria and some yeasts [13]. The gut microbiota has immunoregulatory functions and can affect the host's energy harvest [121], in addition to its effect on lipid metabolism [122] and intestinal barrier integrity [123,124]. Different factors indicate the major role of the gut microbiota in the pathogenesis of atherosclerosis. These factors include the major role of gut microbiota in generating atherogenic substances, such as TMA, the shared bacterial phylotype between atherosclerotic plaque and the oral and gut microbiome, and the specific gut metagenome in atherosclerotic patients [125]. It has been reported that some probiotics can support gut barrier functions, thereby reducing the translocation of bacterial and other immunogenic material from the gut [124,126]. Evidence also associated the use of probiotics with a reduction in different cardiovascular disease risk biomarkers, such as serum LDL and total cholesterol (TC), in addition to systemic inflammation [124,127]. Supplements with adequate probiotics were able to improve major atherosclerotic risk factors such as dyslipidemia, hypercholesterolemia, chronic inflammation, and hypertension [128,129]. Huang et al. reported the positive effect of a Lactobacillus strain in the reduction of atherosclerosis lesion area [130]. In addition, several meta-analyses reported a reduction in TC and LDL-C after the intake of probiotics and, specifically, *Lactobacillus acidophilus* [131,132]. Moreover, it has been reported that probiotics are able to improve the integrity of the epithelial barrier and support the function of tight junctions which can inhibit the translocation of harmful metabolites, such as TMAO and LPS, from entering the peripheral circulation leading to a stable atherosclerotic plaque (Figure 1) [61,133]. VSL-3 is a well-studied probiotic mixture containing eight different probiotic strains: *Bifidobacterium breve, Bifidobacterium longum, Bifidobacterium infantis, Lactobacillus acidophilus, Lactobacillus plantarum, Lactobacillus paracasei, Lactobacillus bulgaricus,* and *Streptococcus thermophilus.* Different studies have shown the beneficial effects of VSL-3 in different diseases, including ulcerative colitis [134], liver disease [135], and Crohn's disease [136]. Moreover, VSL-3 showed a promising potential in the treatment of atherosclerosis. Mencarelli et al. reported that VSL-3 was able to improve insulin signaling and protect against non-alcoholic steatohepatitis and atherosclerosis in ApoE$-/-$ mice with DSS-induced colitis [137]. In addition, Chan et al. compared the effect of VSL-3 with a positive control drug, telmisartan, that has proved to be effective in reducing atherogenesis in ApoE$-/-$ mice [124]. The results showed that VSL-3 was comparable to telmisartan in reducing the biomarkers of vascular inflammation and development of atherosclerosis [138]. The ability of probiotics to support overall gut health has led to more research showing the promising therapeutic effects of probiotics on disease. In fact, probiotics are currently used for the treatment or prevention of irritable bowel syndrome, inflammatory bowel diseases, gluten intolerance, antibiotic-associated diarrhea, and gastroenteritis [139]. Recent evidence has reported the contribution of the gut microbiota in different diseases via the gut–brain axis, gut–liver axis, gut–lung axis, and gut–vascular axis [41,44,126,139]. Furthermore, probiotics can modulate host immune responses [43,140,141], yet the interactions between probiotics, the gut, and the host immune system are very complex and are not fully understood. Studies have reported the positive role of probiotics in inflammation. For example, ApoE$-/-$ mice treated with *Pediococcus acidilactici* R037 showed a reduction in atherosclerotic lesion development via the suppression of pro-inflammatory cytokine production and IF-γ-producing CD4+ T cells [142]. Probiotics are also able to reduce inflammation by increasing the number of Treg cells [143]. A study conducted using VSL-3 showed that the DNA from this consortium was able to limit epithelial pro-inflammatory responses and attenuate the release of TNF-α in response to an *Escherichia coli* DNA injection [144]. Moreover, another study reported the ability of VSL-3 to reduce vascular inflammation in ApoE$-/-$ mice fed a HFD [124]. It has also been shown that the DNA from VSL-3 was able to employ anti-inflammatory effects through TLR9 signaling [145]; the authors also reported that the protective anti-inflammatory effect of probiotics was reconciled through their DNA and not through their metabolites, and that signaling of TLR9 had a major role in mediating this effect [43,146].

Nonetheless, not all probiotics have a proactive role in the treatment of atherosclerosis. For instance, *Lactobacillus reuteri* showed no effect on atherosclerosis in ApoE$-/-$ mice fed a HFD [43,147]. Huang et al. examined the effects of two Lactobacillus strains (*L. acidophilus* ATCC 4356 and 4962) on atherosclerosis development and atherosclerotic lesions in ApoE$-/-$ mice [130]. They reported a dramatic reduction in the atherosclerotic lesion area in the L.4356 group; however, no significant effect was observed in the L.4962 group [130]. In addition, L.4356 was able to significantly reduce plasma cholesterol levels [130]. In another study, Chen et al. reported that L.4356 was able to attenuate the atherosclerotic lesion development in ApoE$-/-$ mice through reducing inflammatory response and oxidative stress [148]. Moreover, Qiu et al. investigated the potential TMAO lowering property of five different probiotics strains, and only *Lactobacillus plantarum* ZDY04 was able to significantly lower the plasma TMAO levels. This was achieved through the remodeling of the gut microbiota, and not by affecting the expression of hepatic FMO3 and metabolizing choline, TMA, and TMAO [149]. Similarly, another study reported the TMAO-lowering potential of *Enterobacter aerogenes* ZDY01 in choline-fed mice; the effect was also attained through gut remodeling [150]. The TMAO-lowering property is strain specific, as a human study investigating the supplementation of *Streptococcus thermophilus* (KB19), *Lactobacillus acidophilus* (KB27), and *Bifidobacteria longum* (KB31) stated that there was no effect on plasma TMAO levels [151]. Another study by Tripolt et al. reported no effect on TMAO levels after 12 weeks of supplementation with *Lactobacillus casei Shirota* in patients with metabolic syndrome [152]. The mechanisms underlying the effects of probiotics on host health are still not fully understood [61]. As a result, the use of probiotics that can directly act on the TMA in the gut might be an alternative approach to reduce serum TMAO levels and prevent the development of atherosclerosis and "fish odor syndrome" [149].

8. Mechanisms Underlying the Therapeutic Effect of Probiotics in Atherosclerosis

The mechanisms underlying the protective effect of probiotics against atherosclerosis are not fully understood. Nevertheless, the action of probiotics at different steps is becoming clear as more studies are being conducted. To start with, it has been reported in previous studies how probiotics can combat gut dysbiosis through strengthening the epithelial tight junctions, preventing the translocation of damaging metabolites, such as LPS and TMAO, into the circulation, which can lead to a stable atherosclerotic plaque [61]. Several studies have demonstrated the hypocholesterolemic effect of probiotics [153,154]. Liong and Shah have pointed out the ability of Lactobacillus strains to reduce cholesterol in an in vitro model, and this was achieved through various mechanisms such as assimilation of cholesterol during growth, incorporation of cholesterol into the membrane of cells, the binding of cholesterol to the cell surface, and co-precipitation with deconjugated bile [155]. These findings were backed up in another study by Zeng et al., who studied *Lactobacillus buchneri* P2 and confirmed the cholesterol removal trait of this bacteria through an assimilation mechanism [156]. Another study by Huang and Zheng reported the cholesterol lowering property of a probiotic strain through the inhibition of the gene expression of NPC1L1 in Caco-2 cells [157,158]. The NPC1L1 protein plays a major role in cholesterol absorption, and it is considered to be a promising target for cholesterol-lowering medication [159]. NPC1L1 has been identified by Duval et al. to be a novel target gene for the liver X receptors (LXRs), which support the crucial role of LXRs in intestinal cholesterol homeostasis [160]. LXRs activation has been reported to reduce whole-body cholesterol levels and reduce atherosclerosis [161]. In addition, VSL-3 was found to improve lipid profiles in mice [162], and this was attained by promoting BA deconjugation and fecal excretion, and by increasing hepatic BA synthesis through the downregulation of the FXR/FGF15 axis [163]. BAs can regulate cholesterol balance, and disruption in the circulation of enterohepatic BAs can lead to gall bladder [164] and gastrointestinal diseases [165]. The metabolism of BAs is also associated with obesity, diabetes, and cardiovascular diseases [166–168]. BAs are synthesized from hepatic cholesterol, and they are further conjugated with amino acids glycine and taurine to form bile salts that are transferred to the intestine. Bile salts' amphiphilic combination is crucial for fat absorption in the intestine, yet excessive bile salts are toxic to the gut bacteria [169]. Bile salt hydrolase (BSH), which is present in the gut microbiome, is responsible for the catalysis of the conjugated bile salts into deconjugated BAs in order to maintain the balance of metabolism of BAs. Deconjugated BAs function as signaling molecules to aid in the secretion of GLP-1 hormone [170], activate other receptors, and impact different metabolic processes involved in various diseases [171,172]. The presence of BSH has been identified in different microbial genera such as *Lactobacillus, Bifidobacterium, Enterococcus, Clostridium spp.*, and *Bacteroides*. It has also been reported that one bacterial strain can possess distinct BSHs that can have different properties [173]. In addition, recent evidence suggested that TMAO can promote atherosclerosis, partially through inhibiting hepatic bile acid synthesis [110].

In addition, probiotics can apply their anti-inflammatory actions through modulating the expression of key transcription factors or microRNAs (miRNAs), which are associated with pro-inflammatory signaling [148,174]. For instance, Chen et al. reported that *L. acidophilus* ATCC 4356 was able to reduce the levels of TNF-α and oxidative stress markers in addition to its ability of reversing the reduction in IL-10 levels via inhibiting the activation of NF-κB and its translocation to the nucleus [148]. In the case of atherosclerosis, T lymphocytes and macrophages accumulate and proliferate at the atherosclerotic lesions, which leads to the secretion of inflammatory cytokines such as TNF-α and IL-10 [148]. These inflammatory cytokines can, in turn, activate intracellular NF-κB signaling pathways which can stimulate the production of more cytokines, leading to further inflammation [175]. Activated NF-κB exists in the fibrotic thickened intima-media and atheromatous areas of the atherosclerotic lesion, smooth muscle cells, macrophages, and endothelial cells; however, little or no activated NF-κB can be detected in vessels lacking atherosclerosis [176,177]. The activation of NF-κB is associated with the phosphorylation of IκB-α and subsequent degradation of IκB-α, which results in the translocation of NF-κB into the nucleus. It has also been reported that TNF-α is one of the inflammatory markers that can promote atherosclerosis. On the other hand, IL-10 is an anti-inflammatory marker that provides a crucial atheroprotective signal [44]. As for the miRNAs, *Lactobacillus acidophilus* has been shown to protect against apoptosis and necrosis in human endothelial cells, which is induced by LPS stimulation, and this in turn was associated with a decrease in the expression of pro-inflammatory miR-155 and increased expression of anti-apoptotic miR-21 [43,174]. Nonetheless, further research is still needed to explain the effect of probiotics on the expression of miRNAs associated with atherosclerosis [43]. More studies showing the atheroprotective effect of different probiotics are described in Table 1.

Table 1. Anti-atherogenic effects of probiotics in human and animal studies.

Reference	Study Design	Study Group	Bacteria Administered	Athero-Protective Outcomes
Chan et al. 2016 [124]	ApoE−/− mice were fed HFD alone or with VSL#3 or a positive control treatment, telmisartan or both for 12 weeks	ApoE−/− mice	VSL#3	VSL#3 reduced pro-inflammatory adhesion molecules and risk factors of plaque rupture, and reduced vascular inflammation and atherosclerosis to a similar extent to telmisartan. Combining both VSL#3 and telmisartan showed no further benefits
Huang et al. 2014 [130]	Eight week-old ApoE−/− mice fed a Western diet with or without *L. acidophilus ATCC 4356* daily for 16 weeks	ApoE−/− mice	*Lactobacillus acidophilus (ATCC 4356)*	*L. acidophilus ATCC 4356* protected ApoE−/− mice from atherosclerosis by reducing their plasma cholesterol levels.
Chen et al. 2013 [148]	Eight week-old ApoE−/− mice treated with *L. acidophilus ATCC 4356* daily for 12 weeks. Wild-type (WT) mice or ApoE−/− mice (control group treated with saline only). Body weight, serum lipid levels, aortic atherosclerotic lesions, and inflammatory status were examined	ApoE−/− mice	*Lactobacillus acidophilus ATCC 4356*	Decreased atherosclerotic lesion size, decreased levels of serum malondialdehyde (MDA), oxLDL, and TNF-α; increased levels of IL-10 and superoxide dismutase (SOD) activity in serum.
Qiu et al. 2018 [149]	Five probiotic strains were investigated for choline-induced TMAO levels in ApoE−/− mice supplemented with 1.3% choline. Only *Lactobacillus plantarum ZDY04 (PLA04)* was subjected for further investigation.	ApoE−/− mice	*Lactobacillus plantarum ZDY01 (PLA01), Lactobacillus rhamnosus ZDY9 (LGG), Lactobacillus plantarum ZDY04 (PLA04), Lactobacillus caseii ZDY8 (CAS), Lactobacillus bulgaricus ZDY5 (BUL)*	*L. plantarum ZDY04* reduced serum TMAO levels and cecal TMA levels, and inhibited atherosclerotic lesion formation. *L. plantarum ZDY04* had no effect on hepatic FMO3 activity.
Qiu et al. 2017 [150]	*Enterobacter aerogenes ZDY01* was administered to choline-fed mice. Serum TMAO and cecal TMA levels were measured	Mice	*Enterobacter aerogenes ZDY01*	Reduction in serum TMAO and cecal TMA levels.

Table 1. *Cont.*

Reference	Study Design	Study Group	Bacteria Administered	Athero-Protective Outcomes
Borges et al. 2019 [151]	21 patients with chronic kidney disease in a double-blind pilot study (3 months duration). A total of 10 patients in the placebo group and 11 patients in the probiotic group. Plasma TMAO, choline, and betaine were measured	Human	*Streptococcus thermophilus* (KB19), *Lactobacillus acidophilus* (KB27), *Bifidobacteria longum* (KB31)	No change in TMAO levels and significant increase in betaine plasma levels after probiotic supplementation. Significant decrease in choline plasma levels in placebo group.
Jones et al. 2012 [178]	114 subjects in a double-blind, placebo-controlled, randomized study received either yogurts containing microencapsulated *L. reuteri* NCIMB 30242 or placebo yogurts	Human	*Lactobacillus reuteri NCIMB 30242*	Reduction in LDL-C, TC, apoB-100, and non-HDL-C.
Rajkumar et al. 2014 [179]	Subjects randomized into four groups: placebo, omega-3 fatty acid, probiotic VSL#3, or both omega-3 and probiotic, for 6 weeks. Blood and fecal samples examined at baseline and after 6 weeks	Human	VSL#3	Reduction in TC, triglyceride(TG), LDL, and VLDL; increased HDL levels, improved insulin sensitivity and decreased hsCRP.
Rerksuppaphol et al. 2015 [180]	Patients diagnosed with hypercholesterolemia received probiotic capsule of *Lactobacillus acidophilus* plus *Bifidobacterium bifidum* three times daily for six weeks. TC, HDL-C, LDL-C, and TG levels were measured	Human	*Lactobacillus acidophilus, Bifidobacterium bifidum*	Decreased TC, HDL-C, and LDL-C levels in probiotic group.
Boutagy et al. 2015 [181]	Nineteen healthy, non-obese males (18–30 years) were randomized to either VSL#3 or placebo during the consumption of a hypercaloric, high-fat diet for 4 weeks. Plasma TMAO, L-carnitine, choline, and betaine (UPLC-MS/MS) were measured at baseline and following a HFD.	Human	VSL#3	Increased plasma TMAO in both the VSL#3 and placebo groups. Plasma L-carnitine, choline, and betaine concentrations did not increase following the HFD in either group. VSL#3 treatment did not influence plasma TMAO concentrations.
Bjerg et al. 2015 [182]	Fecal samples were collected at baseline, after four weeks supplementation, and two weeks after the supplementation was ended; fasting blood samples were collected at baseline and after 4 weeks.	Human	*Lactobacillus paracasei subsp. paracasei, Lactobacillus casei W8®*	Reduced TG
Bernini et al. 2016 [183]	Fifty-one patients with MetS were divided into a control group and a probiotic group. The probiotic group received fermented milk with probiotics for 45 d. The effects of *B. lactis* on lipid profile, glucose metabolism, and pro-inflammatory cytokines were assessed in blood samples.	Human	*Bifidobacterium lactis HN019*	Reduction in body mass index (BMI), TC, LDL-C, TNF-α and IL-6.

Table 1. *Cont.*

Reference	Study Design	Study Group	Bacteria Administered	Athero-Protective Outcomes
Madjd et al. 2016 [184]	Overweight and obese women consumed either a probiotic yogurt (PY) or a standard low-fat yogurt (LF) every day with their main meals for 12 weeks while following a weight-loss program	Human	*Streptococcus thermophiles, Lactobacillus bulgaricus, Lactobacillus acidophilus* LA5, *Bifidobacterium lactis* BB12	Reduction in TC, LDL-C, insulin resistance, postprandial blood glucose, and fasting insulin.
Chan et al. 2016 [185]	12 weeks feeding of HFD as opposed to normal chow diet (ND) in ApoE−/− mice. LGG or TLM supplementation to HFD was studied	ApoE−/− mice	*L. rhamnosus GG (LGG)*	Reduced lesion development; decreased plasma cholesterol, sE-selectin, sICAM-1, sVCAM-1, and endotoxin.
Costabile et al. 2017 [186]	Double-blind, placebo-controlled, randomized design in which subjects received encapsulated *Lactobacillus plantarum ECGC 13110402* twice daily.	Human	*Lactobacillus plantarum* ECGC 13110402	Reduction in LDL. Reduction in systolic blood pressure.
Firouzi et al. 2017 [187]	A randomized, double-blind, parallel-group, controlled clinical trial included 136 participants with type 2 diabetes, aged 30–70 years who received either probiotics or placebo for 12 weeks.	Human	*Lactobacillus acidophilus, Lactobacillus casei, Lactobacillus lactis, Bifidobacterium bifdum, Bifidobacterium longum,* and *Bifidobacterium infantis*	Improved HbA1c and fasting insulin.
Yoshida et al. 2018 [188]	Oral gavage of *Bacteroides vulgatus* and *Bacteroides dorei* in 6-week-old female ApoE−/− mice 5 times per week for 10 weeks. At 16 weeks of age, the mice were euthanized and analyses were performed to evaluate atherosclerosis.	ApoE−/− mice	*Bacteroides vulgatus, Bacteroides dorei*	Reduced plaque inflammation, attenuating atherosclerotic lesion form.
Saika et al. 2018 [189]	Wistar rats fed a high-cholesterol diet received *Saccharomyces cerevisiae ARDMC1*	Wistar rats	*Saccharomyces cerevisiae* ARDMC1	Reduced TC, LDL, and TG.
Huang et al. 2018 [190]	Oral administration with *Enterococcus faecium* to rats for 35 days. The gene transcriptions related to cholesterol metabolism, composition of bile acids in feces, synthesis of TMAO in the liver, and composition of the gut microbiota of rats were examined.	Rats	*Enterococcus faecium* WEFA23	Reduction of cholesterol, upregulation of genes' transcript level relevant to cholesterol decomposition and transportation, and downregulation of genes involved in cholesterol synthesis. Decreased TMAO production followed by increasing the CYP7A1 transcript level.

Table 1. *Cont.*

Reference	Study Design	Study Group	Bacteria Administered	Athero-Protective Outcomes
Szulinska et al. 2018 [191]	81 obese Caucasian women randomly assigned to three groups: a placebo, low dose (LD), and high dose (HD) of lyophilisate powder containing live multispecies probiotic bacteria. The probiotic supplement was administered daily for 12 weeks.	Human	*Bifidobacterium bifidum* W23, *Bifidobacterium lactis* W51, *Bifidobacterium lactis* W52, *Lactobacillus acidophilus* W37, *Lactobacillus brevis* W63, *Lactobacillus casei* W56, *Lactobacillus salivarius* W24, *Lactococcus lactis* W19, *Lactococcus lactis* W58	HD decreased systolic blood pressure, vascular endothelial growth factor, pulse wave analysis systolic pressure, pulse wave analysis pulse pressure, pulse wave analysis augmentation index, pulse wave velocity, IL-6, TNF-α, and thrombomodulin. LD decreased the systolic blood pressure and IL-6 levels.
Tang et al. 2021 [192]	*E.aerogenes* ZDY01 was administered to ApoE−/− mice fed with 1.3% choline	ApoE−/− mice	*Enterobacter aerogenes* ZDY01	Inhibition of choline-induced atherosclerosis. Reduction of cecal TMA and serum TMAO levels and modulation of CDCA-FXR/FGF15 axis.
Wang et al. 2022 [193]	Eight strains of *Bifidobacterium breve* and eight strains of *Bifidobacterium longum* were administered to choline-fed C57BL/6J mice for 6 weeks	C57BL/6J Mice	Eight strains of *Bifidobacterium breve* and eight strains of *Bifidobacterium longum*	*B. breve* Bb4 and *B. longum* BL1 and BL7 significantly reduced plasma TMAO and plasma and cecal TMA concentrations.

9. Different Strategies Involved in Inhibiting TMAO Formation

Considering the composite nature of TMAO formation, different strategies have been predicted to affect the different pathways for TMAO production and reduce the risk of atherosclerosis and cardiovascular disease. Based on this context, a dietary intervention, such as limiting animal-based foods that leads to TMA, could be a straightforward solution; however, this could lead to clinical consequences due to the deficiency of major nutrients that are required for optimal health [194]. Another research interest has focused on identifying the specific enzymes responsible for TMAO production. FMO3 is an enzyme in the liver that is reported to convert TMA into TMAO, and the inhibition of this enzyme would lead to the accumulation of TMA in the gut, which would cause trimethylaminuria, also known as fish malodor syndrome [195,196]. Therefore, the optimal approach for TMAO reduction would be to decrease TMA formation by the gut microbes [197]. Research described the use of 3,3-dimethyl-1-butanol (DMB), the structural analog of choline, as a drug to reduce TMA formation via the inhibition of microbial choline TMA lyase [23]. The results showed that the use of DMB was able to reduce TMAO circulating levels, in addition to its ability to suppress macrophage foam cell formation and ameliorate atherosclerotic lesion formation in vivo [23]. Moreover, DMB can reduce platelet activation and the thrombus formation rate [197]. Despite all these promising results, the effects of DMB were reversible, meaning a direct injection of TMAO could completely reverse the positive outcomes of DMB. Other choline analogues and second-generation TMA lyase inhibitors include fluoromethylcholine (FMC) and iodomethylcholine (IMC). Both FMC and IMC were able to promote the irreversible inhibition of microbial TMA lyase. They were also able to suppress TMA and TMAO levels, and reduce thrombus formation without any noticed toxicity in vivo compared to DMB [197]. Another agent used was the pharmacological product meldonium, which acts as an analogue of γ-butyrobetaine. Evidence has shown that long-term administration of meldonium decreased the levels of circulating L-carnitine in healthy non-vegetarian individuals by suppressing γ-butyrobetaine hydroxylase enzyme [198]. It has also been reported that treatment with meldonium led to a decreased plasma TMAO concentration through increased urinary excretion [199,200]. These different reported treatments could be promising in the prevention of atherosclerosis or cardiovascular disease [201]. Other studies considered that modulating the gut microbiota composition and metabolic function could be an optimal strategy for reducing TMAO levels, as the gut microbiota has been reported to be a major factor in determining the amount of TMA generated. Some literature has demonstrated that antibiotic administration could decrease levels of TMA-generating bacteria, thus decreasing TMAO levels. However, short-term changes in TMAO levels and the high risk of antibiotic resistance are one of the reasons to disregard the use of antibiotics as a therapeutic strategy to modulate TMAO levels. Another studied strategy for modulating the gut microbiota and its metabolic function was repopulating the gut with other microorganisms that are able to cause a decrease in circulating TMA levels. As such, fecal microbial transplant (FMT) has been reported to successfully transmit atherosclerosis susceptibility [202], in addition to increasing the thrombosis potential [203] and leading to higher platelet reactivity in animal models. On the contrary, in a double-blinded controlled pilot study, the FMT from a lean vegan donor did not improve TMAO levels, and it did not have any effect on the parameters of vascular inflammation, despite it causing changes in the intestinal microbiota composition [204]. Furthermore, Kajllmo et al. investigated the effect of FMT from healthy donors used to treat patients with irritable bowel syndrome (IBS) on plasma lipids and LDL/HDL subfractions in a randomized, double-blinded study; they reported no significant effect of FMT on LDL, HDL, and TC levels [205]. However, a recent study by Kim et al. examined the effect of the gut microbiota on the pathogenesis of atherosclerosis in a transgenic atherosclerosis model with C1q/TNF-related protein 9-knockout (CTRP9-KO) mice. CTRP9 plays an important role in cardiovascular homeostasis, promotes endothelial cell function, and improves endothelial-dependent vasorelaxation. They were able to demonstrate that an FMT from wild-type (WT) mice into CTRP9-KO mice was able to decrease atherosclerotic lesions in carotid arteries. In contrast, wild-type (WT) mice transplanted with FMT from CTRP9-KO mice showed progression of atherosclerosis [206]. These results are promising, but more studies are still needed to prove that FMT is a potential treatment for atherosclerosis or CVD [194,207]. Another repopulating strategy to alter the microbiota that has been used is the administration of probiotics. The effects of probiotics on TMAO circulating levels have been reported previously in our review.

10. Conclusions

Recently, gut microbiota and microbial metabolites, including TMA and SCFAs, have attracted the focus of researchers due to their crucial role in the development of atherosclerosis and cardiovascular disease. Gut microbiota composition alters between different age groups, and this compositional shift is associated with immune dysregulation and the onset of aging-associated pathologies such as PAD. Gut dysbiosis in the elderly has been associated with the impairment of intestinal barrier integrity, an increase in gut leakiness, endotoxemia, and subsequent inflammaging. Studies have provided evidence of the beneficial role of probiotics, prebiotics, and SCFAs in the management of several atherogenic risk factors. However, some atheroprotective effects of probiotics were strain-specific, and further research needs to be performed to better understand the mechanisms behind the different effects. Further inspection is also required to confirm the atheroprotective effect of SCFAs, and whether the effect of their supplementation persists over the long term. In addition, targeting TMA and TMAO might act as a potential novel therapeutic strategy to prevent atherosclerosis development, plaque rupture, and cardiovascular disease. Although findings in this regard were promising, the exact mechanisms of gut dysbiosis and microbiota-derived TMAO involved in atherosclerosis are not yet fully understood. More well-conducted studies focusing on molecular mechanisms and precise treatments targeting gut microbiota-dependent metabolites for anti-atherosclerosis and, specifically, in the elderly, remain to be completed in further investigations. Additional studies are also required to better understand the factors leading to TMA-forming bacteria and their consequent therapeutic manipulations.

Author Contributions: Conceptualization and supervision I.H. and R.E.H. Writing original draft, reviewing, and editing the manuscript R.E.H., N.A.-A. and I.H. All authors have read and agreed to the published version of the manuscript.

Funding: Open Access Publication Fund of Charité—Universitätsmedizin Berlin and the German Research Foundation (DFG).

Institutional Review Board Statement: Not applicable.

Informed Consent Statement: Not applicable.

Data Availability Statement: Not applicable.

Acknowledgments: We would like to thank Yingqi Yu for his help with obtaining the publication license for Figure 1, which was created by BioRender.

Conflicts of Interest: The authors declare no conflict of interest.

References

1. Hansson, G.K.; Libby, P.; Schönbeck, U.; Yan, Z.Q. Innate and adaptive immunity in the pathogenesis of atherosclerosis. *Circ. Res.* **2002**, *91*, 281–291. [CrossRef] [PubMed]
2. Komaroff, A.L. The Microbiome and Risk for Atherosclerosis. *JAMA* **2018**, *319*, 2381–2382. [CrossRef]
3. WHO. Diabetes. Available online: https://www.who.int/health-topics/diabetes#tab=tab_1 (accessed on 5 November 2022).
4. Sasor, A.; Ohlsson, B. Microangiopathy is common in submucosal vessels of the colon in patients with diabetes mellitus. *Rev. Diabet. Stud.* **2014**, *11*, 175–180. [CrossRef] [PubMed]
5. Fiordaliso, F.; Clerici, G.; Maggioni, S.; Caminiti, M.; Bisighini, C.; Novelli, D.; Minnella, D.; Corbelli, A.; Morisi, R.; De Iaco, A.; et al. Prospective study on microangiopathy in type 2 diabetic foot ulcer. *Diabetologia* **2016**, *59*, 1542–1548. [CrossRef] [PubMed]
6. Hasheminasabgorji, E.; Jha, J.C. Dyslipidemia, Diabetes and Atherosclerosis: Role of Inflammation and ROS-Redox-Sensitive Factors. *Biomedicines* **2021**, *9*, 1602. [CrossRef] [PubMed]
7. Xiao, Y.; Guo, Z.; Li, Z.; Ling, H.; Song, C. Role and mechanism of action of butyrate in atherosclerotic diseases: A review. *J. Appl. Microbiol.* **2021**, *131*, 543–552. [CrossRef] [PubMed]
8. Koh, A.; De Vadder, F.; Kovatcheva-Datchary, P.; Bäckhed, F. From Dietary Fiber to Host Physiology: Short-Chain Fatty Acids as Key Bacterial Metabolites. *Cell* **2016**, *165*, 1332–1345. [CrossRef]
9. Canfora, E.E.; Jocken, J.W.; Blaak, E.E. Short-chain fatty acids in control of body weight and insulin sensitivity. *Nat. Rev. Endocrinol.* **2015**, *11*, 577–591. [CrossRef] [PubMed]
10. Miyamoto, J.; Kasubuchi, M.; Nakajima, A.; Irie, J.; Itoh, H.; Kimura, I. The role of short-chain fatty acid on blood pressure regulation. *Curr. Opin. Nephrol. Hypertens.* **2016**, *25*, 379–383. [CrossRef]
11. Leung, C.; Rivera, L.; Furness, J.B.; Angus, P.W. The role of the gut microbiota in NAFLD. *Nat. Rev. Gastroenterol. Hepatol.* **2016**, *13*, 412–425. [CrossRef]
12. Brown, J.M.; Hazen, S.L. Microbiome series: Microbial modulation of cardiovascular disease. *Nat. Rev. Microbiol.* **2018**, *16*, 171–181. [CrossRef] [PubMed]
13. El Hage, R.; Hernandez-Sanabria, E.; Van De Wiele, T. Emerging Trends in "Smart Probiotics": Functional Consideration for the Development of Novel Health and Industrial Applications. *Front. Microbiol.* **2017**, *8*, 01889. [CrossRef]
14. Emeline Ragonnaud, A.B. Gut microbiota as the key controllers of "healthy" aging of elderly people. *Immun. Ageing* **2021**, *18*, 2. [CrossRef] [PubMed]
15. Rath, S.; Heidrich, B.; Pieper, D.H.; Vital, M. Uncovering the trimethylamine-producing bacteria of the human gut microbiota. *Microbiome* **2017**, *5*, 54. [CrossRef] [PubMed]
16. Janeiro, M.H.; Ramírez, M.J.; Milagro, F.I.; Martínez, J.A.; Solas, M. Implication of Trimethylamine N-Oxide (TMAO) in Disease: Potential Biomarker or New Therapeutic Target. *Nutrients* **2018**, *10*, 1398. [CrossRef] [PubMed]
17. Rath, S.; Rud, T.; Pieper, D.H.; Vital, M. Potential TMA-Producing Bacteria Are Ubiquitously Found in Mammalia. *Front. Microbiol.* **2020**, *10*, 2966. [CrossRef] [PubMed]
18. Wang, Z.; Klipfell, E.; Bennett, B.J.; Koeth, R.; Levison, B.S.; DuGar, B.; Feldstein, A.E.; Britt, E.B.; Fu, X.; Chung, Y.-M.; et al. Gut flora metabolism of phosphatidylcholine promotes cardiovascular disease. *Nature* **2011**, *472*, 57–63. [CrossRef]
19. Koeth, R.A.; Wang, Z.; Levison, B.S.; Buffa, J.A.; Org, E.; Sheehy, B.T.; Britt, E.B.; Fu, X.; Wu, Y.; Li, L.; et al. Intestinal microbiota metabolism of L-carnitine, a nutrient in red meat, promotes atherosclerosis. *Nat. Med.* **2013**, *19*, 576–585. [CrossRef]
20. Zhu, W.; Gregory, J.C.; Org, E.; Buffa, J.A.; Gupta, N.; Wang, Z.; Li, L.; Fu, X.; Wu, Y.; Mehrabian, M.; et al. Gut Microbial Metabolite TMAO Enhances Platelet Hyperreactivity and Thrombosis Risk. *Cell* **2016**, *165*, 111–124. [CrossRef]
21. Seldin, M.M.; Meng, Y.; Qi, H.; Zhu, W.; Wang, Z.; Hazen, S.L.; Lusis, A.J.; Shih, D.M. Trimethylamine N-Oxide Promotes Vascular Inflammation Through Signaling of Mitogen-Activated Protein Kinase and Nuclear Factor-κB. *J. Am. Heart Assoc.* **2016**, *5*, e002767. [CrossRef] [PubMed]
22. Jia, X.; Osborn, L.J.; Wang, Z. Simultaneous Measurement of Urinary Trimethylamine (TMA) and Trimethylamine N-Oxide (TMAO) by Liquid Chromatography–Mass Spectrometry. *Molecules* **2020**, *25*, 1862. [CrossRef] [PubMed]

23. Wang, Z.; Roberts, A.B.; Buffa, J.A.; Levison, B.S.; Zhu, W.; Org, E.; Gu, X.; Huang, Y.; Zamanian-Daryoush, M.; Culley, M.K.; et al. Non-lethal Inhibition of Gut Microbial Trimethylamine Production for the Treatment of Atherosclerosis. *Cell* **2015**, *163*, 1585–1595. [CrossRef] [PubMed]

24. Warrier, M.; Shih, D.M.; Burrows, A.C.; Ferguson, D.; Gromovsky, A.D.; Brown, A.L.; Marshall, S.; McDaniel, A.; Schugar, R.C.; Wang, Z.; et al. The TMAO Generating Enzyme Flavin Monooxygenase 3 is a Central Regulator of Cholesterol Balance. *Cell Rep.* **2016**, *10*, 326–338. [CrossRef] [PubMed]

25. Miao, J.; Morbid Obesity Study Group; Ling, A.V.; Manthena, P.V.; Gearing, M.E.; Graham, M.J.; Crooke, R.M.; Croce, K.J.; Esquejo, R.M.; Clish, C.; et al. Flavin-containing monooxygenase 3 as a potential player in diabetes-associated atherosclerosis. *Nat. Commun.* **2015**, *6*, 6498. [CrossRef] [PubMed]

26. Chen, K.; Zheng, X.; Feng, M.; Li, D.; Zhang, H. Gut Microbiota-Dependent Metabolite Trimethylamine N-Oxide Contributes to Cardiac Dysfunction in Western Diet-Induced Obese Mice. *Front. Physiol.* **2017**, *8*, 139. [CrossRef] [PubMed]

27. Sun, G.; Yin, Z.; Liu, N.; Bian, X.; Yu, R.; Su, X.; Zhang, B.; Wang, Y. Gut microbial metabolite TMAO contributes to renal dysfunction in a mouse model of diet-induced obesity. *Biochem. Biophys. Res. Commun.* **2017**, *493*, 964–970. [CrossRef]

28. Senthong, V.; Wang, Z.; Fan, Y.; Wu, Y.; Hazen, S.L.; Tang, W.W. Trimethylamine N-Oxide and Mortality Risk in Patients With Peripheral Artery Disease. *J. Am. Heart Assoc.* **2016**, *5*, e004237. [CrossRef]

29. Shih, D.M.; Wang, Z.; Lee, R.; Meng, Y.; Che, N.; Charugundla, S.; Qi, H.; Wu, J.; Pan, C.; Brown, J.M.; et al. Flavin containing monooxygenase 3 exerts broad effects on glucose and lipid metabolism and atherosclerosis. *J. Lipid Res.* **2015**, *56*, 22–37. [CrossRef]

30. Biscetti, F.; Nardella, E.; Cecchini, A.L.; Landolfi, R.; Flex, A. The Role of the Microbiota in the Diabetic Peripheral Artery Disease. *Mediat. Inflamm.* **2019**, *2019*, 4128682. [CrossRef]

31. Pierce, G.L.; Roy, S.J.; Gimblet, C.J. The Gut-Arterial Stiffness Axis: Is TMAO a Novel Target to Prevent Age-Related Aortic Stiffening? *Hypertension* **2021**, *78*, 512–515. [CrossRef]

32. Hui, D.Y. Intestinal phospholipid and lysophospholipid metabolism in cardiometabolic disease. *Curr. Opin. Lipidol.* **2016**, *27*, 507–512. [CrossRef]

33. Tang, W.H.W.; Wang, Z.; Levison, B.S.; Koeth, R.A.; Britt, E.B.; Fu, X.; Wu, Y.; Hazen, S.L. Intestinal microbial metabolism of phosphatidylcholine and cardiovascular risk. *N. Engl. J. Med.* **2013**, *368*, 1575–1584. [CrossRef] [PubMed]

34. Wang, Z.; Bergeron, N.; Levison, B.S.; Li, X.S.; Chiu, S.; Jia, X.; Koeth, R.A.; Li, L.; Wu, Y.; Tang, W.H.W.; et al. Impact of chronic dietary red meat, white meat, or non-meat protein on trimethylamine N-oxide metabolism and renal excretion in healthy men and women. *Eur. Heart J.* **2019**, *40*, 583–594. [CrossRef] [PubMed]

35. Koeth, R.A.; Lam-Galvez, B.R.; Kirsop, J.; Wang, Z.; Levison, B.S.; Gu, X.; Copeland, M.F.; Bartlett, D.; Cody, D.B.; Dai, H.J.; et al. l-Carnitine in omnivorous diets induces an atherogenic gut microbial pathway in humans. *J. Clin. Investig.* **2019**, *129*, 373–387. [CrossRef] [PubMed]

36. Rath, S.; Rud, T.; Karch, A.; Pieper, D.H.; Vital, M. Pathogenic functions of host microbiota. *Microbiome* **2018**, *6*, 174. [CrossRef]

37. Buford, T.W. (Dis)Trust your gut: The gut microbiome in age-related inflammation, health, and disease. *Microbiome* **2017**, *5*, 80. [CrossRef]

38. Rath, S.; Rox, K.; Kleine Bardenhorst, S.; Schminke, U.; Dörr, M.; Mayerle, J.; Frost, F.; Lerch, M.M.; Karch, A.; Brönstrup, M.; et al. Higher Trimethylamine-N-Oxide Plasma Levels with Increasing Age Are Mediated by Diet and Trimethylamine-Forming Bacteria. *mSystems* **2021**, *6*, e00945-21. [CrossRef] [PubMed]

39. Brunt, V.E.; Gioscia-Ryan, R.A.; Richey, J.J.; Zigler, M.C.; Cuevas, L.M.; Gonzalez, A.; Vázquez-Baeza, Y.; Battson, M.L.; Smithson, A.T.; Gilley, A.D.; et al. Suppression of the gut microbiome ameliorates age-related arterial dysfunction and oxidative stress in mice. *J. Physiol.* **2019**, *597*, 2361–2378. [CrossRef]

40. Ke, Y.; Li, D.; Zhao, M.; Liu, C.; Liu, J.; Zeng, A.; Shi, X.; Cheng, S.; Pan, B.; Zheng, L.; et al. Gut flora-dependent metabolite Trimethylamine-N-oxide accelerates endothelial cell senescence and vascular aging through oxidative stress. *Free. Radic. Biol. Med.* **2018**, *116*, 88–100. [CrossRef] [PubMed]

41. Lau, K.; Srivatsav, V.; Rizwan, A.; Nashed, A.; Liu, R.; Shen, R.; Akhtar, M. Bridging the Gap between Gut Microbial Dysbiosis and Cardiovascular Diseases. *Nutrients* **2017**, *9*, 859. [CrossRef]

42. Feng, Q.; Chen, W.D.; Wang, Y.D. Gut Microbiota: An Integral Moderator in Health and Disease. *Front. Microbiol.* **2018**, *9*, 151. [CrossRef] [PubMed]

43. O'Morain, V.L.; Ramji, D.P. The Potential of Probiotics in the Prevention and Treatment of Atherosclerosis. *Mol. Nutr. Food Res.* **2020**, *64*, e1900797. [CrossRef] [PubMed]

44. Tang, W.H.W.; Bäckhed, F.; Landmesser, U.; Hazen, S.L. Intestinal Microbiota in Cardiovascular Health and Disease: JACC State-of-the-Art Review. *J. Am. Coll. Cardiol.* **2019**, *73*, 2089–2105. [CrossRef] [PubMed]

45. Khan, S.S.; Singer, B.D.; Vaughan, D.E. Molecular and physiological manifestations and measurement of aging in humans. *Aging Cell* **2017**, *16*, 624–633. [CrossRef] [PubMed]

46. Ferrucci, L.; Gonzalez-Freire, M.; Fabbri, E.; Simonsick, E.; Tanaka, T.; Moore, Z.; Salimi, S.; Sierra, F.; de Cabo, R. Measuring biological aging in humans: A quest. *Aging Cell* **2020**, *19*, e13080. [CrossRef] [PubMed]

47. Montecino-Rodriguez, E.; Berent-Maoz, B.; Dorshkind, K. Causes, consequences, and reversal of immune system aging. *J. Clin. Investig.* **2013**, *123*, 958–965. [CrossRef]

48. Bosco, N.; Noti, M. The aging gut microbiome and its impact on host immunity. *Genes Immun.* **2021**, *22*, 289–303. [CrossRef]

49. Xu, C.; Zhu, H.; Qiu, P. Aging progression of human gut microbiota. *BMC Microbiol.* **2019**, *19*, 236. [CrossRef]

50. Woodmansey, E.J. Intestinal bacteria and ageing. *J. Appl. Microbiol.* **2007**, *102*, 1178–1186. [CrossRef] [PubMed]
51. Jeffery, I.B.; Lynch, D.B.; O'Toole, P.W. Composition and temporal stability of the gut microbiota in older persons. *ISME J.* **2016**, *10*, 170–182. [CrossRef] [PubMed]
52. O'Toole, P.W.; Jeffery, I.B. Gut microbiota and aging. *Science* **2015**, *350*, 1214–1215. [CrossRef] [PubMed]
53. Mariat, D.; Firmesse, O.; Levenez, F.; Guimarães, V.; Sokol, H.; Doré, J.; Corthier, G.; Furet, J.P. The Firmicutes/Bacteroidetes ratio of the human microbiota changes with age. *BMC Microbiol.* **2009**, *9*, 123. [CrossRef] [PubMed]
54. Schneeberger, M.; Everard, A.; Gómez-Valadés, A.G.; Matamoros, S.; Ramírez, S.; Delzenne, N.M.; Gomis, R.; Claret, M.; Cani, P.D. Akkermansia muciniphila inversely correlates with the onset of inflammation, altered adipose tissue metabolism and metabolic disorders during obesity in mice. *Sci. Rep.* **2015**, *5*, 16643. [CrossRef] [PubMed]
55. Ling, Z.; Liu, X.; Cheng, Y.; Yan, X.; Wu, S. Gut microbiota and aging. *Crit. Rev. Food Sci. Nutr.* **2022**, *62*, 3509–3534. [CrossRef] [PubMed]
56. Yang, T.; Santisteban, M.M.; Rodriguez, V.; Li, E.; Ahmari, N.; Carvajal, J.M.; Zadeh, M.; Gong, M.; Qi, Y.; Zubcevic, J.; et al. Gut Dysbiosis Is Linked to Hypertension. *Hypertension* **2015**, *65*, 1331–1340. [CrossRef] [PubMed]
57. Zhang, L.; Yu, F.; Xia, J. Trimethylamine N-oxide: Role in cell senescence and age-related diseases. *Eur. J. Nutr.* **2022**. [CrossRef] [PubMed]
58. Huang, R.; Yan, L.; Lei, Y. The Gut Microbial-Derived Metabolite Trimethylamine N-Oxide and Atrial Fibrillation: Relationships, Mechanisms, and Therapeutic Strategies. *Clin. Interv. Aging* **2021**, *16*, 1975. [CrossRef] [PubMed]
59. Brunt, V.E.; Gioscia-Ryan, R.A.; Casso, A.G.; VanDongen, N.S.; Ziemba, B.P.; Sapinsley, Z.J.; Richey, J.J.; Zigler, M.C.; Neilson, A.P.; Davy, K.P.; et al. Trimethylamine-N-Oxide Promotes Age-Related Vascular Oxidative Stress and Endothelial Dysfunction in Mice and Healthy Humans. *Hypertension* **2020**, *76*, 101–112. [CrossRef]
60. Morrison, D.J.; Preston, T. Formation of short chain fatty acids by the gut microbiota and their impact on human metabolism. *Gut Microbes* **2016**, *7*, 189–200. [CrossRef] [PubMed]
61. Shen, X.; Li, L.; Sun, Z.; Zang, G.; Zhang, L.; Shao, C.; Wang, Z. Gut Microbiota and Atherosclerosis-Focusing on the Plaque Stability. *Front. Cardiovasc. Med.* **2021**, *8*, 668532. [CrossRef]
62. Chambers, E.S.; Preston, T.; Frost, G.; Morrison, D.J. Role of Gut Microbiota-Generated Short-Chain Fatty Acids in Metabolic and Cardiovascular Health. *Curr. Nutr. Rep.* **2018**, *7*, 198–206. [CrossRef] [PubMed]
63. Vourakis, M.; Mayer, G.; Rousseau, G. The Role of Gut Microbiota on Cholesterol Metabolism in Atherosclerosis. *Int. J. Mol. Sci.* **2021**, *22*, 8074. [CrossRef] [PubMed]
64. D'Souza, W.N.; Douangpanya, J.; Mu, S.; Jaeckel, P.; Zhang, M.; Maxwell, J.R.; Rottman, J.B.; Labitzke, K.; Willee, A.; Beckmann, H.; et al. Differing roles for short chain fatty acids and GPR43 agonism in the regulation of intestinal barrier function and immune responses. *PLoS ONE* **2017**, *12*, e0180190. [CrossRef] [PubMed]
65. Hu, J.; Lin, S.; Zheng, B.; Cheung, P.C.K. Short-chain fatty acids in control of energy metabolism. *Crit. Rev. Food Sci. Nutr.* **2018**, *58*, 1243–1249. [CrossRef] [PubMed]
66. Müller, M.; Hernández, M.A.G.; Goossens, G.H.; Reijnders, D.; Holst, J.J.; Jocken, J.W.E.; van Eijk, H.; Canfora, E.E.; Blaak, E.E. Circulating but not faecal short-chain fatty acids are related to insulin sensitivity, lipolysis and GLP-1 concentrations in humans. *Sci. Rep.* **2019**, *9*, 12515. [CrossRef]
67. Macfarlane, G.T.; Macfarlane, S. Bacteria, Colonic Fermentation, and Gastrointestinal Health. *J. AOAC Int.* **2019**, *95*, 50–60. [CrossRef] [PubMed]
68. Flint, H.J.; Duncan, S.H.; Scott, K.P.; Louis, P. Links between diet, gut microbiota composition and gut metabolism. *Proc. Nutr. Soc.* **2015**, *74*, 13–22. [CrossRef]
69. Hartley, L.; May, M.D.; Loveman, E.; Colquitt, J.L.; Rees, K. Dietary fibre for the primary prevention of cardiovascular disease. *Cochrane Database Syst. Rev.* **2016**, *2016*, CD011472. [CrossRef] [PubMed]
70. Corrêa-Oliveira, R.; Fachi, J.L.; Vieira, A.; Sato, F.T.; Vinolo, M.A.R. Regulation of immune cell function by short-chain fatty acids. *Clin. Transl. Immunol.* **2016**, *5*, e73. [CrossRef]
71. Martin-Gallausiaux, C.; Marinelli, L.; Blottière, H.M.; Larraufie, P.; Lapaque, N. SCFA: Mechanisms and functional importance in the gut. *Proc. Nutr. Soc.* **2021**, *80*, 37–49. [CrossRef]
72. Den Besten, G.; van Eunen, K.; Groen, A.K.; Venema, K.; Reijngoud, D.J.; Bakker, B.M. The role of short-chain fatty acids in the interplay between diet, gut microbiota, and host energy metabolism. *J. Lipid Res.* **2013**, *54*, 2325–2340. [CrossRef] [PubMed]
73. Raqib, R.; Sarker, P.; Bergman, P.; Ara, G.; Lindh, M.; Sack, D.A.; Nasirul Islam, K.M.; Gudmundsson, G.H.; Andersson, J.; Agerberth, B. Improved outcome in shigellosis associated with butyrate induction of an endogenous peptide antibiotic. *Proc. Natl. Acad. Sci. USA* **2006**, *103*, 9178–9183. [CrossRef] [PubMed]
74. Maslowski, K.M.; Vieira, A.T.; Ng, A.; Kranich, J.; Sierro, F.; Yu, D.; Schilter, H.C.; Rolph, M.S.; Mackay, F.; Artis, D.; et al. Regulation of inflammatory responses by gut microbiota and chemoattractant receptor GPR43. *Nature* **2009**, *461*, 1282–1286. [CrossRef] [PubMed]
75. Kim, M.H.; Kang, S.G.; Park, J.H.; Yanagisawa, M.; Kim, C.H. Short-chain fatty acids activate GPR41 and GPR43 on intestinal epithelial cells to promote inflammatory responses in mice. *Gastroenterology* **2013**, *145*, 396–406.e1. [CrossRef] [PubMed]
76. Aguilar, E.C.; Santos, L.C.d.; Leonel, A.J.; de Oliveira, J.S.; Santos, E.A.; Navia-Pelaez, J.M.; da Silva, J.F.; Mendes, B.P.; Capettini, L.S.A.; Teixeira, L.G.; et al. Oral butyrate reduces oxidative stress in atherosclerotic lesion sites by a mechanism involving NADPH oxidase down-regulation in endothelial cells. *J. Nutr. Biochem.* **2016**, *34*, 99–105. [CrossRef]

77. Donohoe, D.R.; Collins, L.B.; Wali, A.; Bigler, R.; Sun, W.; Bultman, S.J. The Warburg effect dictates the mechanism of butyrate-mediated histone acetylation and cell proliferation. *Mol. Cell* **2012**, *48*, 612–626. [CrossRef] [PubMed]
78. Vinolo, M.A.; Rodrigues, H.G.; Nachbar, R.T.; Curi, R. Regulation of inflammation by short chain fatty acids. *Nutrients* **2011**, *3*, 858–876. [CrossRef]
79. Vinolo, M.A.; Hirabara, S.M.; Curi, R. G-protein-coupled receptors as fat sensors. *Curr. Opin. Clin. Nutr. Metab. Care* **2012**, *15*, 112–116. [CrossRef]
80. Tan, J.; McKenzie, C.; Potamitis, M.; Thorburn, A.N.; Mackay, C.R.; Macia, L. Chapter Three—The Role of Short-Chain Fatty Acids in Health and Disease. In *Advances in Immunology*; Alt, F.W., Ed.; Academic Press: Cambridge, MA, USA, 2014; Volume 121, pp. 91–119.
81. Kelly, C.J.; Zheng, L.; Campbell, E.L.; Saeedi, B.; Scholz, C.C.; Bayless, A.J.; Wilson, K.E.; Glover, L.E.; Kominsky, D.J.; Magnuson, A. Crosstalk between microbiota-derived short-chain fatty acids and intestinal epithelial HIF augments tissue barrier function. *Cell Host Microbe* **2015**, *17*, 662–671. [CrossRef]
82. Osto, E. The promise of the gut metabolite propionate for a novel and personalized lipid-lowering treatment. *Eur. Heart J.* **2022**, *43*, 534. [CrossRef]
83. Haghikia, A.; Zimmermann, F.; Schumann, P.; Jasina, A.; Roessler, J.; Schmidt, D.; Heinze, P.; Kaisler, J.; Nageswaran, V.; Aigner, A. Propionate attenuates atherosclerosis by immune-dependent regulation of intestinal cholesterol metabolism. *Eur. Heart J.* **2022**, *43*, 518–533. [CrossRef] [PubMed]
84. Bartolomaeus, H.; Balogh, A.; Yakoub, M.; Homann, S.; Markó, L.; Höges, S.; Tsvetkov, D.; Krannich, A.; Wundersitz, S.; Avery, E.G.; et al. Short-Chain Fatty Acid Propionate Protects From Hypertensive Cardiovascular Damage. *Circulation* **2019**, *139*, 1407–1421. [CrossRef]
85. Kumar, J.; Rani, K.; Datt, C. Molecular link between dietary fibre, gut microbiota and health. *Mol. Biol. Rep.* **2020**, *47*, 6229–6237. [CrossRef] [PubMed]
86. Chambers, E.S.; Viardot, A.; Psichas, A.; Morrison, D.J.; Murphy, K.G.; Zac-Varghese, S.E.; MacDougall, K.; Preston, T.; Tedford, C.; Finlayson, G.S.; et al. Effects of targeted delivery of propionate to the human colon on appetite regulation, body weight maintenance and adiposity in overweight adults. *Gut* **2015**, *64*, 1744–1754. [CrossRef] [PubMed]
87. Li, Z.; Yi, C.X.; Katiraei, S.; Kooijman, S.; Zhou, E.; Chung, C.K.; Gao, Y.; van den Heuvel, J.K.; Meijer, O.C.; Berbée, J.F.P.; et al. Butyrate reduces appetite and activates brown adipose tissue via the gut-brain neural circuit. *Gut* **2018**, *67*, 1269–1279. [CrossRef] [PubMed]
88. Hu, S.; Kuwabara, R.; de Haan, B.J.; Smink, A.M.; de Vos, P. Acetate and Butyrate Improve β-cell Metabolism and Mitochondrial Respiration under Oxidative Stress. *Int. J. Mol. Sci.* **2020**, *21*, 1542. [CrossRef] [PubMed]
89. Arora, T.; Tremaroli, V. Therapeutic Potential of Butyrate for Treatment of Type 2 Diabetes. *Front. Endocrinol.* **2021**, *12*, 761834. [CrossRef] [PubMed]
90. Aguilar, E.C.; Leonel, A.J.; Teixeira, L.G.; Silva, A.R.; Silva, J.F.; Pelaez, J.M.; Capettini, L.S.; Lemos, V.S.; Santos, R.A.; Alvarez-Leite, J.I. Butyrate impairs atherogenesis by reducing plaque inflammation and vulnerability and decreasing NFκB activation. *Nutr. Metab. Cardiovasc. Dis.* **2014**, *24*, 606–613. [CrossRef] [PubMed]
91. Säemann, M.D.; Böhmig, G.A.; Osterreicher, C.H.; Burtscher, H.; Parolini, O.; Diakos, C.; Stöckl, J.; Hörl, W.H.; Zlabinger, G.J. Anti-inflammatory effects of sodium butyrate on human monocytes: Potent inhibition of IL-12 and up-regulation of IL-10 production. *FASEB J* **2000**, *14*, 2380–2382. [CrossRef]
92. Verhaar, B.J.H.; Prodan, A.; Nieuwdorp, M.; Muller, M. Gut Microbiota in Hypertension and Atherosclerosis: A Review. *Nutrients* **2020**, *12*, 2982. [CrossRef]
93. Zapolska-Downar, D.; Siennicka, A.; Kaczmarczyk, M.; Kołodziej, B.; Naruszewicz, M. Butyrate inhibits cytokine-induced VCAM-1 and ICAM-1 expression in cultured endothelial cells: The role of NF-kappaB and PPARalpha. *J. Nutr. Biochem.* **2004**, *15*, 220–228. [CrossRef] [PubMed]
94. Du, Y.; Li, X.; Su, C.; Xi, M.; Zhang, X.; Jiang, Z.; Wang, L.; Hong, B. Butyrate protects against high-fat diet-induced atherosclerosis via up-regulating ABCA1 expression in apolipoprotein E-deficiency mice. *Br. J. Pharmacol.* **2020**, *177*, 1754–1772. [CrossRef]
95. Zhou, K. Strategies to promote abundance of Akkermansia muciniphila, an emerging probiotics in the gut, evidence from dietary intervention studies. *J. Funct. Foods* **2017**, *33*, 194–201. [CrossRef]
96. Karlsson, C.L.; Onnerfält, J.; Xu, J.; Molin, G.; Ahrné, S.; Thorngren-Jerneck, K. The microbiota of the gut in preschool children with normal and excessive body weight. *Obesity* **2012**, *20*, 2257–2261. [CrossRef]
97. Santacruz, A.; Collado, M.C.; García-Valdés, L.; Segura, M.T.; Martín-Lagos, J.A.; Anjos, T.; Martí-Romero, M.; Lopez, R.M.; Florido, J.; Campoy, C.; et al. Gut microbiota composition is associated with body weight, weight gain and biochemical parameters in pregnant women. *Br. J. Nutr.* **2010**, *104*, 83–92. [CrossRef]
98. Everard, A.; Belzer, C.; Geurts, L.; Ouwerkerk, J.P.; Druart, C.; Bindels, L.B.; Guiot, Y.; Derrien, M.; Muccioli, G.G.; Delzenne, N.M.; et al. Cross-talk between Akkermansia muciniphila and intestinal epithelium controls diet-induced obesity. *Proc. Natl. Acad. Sci. USA* **2013**, *110*, 9066–9071. [CrossRef] [PubMed]
99. Turner, J.R. Intestinal mucosal barrier function in health and disease. *Nat. Rev. Immunol.* **2009**, *9*, 799–809. [CrossRef] [PubMed]
100. Li, J.; Lin, S.; Vanhoutte, P.M.; Woo, C.W.; Xu, A. Akkermansia Muciniphila Protects Against Atherosclerosis by Preventing Metabolic Endotoxemia-Induced Inflammation in Apoe−/− Mice. *Circulation* **2016**, *133*, 2434–2446. [CrossRef]

101. Collado, M.C.; Derrien, M.; Isolauri, E.; de Vos, W.M.; Salminen, S. Intestinal integrity and Akkermansia muciniphila, a mucin-degrading member of the intestinal microbiota present in infants, adults, and the elderly. *Appl. Environ. Microbiol.* **2007**, *73*, 7767–7770. [CrossRef] [PubMed]

102. Hasani, A.; Ebrahimzadeh, S.; Hemmati, F.; Khabbaz, A.; Hasani, A.; Gholizadeh, P. The role of Akkermansia muciniphila in obesity, diabetes and atherosclerosis. *J. Med. Microbiol.* **2021**, *70*, 001435. [CrossRef]

103. Bodogai, M.; O'Connell, J.; Kim, K.; Kim, Y.; Moritoh, K.; Chen, C.; Gusev, F.; Vaughan, K.; Shulzhenko, N.; Mattison, J.A. Commensal bacteria contribute to insulin resistance in aging by activating innate B1a cells. *Sci. Transl. Med.* **2018**, *10*, eaat4271. [CrossRef] [PubMed]

104. Nie, K.; Ma, K.; Luo, W.; Shen, Z.; Yang, Z.; Xiao, M.; Tong, T.; Yang, Y.; Wang, X. Roseburia intestinalis: A Beneficial Gut Organism From the Discoveries in Genus and Species. *Front. Cell Infect. Microbiol.* **2021**, *11*, 757718. [CrossRef] [PubMed]

105. Kasahara, K.; Krautkramer, K.A.; Org, E.; Romano, K.A.; Kerby, R.L.; Vivas, E.I.; Mehrabian, M.; Denu, J.M.; Bäckhed, F.; Lusis, A.J.; et al. Interactions between Roseburia intestinalis and diet modulate atherogenesis in a murine model. *Nat. Microbiol.* **2018**, *3*, 1461–1471. [CrossRef] [PubMed]

106. Quévrain, E.; Maubert, M.A.; Michon, C.; Chain, F.; Marquant, R.; Tailhades, J.; Miquel, S.; Carlier, L.; Bermúdez-Humarán, L.G.; Pigneur, B.; et al. Identification of an anti-inflammatory protein from Faecalibacterium prausnitzii, a commensal bacterium deficient in Crohn's disease. *Gut* **2016**, *65*, 415–425. [CrossRef] [PubMed]

107. Breyner, N.M.; Michon, C.; de Sousa, C.S.; Vilas Boas, P.B.; Chain, F.; Azevedo, V.A.; Langella, P.; Chatel, J.M. Microbial Anti-Inflammatory Molecule (MAM) from Faecalibacterium prausnitzii Shows a Protective Effect on DNBS and DSS-Induced Colitis Model in Mice through Inhibition of NF-κB Pathway. *Front. Microbiol.* **2017**, *8*, 114. [CrossRef]

108. Brown, A.J.; Goldsworthy, S.M.; Barnes, A.A.; Eilert, M.M.; Tcheang, L.; Daniels, D.; Muir, A.I.; Wigglesworth, M.J.; Kinghorn, I.; Fraser, N.J.; et al. The Orphan G protein-coupled receptors GPR41 and GPR43 are activated by propionate and other short chain carboxylic acids. *J. Biol. Chem.* **2003**, *278*, 11312–11319. [CrossRef]

109. Maioli, T.U.; Borras-Nogues, E.; Torres, L.; Barbosa, S.C.; Martins, V.D.; Langella, P.; Azevedo, V.A.; Chatel, J.M. Possible Benefits of Faecalibacterium prausnitzii for Obesity-Associated Gut Disorders. *Front. Pharmacol.* **2021**, *12*, 740636. [CrossRef] [PubMed]

110. Chen, M.-l.; Yi, L.; Zhang, Y.; Zhou, X.; Ran, L.; Yang, J.; Zhu, J.-d.; Zhang, Q.-y.; Mi, M.-t. Resveratrol attenuates trimethylamine-N-oxide (TMAO)-induced atherosclerosis by regulating TMAO synthesis and bile acid metabolism via remodeling of the gut microbiota. *MBio* **2016**, *7*, e02210–e02215. [CrossRef] [PubMed]

111. Chung, J.H.; Manganiello, V.; Dyck, J.R. Resveratrol as a calorie restriction mimetic: Therapeutic implications. *Trends Cell Biol.* **2012**, *22*, 546–554. [CrossRef] [PubMed]

112. Shin, N.R.; Lee, J.C.; Lee, H.Y.; Kim, M.S.; Whon, T.W.; Lee, M.S.; Bae, J.W. An increase in the Akkermansia spp. population induced by metformin treatment improves glucose homeostasis in diet-induced obese mice. *Gut* **2014**, *63*, 727–735. [CrossRef]

113. Anhê, F.F.; Roy, D.; Pilon, G.; Dudonné, S.; Matamoros, S.; Varin, T.V.; Garofalo, C.; Moine, Q.; Desjardins, Y.; Levy, E.; et al. A polyphenol-rich cranberry extract protects from diet-induced obesity, insulin resistance and intestinal inflammation in association with increased Akkermansia spp. population in the gut microbiota of mice. *Gut* **2015**, *64*, 872–883. [CrossRef]

114. Larrosa, M.; Yañéz-Gascón, M.J.; Selma, M.a.V.; Gonzalez-Sarrias, A.; Toti, S.; Cerón, J.J.; Tomas-Barberan, F.; Dolara, P.; Espín, J.C. Effect of a low dose of dietary resveratrol on colon microbiota, inflammation and tissue damage in a DSS-induced colitis rat model. *J. Agric. Food Chem.* **2009**, *57*, 2211–2220. [CrossRef]

115. Jung, C.M.; Heinze, T.M.; Schnackenberg, L.K.; Mullis, L.B.; Elkins, S.A.; Elkins, C.A.; Steele, R.S.; Sutherland, J.B. Interaction of dietary resveratrol with animal-associated bacteria. *FEMS Microbiol. Lett.* **2009**, *297*, 266–273. [CrossRef] [PubMed]

116. Queipo-Ortuño, M.I.; Boto-Ordóñez, M.; Murri, M.; Gomez-Zumaquero, J.M.; Clemente-Postigo, M.; Estruch, R.; Cardona Diaz, F.; Andres-Lacueva, C.; Tinahones, F.J. Influence of red wine polyphenols and ethanol on the gut microbiota ecology and biochemical biomarkers. *Am. J. Clin. Nutr.* **2012**, *95*, 1323–1334. [CrossRef]

117. Qiao, Y.; Sun, J.; Xia, S.; Tang, X.; Shi, Y.; Le, G. Effects of resveratrol on gut microbiota and fat storage in a mouse model with high-fat-induced obesity. *Food Funct.* **2014**, *5*, 1241–1249. [CrossRef] [PubMed]

118. Etxeberria, U.; Arias, N.; Boqué, N.; Macarulla, M.; Portillo, M.; Martínez, J.; Milagro, F. Reshaping faecal gut microbiota composition by the intake of trans-resveratrol and quercetin in high-fat sucrose diet-fed rats. *J. Nutr. Biochem.* **2015**, *26*, 651–660. [CrossRef]

119. Hotel, A.C.P.; Cordoba, A. Health and nutritional properties of probiotics in food including powder milk with live lactic acid bacteria. *Prevention* **2001**, *5*, 1–10.

120. Hill, C.; Guarner, F.; Reid, G.; Gibson, G.R.; Merenstein, D.J.; Pot, B.; Morelli, L.; Canani, R.B.; Flint, H.J.; Salminen, S. Expert consensus document: The International Scientific Association for Probiotics and Prebiotics consensus statement on the scope and appropriate use of the term probiotic. *Nat. Rev. Gastroenterol. Hepatol.* **2014**, *11*, 506–514. [CrossRef]

121. Cani, P.D.; Delzenne, N.M.; Amar, J.; Burcelin, R. Role of gut microflora in the development of obesity and insulin resistance following high-fat diet feeding. *Pathol. Biol.* **2008**, *56*, 305–309. [CrossRef]

122. Velagapudi, V.R.; Hezaveh, R.; Reigstad, C.S.; Gopalacharyulu, P.; Yetukuri, L.; Islam, S.; Felin, J.; Perkins, R.; Borén, J.; Oresic, M.; et al. The gut microbiota modulates host energy and lipid metabolism in mice. *J. Lipid Res.* **2010**, *51*, 1101–1112. [CrossRef]

123. Sekirov, I.; Russell, S.L.; Antunes, L.C.; Finlay, B.B. Gut microbiota in health and disease. *Physiol. Rev.* **2010**, *90*, 859–904. [CrossRef] [PubMed]

124. Chan, Y.K.; El-Nezami, H.; Chen, Y.; Kinnunen, K.; Kirjavainen, P.V. Probiotic mixture VSL#3 reduce high fat diet induced vascular inflammation and atherosclerosis in ApoE(−/−) mice. *AMB Express* **2016**, *6*, 61. [CrossRef] [PubMed]
125. Karlsson, C.; Ahrné, S.; Molin, G.; Berggren, A.; Palmquist, I.; Fredrikson, G.N.; Jeppsson, B. Probiotic therapy to men with incipient arteriosclerosis initiates increased bacterial diversity in colon: A randomized controlled trial. *Atherosclerosis* **2010**, *208*, 228–233. [CrossRef]
126. Mennigen, R.; Bruewer, M. Effect of probiotics on intestinal barrier function. *Ann. N. Y. Acad. Sci* **2009**, *1165*, 183–189. [CrossRef]
127. Kekkonen, R.A.; Lummela, N.; Karjalainen, H.; Latvala, S.; Tynkkynen, S.; Jarvenpaa, S.; Kautiainen, H.; Julkunen, I.; Vapaatalo, H.; Korpela, R. Probiotic intervention has strain-specific anti-inflammatory effects in healthy adults. *World J. Gastroenterol.* **2008**, *14*, 2029–2036. [CrossRef] [PubMed]
128. Vasquez, E.C.; Pereira, T.M.C.; Peotta, V.A.; Baldo, M.P.; Campos-Toimil, M. Probiotics as Beneficial Dietary Supplements to Prevent and Treat Cardiovascular Diseases: Uncovering Their Impact on Oxidative Stress. *Oxidative Med. Cell. Longev.* **2019**, *2019*, 3086270. [CrossRef]
129. Zeng, W.; Shen, J.; Bo, T.; Peng, L.; Xu, H.; Nasser, M.I.; Zhuang, Q.; Zhao, M. Cutting Edge: Probiotics and Fecal Microbiota Transplantation in Immunomodulation. *J. Immunol. Res.* **2019**, *2019*, 1603758. [CrossRef]
130. Huang, Y.; Wang, J.; Quan, G.; Wang, X.; Yang, L.; Zhong, L. Lactobacillus acidophilus ATCC 4356 prevents atherosclerosis via inhibition of intestinal cholesterol absorption in apolipoprotein E-knockout mice. *Appl. Environ. Microbiol.* **2014**, *80*, 7496–7504. [CrossRef]
131. Wu, Y.; Zhang, Q.; Ren, Y.; Ruan, Z. Effect of probiotic Lactobacillus on lipid profile: A systematic review and meta-analysis of randomized, controlled trials. *PLoS ONE* **2017**, *12*, e0178868. [CrossRef]
132. Yan, S.; Tian, Z.; Li, M.; Li, B.; Cui, W. Effects of probiotic supplementation on the regulation of blood lipid levels in overweight or obese subjects: A meta-analysis. *Food Funct.* **2019**, *10*, 1747–1759. [CrossRef]
133. Deng, X.; Ma, J.; Song, M.; Jin, Y.; Ji, C.; Ge, W.; Guo, C. Effects of products designed to modulate the gut microbiota on hyperlipidaemia. *Eur. J. Nutr.* **2019**, *58*, 2713–2729. [CrossRef] [PubMed]
134. Mardini, H.E.; Grigorian, A.Y. Probiotic mix VSL#3 is effective adjunctive therapy for mild to moderately active ulcerative colitis: A meta-analysis. *Inflamm. Bowel Dis.* **2014**, *20*, 1562–1567. [CrossRef] [PubMed]
135. Dhiman, R.K.; Rana, B.; Agrawal, S.; Garg, A.; Chopra, M.; Thumburu, K.K.; Khattri, A.; Malhotra, S.; Duseja, A.; Chawla, Y.K. Probiotic VSL#3 reduces liver disease severity and hospitalization in patients with cirrhosis: A randomized, controlled trial. *Gastroenterology* **2014**, *147*, 1327–1337.e1323. [CrossRef] [PubMed]
136. Fedorak, R.N.; Feagan, B.G.; Hotte, N.; Leddin, D.; Dieleman, L.A.; Petrunia, D.M.; Enns, R.; Bitton, A.; Chiba, N.; Paré, P.; et al. The probiotic VSL#3 has anti-inflammatory effects and could reduce endoscopic recurrence after surgery for Crohn's disease. *Clin. Gastroenterol. Hepatol.* **2015**, *13*, 928–935.e2. [CrossRef] [PubMed]
137. Mencarelli, A.; Cipriani, S.; Renga, B.; Bruno, A.; D'Amore, C.; Distrutti, E.; Fiorucci, S. VSL#3 resets insulin signaling and protects against NASH and atherosclerosis in a model of genetic dyslipidemia and intestinal inflammation. *PLoS ONE* **2012**, *7*, e45425. [CrossRef]
138. Blessing, E.; Preusch, M.; Kranzhöfer, R.; Kinscherf, R.; Marx, N.; Rosenfeld, M.E.; Isermann, B.; Weber, C.M.; Kreuzer, J.; Gräfe, J.; et al. Anti-atherosclerotic properties of telmisartan in advanced atherosclerotic lesions in apolipoprotein E deficient mice. *Atherosclerosis* **2008**, *199*, 295–303. [CrossRef]
139. Rao, R.K.; Samak, G. Protection and Restitution of Gut Barrier by Probiotics: Nutritional and Clinical Implications. *Curr. Nutr. Food Sci.* **2013**, *9*, 99–107. [CrossRef] [PubMed]
140. Li, S.-C.; Hsu, W.-F.; Chang, J.-S.; Shih, C.-K. Combination of Lactobacillus acidophilus and Bifidobacterium animalis subsp. lactis shows a stronger anti-inflammatory effect than individual strains in HT-29 cells. *Nutrients* **2019**, *11*, 969. [CrossRef]
141. Moludi, J.; Kafil, H.S.; Qaisar, S.A.; Gholizadeh, P.; Alizadeh, M.; Vayghyan, H.J. Effect of probiotic supplementation along with calorie restriction on metabolic endotoxemia, and inflammation markers in coronary artery disease patients: A double blind placebo controlled randomized clinical trial. *Nutr. J.* **2021**, *20*, 47. [CrossRef]
142. Mizoguchi, T.; Kasahara, K.; Yamashita, T.; Sasaki, N.; Yodoi, K.; Matsumoto, T.; Emoto, T.; Hayashi, T.; Kitano, N.; Yoshida, N.; et al. Oral administration of the lactic acid bacterium Pediococcus acidilactici attenuates atherosclerosis in mice by inducing tolerogenic dendritic cells. *Heart Vessels* **2017**, *32*, 768–776. [CrossRef] [PubMed]
143. Smits, H.H.; Engering, A.; van der Kleij, D.; de Jong, E.C.; Schipper, K.; van Capel, T.M.; Zaat, B.A.; Yazdanbakhsh, M.; Wierenga, E.A.; van Kooyk, Y.; et al. Selective probiotic bacteria induce IL-10-producing regulatory T cells in vitro by modulating dendritic cell function through dendritic cell-specific intercellular adhesion molecule 3-grabbing nonintegrin. *J. Allergy Clin. Immunol.* **2005**, *115*, 1260–1267. [CrossRef] [PubMed]
144. Jijon, H.; Backer, J.; Diaz, H.; Yeung, H.; Thiel, D.; McKaigney, C.; De Simone, C.; Madsen, K. DNA from probiotic bacteria modulates murine and human epithelial and immune function. *Gastroenterology* **2004**, *126*, 1358–1373. [CrossRef]
145. Rachmilewitz, D.; Katakura, K.; Karmeli, F.; Hayashi, T.; Reinus, C.; Rudensky, B.; Akira, S.; Takeda, K.; Lee, J.; Takabayashi, K.; et al. Toll-like receptor 9 signaling mediates the anti-inflammatory effects of probiotics in murine experimental colitis. *Gastroenterology* **2004**, *126*, 520–528. [CrossRef]
146. Robbins, C.S.; Hilgendorf, I.; Weber, G.F.; Theurl, I.; Iwamoto, Y.; Figueiredo, J.-L.; Gorbatov, R.; Sukhova, G.K.; Gerhardt, L.; Smyth, D. Local proliferation dominates lesional macrophage accumulation in atherosclerosis. *Nat. Med.* **2013**, *19*, 1166–1172. [CrossRef] [PubMed]

147. Fåk, F.; Bäckhed, F. Lactobacillus reuteri prevents diet-induced obesity, but not atherosclerosis, in a strain dependent fashion in Apoe−/− mice. *PLoS ONE* **2012**, *7*, e46837. [CrossRef]

148. Chen, L.; Liu, W.; Li, Y.; Luo, S.; Liu, Q.; Zhong, Y.; Jian, Z.; Bao, M. Lactobacillus acidophilus ATCC 4356 attenuates the atherosclerotic progression through modulation of oxidative stress and inflammatory process. *Int. Immunopharmacol.* **2013**, *17*, 108–115. [CrossRef] [PubMed]

149. Qiu, L.; Tao, X.; Xiong, H.; Yu, J.; Wei, H. Lactobacillus plantarum ZDY04 exhibits a strain-specific property of lowering TMAO via the modulation of gut microbiota in mice. *Food Funct.* **2018**, *9*, 4299–4309. [CrossRef]

150. Qiu, L.; Yang, D.; Tao, X.; Yu, J.; Xiong, H.; Wei, H. Enterobacter aerogenes ZDY01 Attenuates Choline-Induced Trimethylamine N-Oxide Levels by Remodeling Gut Microbiota in Mice. *J. Microbiol. Biotechnol.* **2017**, *27*, 1491–1499. [CrossRef] [PubMed]

151. Borges, N.A.; Stenvinkel, P.; Bergman, P.; Qureshi, A.; Lindholm, B.; Moraes, C.; Stockler-Pinto, M.; Mafra, D. Effects of probiotic supplementation on trimethylamine-N-oxide plasma levels in hemodialysis patients: A pilot study. *Probiotics Antimicrob. Proteins* **2019**, *11*, 648–654. [CrossRef]

152. Tripolt, N.J.; Leber, B.; Triebl, A.; Köfeler, H.; Stadlbauer, V.; Sourij, H. Effect of Lactobacillus casei Shirota supplementation on trimethylamine-N-oxide levels in patients with metabolic syndrome: An open-label, randomized study. *Atherosclerosis* **2015**, *242*, 141–144. [CrossRef]

153. Agerholm-Larsen, L.; Raben, A.; Haulrik, N.; Hansen, A.S.; Manders, M.; Astrup, A. Effect of 8 week intake of probiotic milk products on risk factors for cardiovascular diseases. *Eur. J. Clin. Nutr.* **2000**, *54*, 288–297. [CrossRef] [PubMed]

154. Ooi, L.G.; Liong, M.T. Cholesterol-lowering effects of probiotics and prebiotics: A review of in vivo and in vitro findings. *Int. J. Mol. Sci.* **2010**, *11*, 2499–2522. [CrossRef] [PubMed]

155. Liong, M.T.; Shah, N.P. Acid and bile tolerance and cholesterol removal ability of lactobacilli strains. *J. Dairy Sci.* **2005**, *88*, 55–66. [CrossRef] [PubMed]

156. Zeng, X.Q.; Pan, D.D.; Guo, Y.X. The probiotic properties of Lactobacillus buchneri P2. *J. Appl. Microbiol.* **2010**, *108*, 2059–2066. [CrossRef] [PubMed]

157. Hassan, A.; Din, A.U.; Zhu, Y.; Zhang, K.; Li, T.; Wang, Y.; Luo, Y.; Wang, G. Updates in understanding the hypocholesterolemia effect of probiotics on atherosclerosis. *Appl. Microbiol. Biotechnol.* **2019**, *103*, 5993–6006. [CrossRef] [PubMed]

158. Huang, Y.; Zheng, Y. The probiotic Lactobacillus acidophilus reduces cholesterol absorption through the down-regulation of Niemann-Pick C1-like 1 in Caco-2 cells. *Br. J. Nutr.* **2010**, *103*, 473–478. [CrossRef]

159. Miura, S.; Saku, K. Ezetimibe, a selective inhibitor of the transport of cholesterol. *Intern. Med.* **2008**, *47*, 1165–1170. [CrossRef]

160. Duval, C.; Touche, V.; Tailleux, A.; Fruchart, J.C.; Fievet, C.; Clavey, V.; Staels, B.; Lestavel, S. Niemann-Pick C1 like 1 gene expression is down-regulated by LXR activators in the intestine. *Biochem. Biophys. Res. Commun.* **2006**, *340*, 1259–1263. [CrossRef]

161. Bradley, M.N.; Hong, C.; Chen, M.; Joseph, S.B.; Wilpitz, D.C.; Wang, X.; Lusis, A.J.; Collins, A.; Hseuh, W.A.; Collins, J.L.; et al. Ligand activation of LXR beta reverses atherosclerosis and cellular cholesterol overload in mice lacking LXR alpha and apoE. *J. Clin. Investig.* **2007**, *117*, 2337–2346. [CrossRef]

162. Kriaa, A.; Bourgin, M.; Potiron, A.; Mkaouar, H.; Jablaoui, A.; Gérard, P.; Maguin, E.; Rhimi, M. Microbial impact on cholesterol and bile acid metabolism: Current status and future prospects. *J. Lipid Res.* **2019**, *60*, 323–332. [CrossRef]

163. Degirolamo, C.; Rainaldi, S.; Bovenga, F.; Murzilli, S.; Moschetta, A. Microbiota modification with probiotics induces hepatic bile acid synthesis via downregulation of the Fxr-Fgf15 axis in mice. *Cell Rep.* **2014**, *7*, 12–18. [CrossRef] [PubMed]

164. Cai, J.S.; Chen, J.H. The mechanism of enterohepatic circulation in the formation of gallstone disease. *J. Membr. Biol.* **2014**, *247*, 1067–1082. [CrossRef] [PubMed]

165. Copaci, I.; Micu, L.; Iliescu, L.; Voiculescu, M. New therapeutical indications of ursodeoxycholic acid. *Rom. J. Gastroenterol.* **2005**, *14*, 259–266. [PubMed]

166. Cariou, B.; Chetiveaux, M.; Zaïr, Y.; Pouteau, E.; Disse, E.; Guyomarc'h-Delasalle, B.; Laville, M.; Krempf, M. Fasting plasma chenodeoxycholic acid and cholic acid concentrations are inversely correlated with insulin sensitivity in adults. *Nutr. Metab.* **2011**, *8*, 48. [CrossRef] [PubMed]

167. Joyce, S.A.; MacSharry, J.; Casey, P.G.; Kinsella, M.; Murphy, E.F.; Shanahan, F.; Hill, C.; Gahan, C.G. Regulation of host weight gain and lipid metabolism by bacterial bile acid modification in the gut. *Proc. Natl. Acad. Sci. USA* **2014**, *111*, 7421–7426. [CrossRef]

168. Charach, G.; Argov, O.; Geiger, K.; Charach, L.; Rogowski, O.; Grosskopf, I. Diminished bile acids excretion is a risk factor for coronary artery disease: 20-year follow up and long-term outcome. *Therap. Adv. Gastroenterol.* **2018**, *11*, 1756283x17743420. [CrossRef]

169. Ignacio Barrasa, J.; Olmo, N.; Pérez-Ramos, P.; Santiago-Gómez, A.; Lecona, E.; Turnay, J.; Antonia Lizarbe, M. Deoxycholic and chenodeoxycholic bile acids induce apoptosis via oxidative stress in human colon adenocarcinoma cells. *Apoptosis* **2011**, *16*, 1054–1067. [CrossRef]

170. Parker, H.E.; Wallis, K.; le Roux, C.W.; Wong, K.Y.; Reimann, F.; Gribble, F.M. Molecular mechanisms underlying bile acid-stimulated glucagon-like peptide-1 secretion. *Br. J. Pharmacol.* **2012**, *165*, 414–423. [CrossRef] [PubMed]

171. Haeusler, R.A.; Astiarraga, B.; Camastra, S.; Accili, D.; Ferrannini, E. Human insulin resistance is associated with increased plasma levels of 12α-hydroxylated bile acids. *Diabetes* **2013**, *62*, 4184–4191. [CrossRef]

172. Cipriani, S.; Mencarelli, A.; Chini, M.G.; Distrutti, E.; Renga, B.; Bifulco, G.; Baldelli, F.; Donini, A.; Fiorucci, S. The bile acid receptor GPBAR-1 (TGR5) modulates integrity of intestinal barrier and immune response to experimental colitis. *PLoS ONE* **2011**, *6*, e25637. [CrossRef]

173. Song, Z.; Cai, Y.; Lao, X.; Wang, X.; Lin, X.; Cui, Y.; Kalavagunta, P.K.; Liao, J.; Jin, L.; Shang, J. Taxonomic profiling and populational patterns of bacterial bile salt hydrolase (BSH) genes based on worldwide human gut microbiome. *Microbiome* **2019**, *7*, 1–16. [CrossRef] [PubMed]

174. Kalani, M.; Hodjati, H.; Sajedi Khanian, M.; Doroudchi, M. Lactobacillus acidophilus Increases the Anti-apoptotic Micro RNA-21 and Decreases the Pro-inflammatory Micro RNA-155 in the LPS-Treated Human Endothelial Cells. *Probiotics Antimicrob. Proteins* **2016**, *8*, 61–72. [CrossRef] [PubMed]

175. Amaretti, A.; di Nunzio, M.; Pompei, A.; Raimondi, S.; Rossi, M.; Bordoni, A. Antioxidant properties of potentially probiotic bacteria: In vitro and in vivo activities. *Appl. Microbiol. Biotechnol.* **2013**, *97*, 809–817. [CrossRef] [PubMed]

176. Ohira, H.; Tsutsui, W.; Fujioka, Y. Are Short Chain Fatty Acids in Gut Microbiota Defensive Players for Inflammation and Atherosclerosis? *J. Atheroscler Thromb.* **2017**, *24*, 660–672. [CrossRef]

177. Bultman, S.J. Bacterial butyrate prevents atherosclerosis. *Nat. Microbiol.* **2018**, *3*, 1332–1333. [CrossRef]

178. Jones, M.L.; Martoni, C.J.; Parent, M.; Prakash, S. Cholesterol-lowering efficacy of a microencapsulated bile salt hydrolase-active Lactobacillus reuteri NCIMB 30242 yoghurt formulation in hypercholesterolaemic adults. *Br. J. Nutr.* **2012**, *107*, 1505–1513. [CrossRef]

179. Rajkumar, H.; Mahmood, N.; Kumar, M.; Varikuti, S.R.; Challa, H.R.; Myakala, S.P. Effect of probiotic (VSL#3) and omega-3 on lipid profile, insulin sensitivity, inflammatory markers, and gut colonization in overweight adults: A randomized, controlled trial. *Mediators Inflamm.* **2014**, *2014*, 348959. [CrossRef] [PubMed]

180. Rerksuppaphol, S.; Rerksuppaphol, L. A Randomized Double-blind Controlled Trial of Lactobacillus acidophilus Plus Bifidobacterium bifidum versus Placebo in Patients with Hypercholesterolemia. *J. Clin. Diagn. Res.* **2015**, *9*, Kc01–Kc04. [CrossRef]

181. Boutagy, N.E.; Neilson, A.P.; Osterberg, K.L.; Smithson, A.T.; Englund, T.R.; Davy, B.M.; Hulver, M.W.; Davy, K.P. Probiotic supplementation and trimethylamine-N-oxide production following a high-fat diet. *Obesity* **2015**, *23*, 2357–2363. [CrossRef] [PubMed]

182. Bjerg, A.T.; Sørensen, M.B.; Krych, L.; Hansen, L.H.; Astrup, A.; Kristensen, M.; Nielsen, D.S. The effect of Lactobacillus paracasei subsp. paracasei L. casei W8® on blood levels of triacylglycerol is independent of colonisation. *Benef Microbes* **2015**, *6*, 263–269. [CrossRef] [PubMed]

183. Bernini, L.J.; Simão, A.N.; Alfieri, D.F.; Lozovoy, M.A.; Mari, N.L.; de Souza, C.H.; Dichi, I.; Costa, G.N. Beneficial effects of Bifidobacterium lactis on lipid profile and cytokines in patients with metabolic syndrome: A randomized trial. Effects of probiotics on metabolic syndrome. *Nutrition* **2016**, *32*, 716–719. [CrossRef]

184. Madjd, A.; Taylor, M.A.; Mousavi, N.; Delavari, A.; Malekzadeh, R.; Macdonald, I.A.; Farshchi, H.R. Comparison of the effect of daily consumption of probiotic compared with low-fat conventional yogurt on weight loss in healthy obese women following an energy-restricted diet: A randomized controlled trial. *Am. J. Clin. Nutr.* **2016**, *103*, 323–329. [CrossRef] [PubMed]

185. Chan, Y.K.; Brar, M.S.; Kirjavainen, P.V.; Chen, Y.; Peng, J.; Li, D.; Leung, F.C.; El-Nezami, H. High fat diet induced atherosclerosis is accompanied with low colonic bacterial diversity and altered abundances that correlates with plaque size, plasma A-FABP and cholesterol: A pilot study of high fat diet and its intervention with Lactobacillus rhamnosus GG (LGG) or telmisartan in ApoE(−/−) mice. *BMC Microbiol.* **2016**, *16*, 264. [CrossRef]

186. Costabile, A.; Buttarazzi, I.; Kolida, S.; Quercia, S.; Baldini, J.; Swann, J.R.; Brigidi, P.; Gibson, G.R. An in vivo assessment of the cholesterol-lowering efficacy of Lactobacillus plantarum ECGC 13110402 in normal to mildly hypercholesterolaemic adults. *PLoS ONE* **2017**, *12*, e0187964. [CrossRef]

187. Firouzi, S.; Majid, H.A.; Ismail, A.; Kamaruddin, N.A.; Barakatun-Nisak, M.Y. Effect of multi-strain probiotics (multi-strain microbial cell preparation) on glycemic control and other diabetes-related outcomes in people with type 2 diabetes: A randomized controlled trial. *Eur. J. Nutr.* **2017**, *56*, 1535–1550. [CrossRef]

188. Yoshida, N.; Emoto, T.; Yamashita, T.; Watanabe, H.; Hayashi, T.; Tabata, T.; Hoshi, N.; Hatano, N.; Ozawa, G.; Sasaki, N. Bacteroides vulgatus and Bacteroides dorei reduce gut microbial lipopolysaccharide production and inhibit atherosclerosis. *Circulation* **2018**, *138*, 2486–2498. [CrossRef]

189. Saikia, D.; Manhar, A.K.; Deka, B.; Roy, R.; Gupta, K.; Namsa, N.D.; Chattopadhyay, P.; Doley, R.; Mandal, M. Hypocholesterolemic activity of indigenous probiotic isolate Saccharomyces cerevisiae ARDMC1 in a rat model. *J. Food Drug Anal.* **2018**, *26*, 154–162. [CrossRef]

190. Huang, F.; Zhang, F.; Xu, D.; Zhang, Z.; Xu, F.; Tao, X.; Qiu, L.; Wei, H. Enterococcus faecium WEFA23 from infants lessens high-fat-diet-induced hyperlipidemia via cholesterol 7-alpha-hydroxylase gene by altering the composition of gut microbiota in rats. *J. Dairy Sci.* **2018**, *101*, 7757–7767. [CrossRef]

191. Szulińska, M.; Łoniewski, I.; Skrypnik, K.; Sobieska, M.; Korybalska, K.; Suliburska, J.; Bogdański, P. Multispecies probiotic supplementation favorably affects vascular function and reduces arterial stiffness in obese postmenopausal women—A 12-week placebo-controlled and randomized clinical study. *Nutrients* **2018**, *10*, 1672. [CrossRef]

192. Tang, J.; Qin, M.; Tang, L.; Shan, D.; Zhang, C.; Zhang, Y.; Wei, H.; Qiu, L.; Yu, J. Enterobacter aerogenes ZDY01 inhibits choline-induced atherosclerosis through CDCA-FXR-FGF15 axis. *Food Funct.* **2021**, *12*, 9932–9946. [CrossRef]

193. Wang, Q.; Guo, M.; Liu, Y.; Xu, M.; Shi, L.; Li, X.; Zhao, J.; Zhang, H.; Wang, G.; Chen, W. Bifidobacterium breve and Bifidobacterium longum Attenuate Choline-Induced Plasma Trimethylamine N-Oxide Production by Modulating Gut Microbiota in Mice. *Nutrients* **2022**, *14*, 1222. [CrossRef]
194. Simó, C.; García-Cañas, V. Dietary bioactive ingredients to modulate the gut microbiota-derived metabolite TMAO. New opportunities for functional food development. *Food Funct.* **2020**, *11*, 6745–6776. [CrossRef] [PubMed]
195. Messenger, J.; Clark, S.; Massick, S.; Bechtel, M. A review of trimethylaminuria:(fish odor syndrome). *J. Clin. Aesthetic Dermatol.* **2013**, *6*, 45.
196. Cashman, J.R.; Camp, K.; Fakharzadeh, S.S.; Fennessey, P.V.; Hines, R.N.; Mamer, O.A.; Mitchell, S.C.; Preti, G.; Schlenk, D.; Smith, R.L. Biochemical and clinical aspects of the human flavin-containing monooxygenase form 3 (FMO3) related to trimethylaminuria. *Curr. Drug Metab.* **2003**, *4*, 151–170. [CrossRef] [PubMed]
197. Roberts, A.B.; Gu, X.; Buffa, J.A.; Hurd, A.G.; Wang, Z.; Zhu, W.; Gupta, N.; Skye, S.M.; Cody, D.B.; Levison, B.S. Development of a gut microbe–targeted nonlethal therapeutic to inhibit thrombosis potential. *Nat. Med.* **2018**, *24*, 1407–1417. [CrossRef] [PubMed]
198. Liepinsh, E.; Konrade, I.; Skapare, E.; Pugovics, O.; Grinberga, S.; Kuka, J.; Kalvinsh, I.; Dambrova, M. Mildronate treatment alters γ-butyrobetaine and l-carnitine concentrations in healthy volunteers. *J. Pharm. Pharmacol.* **2011**, *63*, 1195–1201. [CrossRef] [PubMed]
199. Kuka, J.; Liepinsh, E.; Makrecka-Kuka, M.; Liepins, J.; Cirule, H.; Gustina, D.; Loza, E.; Zharkova-Malkova, O.; Grinberga, S.; Pugovics, O. Suppression of intestinal microbiota-dependent production of pro-atherogenic trimethylamine N-oxide by shifting L-carnitine microbial degradation. *Life Sci.* **2014**, *117*, 84–92. [CrossRef]
200. Dambrova, M.; Skapare-Makarova, E.; Konrade, I.; Pugovics, O.; Grinberga, S.; Tirzite, D.; Petrovska, R.; Kalvins, I.; Liepins, E. Meldonium decreases the diet-increased plasma levels of trimethylamine N-oxide, a metabolite associated with atherosclerosis. *J. Clin. Pharmacol.* **2013**, *53*, 1095–1098. [CrossRef]
201. Zhu, Y.; Li, Q.; Jiang, H. Gut microbiota in atherosclerosis: Focus on trimethylamine N-oxide. *Apmis* **2020**, *128*, 353–366. [CrossRef]
202. Gregory, J.C.; Buffa, J.A.; Org, E.; Wang, Z.; Levison, B.S.; Zhu, W.; Wagner, M.A.; Bennett, B.J.; Li, L.; DiDonato, J.A. Transmission of atherosclerosis susceptibility with gut microbial transplantation. *J. Biol. Chem.* **2015**, *290*, 5647–5660. [CrossRef]
203. Skye, S.M.; Zhu, W.; Romano, K.A.; Guo, C.-J.; Wang, Z.; Jia, X.; Kirsop, J.; Haag, B.; Lang, J.M.; DiDonato, J.A. Microbial transplantation with human gut commensals containing CutC is sufficient to transmit enhanced platelet reactivity and thrombosis potential. *Circ. Res.* **2018**, *123*, 1164–1176. [CrossRef] [PubMed]
204. Smits, L.P.; Kootte, R.S.; Levin, E.; Prodan, A.; Fuentes, S.; Zoetendal, E.G.; Wang, Z.; Levison, B.S.; Cleophas, M.C.; Kemper, E.M. Effect of vegan fecal microbiota transplantation on carnitine-and choline-derived trimethylamine-N-oxide production and vascular inflammation in patients with metabolic syndrome. *J. Am. Heart Assoc.* **2018**, *7*, e008342. [CrossRef] [PubMed]
205. Kjellmo, C.A.; Johnsen, P.H.; Hovland, A.; Lappegård, K.T.; Mathisen, M. Fecal microbiota transplantation in irritable bowel syndrome does not affect plasma lipids or LDL and HDL subfractions. *Atherosclerosis* **2017**, *263*, e206. [CrossRef]
206. Kim, E.S.; Yoon, B.H.; Lee, S.M.; Choi, M.; Kim, E.H.; Lee, B.W.; Kim, S.Y.; Pack, C.G.; Sung, Y.H.; Baek, I.J.; et al. Fecal microbiota transplantation ameliorates atherosclerosis in mice with C1q/TNF-related protein 9 genetic deficiency. *Exp. Mol. Med.* **2022**, *54*, 103–114. [CrossRef] [PubMed]
207. Zhou, Y.; Xu, H.; Huang, H.; Li, Y.; Chen, H.; He, J.; Du, Y.; Chen, Y.; Zhou, Y.; Nie, Y. Are there potential applications of fecal microbiota transplantation beyond intestinal disorders? *BioMed Res. Int.* **2019**, *2019*, 3469754. [CrossRef] [PubMed]

International Journal of
Molecular Sciences

Review

The Role of Oxidative Stress Enhanced by Adiposity in Cardiometabolic Diseases

Iwona Świątkiewicz [1,2,*], Marcin Wróblewski [3], Jarosław Nuszkiewicz [3], Paweł Sutkowy [3], Joanna Wróblewska [3] and Alina Woźniak [3]

[1] Department of Cardiology and Internal Medicine, Collegium Medicum, Nicolaus Copernicus University, 85-094 Bydgoszcz, Poland
[2] Division of Cardiovascular Medicine, University of California San Diego, La Jolla, CA 92037, USA
[3] Department of Medical Biology and Biochemistry, Collegium Medicum, Nicolaus Copernicus University, 85-092 Bydgoszcz, Poland
* Correspondence: iwona.swiatkiewicz@gmail.com

Abstract: Cardiometabolic diseases (CMDs), including cardiovascular disease (CVD), metabolic syndrome (MetS), and type 2 diabetes (T2D), are associated with increased morbidity and mortality. The growing prevalence of CVD is mostly attributed to the aging population and common occurrence of risk factors, such as high systolic blood pressure, elevated plasma glucose, and increased body mass index, which led to a global epidemic of obesity, MetS, and T2D. Oxidant–antioxidant balance disorders largely contribute to the pathogenesis and outcomes of CMDs, such as systemic essential hypertension, coronary artery disease, stroke, and MetS. Enhanced and disturbed generation of reactive oxygen species in excess adipose tissue during obesity may lead to increased oxidative stress. Understanding the interplay between adiposity, oxidative stress, and cardiometabolic risks can have translational impacts, leading to the identification of novel effective strategies for reducing the CMDs burden. The present review article is based on extant results from basic and clinical studies and specifically addresses the various aspects associated with oxidant–antioxidant balance disorders in the course of CMDs in subjects with excess adipose tissue accumulation. We aim at giving a comprehensive overview of existing knowledge, knowledge gaps, and future perspectives for further basic and clinical research. We provide insights into both the mechanisms and clinical implications of effects related to the interplay between adiposity and oxidative stress for treating and preventing CMDs. Future basic research and clinical trials are needed to further examine the mechanisms of adiposity-enhanced oxidative stress in CMDs and the efficacy of antioxidant therapies for reducing risk and improving outcome of patients with CMDs.

Keywords: oxidative stress; obesity; cardiovascular disease; cardiometabolic diseases; coronary artery disease; metabolic syndrome; type 2 diabetes

Citation: Świątkiewicz, I.; Wróblewski, M.; Nuszkiewicz, J.; Sutkowy, P.; Wróblewska, J.; Woźniak, A. The Role of Oxidative Stress Enhanced by Adiposity in Cardiometabolic Diseases. *Int. J. Mol. Sci.* **2023**, *24*, 6382. https://doi.org/10.3390/ijms24076382

Academic Editors: Yutang Wang and Dianna Magliano

Received: 12 February 2023
Revised: 23 March 2023
Accepted: 24 March 2023
Published: 28 March 2023

1. Introduction

Cardiometabolic diseases (CMDs), such as cardiovascular disease (CVD), metabolic syndrome (MetS), and type 2 diabetes (T2D), are associated with increased morbidity and mortality [1,2]. CVD, including coronary artery disease (CAD) and systemic essential hypertension (HTN), are among the main causes of premature and excess mortality in developed countries [1]. CAD is a leading single cause of death in people over 50 years of age [1]. HTN remains a major cardiovascular risk factor, almost doubling the risk of death, with a rising systolic and diastolic blood pressure (BP) of as much as 20 and 10 mmHg, respectively [3,4]. Moreover, while the overall prevalence of HTN in the adult population is ~30–45% globally, HTN becomes progressively more common with advancing age, reaching >60% in people aged >60 years [3]. MetS, which occurs in approximately 25–30% of adults, doubles the long-term risk of developing CVD and is associated with a 5-fold increase in the risk of T2D [2,5]. The incidence of T2D is constantly growing, with an increase in deaths from T2D by 70% globally between 2000 and 2019 [1]. The increasing prevalence of CMDs has been attributed to the aging population and the common occurrence of risk factors, such as high systolic BP, elevated plasma glucose, and increased body mass index (BMI) [1]. The global epidemic of obesity, MetS, and T2D in adult and children populations developed over the last decades [1]. Notably, elevated fasting plasma glucose and high BMI are among the leading risk factors that displayed the largest increases in risk exposure over the period from 1990 to 2019 [6].

The CMDs are characterized by coexistence of multiple risk factors, including excess weight and adiposity, dyslipidemia, insulin resistance, high fasting plasma glucose, impaired glucose tolerance or T2D, cigarette smoking, elevated BP, physical inactivity, and erratic dietary patterns, all of which contribute to the pathogenesis of CMDs and impact patient outcomes [2,7–10]. The leading risk factor globally for attributable deaths is high systolic BP, which accounts for ~10.8 million deaths (i.e., ~19% of all deaths) and has emerged as the most important risk factor in older people [6]. Specifically, high systolic BP accounts for 9.3% of the disability-adjusted life years (DALYs) in the entire population and 6.0%, 16.1%, and 19.5% for subgroups with ages 25–49 years, 50–74 years, and ≥75 years, respectively. The marked rise of prevalence of high fasting plasma glucose and high BMI, and their large contribution to CMDs burden, is particularly alarming, which is reinforced by insufficient understanding of underlying mechanisms. For example, the prevalence of high BMI is rising significantly faster than prevalence of low physical activity, excessive caloric intake, and poor diet quality, all of which contribute to high BMI [6]. Identifying and addressing risk factors can reduce cardiometabolic risks; however, the efficacy of currently used preventive and therapeutic strategies is insufficient [8].

Obesity is associated with an increased risk of metabolic disorders, including MetS and T2D, and is among the major risk factors for CVD [1,2,9,11]. Obesity-related mortality and disability are caused mainly by CVD [1]. The prevalence of obesity increased during the past three decades at a faster pace than the related disease burden [1,12]. In the adult US population, the prevalence of overweight and obese was reported as high as 71% and 40%, respectively [2]. Overweight or obesity occurs in ~83%, ~76%, and ~74% of subjects with T2D, HTN, and dyslipidemia, respectively [13]. Importantly, ~33% of overweight and ~65% of obese individuals fulfilled the criteria for MetS [14]. The prevalence of abdominal obesity manifesting as an elevated waist circumference, which is a typical feature of MetS, was reported in the adult US population at ~56% [15]. Additionally, an increased waist circumference was the most common abnormality in 34,821 subjects with MetS, mostly from the European countries enrolled in the Metabolic syndrome and Arteries Research (MARE) Consortium [9].

Understanding the underlying mechanisms and identifying risk factors for CMDs plays a significant role in reducing cardiometabolic risk and improving patient outcome. Oxidative stress, which results from a lack of balance between oxygen derivatives generation and their removal by the antioxidant defense system, contributes to the pathophysiology of obesity, atherosclerosis, and CMDs [10,16–19]. The significance of oxidative stress relates to the fundamental role of reactive oxygen species (ROS) and redox signaling in molecular, cellular, and systems processes [20,21]. In obesity, enhanced and disturbed generation of reactive oxygen species (ROS) in excess adipose tissue (AT) may lead to increased oxidative stress. Oxidant–antioxidant imbalance is a common feature of various CMDs, including HTN, CAD, stroke, and MetS. Oxidative stress can result in various disorders, such as endothelial damage, vascular dysfunction, cardiovascular remodeling, and systemic inflammation [17,19]. In addition, oxidative stress is shown to be associated with impaired insulin signaling pathways and insulin resistance [17–19,22].

The mechanisms and clinical implications related to the interplay between adiposity, oxidative stress, and CMDs have not been comprehensively addressed [23–27]. The present review article is based on extant results from basic and clinical studies, and specifically addresses the various aspects associated with oxidant–antioxidant balance disorders in the course of CMDs in subjects with excessive accumulation of AT and obesity. We aim to give a comprehensive overview of existing knowledge about associations between adiposity-enhanced oxidative stress and cardiometabolic risks to indicate knowledge gaps and offer future perspectives for further basic and clinical research. We provide insights into both the mechanisms and clinical implications of effects related to the interplay between adiposity and oxidative stress for treating and preventing CMDs.

2. Characteristics of Adipose Tissue

Adipose tissue (AT) is one of the main types of loose connective tissue [28]. Adipocytes constitute the main fraction of AT-building cells [29]. In addition to adipocytes, there are also stromal vascular fraction, adipose-derived stem cells, preadipocytes, macrophages, lymphocytes, eosinophils, mast cells, fibroblasts, and nerve cells [27,30,31]. There are four types of AT differentiated by histological structure and function: white adipose tissue (WAT), brown adipose tissue (BAT), beige adipose tissue, and pink adipose tissue [27,32]. WAT, one of the largest organs, is the main energy store of the organism, which captures and accumulates lipids [33]. By collecting triacylglycerols and glucose, WAT protects other tissues [33]. Adipocytes that make up WAT are characterized by a significant lipid content [34]. High lipid content in WAT acts as a thermal insulator and helps maintain internal body temperature [35]. WAT also produces and releases a variety of bioactive molecules, including the adipokines, which are biologically active proteins with a low molecular weight synthetized and secreted mainly by WAT [36]. These biomolecules exhibit autocrine, paracrine, and endocrine effects on tissues [37]. So far, over 600 adipokines were detected and described in scientific literature [38]. The main role of adipokines is to regulate metabolism and bioenergetic homeostasis [39]. Moreover, adipokines have immunomodulatory properties [40]. In the course of obesity, a change in the adipokine profile is observed in favor of the increased secretion of pro-inflammatory adipokines with a simultaneous reduction in the level of anti-inflammatory adipokines [41,42]. This leads to chronic low-grade inflammation, which affects not only AT, but also other tissues [43]. BAT is made up of adipocytes containing many fat droplets of varying sizes [44]. Compared to WAT, BAT is characterized by a large number of mitochondria in adipocytes [45]. A significant number of mitochondria enables the implementation of the main BAT function: non-shivering thermogenesis [46]. Beige AT is a transition form between WAT and BAT [47]. It is formed as a result of the beiging of WAT adipocytes [48]. Beige adipocytes acquire the properties that are typical for BAT, and their role is also changed, from cells constituting an energy store to energy-releasing adipocytes [49]. The main factor leading to the formation of beige AT is chronic exposure to low temperatures [46]. Pink AT is formed in mammary gland alveolar epithelial cells [50,51]. This tissue is involved in the production and secretion of milk during lactation [52].

AT is characterized by high plasticity and adaptation to changing conditions [46]. Not only are the type and volume of AT important, but also the location. In clinical terms, visceral AT is extremely significant. The increase in visceral AT volume results in abdominal obesity and an increased risk of CMDs [53]. Visceral AT is formed mainly by WAT and is a source of adipokines [54]. Epicardial AT, which is a particular form of visceral AT, participates in the pathogenesis of CAD, atrial fibrillation (AF), and heart failure (HF) with preserved left ventricular (LV) ejection fraction [55,56]. Additionally, the association between perivascular AT (PVAT) and the occurrence of CVD was found [57]. In the course of obesity, PVAT hypertrophy and hyperplasia are observed [58]. PVAT expansion leads to atheromatous plaque development and vascular calcification [59]. Under physiological conditions, a positive effect of PVAT on cardiovascular homeostasis is observed in patients with an AT amount within normal limits [60,61]. Subcutaneous AT is the second largest depot of fat in the human body [62]. The amount of subcutaneous AT is proportional to visceral AT and increases in the course of obesity [63]. Recent studies indicated that subcutaneous AT participates in the regulation of lipid-carbohydrate metabolism [64].

Obesity is defined as a state of excessive accumulation of AT that exceeds the adaptive abilities of the organism and increases the risk of developing other diseases [65,66]. Obesity is the global epidemic affecting more than 2.3 billion people worldwide, both adults and children [1,9].

In clinical practice, the calculation of BMI is the most common method of diagnosing excess weight, including obesity [67]. A BMI value of 18.5–24.9 kg/m^2 was determined as a normal value, while values of 25.0–29.9 kg/m^2 and $\geq$30.0 kg/m^2 indicate overweight and obesity, respectively [68]. Patients with BMI values of 30.0–34.9 kg/m^2 are diagnosed with obesity class I, BMI of 35.0–39.9 kg/m^2 is considered as obesity class II, and BMI $\geq$ 40 kg/m^2 is defined as obesity class III [68]. However, population- and country-specific criteria should be considered. For example, the optimal cut-off point for the identification of metabolic disorders in the Polish population is 27.2 kg/m^2 [69]. Additionally, for an equivalent age-adjusted and sex-adjusted obesity-based risk of T2D at a BMI of 30.0 kg/m^2 in White populations, the lower BMI cutoffs for South Asian (23.9 kg/m^2), Black (28.1 kg/m^2), Chinese (26.9 kg/m^2), and Arab (26.6 kg/m^2) populations were found [70].

Increased waist circumference, which indicates the presence of excess central (abdominal) obesity, is a typical finding in subjects with MetS and is common in patients with other CMDs [5,71]. For the increased waist circumference, the population-, ethnic-, gender-, and country-specific definitions should be used. The cut-off values of increased waist circumference for different populations are provided in Table 1 [5,71]. The waist circumference measurement is recommended for those with a BMI of 25 to 34.9 kg/m^2 to provide additional information on CVD risk; however, if BMI is >30 kg/m^2, central obesity can be assumed and waist circumference does not need to be measured [5,71]. Nevertheless, owing to the need for screening of individuals with a metabolically obese normal weight, the measuring of waist circumference should be considered when BMI is $\geq$22.5 kg/m^2 in females and $\geq$23.8 kg/m^2 in males [14,69,72].

Table 1. Population- and gender-specific cut-off values of waist circumference for the diagnosis of metabolic syndrome.

Population (Country/Ethnic Group)	Waist Circumference [cm]	
	Male	Female
Europid origin, Sub-Saharan Africa, Eastern Mediterranean, Middle East	94	80
United States of America	102	88
Asia, South and Central America	90	80

Waist–hip ratio (WHR), another indicator of abdominal obesity, is calculated as the ratio of the waist circumference to the hip circumference [73]. The WHR reference values are gender specific. For males, physiologically WHR is >0.90 and for females >0.85 [74].

BMI, waist circumference, and WHR are often used in clinical practice due to the simplicity of measurement and calculations, while other methods, such as bioelectrical impedance, are used to determine the content of AT in the body [75]. It is estimated that in healthy adult males and healthy adult females, fat should account for 17.6–25.3% and 28.8–35.7% of body mass, respectively [76].

Obesity is closely related to the MetS that is associated with adverse outcome [2,5,9,71,77,78]. MetS is defined as a set of interrelated factors that significantly increase the risk of other CMDs, including T2D and CVD [5,9,10,71,78–81]. Abdominal obesity is among the diagnostic criteria for MetS [80]. According to the International Diabetes Federation (IDF) criteria, MetS is defined in the presence of $\geq$3 of the following five risk factors: increased waist circumference (population- and country-specific definitions should be used), elevated fasting plasma glucose ($\geq$100 mg/dL or drug treatment for this disorder), elevated BP (systolic $\geq$130 and/or diastolic $\geq$85 mmHg or antihypertensive drug treatment), hypertriglyceridemia ($\geq$150 mg/dL or drug treatment for this disorder), and reduced high-density lipoprotein cholesterol (HDL-C) (<40 mg/dL in males and <50 mg/dL in females or drug treatment for this disorder) [5,71]. Most T2D patients met the MetS criteria [5,71]. MetS occurs in ~25% of adults depending on age, gender, race, country of origin, and diagnostic criteria [2,5,9,14,71,82–84]. For example, in the National Health and Nutrition Examination Survey (NHANES) [14,82,83], MetS was diagnosed in ~34% of US adults 20 years of age and over. Importantly, while MetS occurred in 16% of females and 20% of males under 40 years of age, 52% of males and 54% of females 60 years of age and over met the MetS criteria [14]. The pathogenesis of MetS is influenced by genetic and lifestyle factors [85]. In patients with MetS, lifestyle modifications, and in some cases, pharmacological treatment, are required for improving patient outcome [3,74]. A lack of effective intervention in patients with MetS may lead to developing other CMDs, such as CVD, disability, and premature death [84].

3. The Oxidant–Antioxidant Balance and Oxidative Stress

Reactive oxygen species (ROS) and reactive nitrogen species (RNS) are the products of normal cellular metabolism [20,21]. The main source of ROS in the cell is a mitochondrial respiratory chain. ROS are also generated in peroxisomes, endoplasmic reticulum, and during reactions catalyzed by xanthine oxidase, endothelial oxidases, and phagocyte-reduced nicotinamide adenine dinucleotide phosphate (NADPH) oxidase (NOX) [86–88]. Nitric oxide ($\cdot$NO) is, in turn, generated from L-Arginine with the participation of nitric oxide synthases (NOS) [89]. ROS/RNS in the living system play a double role. They are not only deleterious species, but also act as "second messengers" by participating in a number of normal regulatory processes [89–91]. It was shown that redox signaling may control cellular functions through modifications of the activity of enzymes participating in metabolic processes, regulation of gene expression, transcription factors, and through the impact on the nature of the epigenetic modifications [92]. The key redox signaling agents are hydrogen peroxide (H_2O_2) and the superoxide anion radical (O_2^-) [93]. O_2^- is considered a primary ROS, participating in both signaling and in cell injury [94]. As a result of the reactions with the participation of this radical, other ROS are generated, such as H_2O_2 and hydroxyl radical ($\cdot$OH) [95]. The manner of ROS/RNS activity depends mainly on their concentration [86]. Occurring in low/moderate concentrations, ROS show beneficial effects [20]. At high concentrations, however, they may damage all major cellular components [96]. Specifically, they participate in the oxidation of proteins, carbohydrates, lipids, and DNA, causing damage to DNA, cellular membranes, and organelles [97].

Under physiological conditions, intracellular ROS homeostasis is subject to strict control, resulting in exceptionally low levels of ROS in the cell [90]. Complex mechanisms, with which aerobic organisms are equipped, protect against excessive ROS generation [98–101]. In order to maintain redox balance, their activity encompasses prevention, interception, and repair [102]. It was proven that the main role in antioxidant protection is played by enzymes [103], such as superoxide dismutase (SOD), glutathione peroxidase (GPx), catalase (CAT), glutathione reductase (GR), and xanthine oxidoreductase (XOR) [104]. Redox homeostasis of the cell is also maintained by non-enzymatic ROS/RNS scavengers, including both endogenous antioxidants, such as glutathione, ferritin, ceruloplasmin, uric acid, and coenzyme Q and exogenous antioxidants, such as carotenoids, polyphenolic compounds, vitamin C, and vitamin E [105–107].

Lack of balance between ROS/RNS generation and their removal by the antioxidant defense system leads to oxidative stress [16]. It can result from excessive generation of these oxygen and/or nitrogen derivatives or from weakening activity of antioxidant mechanisms [105]. More recent publications suggest two ways of classifying oxidative stress: time based and concentration/intensity based [90]. Based on time criterion, we can distinguish "acute" and "chronic" oxidative stress [90]. The classification based on its intensity indicates basal, low intensity, intermediate intensity, and high intensity oxidative stress [108]. One of the most significant consequences of oxidative stress is enhancement of lipid peroxidation, which results in the oxidation of polyunsaturated fatty acids (PUFAs), which are part of cellular membranes [109]. The products of this process are conjugated dienes (CD) and lipid peroxides (ROOH) [109], and also the so-called secondary products of lipid peroxidation, such as malondialdehyde (MDA), 4-hydroxy-2-nonenal (4-HNE) [110], and isoprostanes (IsoPs) [111]. It was shown that oxidative stress may take part in the etiopathogenesis of numerous systemic diseases, including CMDs such as CVD [16,86,97] and MetS [112,113].

4. Adipose Tissue as a Source of Free Radicals

The main endogenous sources of ROS in AT are mitochondria and NOX. Key importance is also given to NOS, Fenton's reaction, microsomal cytochrome P450, peroxisomal β-oxidation, as well as lipoxygenases and cyclooxygenases [114].

The results of in vitro studies show that during the basic state of mitochondria activity, 0.2–2% of oxygen is converted in a respiratory chain into ROS [115], mainly from complex I and complex III (originally O_2^-) [116]. Complex I receives electrons from reduced nicotinamide adenine dinucleotide (NADH), which leads to a premature leak of electrons and penetration of ROS to the mitochondrial matrix. In complex III, however, ROS are generated on both sides of the inner mitochondrial membrane [117]. In obesity, mitochondria are particularly susceptible to ROS generation because when an excess of nutrients occurs in adipocytes and mitochondrial substrates, the ROS concentration increases [118]. In a cell culture of 3T3-L1 adipocytes, it was shown that high concentrations of glucose or free fatty acids increase ROS generation in mitochondria [119]. It is considered that the excess of ROS in mitochondria leads to their dysfunction, which is the cause of T2D, non-alcoholic fatty liver disease, HF, and MetS. Other studies also suggest a possibility of developing myocarditis caused by changes in immunological response due to mitochondrial damage to DNA by free radicals [120].

NOX is a membranous protein that transports electrons from NADPH to O_2, the side effect of which is ROS generation in the cytoplasm. Among seven NOX isoforms identified in mammalian adipocytes, the most numerous isoform is NOX4 [114], although its exceptionally strong expression was also noted in the cardiovascular system (together with NOX1 and NOX5 isoforms) [121]. In cell cultures, it was shown that NOX4 expression and the resulting ROS generation increased in adipocytes exposed to excess glucose or palmitic acid salts [122,123]. NOX4 is also distinguished by the fact that it generates H_2O_2, which penetrates through membranes and is a more durable form of ROS in comparison with O_2^-, produced by the other NOX isoforms. What is also significant for AT is NOX2 isoform, because it dominates in the cellular membrane of macrophages of AT and generates O_2^- in response to lipopolysaccharides or saturated fatty acids [124].

In obese individuals, ROS sources in AT can change from NOX4 at an early stage to NOX2 at a medium stage, and then passing to the late stage mainly into impaired oxidative phosphorylation in mitochondria [125]. With reference to vascular stroma cells, macrophages seem to be of greatest significance for ROS sources in AT. They are even considered to be the main factor regulating activity of AT in free radical signaling pathways [124].

A significant endogenous ROS source in adipocytes can also be endoplasmic reticular oxidoreductin 1 (ERO1) and the xanthine dehydrogenase (XDH)/oxidoreductase (XOD) system (Figure 1). ERO1 is a protein disulfide oxidase of endoplasmic reticulum, which generates H_2O_2 as a result of protein folding and secretion [124]. The XDH/XOD system becomes an ROS source under oxidative stress conditions, which is, for example, observed during obesity. XOD then changes into xanthine oxidase (XO) and generates O_2^- and H_2O_2 in a series of catabolic reactions of purine conversion into uric acid [23].

Fisher-Wellman and Neufer [116], moreover, propounded a hypothesis that under insulin resistance conditions, when a decreased concentration of glutathione (GSH) is observed, pyruvate dehydrogenase and nicotinamide nucleotide transhydrogenase can be a significant source of H_2O_2 in adipocytes. Enhanced oxidative redox signaling in AT, as it was already mentioned mainly in obese individuals, results in further peroxidative consequences for the organism, being a result of this tissue disorder [124]. Apart from the abovementioned mechanisms, it is mainly about abnormal adipokine secretion into the bloodstream. In the obese state, leptin [126] and chemerin [127] are secreted in excess (see Figure 1), whereas adiponectin [128] and omentin-1 [129] are subject to decreased secretive activity of AT. This leads to oxidative stress and an inflammatory state in vascular endothelium, and by the same token, dysfunction of blood vessels (HTN, atherosclerosis-typical for MetS). In the context of the present paper, what is of particular significance is PVAT; since it is adjacent to blood vessels, it affects them directly. PVAT in obese individuals has a vasoconstrictive activity, whereas in a healthy organism, it relaxes the smooth muscle of blood vessels [23].

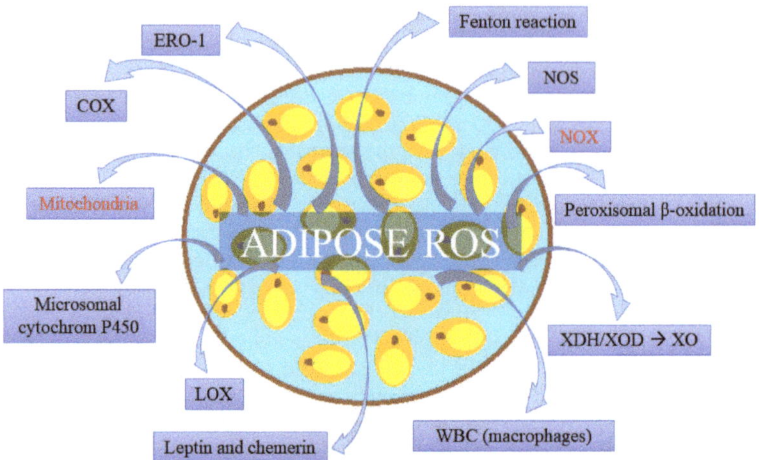

Figure 1. Adipose tissue as a source of reactive oxygen species (ROS). The main sources are mitochondria and reduced nicotinamide adenine dinucleotide phosphate (NADPH) oxidase (NOX). (ERO-1) endoplasmic reticular oxidoreductin 1, (COX) cyclooxygenase, (LOX) lipoxygenase, (WBC) white blood cells, (XDH) xanthine dehydrogenase, (XOR) xanthine oxidoreductase, (XO) xanthine oxidase, and (NOS) nitric oxide synthase.

5. Oxidative Stress in Cardiometabolic Disorders in Subjects with Obesity

Oxidative stress is a common feature of CMDs including various types of CVD, such as HTN and atherosclerotic CVD, and MetS (Figure 2). The interplay between obesity, oxidative stress, and low-grade chronic inflammation in the course of CMDs is linked to adverse outcomes, such as premature and accelerated process of significant atherosclerosis, CAD events including acute myocardial infarction, post-infarct LV dysfunction and remodeling, and HF in long-term follow-up [130–137]. Several clinical studies were conducted in the populations of overweight or obese adults and children to examine associations between oxidative stress markers and cardiometabolic parameters, as well as the presence of CMDs or a risk of developing CMD, e.g., [131–139]. In general, the levels of oxidative stress markers were abnormal in obese individuals, both healthy and those with CMDs, compared to non-obese controls [138–146]. Moreover, measurement of various oxidative stress markers can capture different stages of oxidative stress development, which may result in various types and varying severity of tissue damage. While some markers are more specific for early stages of oxidative stress development, which are characterized by the production of ROS (e.g., H_2O_2), the others are related to later and more severe stages, such as ROS-mediated lipid peroxidation (e.g., IsoPs) or DNA damage (e.g., 8-hydroxy-2′-deoxyguanosine) [97,109,111,139–143,145,147,148]. More advanced and intense oxidative stress may be associated with a more likely occurrence of endothelial dysfunction, atherosclerotic vascular changes, or symptomatic CMDs. Importantly, the findings of some studies suggest a usefulness of oxidative stress markers for identifying individuals with increased risk of development or progression of CMDs. Specifically, IsoPs, which are markers of lipid peroxidation and have vasoconstricting and inflammatory properties, were shown to be useful for predicting HTN [139,145,147]. An increase in prevalence of childhood obesity over recent years, which may contribute to increased cardiometabolic risk in children populations, resulted in expanding clinical research addressing an occurrence of oxidative stress in children populations, especially those with obesity [1,139,140].

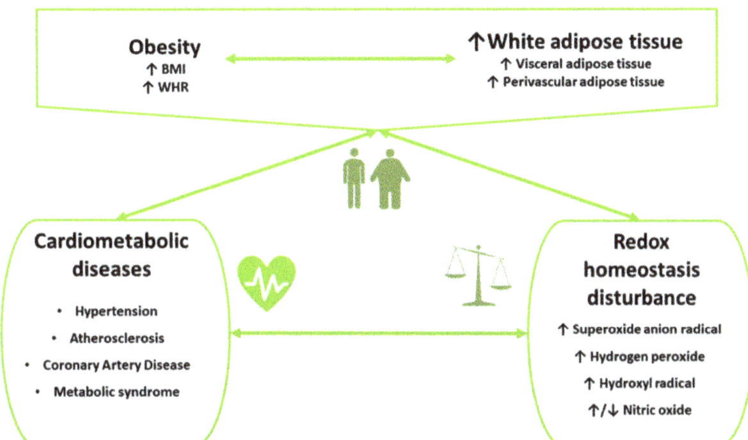

Figure 2. Demonstrated and potential associations between adiposity, redox homeostasis disorders, and cardiometabolic diseases.

5.1. Systemic Essential Hypertension

The presence of oxidative stress was demonstrated in both animal and human models of HTN [149,150]. Associations between oxidant–antioxidant balance markers, adiposity indices, and BP values were observed in obese healthy subjects, obese individuals with elevated BP, and patients with HTN [139–144,147,150,151]. Additionally, urinary and plasma IsoPs were significantly lower in treated hypertensive men compared with the untreated men [152].

In the study of Atabek et al. [140], positive correlations between plasma concentrations of peroxy radicals and systolic BP, as well as between peroxy radicals and total cholesterol level, were found in the group of obese children, 25% of whom had hyperlipidemia and HTN. Importantly, in the control group of healthy and non-obese children, the levels of peroxy radicals were significantly lower compared to obese children. In addition, no correlations between oxidative stress markers and cardiometabolic parameters including BP values were found in the control group [140]. Furthermore, in the study of Morandi et al. [151], which included children and adolescents with obesity, total systemic antioxidant capacity (TAC) was inversely associated with systolic BP and pulse pressure, which is a marker of arterial stiffness and subclinical vascular damage. In addition, a negative association between TAC and a risk of systolic HTN was observed, which was independent of another significant predictor of HTN, i.e., BMI z-score. Importantly, in this study, the participants with HTN and the composite measure defined as "elevated systolic BP + HTN" accounted for 25% and 48% of the studied population, respectively. The relationship between oxidative stress and systolic BP could be explained by hypothesizing that oxidative stress would increase the BP by producing endothelial dysfunction, which was shown previously to be strongly correlated with systolic BP [151,153].

Several authors reported higher plasma and urine concentrations of IsoPs in obese subjects with elevated BP compared to non-obese individuals and obese subjects with normal BP [139,141,145,147,151]. Additionally, positive correlations between IsoPs and various cardiometabolic markers, such as BP, central adiposity indices (BMI derivatives and waist circumference), insulin resistance markers, body fat, total cholesterol, triglycerides

(TGs), high sensitivity-C-reactive protein (hs-CRP), T2D diagnosis, and cigarette smoking were found [139,141,145,147,151].

In the study of Ostrow et al. [139], which included children with obesity, 8-IsoP correlated with mean 24 h systolic BP and was higher in subjects with "masked HTN", defined as elevated mean 24 h systolic BP, compared to normotensive subjects. Importantly, the participants with "masked HTN" accounted for 16% of the study group [136]. In another study including 897 premenopausal overweight women with an average BMI of 27 kg/m^2 and without a history of CVD, the urine levels of F2-IsoP and F2-IsoP metabolite (15-F2t-IsoP-M) were positively correlated with diastolic BP, central adiposity indices, T2D diagnosis, and cigarette smoking [147]. Moreover, the findings of this study suggest that elevated F2-IsoP metabolite may be considered as a predictor of an incident HTN in the long-term follow-up with a maximum of 11.5 years. Additionally, F2-IsoP was shown to be associated with diastolic BP in African American, but not in White American healthy obese youth, although positive correlations between oxidative stress and body fat were found in obese healthy young subjects of both races [145]. These findings suggest that oxidative stress may be a mechanistic link between key risk factors and occurrence of CMDs and indicate a usefulness of oxidative stress markers, such as markers of lipid peroxidation (e.g., IsoPs), for predicting the development of CVD, such as HTN [139,145,147].

However, some authors did not observe positive correlations between BP values, such as daytime mean BP, night-time mean BP, and systolic BP, and oxidative stress markers, such as IsoPs levels and NO production/metabolism markers [141,154]. Specifically, in the study of a large cohort of obese children characterized by a relatively higher composite prevalence of HTN, consisting of sustained HTN and "masked" HTN [155] (i.e., 17% in obese children vs. ~6% in the group of children with normal weight), no correlations between BP values and oxidative stress markers were observed [141]. However, in this study, oxidative stress markers were correlated with measures of obesity and insulin resistance, independently of BMI [141]. The authors concluded that elevated IsoPs levels may represent an early marker of cardiometabolic dysfunction, even in the absence of established HTN [141]. Several IsoPs actions related to vascular dysfunction, such as vasoconstriction, induction of platelet aggregation, and enhancement of adhesion of neutrophils and monocytes to endothelial cells may be involved in the pathogenesis of cardiometabolic dysfunction. Additionally, the positive associations between IsoPs and cardiometabolic risk factors, such as insulin resistance indices, inflammation markers, and atherogenic lipids were found. However, no differences in IsoPs levels between groups of overweight treated hypertensive patients, overweight untreated hypertensive patients, and normotensive controls were reported [152].

Moreover, Baráth et al. [143] observed significantly higher concentrations of MDA, another marker of lipid peroxidation, in obese patients with HTN compared to non-obese hypertensive patients, obese normotensive subjects, and healthy controls. Additionally, Minuz et al. [156] demonstrated enhanced oxidative stress and persistent platelet activation in patients with HTN and advanced vascular lesions in the course of severe hypertensive retinopathy. These findings indicate that oxidative stress markers might be useful for identifying those hypertensive patients who are at an increased risk of cardiovascular events and who might benefit from a long-term antiplatelet therapy [156]. Furthermore, the associations between the oxidative stress to DNA and BP measures were also investigated [142,148]. In the study of Yavuzer et al. [148], an increased urinary level of the 8-oxo-7,8-dihydro-2′-deoxyguanosine (8-oxodG), the oxidative stress marker of DNA damage, was found in elderly individuals with HTN compared to controls. However, no differences in the 8-oxodG between the groups of obese men with and without HTN were found in the study of Cejvanovic et al. [142]. Nevertheless, statistically significant associations between 8-oxodG and mean 24 h systolic and diastolic BP were observed.

5.2. Atherosclerotic Risk Factors, Atherosclerosis, and Metabolic Syndrome

Atherosclerosis is a chronic vascular disease characterized by the formation of an atherosclerotic plaque in the vessel wall of medium- or large-sized arteries, which results in the development of atherosclerotic CVD. While the pathogenesis of atherosclerosis is not well understood, extensive research indicates that atherosclerosis is a consequence of various inflammatory, oxidative, and mechanical processes. The atherosclerotic process is initiated by damage of endothelial cells, which can be triggered by oxidative stress associated with atherosclerotic risk factors, such as obesity (especially abdominal obesity), insulin resistance, elevated glucose levels in prediabetes and T2D, HTN, dyslipidemia, and cigarette smoking, all of which often coexist [157–160]. The MetS reflects the clustering of individual cardiometabolic risk factors related to abdominal obesity, insulin resistance, dyslipidemia, and elevated BP [5,10]. Accumulation of visceral fat leading to abdominal obesity results in disturbances in the production of inflammatory and anti-inflammatory cytokines, followed by the development of chronic low-grade inflammation [146,161]. Central obesity plays a significant role in the pathogenesis of atherosclerotic CVD. The oxidative theory of atherosclerosis assumes that the process of atherogenesis is significantly influenced by the oxidative modification of lipoproteins, which leads to lipid peroxidation followed by formation of foam cells and damage of endothelial cells of the vessel wall [160]. Additionally, IsoPs exert several actions that may be involved in the pathogenesis of vascular dysfunction. The IsoPs are potent vasoconstrictors in most vascular beds, induce platelet aggregation, and enhance the adhesion of neutrophils and monocytes to endothelial cells, all of which may contribute to atherosclerosis [141].

Enhanced systemic oxidative stress can be a significant mechanism linking obesity, especially central obesity, to atherosclerotic CMDs, and may contribute directly or indirectly to the development of atherosclerosis [136,146,157]. Oxidative stress may also interact with inflammatory processes in the early stages of atherosclerosis [146,161]. Positive correlations were found in healthy young subjects between oxidative stress markers (such as MDA and CD), inflammatory markers (such as CRP), measures of abdominal obesity (except for BMI), and the presence of MetS [161]. Furthermore, oxidative stress in those with obesity is involved in various interactions with metabolism and actions of the atherogenic lipids. Those interactions include a postprandial increase in TGs concentrations, oxidative modification of low-density lipoproteins (LDL) resulting in the formation of oxidized LDL (oxLDL), which has significant atherogenic properties, and a decrease in circulating high-density lipoproteins (HDL), which also have a protective effect against atherosclerosis [157,162]. These interactions may lead to further production of ROS in the endothelium, promotion of proinflammatory vascular processes, increase in endothelial damage, and development of endothelial dysfunction, which can initiate atherosclerotic processes [157]. It was shown that peroxy radical levels were higher in subjects with hyperlipidemia compared to those with normal lipid levels, while both groups had similar BMI [140]. Significant unfavorable correlations between lipid atherosclerotic risk factors were observed in overweight females with hypothyroidism across different BMI ranges [163]. These correlations include non-high-density lipoprotein cholesterol (non-HDL-C) level and the ratios of TGs/HDL-C, total cholesterol/ high-density lipoprotein cholesterol (HDL-C) and low-density lipoprotein cholesterol (LDL-C)/HDL-C, and oxidative stress markers, such as thiobarbituric acid reactive substances (TBARS) and protein carbonyls, as well as markers of antioxidant defense, such as glutathione (GSH), CAT, and GPx [163]. In addition, elevated levels of oxLDL are involved in the formation of foam cells, both development and destabilization of atherosclerotic plaque, have an association with cytotoxic and procoagulant activity, and increase expression of adhesion molecules in endothelial cells. These elevated levels of oxLDL were observed in patients with advanced atherosclerosis, such as patients with CAD and ischemic stroke [133,164,165]. Additionally, in obese men compared to slim and physically active ones, the concentration of paraoxonase-1 (PON-1) that inhibits formation of oxLDL was lower, while the levels of MDA, the lectin-like receptor of oxidized LDL type I (LOX-1), and pro-inflammatory cytokine tumor necrosis factor α were higher [146]. Additionally, in the study of patients with morbid obesity treated with bariatric surgery, the levels of oxLDL and MDA were higher in subjects with carotid atheromatous plaques detected by ultrasound examination compared to patients with no visible carotid atheroma [155]. After bariatric surgery and lowering body mass, oxLDL levels decreased in both groups but were still higher in patients with atheromatous plaques, while the levels of PON-1 and CAT were higher in patients without atheromatous plaques. Importantly, MDA levels decreased significantly in both groups of patients after the bariatric surgery, which indicated that bariatric surgery reduced lipid peroxidation independently of the presence of atherosclerotic lesions.

Impairment of endothelial function is an early disorder involved in the pathogenesis of atherosclerosis [166]. A characteristic feature of endothelial dysfunction is lowered bioaccessibility of endothelial-dependent blood vessel dilating substances, in particular NO [167,168]. An increase in total AT and abdominal fat is connected with an impairment of endothelial function dependent on NO, development of oxidative stress, and increased production of vasoconstrictor proteins [169]. Vascular NADPH oxidases are multi-subunit enzymatic complexes occurring in myocytes of blood vessels and endothelial cells, which are a major source of superoxide anion radical within the walls of blood vessels [133]. NADPH oxidases play a key role in the pathogenesis of vascular diseases including cerebrovascular accident [170]. The findings of Silver et al. [169] demonstrated that vascular endothelial cell protein expression of NADPH oxidase-p47(phox), CAT, nitrotyrosine, and phosphorylation of eNOS were greater in the overweight/obese patients than in the normal-weight patients. This may provide novel insight into the molecular mechanisms linking obesity to oxidative stress and increased risk of atherosclerotic CVD. In the study by López-Domènech et al. [171], after losing weight, a decrease in BP, improvements in metabolic parameters, reduction in inflammatory response and oxidative stress parameters in leukocytes, such as decrease in superoxide and protein carbonyl, and enhancement of antioxidant systems activity were observed, even in subjects with morbid obesity undergoing dietary intervention with calorie restriction. In addition, a significant reduction in subclinical markers of atherosclerosis, such as small and dense LDL particles, myeloperoxidase (MPO), sP-selectin, and leukocyte adhesion, was reported. These findings suggest that the improvement of oxidative stress and inflammation may be underlying mechanisms responsible for reducing the risk of CVD in obese subjects after losing weight resulting from a calorie restriction diet.

5.3. Coronary Artery Disease

Elevated levels of free fatty acids and triacylglycerols in obese persons favor ectopic lipid accumulation, especially in the heart [172]. The amount of epicardial AT is strongly correlated with visceral obesity. During CAD development, epicardial and perivascular AT accumulates in the myocardium around medium and small coronary arteries. This results in compression, local delivery of free fatty acids and cardioactive hormones, release of pro-inflammatory adipokines, apoptosis, coronary calcification, and atheromatous plaque formation [172]. Obese hearts are less metabolically flexible and cardiometabolic changes in the heart favor ROS generation [173]. Increased fatty acid uptake and fatty acid oxidation, mitochondria dysfunction, augmented NOX activity, and decreased antioxidant capacity enhance oxidative stress in the heart and cardiomyocytes [174]. It seems that mitochondrial dysfunction plays a key role in MetS and enhances progression of metabolic disorders [175]. Little data are available to provide evidence on elevated level of oxidative stress in the heart tissue of CAD patients. In the study of obese males aged <55 years undergoing coronary artery bypass graft surgery (CABG), Niemann et al. [175] found increased levels of oxidative stress markers, such as 8-OHdG and protein carbonyls, in cardiomyocytes, as well as disorders of mitochondrial function. These abnormalities were comparable to changes observed in older patients. Importantly, young obese patients demonstrated signs of disturbed mitochondrial function and biogenesis otherwise seen only in old patients (obese or normal-weight). In addition, the length of telomeres was significantly reduced in cardiomyocytes of obese CAD patients compared to slim ones in the same age group [175]. Additionally, mean telomere length was reduced up to 30% in cardiomyocytes of old (normal-weight and obese) and young obese subjects, with the shortest telomeres in the old obese patients. Shortening the length of telomeres constitutes a sensitive indicator of increased oxidative stress in post-mitotic cells, such as cardiomyocytes [174]. In the group of obese patients undergoing CABG, Gramlich et al. [176] found increased levels of ROS and an enhanced expression of ROS-producing enzymes (i.e., p47phox, XO), decreased antioxidant defense mechanisms, and elevated inflammatory markers (such as vascular cell adhesion molecule-1) in the right atrial myocardial tissue and in serum, which were more pronounced with increasing BMI. These findings may indicate an increased risk of developing cardiomyopathy and cardiac dysfunction in obese patients undergoing CABG due to ongoing ischemia and reperfusion related to CABG.

5.4. Obesity and Oxidative Stress: Direct Link to CVD Outcomes and Mortality

The interactions between obesity and oxidative stress promote endothelial dysfunction, atherosclerotic coronary lesions, LV hypertrophy (LVH), myocardial fibrosis, cardiac remodeling, and LV diastolic and systolic dysfunction. These all contribute to the pathophysiology, symptoms, and outcomes of CVD including various arrhythmias, such as AF, myocardial infarction, and HF (particularly HF with preserved LV ejection fraction) [172,173]. Associations between oxidative stress and CVD outcomes including CVD-related mortality were examined in a few clinical studies [177,178]. In a large population-based cohort study including 9949 older adults (aged 50–75 years) from Germany, the urinary levels of oxidation end products of lipids and DNA, i.e., 8-IsoP and oxidized guanine/guanosine (OxGua), were shown to be independent predictors for CVD-related mortality and stroke incidence (8-IsoP was also a predictor of fatal stroke) in a 14-year follow-up [177]. Moreover, adding these biomarkers to the European Society of Cardiology SCORE scale improved its abilities for the prediction of CVD mortality. Furthermore, both biomarkers were associated with an incidence of myocardial infarction only in obese subjects (i.e., those with BMI ≥ 30 kg/m^2), but not in the total population. These findings provided strong evidence of the involvement of oxidative stress in the pathophysiology and outcome of CAD, as well as indicated a usefulness of 8-IsoP and OxGua measurements for prediction of myocardial infarction in obese older subjects. Several biological mechanisms might contribute to these results, such as the significant role of oxidative stress in the initiation and progression of atherosclerosis, including atheromatous plaque rupture and the enhancement of systemic oxidative stress related to the accumulation of AT during obesity. However, in another prospective cohort study conducted by Godreau et al. [178], no association between the serum level of 8-IsoP (which is not so stable as urine level) and mortality was observed in the subjects with elevated BMI in contrast to individuals with low to normal BMI.

Table 2 displays the relationships between oxidative stress markers and cardiometabolic parameters in obese subjects.

Table 2. Relationships between selected oxidative stress markers and cardiometabolic parameters in obese subjects with cardiometabolic disorders.

Oxidative Stress Marker	Results of Clinical Studies	Refs.
Isoprostanes	Higher concentration in hypertensive than normotensive subjects Positive correlations with: mean 24 h systolic BP, systolic BP, diastolic BP (in African Americans but not in White Americans), central adiposity indices, body fat, total cholesterol, TGs, total cholesterol/HDL-C, T2D diagnosis, insulin resistance markers, hs-CRP, cigarette smoking Negative correlations with: peak oxygen consumption (VO_2 max) Predictor for: HTN in a long-term follow-up, myocardial infarction in a long-term follow-up	[139,141,146,147,151,154,177]
H_2O_2	Positive correlations with: central adiposity indices, interleukin-6	[139]
Peroxy radicals	Higher concentration in subjects with hyperlipidemia than without Positive correlations with: systolic BP, total cholesterol	[140]
TAC	Negative correlation with: systolic BP, pulse pressure Positive correlations with: pulse wave velocity	[141,151]
MDA	Higher concentration in obese hypertensive patients compared to obese normotensives, non-obese hypertensives, and healthy non-obese controls	[143,179]
TBARS	Positive correlations with: non-HDL-C, TGs/HDL-C, total cholesterol/HDL-C, LDL-C/HDL-C	[163]
Protein Carbonyls	Higher concentration in CAD obese patients than in patients with normal weight Positive correlations with: non-HDL-C, TGs/HDL-C, total cholesterol/HDL-C, LDL-C/HDL-C	[163,175]
RBC GPx	Negative correlations with: non-HDL-C, TGs/HDL-C, total cholesterol/HDL-C, LDL-C/HDL-C	[163]
8-OHdG	Higher concentration in CAD obese patients than in patients with normal weight Positive correlations with: mean 24 h systolic and diastolic BP Predictor for: myocardial infarction in a long-term follow-up	[142,175,177]

Abbreviations: (BP) blood pressure, (CAD) coronary artery disease, (HDL-C) high-density lipoprotein cholesterol, (hs-CRP) high-sensitivity C-reactive protein, (HTN) systemic essential hypertension, (H_2O_2) hydrogen peroxide, (LDL-C) low-density lipoprotein cholesterol, (MDA) malondialdehyde, (RBC GPx) glutathione peroxidase in red blood cells, (TAC) total anti-oxidant capacity, (TBARS) thiobarbituric acid reactive substances, (TGs) triglycerides, (T2D) type 2 diabetes, and (8-OHdG) 8-hydroxy-2′-deoxyguanosine.

5.5. Oxidative Stress and Cardiometabolic Risks: Clinical Perspectives

The management of patients with CMDs is challenging yet critical. Preventative and therapeutic approaches, which aim at mitigating oxidative stress through lifestyle and pharmacological interventions, represent promising strategies for patients with a diagnosis or high risk of CMDs [113,180–185].

The use of natural antioxidant compounds, such as vitamins, flavonoids, and polyphenols, as well various diet modifications, may improve systemic oxidative status and be helpful for the treatment and prevention of CVD and MetS [180,186]. For example, there is some evidence that the antioxidant capacity of green tea, cocoa, and extra virgin olive oil is associated with cardioprotection and a decrease in the BP, making these plant-derived nutraceuticals interesting potential tools against HTN and other types of CVD [180,187]. Additionally, small clinical studies show that increasing antioxidant capacity by vitamin E supplementation improves cholesterol levels, markers of oxidative stress, arterial compliance, increases insulin sensitivity, and decreases the systolic BP and, to a much lesser extent, diastolic BP in mildly hypertensive adults [188–190]. However, the results of large randomized trials do not support the use of vitamins E and C for reducing cardiovascular risk in patients at high risk for cardiovascular events, i.e., patients with CVD or T2D including patients surviving recent myocardial infarction [181,191,192]. A possible explanation may be that these vitamins are not specifically targeted to the sites of ROS generation (e.g., mitochondria) and that vitamins react more slowly with ROS than ROS can interact with NO [181]. Based on preclinical studies and small human studies, numerous other antioxidant compounds, such as resveratrol, quercetin, catechins, and several others, might exert beneficial effects for prevention and treatment of CMDs. However, large cohort randomized controlled clinical trials with adequate methodology, such as rigorous inclusion and exclusion criteria, sufficient duration of intervention, and long-term follow up, are needed to provide sufficient clinical evidence for improving cardiometabolic outcomes in subjects at high cardiometabolic risk and patients with CMDs [180,181,193,194].

Weight loss interventions, both dietary and surgical, were shown to be associated with a reduction in oxidative stress and improvements of subclinical atherosclerotic markers, suggesting that these mechanisms may contribute to the reduced risk of CVD in obese subjects after losing weight [171,179]. Structured lifestyle interventions including comprehensive multicomponent intensive cardiac rehabilitation, which can be enhanced by a plant-based diet with antioxidant capacity, are critical to improving the outcome of patients with CVD and cardiometabolic risk factors [182]. Some data indicate the beneficial effects of mitigating circadian disruption on reducing cardiometabolic risks [113,183,184,195,196]. Circadian disruption itself may be secondary to various factors, such as a prolonged daily eating period or sleep disruption that may also be associated with increased oxidative stress. While a few studies on time-restricted eating (TRE), an intervention based on modifying timing and duration of daily food intake, suggest a reduction in lipid peroxidation in obese subjects, including individuals with prediabetes; the results of ongoing studies on the effects of TRE in patients with CMDs, including MetS, are warranted [183–185]. Available data suggest that the benefits of lifestyle modifications, including calorie restriction diets, plant diets, or TRE, go beyond the benefits of caloric restriction and weight loss; however, feasibility and sustainability of these therapeutic interventions in both clinical studies and real-world clinical practice may be limited [197]. There is a need for further clinical research including large-scale randomized controlled trials with longer duration of TRE intervention, long-term follow-up, measurement of circadian rhythms, and additional tools for recording food intake and chrono-nutrition assessment to determine the efficacy of TRE for reducing long-term cardiometabolic risk. Additionally, future clinical trials are warranted to establish the optimal protocols of intensive cardiac rehabilitation and provide tools for sustained lifestyle changes.

Pharmacological scavenging and/or preventing the generation of ROS may both be other preventative or therapeutic options to reduce deleterious effects of oxidative stress in CMDs; however, current clinical evidence on specific antioxidant pharmacotherapies is limited [181]. Nevertheless, given that angiotensin II is a key upstream trigger of ROS formation, angiotensin-converting enzyme (ACE) inhibitors, one of the most common evidence-based pharmacotherapies for CMDs, exert their beneficial clinical cardiovascular effects in part through antioxidative mechanisms. Thus, although vitamin E therapy can be regarded as a secondary therapy that scavenges already-formed ROS, ACE inhibitors can be considered as a primary therapy that blocks ROS production at the enzymatic source [181].

The antioxidant effects of antidiabetic and antiobesity medications, such as glucagon-like peptide-1 receptor agonists (GLP-1RA) and metformin, may be useful for the management of CMDs. In addition to the glucose-lowering and weight-decreasing effects, GLP-1RA affect cellular pathways involved in redox homeostasis [198]. Several in vitro and in vivo studies proved that GLP-1RA reduce ROS and protect against oxidative stress-related cell damage induced by various stress factors, such as high glucose and fatty acids, through various mechanisms, such as activating the Nrf2 signaling pathway and enhancing the expression of antioxidant and detoxification enzymes [198–200]. GLP-1 protects endothelial cells from oxidant injury by reducing intracellular ROS and preventing both endothelial dysfunction and excessively stimulated autophagy, possibly by restoring HDAC6 through a GLP-1RA-ERK1/2-dependent manner [200]. The antioxidant effects of GLP-1RA may be involved in protection against atherosclerosis and diabetes complications, such as diabetic cardiomyopathy and nephropathy [198,201]. GLP-1RA also reduce glycemic variability, which has emerged as a risk factor for diabetic and cardiovascular complications, possibly through enhancing oxidative stress [201,202]. Additionally, in recent years, the utility of metformin was expanded beyond the first-line treatment for T2D due to various effects related to pleiotropic mechanisms of action, including AMPK-dependent and AMPK-independent pathways [203,204]. In addition to affecting glucose and lipid metabolism, as well as improving insulin resistance and obesity, metformin was shown to restore the cellular redox balance and affect mitochondrial function. Moreover, recent in vitro and in vivo studies showed that metformin inhibits hepatic gluconeogenesis in a substrate-selective manner via a redox-dependent mechanism of action [204]. While clinical data on antioxidant effects of GLP-1RA and metformin in CMDs are limited, further studies are needed to better understand the mechanisms of oxidative stress protection that are independent of the effects on glucose metabolism or body weight.

Despite a large amount of evidence on unfavorable effects of oxidative stress in obesity and CMDs and beneficial effects of antioxidant therapies in preclinical studies, further basic and clinical research is needed to investigate the oxidative stress-related molecular mechanisms involved in the pathophysiology of CMDs, and demonstrate established benefits of antioxidant therapies for the prevention and treatment of CMDs. There is a need to address various aspects associated with a translational gap between the preclinical and clinical phases of developing and implementing antioxidant therapies in CMDs. This includes the pathophysiological complexity of CMDs, singular molecular targets of antioxidant agents, low bioavailability of natural antioxidants, clinically irrelevant dosages of compounds in experimental studies, design of experimental studies that do not adequately reflect human populations including subjects with various comorbidities, and a lack of established knowledge on mechanisms of the switch from protective oxidative signaling to deleterious oxidative stress [180]. Antioxidant therapies represent an approach that can potentially have translational impacts leading to improvements in health and a reduction in risks for CMDs, disability, and premature death; however, more clinical evidence on benefits provided by such therapies is warranted. Additionally, the mechanisms of the beneficial effects of antioxidant therapies are still poorly understood. Extensions of study protocols adding the broader spectrum of relevant biomarkers and mitochondrial function evaluation are desirable.

6. Conclusions

Cardiometabolic diseases (CMDs), such as CVD, MetS, and T2D are associated with increased morbidity and mortality. The growing prevalence of CMDs is mostly attributed to the aging population and common occurrence of risk factors, such as high systolic BP, elevated plasma glucose, and increased BMI, which lead to the global epidemic of obesity, MetS, and T2D. Oxidant–antioxidant balance disorders largely contribute to the pathogenesis and outcomes of CMDs, such as HTN, atherosclerosis, CAD, cerebrovascular disease, and MetS. Enhanced and disturbed generation of ROS in excess AT during obesity may lead to increased oxidative stress. Understanding the mechanisms linking adiposity and oxidant–antioxidant balance disorders to the pathogenesis and clinical outcome of CMDs is of great importance to improve the management of patients with CMDs and guide further basic and clinical research.

Expanding the knowledge on adiposity-enhanced oxidative stress related to cardiometabolic disorders can have translational impacts leading to the identification of beneficial lifestyle interventions and the development of novel effective pharmacotherapies, which can reduce the CMDs burden. Future basic research and clinical trials are needed to further examine the molecular mechanisms of adiposity-enhanced oxidative stress in CMDs and efficacy of antioxidant therapies for reducing risk and improving the outcome of patients with CMDs.

Author Contributions: Design and the methodology, A.W., J.N. and I.Ś.; writing—original draft preparation, I.Ś., M.W., P.S., J.N., J.W. and A.W.; writing—review and editing, I.Ś., A.W. and M.W.; visualization, P.S., M.W., J.N. and I.Ś.; supervision, I.Ś. and A.W. All authors have read and agreed to the published version of the manuscript.

Funding: This research received no external funding.

Institutional Review Board Statement: Not applicable.

Informed Consent Statement: Not applicable.

Data Availability Statement: Not applicable.

Acknowledgments: We thank MaryAnn Villarreal-Gonzalez and two anonymous reviewers for valuable comments.

Conflicts of Interest: The authors declare no conflict of interest.

References

1. Vos, T.; Lim, S.S.; Abbafati, C.; Abbas, K.M.; Abbasi, M.; Abbasifard, M.; Abbasi-Kangevari, M.; Abbastabar, H.; Abd-Allah, F.; Abdelalim, A.; et al. Global burden of 369 diseases and injuries in 204 countries and territories, 1990–2019: A systematic analysis for the Global Burden of Disease Study 2019. *Lancet* **2020**, *396*, 1204–1222. [CrossRef] [PubMed]
2. Virani, S.S.; Alonso, A.; Benjamin, E.J.; Bittencourt, M.S.; Callaway, C.W.; Carson, A.P.; Chamberlain, A.M.; Chang, A.R.; Cheng, S.; Delling, F.; et al. Heart Disease and Stroke Statistics—2020 Update. A Report From The American Heart Association. *Circulation* **2020**, *141*, e139–e596. [CrossRef] [PubMed]
3. Williams, B.; Mancia, G.; Spiering, W.; Rosei, E.A.; Azizi, M.; Burnier, M.; Clement, D.L.; Coca, A.; de Simone, G.; Dominiczak, A.; et al. 2018 ESC/ESH Guidelines for the management of arterial hypertension. *Eur. Heart J.* **2018**, *39*, 3021–3104. [CrossRef] [PubMed]
4. Lewington, S.; Clarke, R.; Qizilbash, N.; Peto, R.; Collins, R.; Collaboration, P.S. Age-specific relevance of usual blood pressure to vascular mortality: A meta-analysis of individual data for one million adults in 61 prospective studies. *Lancet* **2002**, *360*, 1903–1913. [CrossRef]
5. Alberti, K.G.M.M.; Eckel, R.H.; Grundy, S.M.; Zimmet, P.Z.; Cleeman, J.I.; Donato, K.A.; Fruchart, J.C.; James, W.P.T.; Loria, C.M.; Smith, S.C. Harmonizing the metabolic syndrome: A joint interim statement of the international diabetes federation task force on epidemiology and prevention; National heart, lung, and blood institute; American heart association; World heart federation; International. *Circulation* **2009**, *120*, 1640–1645. [CrossRef]
6. Abbafati, C.; Abbas, K.M.; Abbasi-Kangevari, M.; Abd-Allah, F.; Abdelalim, A.; Abdollahi, M.; Abdollahpour, I.; Abegaz, K.H.; Abolhassani, H.; Aboyans, V.; et al. Global burden of 87 risk factors in 204 countries and territories, 1990–2019: A systematic analysis for the Global Burden of Disease Study 2019. *Lancet* **2020**, *396*, 1223–1249. [CrossRef]
7. Yusuf, P.S.; Hawken, S.; Ôunpuu, S.; Dans, T.; Avezum, A.; Lanas, F.; McQueen, M.; Budaj, A.; Pais, P.; Varigos, J.; et al. Effect of potentially modifiable risk factors associated with myocardial infarction in 52 countries (the INTERHEART study): Case-control study. *Lancet* **2004**, *364*, 937–952. [CrossRef]
8. Sperling, L.S.; Mechanick, J.I.; Neeland, I.J.; Herrick, C.J.; Després, J.P.; Ndumele, C.E.; Vijayaraghavan, K.; Handelsman, Y.; Puckrein, G.A.; Araneta, M.R.G.; et al. The CardioMetabolic Health Alliance Working toward a New Care Model for the Metabolic Syndrome. *J. Am. Coll. Cardiol.* **2015**, *66*, 1050–1067. [CrossRef]
9. Scuteri, A.; Laurent, S.; Cucca, F.; Cockcroft, J.; Guimaraes Cunha, P.; Rodriguez Mañas, L.; Mattace Raso, F.U.; Lorenza Muiesan, M.; Ryliškytė, L.; Rietzschel, E.; et al. Metabolic Syndrome Across Europe: Different Clusters of Risk Factors. *Eur. J. Prev. Cardiol.* **2015**, *22*, 486–491. [CrossRef]
10. Eckel, R.H.; Alberti, K.G.M.M.; Grundy, S.M.; Zimmet, P.Z. The metabolic syndrome. *Lancet* **2010**, *375*, 181–183. [CrossRef]

11. Khan, S.S.; Ning, H.; Wilkins, J.T.; Allen, N.; Carnethon, M.; Berry, J.D.; Sweis, R.N.; Lloyd-Jones, D.M. Association of body mass index with lifetime risk of cardiovascular disease and compression of morbidity. *JAMA Cardiol.* **2018**, *3*, 280–287. [CrossRef] [PubMed]

12. Smith, K.B.; Smith, M.S. Obesity Statistics. *Prim. Care Clin. Off. Pract.* **2016**, *43*, 121–135. [CrossRef] [PubMed]

13. Bays, H.E.; Toth, P.P.; Kris-Etherton, P.M.; Abate, N.; Aronne, L.J.; Brown, W.V.; Gonzalez-Campoy, J.M.; Jones, S.R.; Kumar, R.; La Forge, R.; et al. Obesity, adiposity, and dyslipidemia: A consensus statement from the National Lipid Association. *J. Clin. Lipidol.* **2013**, *7*, 304–383. [CrossRef] [PubMed]

14. Ervin, R.B. Prevalence of metabolic syndrome among adults 20 years of age and over, by sex, age, race and ethnicity, and body mass index: United States, 2003–2006. *Natl. Health Stat. Rep.* **2009**, *13*, 1–8.

15. Beltrán-Sánchez, H.; Harhay, M.O.; Harhay, M.M.; McElligott, S. Prevalence and Trends of Metabolic Syndrome in the Adult U.S. Population, 1999–2010. *J. Am. Coll. Cardiol.* **2013**, *62*, 697–703. [CrossRef] [PubMed]

16. Sinha, N.; Dabla, P.K. Oxidative stress and antioxidants in hypertension-a current review. *Curr. Hypertens. Rev.* **2015**, *11*, 132–142. [CrossRef]

17. Roberts, C.K.; Sindhu, K.K. Oxidative stress and metabolic syndrome. *Life Sci.* **2009**, *84*, 705–712. [CrossRef]

18. Rains, J.L.; Jain, S.K. Oxidative stress, insulin signaling, and diabetes. *Free Radic. Biol. Med.* **2011**, *50*, 567–575. [CrossRef]

19. Rani, V.; Deep, G.; Singh, R.K.; Palle, K.; Yadav, U.C.S. Oxidative stress and metabolic disorders: Pathogenesis and therapeutic strategies. *Life Sci.* **2016**, *148*, 183–193. [CrossRef]

20. Valko, M.; Leibfritz, D.; Moncol, J.; Cronin, M.T.D.; Mazur, M.; Telser, J. Free radicals and antioxidants in normal physiological functions and human disease. *Int. J. Biochem. Cell Biol.* **2007**, *39*, 44–84. [CrossRef]

21. Bielli, A.; Scioli, M.G.; Mazzaglia, D.; Doldo, E.; Orlandi, A. Antioxidants and vascular health. *Life Sci.* **2015**, *143*, 209–216. [CrossRef]

22. Houstis, N.; Rosen, E.D.; Lander, E.S. Reactive oxygen species have a causal role in multiple forms of insulin resistance. *Nature* **2006**, *440*, 944–948. [CrossRef] [PubMed]

23. Zhou, Y.; Li, H.; Xia, N. The Interplay Between Adipose Tissue and Vasculature: Role of Oxidative Stress in Obesity. *Front. Cardiovasc. Med.* **2021**, *8*, 650214. [CrossRef] [PubMed]

24. Akoumianakis, I.; Antoniades, C. The interplay between adipose tissue and the cardiovascular system: Is fat always bad? *Cardiovasc. Res.* **2017**, *113*, 999–1008. [CrossRef] [PubMed]

25. Bini, S.; D'Erasmo, L.; Di Costanzo, A.; Minicocci, I.; Pecce, V.; Arca, M. The interplay between angiopoietin-like proteins and adipose tissue: Another piece of the relationship between adiposopathy and cardiometabolic diseases? *Int. J. Mol. Sci.* **2021**, *22*, 742. [CrossRef]

26. Berg, A.H.; Scherer, P.E. Adipose tissue, inflammation, and cardiovascular disease. *Circ. Res.* **2005**, *96*, 939–949. [CrossRef]

27. Koenen, M.; Hill, M.A.; Cohen, P.; Sowers, J.R. Obesity, Adipose Tissue and Vascular Dysfunction. *Circ. Res.* **2021**, *128*, 951–968. [CrossRef]

28. Man, A.W.C.; Zhou, Y.; Xia, N.; Li, H. Perivascular Adipose Tissue as a Target for Antioxidant Therapy for Cardiovascular Complications. *Antioxidants* **2020**, *9*, 574. [CrossRef]

29. Schrover, I.; Spiering, W.; Leiner, T.; Visseren, F. Adipose Tissue Dysfunction: Clinical Relevance and Diagnostic Possibilities. *Horm. Metab. Res.* **2016**, *48*, 213–225. [CrossRef]

30. Zhao, R.; Zhao, T.; He, Z.; Cai, R.; Pang, W. Composition, isolation, identification and function of adipose tissue-derived exosomes. *Adipocyte* **2021**, *10*, 587–604. [CrossRef]

31. Reyes-Farias, M.; Fos-Domenech, J.; Serra, D.; Herrero, L.; Sánchez-Infantes, D. White adipose tissue dysfunction in obesity and aging. *Biochem. Pharmacol.* **2021**, *192*, 114723. [CrossRef] [PubMed]

32. Wróblewski, M.; Szewczyk-Golec, K.; Hołyńska-Iwan, I.; Wróblewska, J.; Woźniak, A. Characteristics of Selected Adipokines in Ascites and Blood of Ovarian Cancer Patients. *Cancers* **2021**, *13*, 4702. [CrossRef]

33. Heinonen, S.; Jokinen, R.; Rissanen, A.; Pietiläinen, K.H. White adipose tissue mitochondrial metabolism in health and in obesity. *Obes. Rev.* **2020**, *21*, e12958. [CrossRef] [PubMed]

34. Corrêa, L.H.; Heyn, G.S.; Magalhaes, K.G. The Impact of the Adipose Organ Plasticity on Inflammation and Cancer Progression. *Cells* **2019**, *8*, 662. [CrossRef] [PubMed]

35. Alexander, C.M.; Kasza, I.; Yen, C.-L.E.; Reeder, S.B.; Hernando, D.; Gallo, R.L.; Jahoda, C.A.B.; Horsley, V.; MacDougald, O.A. Dermal white adipose tissue: A new component of the thermogenic response. *J. Lipid Res.* **2015**, *56*, 2061–2069. [CrossRef] [PubMed]

36. Unamuno, X.; Gómez-Ambrosi, J.; Rodríguez, A.; Becerril, S.; Frühbeck, G.; Catalán, V. Adipokine dysregulation and adipose tissue inflammation in human obesity. *Eur. J. Clin. Investig.* **2018**, *48*, e12997. [CrossRef]

37. Kovács, D.; Fazekas, F.; Oláh, A.; Törőcsik, D. Adipokines in the Skin and in Dermatological Diseases. *Int. J. Mol. Sci.* **2020**, *21*, 9048. [CrossRef]

38. Farkhondeh, T.; Llorens, S.; Pourbagher-Shahri, A.M.; Ashrafizadeh, M.; Talebi, M.; Shakibaei, M.; Samarghandian, S. An Overview of the Role of Adipokines in Cardiometabolic Diseases. *Molecules* **2020**, *25*, 5218. [CrossRef]

39. Francisco, V.; Pino, J.; Gonzalez-Gay, M.A.; Mera, A.; Lago, F.; Gómez, R.; Mobasheri, A.; Gualillo, O. Adipokines and inflammation: Is it a question of weight? *Br. J. Pharmacol.* **2018**, *175*, 1569–1579. [CrossRef]

40. Weidinger, C.; Ziegler, J.F.; Letizia, M.; Schmidt, F.; Siegmund, B. Adipokines and Their Role in Intestinal Inflammation. *Front. Immunol.* **2018**, *9*, 1974. [CrossRef]

41. Zorena, K.; Jachimowicz-Duda, O.; Ślęzak, D.; Robakowska, M.; Mrugacz, M. Adipokines and Obesity. Potential Link to Metabolic Disorders and Chronic Complications. *Int. J. Mol. Sci.* **2020**, *21*, 3570. [CrossRef] [PubMed]

42. Taylor, E.B. The complex role of adipokines in obesity, inflammation, and autoimmunity. *Clin. Sci.* **2021**, *135*, 731–752. [CrossRef] [PubMed]

43. Kawai, T.; Autieri, M.V.; Scalia, R. Adipose tissue inflammation and metabolic dysfunction in obesity. *Am. J. Physiol. Physiol.* **2021**, *320*, C375–C391. [CrossRef] [PubMed]

44. Marlatt, K.L.; Ravussin, E. Brown Adipose Tissue: An Update on Recent Findings. *Curr. Obes. Rep.* **2017**, *6*, 389–396. [CrossRef] [PubMed]

45. Saito, M.; Matsushita, M.; Yoneshiro, T.; Okamatsu-Ogura, Y. Brown Adipose Tissue, Diet-Induced Thermogenesis, and Thermogenic Food Ingredients: From Mice to Men. *Front. Endocrinol.* **2020**, *11*, 222. [CrossRef]

46. Scheele, C.; Wolfrum, C. Brown Adipose Crosstalk in Tissue Plasticity and Human Metabolism. *Endocr. Rev.* **2020**, *41*, 53–65. [CrossRef]

47. Zwick, R.K.; Guerrero-Juarez, C.F.; Horsley, V.; Plikus, M.V. Anatomical, Physiological, and Functional Diversity of Adipose Tissue. *Cell Metab.* **2018**, *27*, 68–83. [CrossRef]

48. Cheng, L.; Wang, J.; Dai, H.; Duan, Y.; An, Y.; Shi, L.; Lv, Y.; Li, H.; Wang, C.; Ma, Q.; et al. Brown and beige adipose tissue: A novel therapeutic strategy for obesity and type 2 diabetes mellitus. *Adipocyte* **2021**, *10*, 48–65. [CrossRef]

49. Pilkington, A.-C.; Paz, H.A.; Wankhade, U.D. Beige Adipose Tissue Identification and Marker Specificity—Overview. *Front. Endocrinol.* **2021**, *12*, 599134. [CrossRef]

50. Giordano, A.; Smorlesi, A.; Frontini, A.; Barbatelli, G.; Cinti, S. White, brown and pink adipocytes: The extraordinary plasticity of the adipose organ. *Eur. J. Endocrinol.* **2014**, *170*, R159–R171. [CrossRef]

51. Cinti, S. Pink Adipocytes. *Trends Endocrinol. Metab.* **2018**, *29*, 651–666. [CrossRef] [PubMed]

52. Valencak, T.G.; Osterrieder, A.; Schulz, T.J. Sex matters: The effects of biological sex on adipose tissue biology and energy metabolism. *Redox Biol.* **2017**, *12*, 806–813. [CrossRef]

53. Tchernof, A.; Després, J.-P. Pathophysiology of Human Visceral Obesity: An Update. *Physiol. Rev.* **2013**, *93*, 359–404. [CrossRef] [PubMed]

54. Longo, M.; Zatterale, F.; Naderi, J.; Parrillo, L.; Formisano, P.; Raciti, G.A.; Beguinot, F.; Miele, C. Adipose Tissue Dysfunction as Determinant of Obesity-Associated Metabolic Complications. *Int. J. Mol. Sci.* **2019**, *20*, 2358. [CrossRef] [PubMed]

55. Nalliah, C.J.; Bell, J.R.; Raaijmakers, A.J.A.; Waddell, H.M.; Wells, S.P.; Bernasochi, G.B.; Montgomery, M.K.; Binny, S.; Watts, T.; Joshi, S.B.; et al. Epicardial Adipose Tissue Accumulation Confers Atrial Conduction Abnormality. *J. Am. Coll. Cardiol.* **2020**, *76*, 1197–1211. [CrossRef] [PubMed]

56. Villasante Fricke, A.C.; Iacobellis, G. Epicardial Adipose Tissue: Clinical Biomarker of Cardio-Metabolic Risk. *Int. J. Mol. Sci.* **2019**, *20*, 5989. [CrossRef] [PubMed]

57. Nosalski, R.; Guzik, T.J. Perivascular adipose tissue inflammation in vascular disease. *Br. J. Pharmacol.* **2017**, *174*, 3496–3513. [CrossRef] [PubMed]

58. Saxton, S.N.; Clark, B.J.; Withers, S.B.; Eringa, E.C.; Heagerty, A.M. Mechanistic Links Between Obesity, Diabetes, and Blood Pressure: Role of Perivascular Adipose Tissue. *Physiol. Rev.* **2019**, *99*, 1701–1763. [CrossRef]

59. Kim, H.W.; Shi, H.; Winkler, M.A.; Lee, R.; Weintraub, N.L. Perivascular Adipose Tissue and Vascular Perturbation/Atherosclerosis. *Arterioscler. Thromb. Vasc. Biol.* **2020**, *40*, 2569–2576. [CrossRef]

60. Sowka, A.; Dobrzyn, P. Role of Perivascular Adipose Tissue-Derived Adiponectin in Vascular Homeostasis. *Cells* **2021**, *10*, 1485. [CrossRef]

61. Chang, L.; Garcia-Barrio, M.T.; Chen, Y.E. Perivascular Adipose Tissue Regulates Vascular Function by Targeting Vascular Smooth Muscle Cells. *Arterioscler. Thromb. Vasc. Biol.* **2020**, *40*, 1094–1109. [CrossRef] [PubMed]

62. Brüggen, M.; Stingl, G. Subcutaneous white adipose tissue: The deepest layer of the cutaneous immune barrier. *JDDG J. Der Dtsch. Dermatol. Ges.* **2020**, *18*, 1225–1227. [CrossRef] [PubMed]

63. Koh, H.-C.E.; van Vliet, S.; Pietka, T.A.; Meyer, G.A.; Razani, B.; Laforest, R.; Gropler, R.J.; Mittendorfer, B. Subcutaneous Adipose Tissue Metabolic Function and Insulin Sensitivity in People With Obesity. *Diabetes* **2021**, *70*, 2225–2236. [CrossRef] [PubMed]

64. Booth, A.D.; Magnuson, A.M.; Fouts, J.; Wei, Y.; Wang, D.; Pagliassotti, M.J.; Foster, M.T. Subcutaneous adipose tissue accumulation protects systemic glucose tolerance and muscle metabolism. *Adipocyte* **2018**, *7*, 261–272. [CrossRef]

65. Fassio, A.; Idolazzi, L.; Rossini, M.; Gatti, D.; Adami, G.; Giollo, A.; Viapiana, O. The obesity paradox and osteoporosis. *Eat. Weight Disord.—Stud. Anorex. Bulim. Obes.* **2018**, *23*, 293–302. [CrossRef] [PubMed]

66. Vecchié, A.; Dallegri, F.; Carbone, F.; Bonaventura, A.; Liberale, L.; Portincasa, P.; Frühbeck, G.; Montecucco, F. Obesity phenotypes and their paradoxical association with cardiovascular diseases. *Eur. J. Intern. Med.* **2018**, *48*, 6–17. [CrossRef] [PubMed]

67. Antonopoulos, A.S.; Oikonomou, E.K.; Antoniades, C.; Tousoulis, D. From the BMI paradox to the obesity paradox: The obesity-mortality association in coronary heart disease. *Obes. Rev.* **2016**, *17*, 989–1000. [CrossRef]

68. Elagizi, A.; Kachur, S.; Lavie, C.J.; Carbone, S.; Pandey, A.; Ortega, F.B.; Milani, R.V. An Overview and Update on Obesity and the Obesity Paradox in Cardiovascular Diseases. *Prog. Cardiovasc. Dis.* **2018**, *61*, 142–150. [CrossRef]

69. Głuszek, S.; Ciesla, E.; Głuszek-Osuch, M.; Kozieł, D.; Kiebzak, W.; Wypchło, Ł.; Suliga, E. Anthropometric indices and cut-off points in the diagnosis of metabolic disorders. *PLoS ONE* **2020**, *15*, e0235121. [CrossRef]
70. Caleyachetty, R.; Barber, T.M.; Mohammed, N.I.; Cappuccio, F.P.; Hardy, R.; Mathur, R.; Banerjee, A.; Gill, P. Ethnicity-specific BMI cutoffs for obesity based on type 2 diabetes risk in England: A population-based cohort study. *Lancet Diabetes Endocrinol.* **2021**, *9*, 419–426. [CrossRef]
71. International Diabetes Federation Consensus Worldwide Definition of the Metabolic Syndrome. Available online: https://idf. org/e-library/consensus-statements.html (accessed on 10 February 2023).
72. Suliga, E.; Kozieł, D.; Głuszek, S. Prevalence of metabolic syndrome in normal weight individuals. *Ann. Agric. Environ. Med.* **2016**, *23*, 631–635. [CrossRef]
73. Burton, R.F. The waist-hip ratio: A flawed index. *Ann. Hum. Biol.* **2020**, *47*, 629–631. [CrossRef] [PubMed]
74. Li, Y.; He, Y.; Yang, L.; Liu, Q.; Li, C.; Wang, Y.; Yang, P.; Wang, J.; Chen, Z.; Huang, X. Body Roundness Index and Waist–Hip Ratio Result in Better Cardiovascular Disease Risk Stratification: Results From a Large Chinese Cross-Sectional Study. *Front. Nutr.* **2022**, *9*, 801582. [CrossRef] [PubMed]
75. Borga, M.; West, J.; Bell, J.D.; Harvey, N.C.; Romu, T.; Heymsfield, S.B.; Dahlqvist Leinhard, O. Advanced body composition assessment: From body mass index to body composition profiling. *J. Investig. Med.* **2018**, *66*, 1–9. [CrossRef] [PubMed]
76. Branco, B.H.M.; Bernuci, M.P.; Marques, D.C.; Carvalho, I.Z.; Barrero, C.A.L.; de Oliveira, F.M.; Ladeia, G.F.; Júnior, N.N. Proposal of a normative table for body fat percentages of Brazilian young adults through bioimpedanciometry. *J. Exerc. Rehabil.* **2018**, *14*, 974–979. [CrossRef] [PubMed]
77. Mottillo, S.; Filion, K.B.; Genest, J.; Joseph, L.; Pilote, L.; Poirier, P.; Rinfret, S.; Schiffrin, E.L.; Eisenberg, M.J. The metabolic syndrome and cardiovascular risk: A systematic review and meta-analysis. *J. Am. Coll. Cardiol.* **2010**, *56*, 1113–1132. [CrossRef]
78. McCracken, E.; Monaghan, M.; Sreenivasan, S. Pathophysiology of the metabolic syndrome. *Clin. Dermatol.* **2018**, *36*, 14–20. [CrossRef]
79. Castro-Barquero, S.; Ruiz-León, A.M.; Sierra-Pérez, M.; Estruch, R.; Casas, R. Dietary Strategies for Metabolic Syndrome: A Comprehensive Review. *Nutrients* **2020**, *12*, 2983. [CrossRef]
80. Rochlani, Y.; Pothineni, N.V.; Kovelamudi, S.; Mehta, J.L. Metabolic syndrome: Pathophysiology, management, and modulation by natural compounds. *Ther. Adv. Cardiovasc. Dis.* **2017**, *11*, 215–225. [CrossRef]
81. Iqbal, J.; Al Qarni, A.; Hawwari, A.; Alghanem, A.F.; Ahmed, G. Metabolic Syndrome, Dyslipidemia and Regulation of Lipoprotein Metabolism. *Curr. Diabetes Rev.* **2018**, *14*, 427–433. [CrossRef]
82. Moore, J.X.; Chaudhary, N.; Akinyemiju, T. Metabolic Syndrome Prevalence by Race/Ethnicity and Sex in the United States, National Health and Nutrition Examination Survey, 1988–2012. *Prev. Chronic Dis.* **2017**, *14*, 160287. [CrossRef] [PubMed]
83. Hirode, G.; Wong, R.J. Trends in the Prevalence of Metabolic Syndrome in the United States, 2011–2016. *JAMA* **2020**, *323*, 2526–2528. [CrossRef] [PubMed]
84. Saklayen, M.G. The Global Epidemic of the Metabolic Syndrome. *Curr. Hypertens. Rep.* **2018**, *20*, 12. [CrossRef]
85. Bovolini, A.; Garcia, J.; Andrade, M.A.; Duarte, J.A. Metabolic Syndrome Pathophysiology and Predisposing Factors. *Int. J. Sports Med.* **2021**, *42*, 199–214. [CrossRef] [PubMed]
86. Sarniak, A.; Lipińska, J.; Tytman, K.; Lipińska, S. Endogenne mechanizmy powstawania reaktywnych form tlenu (ROS). *Postep. Hig. Med. Dosw.* **2016**, *70*, 1150–1165. [CrossRef] [PubMed]
87. Herb, M.; Schramm, M. Functions of ros in macrophages and antimicrobial immunity. *Antioxidants* **2021**, *10*, 313. [CrossRef] [PubMed]
88. Konno, T.; Melo, E.P.; Chambers, J.E.; Avezov, E. Intracellular sources of ROS/H2O2 in health and neurodegeneration: Spotlight on endoplasmic reticulum. *Cells* **2021**, *10*, 233. [CrossRef]
89. Di Meo, S.; Reed, T.T.; Venditti, P.; Victor, V.M. Role of ROS and RNS Sources in Physiological and Pathological Conditions. *Oxid. Med. Cell. Longev.* **2016**, *2016*, 1245049. [CrossRef]
90. Lushchak, V.I.; Storey, K.B. Oxidative stress concept updated: Definitions, classifications, and regulatory pathways implicated. *EXCLI J.* **2021**, *20*, 956–967. [CrossRef]
91. Sadiq, I.Z. Free Radicals and Oxidative Stress: Signaling Mechanisms, Redox Basis for Human Diseases, and Cell Cycle Regulation. *Curr. Mol. Med.* **2021**, *23*, 13–35. [CrossRef]
92. Lennicke, C.; Cochemé, H.M. Redox metabolism: ROS as specific molecular regulators of cell signaling and function. *Mol. Cell* **2021**, *81*, 3691–3707. [CrossRef]
93. Sies, H.; Jones, D.P. Reactive oxygen species (ROS) as pleiotropic physiological signalling agents. *Nat. Rev. Mol. Cell Biol.* **2020**, *21*, 363–383. [CrossRef] [PubMed]
94. Xia, Y. Superoxide generation from nitric oxide synthases. *Antioxid. Redox Signal.* **2007**, *9*, 1773–1778. [CrossRef]
95. Woźniak, A. Signs of oxidative stress after exercise. *Biol. Sport* **2003**, *20*, 93–112.
96. Dröge, W. Free radicals in the physiological control of cell function. *Physiol. Rev.* **2002**, *82*, 47–95. [CrossRef] [PubMed]
97. Baysal, S.S.; Koc, S. Oxidant-antioxidant balance in patients with coronary slow flow. *Pak. J. Med. Sci.* **2019**, *35*, 786–792. [CrossRef] [PubMed]
98. Woźniak, B.; Woźniak, A.; Kasprzak, H.A.; Drewa, G.; Mila-Kierzenkowska, C.; Drewa, T.; Planutis, G. Lipid peroxidation and activity of some antioxidant enzymes in patients with glioblastoma and astrocytoma. *J. Neurooncol.* **2007**, *81*, 21–26. [CrossRef] [PubMed]

99. Woźniak, A.; Górecki, D.; Szpinda, M.; Mila-Kierzenkowska, C.; Woźniak, B. Oxidant-antioxidant balance in the blood of patients with chronic obstructive pulmonary disease after smoking cessation. *Oxid. Med. Cell. Longev.* **2013**, *2013*, 897075. [CrossRef]

100. Ahmadinejad, F.; Møller, S.G.; Hashemzadeh-Chaleshtori, M.; Bidkhori, G.; Jami, M.S. Molecular mechanisms behind free radical scavengers function against oxidative stress. *Antioxidants* **2017**, *6*, 51. [CrossRef]

101. Nuszkiewicz, J.; Czuczejko, J.; Maruszak, M.; Pawłowska, M.; Woźniak, A.; Małkowski, B.; Szewczyk-Golec, K. Parameters of Oxidative Stress, Vitamin D, Osteopontin, and Melatonin in Patients with Lip, Oral Cavity, and Pharyngeal Cancer. *Oxid. Med. Cell. Longev.* **2021**, *2021*, 2364931. [CrossRef]

102. Sies, H.; Berndt, C.; Jones, D.P. Oxidative Stress. *Annu. Rev. Biochem.* **2017**, *86*, 715–748. [CrossRef] [PubMed]

103. Sies, H. Oxidative stress: A concept in redox biology and medicine. *Redox Biol.* **2015**, *4*, 180–183. [CrossRef] [PubMed]

104. Cecerska-Heryć, E.; Surowska, O.; Heryć, R.; Serwin, N.; Napiontek-Balińska, S.; Dołęgowska, B. Are antioxidant enzymes essential markers in the diagnosis and monitoring of cancer patients—A review. *Clin. Biochem.* **2021**, *93*, 1–8. [CrossRef] [PubMed]

105. Pisoschi, A.M.; Pop, A. The role of antioxidants in the chemistry of oxidative stress: A review. *Eur. J. Med. Chem.* **2015**, *97*, 55–74. [CrossRef] [PubMed]

106. Jakubczyk, K.; Kałduńska, J.; Dec, K.; Kawczuga, D.; Janda, K. Antioxidant properties of small-molecule non-enzymatic compounds. *Pol. Merkur. Lek.* **2020**, *48*, 128–132.

107. Nuszkiewicz, J.; Woźniak, A.; Szewczyk-Golec, K. Ionizing radiation as a source of oxidative stress—The protective role of melatonin and vitamin d. *Int. J. Mol. Sci.* **2020**, *21*, 5804. [CrossRef]

108. Lushchak, V.I. Free radicals, reactive oxygen species, oxidative stress and its classification. *Chem. Biol. Interact.* **2014**, *224*, 164–175. [CrossRef]

109. Gaschler, M.M.; Stockwell, B.R. Lipid Peroxidation in Cell Death. *Biochem. Biophys. Res. Commun.* **2017**, *482*, 419–425. [CrossRef]

110. Ayala, A.; Muñoz, M.F.; Argüelles, S. Lipid peroxidation: Production, metabolism, and signaling mechanisms of malondialdehyde and 4-hydroxy-2-nonenal. *Oxid. Med. Cell. Longev.* **2014**, *2014*, 360438. [CrossRef]

111. Janicka, M.; Kot-Wasik, A.; Kot, J.; Namieśnik, J. Isoprostanes-biomarkers of lipid peroxidation: Their utility in evaluating oxidative stress and analysis. *Int. J. Mol. Sci.* **2010**, *11*, 4631–4659. [CrossRef]

112. Kargar, B.; Zamanian, Z.; Hosseinabadi, M.B.; Gharibi, V.; Moradi, M.S.; Cousins, R. Understanding the role of oxidative stress in the incidence of metabolic syndrome and obstructive sleep apnea. *BMC Endocr. Disord.* **2021**, *21*, 77. [CrossRef] [PubMed]

113. Świątkiewicz, I.; Woźniak, A.; Taub, P.R. Time-Restricted Eating and Metabolic Syndrome: Current Status and Future Perspectives. *Nutrients* **2021**, *13*, 221. [CrossRef] [PubMed]

114. Castro, J.P.; Grune, T.; Speckmann, B. The two faces of reactive oxygen species (ROS) in adipocyte function and dysfunction. *Biol. Chem.* **2016**, *397*, 709–724. [CrossRef]

115. Brand, M.D. The sites and topology of mitochondrial superoxide production. *Exp. Gerontol.* **2010**, *45*, 466–472. [CrossRef]

116. Fisher-Wellman, K.H.; Neufer, P.D. Linking mitochondrial bioenergetics to insulin resistance via redox biology. *Trends Endocrinol. Metab.* **2012**, *23*, 142–153. [CrossRef] [PubMed]

117. Cadenas, S. Mitochondrial uncoupling, ROS generation and cardioprotection. *Biochim. Biophys. Acta—Bioenerg.* **2018**, *1859*, 940–950. [CrossRef]

118. Le Lay, S.; Simard, G.; Martinez, M.C.; Andriantsitohaina, R. Oxidative stress and metabolic pathologies: From an adipocentric point of view. *Oxid. Med. Cell. Longev.* **2014**, *2014*, 908539. [CrossRef]

119. Gao, C.L.; Zhu, C.; Zhao, Y.P.; Chen, X.H.; Ji, C.B.; Zhang, C.M.; Zhu, J.G.; Xia, Z.K.; Tong, M.L.; Guo, X.R. Mitochondrial dysfunction is induced by high levels of glucose and free fatty acids in 3T3-L1 adipocytes. *Mol. Cell. Endocrinol.* **2010**, *320*, 25–33. [CrossRef]

120. Yao, X.; Carlson, D.; Sun, Y.; Ma, L.; Wolf, S.E.; Minei, J.P.; Zang, Q.S. Mitochondrial ROS induces cardiac inflammation via a pathway through mtDNA damage in a pneumonia-related sepsis model. *PLoS ONE* **2015**, *10*, e0139416. [CrossRef]

121. Forrester, S.J.; Kikuchi, D.S.; Hernandes, M.S.; Xu, Q.; Griendling, K.K. Reactive oxygen species in metabolic and inflammatory signaling. *Circ. Res.* **2018**, *122*, 877–902. [CrossRef]

122. Han, C.Y.; Umemoto, T.; Omer, M.; Den Hartigh, L.J.; Chiba, T.; LeBoeuf, R.; Buller, C.L.; Sweet, I.R.; Pennathur, S.; Abel, E.D.; et al. NADPH oxidase-derived reactive oxygen species increases expression of monocyte chemotactic factor genes in cultured adipocytes. *J. Biol. Chem.* **2012**, *287*, 10379–10393. [CrossRef] [PubMed]

123. Mahadev, K.; Wu, X.; Zilbering, A.; Zhu, L.; Lawrence, J.T.R.; Goldstein, B.J. Hydrogen Peroxide Generated during Cellular Insulin Stimulation Is Integral to Activation of the Distal Insulin Signaling Cascade in 3T3-L1 Adipocytes. *J. Biol. Chem.* **2001**, *276*, 48662–48669. [CrossRef]

124. Hauck, A.K.; Huang, Y.; Hertzel, A.V.; Bernlohr, D.A. Adipose oxidative stress and protein carbonylation. *J. Biol. Chem.* **2019**, *294*, 1083–1088. [CrossRef]

125. Han, C.Y. Roles of reactive oxygen species on insulin resistance in adipose tissue. *Diabetes Metab. J.* **2016**, *40*, 272–279. [CrossRef]

126. Knudson, J.D.; Dincer, Ü.D.; Zhang, C.; Swafford, A.N.; Koshida, R.; Picchi, A.; Focardi, M.; Dick, G.M.; Tune, J.D. Leptin receptors are expressed in coronary arteries, and hyperleptinemia causes significant coronary endothelial dysfunction. *Am. J. Physiol.—Hear. Circ. Physiol.* **2005**, *289*, 48–57. [CrossRef] [PubMed]

127. Neves, K.B.; Lobato, N.S.; Lopes, R.A.M.; Filgueira, F.P.; Zanotto, C.Z.; Oliveira, A.M.; Tostes, R.C. Chemerin reduces vascular nitric oxide/cGMP signalling in rat aorta: A link to vascular dysfunction in obesity? *Clin. Sci.* **2014**, *127*, 111–122. [CrossRef] [PubMed]

128. Antonopoulos, A.S.; Margaritis, M.; Coutinho, P.; Shirodaria, C.; Psarros, C.; Herdman, L.; Sanna, F.; De Silva, R.; Petrou, M.; Sayeed, R.; et al. Adiponectin as a link between type 2 diabetes and vascular NADPH oxidase activity in the human arterial wall: The regulatory role of perivascular adipose tissue. *Diabetes* **2015**, *64*, 2207–2219. [CrossRef]

129. Leandro, A.; Queiroz, M.; Azul, L.; Seiça, R.; Sena, C.M. Omentin: A novel therapeutic approach for the treatment of endothelial dysfunction in type 2 diabetes. *Free Radic. Biol. Med.* **2021**, *162*, 233–242. [CrossRef]

130. Świątkiewicz, I.; Kozinski, M.; Magielski, P.; Gierach, J.; Fabiszak, T.; Kubica, A.; Sukiennik, A.; Navarese, E.P.; Odrowaz-Sypniewska, G.; Kubica, J. Usefulness of C-reactive protein as a marker of early post-infarct left ventricular systolic dysfunction. *Inflamm. Res.* **2012**, *61*, 725–734. [CrossRef]

131. Świątkiewicz, I.; Kozinski, M.; Magielski, P.; Fabiszak, T.; Sukiennik, A.; Navarese, E.P.; Odrowaz-Sypniewska, G.; Kubica, J. Value of C-reactive protein in predicting left ventricular remodelling in patients with a first ST-segment elevation myocardial infarction. *Mediat. Inflamm.* **2012**, *2012*, 250867. [CrossRef]

132. Świątkiewicz, I.; Magielski, P.; Kubica, J.; Zadourian, A.; Demaria, A.N.; Taub, P.R. Enhanced inflammation is a marker for risk of post-infarct ventricular dysfunction and heart failure. *Int. J. Mol. Sci.* **2020**, *21*, 807. [CrossRef] [PubMed]

133. Qi-An, S.; Marschall, S.R.; Nageswara, R.M. Oxidative Stress, NADPH Oxidases, and Arteries. *Hameostaseologie* **2016**, *36*, 77–88. [CrossRef]

134. McGill, H.C.; McMahan, C.A.; Herderick, E.E.; Zieske, A.W.; Malcom, G.T.; Tracy, R.E.; Strong, J.P. Obesity accelerates the progression of coronary atherosclerosis in young men. *Circulation* **2002**, *105*, 2712–2718. [CrossRef]

135. Hajjar, D.P.; Gotto, A.M. Biological relevance of inflammation and oxidative stress in the pathogenesis of arterial diseases. *Am. J. Pathol.* **2013**, *182*, 1474–1481. [CrossRef]

136. Morrow, J.D. Is oxidant stress a connection between obesity and atherosclerosis? *Arterioscler. Thromb. Vasc. Biol.* **2003**, *23*, 368–370. [CrossRef] [PubMed]

137. Świątkiewicz, I.; Magielski, P.; Kubica, J. C-reactive protein as a risk marker for post-infarct heart failure over a multi-year period. *Int. J. Mol. Sci.* **2021**, *22*, 3169. [CrossRef] [PubMed]

138. Adenan, D.M.; Jaafar, Z.; Jayapalan, J.J.; Aziz, A.A. Plasma antioxidants and oxidative stress status in obese women: Correlation with cardiopulmonary response. *PeerJ* **2020**, *8*, e9230. [CrossRef] [PubMed]

139. Ostrow, V.; Wu, S.; Aguilar, A.; Bonner, R.; Suarez, E.; De Luca, F. Association between oxidative stress and masked hypertension in a multi-ethnic population of obese children and adolescents. *J. Pediatr.* **2011**, *158*, 628–633.e1. [CrossRef]

140. Atabek, M.E.; Vatansev, H.; Erkul, I. Oxidative stress in childhood obesity. *J. Pediatr. Endocrinol. Metab.* **2004**, *17*, 1063–1068. [CrossRef]

141. Correia-Costa, L.; Sousa, T.; Morato, M.; Cosme, D.; Afonso, J.; Areias, J.C.; Schaefer, F.; Guerra, A.; Afonso, A.C.; Azevedo, A.; et al. Oxidative stress and nitric oxide are increased in obese children and correlate with cardiometabolic risk and renal function. *Br. J. Nutr.* **2016**, *116*, 805–815. [CrossRef]

142. Cejvanovic, V.; Asferg, C.; Kjær, L.K.; Andersen, U.B.; Linneberg, A.; Frystyk, J.; Henriksen, T.; Flyvbjerg, A.; Christiansen, M.; Weimann, A.; et al. Markers of oxidative stress in obese men with and without hypertension. *Scand. J. Clin. Lab. Investig.* **2016**, *76*, 620–625. [CrossRef] [PubMed]

143. Baráth, Á.; Németh, I.; Karg, E.; Endreffy, E.; Bereczki, C.; Gellén, B.; Haszon, I.; Túri, S. Roles of paraoxonase and oxidative stress in adolescents with uraemic, essential or obesity-induced hypertension. *Kidney Blood Press. Res.* **2006**, *29*, 144–151. [CrossRef] [PubMed]

144. Gaman, M.; Epingeac, M.E.; Gad, M.; Diaconu, C.C.; Gaman, A.M. Oxidative Stress Levels are Increased in Obesity and Obesity-Related Complications. *J. Hypertens.* **2019**, *37*, e206. [CrossRef]

145. Warolin, J.; Coenen, K.R.; Kantor, J.L.; Whitaker, L.E.; Wang, L.; Acra, S.A.; Roberts, L.J.; Buchowski, M.S. The relationship of oxidative stress, adiposity and metabolic risk factors in healthy Black and White American youth. *Pediatr. Obes.* **2014**, *9*, 43–52. [CrossRef] [PubMed]

146. Kupczyk, D.; Bilski, R.; Sokołowski, K.; Pawłowska, M.; Woźniak, A.; Szewczyk-Golec, K. Paraoxonase 1: The lectin-like oxidized ldl receptor type i and oxidative stress in the blood of men with type ii obesity. *Dis. Markers* **2019**, *2019*, 6178017. [CrossRef]

147. Anderson, C.; Milne, G.L.; Park, Y.M.M.; Sandler, D.P.; Nichols, H.B. Cardiovascular disease risk factors and oxidative stress among premenopausal women. *Free Radic. Biol. Med.* **2018**, *115*, 246–251. [CrossRef]

148. Yavuzer, S.; Yavuzer, H.; Cengiz, M.; Erman, H.; Demirdag, F.; Doventas, A.; Balci, H.; Erdincler, D.S.; Uzun, H. The role of protein oxidation and DNA damage in elderly hypertension. *Aging Clin. Exp. Res.* **2016**, *28*, 625–632. [CrossRef]

149. Griendling, K.K.; Camargo, L.L.; Rios, F.J.; Alves-Lopes, R.; Montezano, A.C.; Touyz, R.M. Oxidative Stress and Hypertension. *Circ. Res.* **2021**, *128*, 993–1020. [CrossRef]

150. Montezano, A.C.; Dulak-Lis, M.; Tsiropoulou, S.; Harvey, A.; Briones, A.M.; Touyz, R.M. Oxidative stress and human hypertension: Vascular mechanisms, biomarkers, and novel therapies. *Can. J. Cardiol.* **2015**, *31*, 631–641. [CrossRef]

151. Morandi, A.; Corradi, M.; Piona, C.; Fornari, E.; Puleo, R.; Maffeis, C. Systemic anti-oxidant capacity is inversely correlated with systolic blood pressure and pulse pressure in children with obesity. *Nutr. Metab. Cardiovasc. Dis.* **2020**, *30*, 508–513. [CrossRef]

152. Ward, N.C.; Hodgson, J.M.; Puddey, I.B.; Mori, T.A.; Beilin, L.J.; Croft, K.D. Oxidative stress in human hypertension: Association with antihypertensive treatment, gender, nutrition, and lifestyle. *Free Radic. Biol. Med.* **2004**, *36*, 226–232. [CrossRef] [PubMed]

153. Benjamin, E.J.; Larson, M.G.; Keyes, M.J.; Mitchell, G.F.; Vasan, R.S.; Keaney, J.F.; Lehman, B.T.; Fan, S.; Osypiuk, E.; Vita, J.A. Clinical Correlates and Heritability of Flow-Mediated Dilation in the Community: The Framingham Heart Study. *Circulation* **2004**, *109*, 613–619. [CrossRef] [PubMed]
154. Dennis, B.A.; Ergul, A.; Gower, B.A.; Allison, J.D.; Davis, C.L. Oxidative stress and cardiovascular risk in overweight children in an exercise intervention program. *Child. Obes.* **2013**, *9*, 15–21. [CrossRef] [PubMed]
155. Flynn, J.T.; Daniels, S.R.; Hayman, L.L.; Maahs, D.M.; McCrindle, B.W.; Mitsnefes, M.; Zachariah, J.P.; Urbina, E.M. Update: Ambulatory blood pressure monitoring in children and adolescents: A scientific statement from the American Heart Association. *Hypertension* **2014**, *63*, 1116–1135. [CrossRef] [PubMed]
156. Minuz, P.; Patrignani, P.; Gaino, S.; Seta, F.; Capone, M.L.; Tacconelli, S.; Degan, M.; Faccini, G.; Fornasiero, A.; Talamini, G.; et al. Determinants of Platelet Activation in Human Essential Hypertension. *Hypertension* **2004**, *43*, 64–70. [CrossRef] [PubMed]
157. Matsuda, M.; Shimomura, I. Increased oxidative stress in obesity: Implications for metabolic syndrome, diabetes, hypertension, dyslipidemia, atherosclerosis, and cancer. *Obes. Res. Clin. Pract.* **2013**, *7*, e330–e341. [CrossRef]
158. Meigs, J.B.; Larson, M.G.; Fox, C.S.; Keaney, J.F.; Vasan, R.S.; Benjamin, E.J. Association of Oxidative Stress, Insulin Resistance, and Diabetes Risk Phenotypes. *Diabetes Care* **2007**, *30*, 2529–2535. [CrossRef]
159. Kar, K.; Bhattacharyya, A.; Paria, B. Elevated mda level correlates with insulin resistance in prediabetes. *J. Clin. Diagn. Res.* **2018**, *12*, BC22–BC24. [CrossRef]
160. Chisolm, G.M.; Steinberg, D. The oxidative modification hypothesis of atherogenesis: An overview. *Free Radic. Biol. Med.* **2000**, *28*, 1815–1826. [CrossRef]
161. Kelishadi, R.; Sharifi, M.; Khosravi, A.; Adeli, K. Relationship between C-reactive protein and atherosclerotic risk factors and oxidative stress markers among young persons 10–18 years old. *Clin. Chem.* **2007**, *53*, 456–464. [CrossRef]
162. Kunitomo, M. Oxidative Stress and Atherosclerosis. *Yakugaku Zasshi* **2007**, *127*, 1997–2014. [CrossRef] [PubMed]
163. Nanda, N.; Bobby, Z.; Hamide, A.; Koner, B.C.; Sridhar, M.G. Association between oxidative stress and coronary lipid risk factors in hypothyroid women is independent of body mass index. *Metabolism* **2007**, *56*, 1350–1355. [CrossRef] [PubMed]
164. Pizzino, G.; Irrera, N.; Cucinotta, M.; Pallio, G.; Mannino, F.; Arcoraci, V.; Squadrito, F.; Altavilla, D.; Bitto, A. Oxidative Stress: Harms and Benefits for Human Health. *Oxid. Med. Cell. Longev.* **2017**, *2017*, 8416763. [CrossRef] [PubMed]
165. Ishigaki, Y.; Oka, Y.; Katagiri, H. Circulating oxidized LDL: A biomarker and a pathogenic factor. *Curr. Opin. Lipidol.* **2009**, *20*, 363–369. [CrossRef]
166. Zapolska, D.D.; Bryk, D.; Olejarz, W. Trans Fatty Acids and Atherosclerosis-effects on Inflammation and Endothelial Function. *J. Nutr. Food Sci.* **2015**, *5*, 6. [CrossRef]
167. Lai, W.K.C.; Kan, M.Y. Homocysteine-induced endothelial dysfunction. *Ann. Nutr. Metab.* **2015**, *67*, 1–12. [CrossRef]
168. Hadi, H.A.R.; Carr, C.S.; Al Suwaidi, J. Endothelial dysfunction: Cardiovascular risk factors, therapy, and outcome. *Vasc. Health Risk Manag.* **2005**, *1*, 183–198.
169. Silver, A.E.; Beske, S.D.; Christou, D.D.; Donato, A.J.; Moreau, K.L.; Eskurza, I.; Gates, P.E.; Seals, D.R. Overweight and obese humans demonstrate increased vascular endothelial NAD(P)H oxidase-p47phox expression and evidence of endothelial oxidative stress. *Circulation* **2007**, *115*, 627–637. [CrossRef]
170. Saeed, R.K. Is oxidative stress, a link between nephrolithiasis and obesity, hypertension, diabetes, chronik kidney disease, metabolic syndrome? *Urol. Res.* **2012**, *40*, 95–112. [CrossRef]
171. López-Domènech, S.; Martínez-Herrera, M.; Abad-Jiménez, Z.; Morillas, C.; Escribano-López, I.; Díaz-Morales, N.; Bañuls, C.; Víctor, V.M.; Rocha, M. Dietary weight loss intervention improves subclinical atherosclerosis and oxidative stress markers in leukocytes of obese humans. *Int. J. Obes.* **2019**, *43*, 2200–2209. [CrossRef]
172. Gutiérrez-cuevas, J.; Sandoval-rodriguez, A.; Meza-rios, A.; Monroy-ramírez, H.C.; Galicia-moreno, M.; García-bañuelos, J.; Santos, A.; Armendariz-borunda, J. Molecular mechanisms of obesity-linked cardiac dysfunction: An up-date on current knowledge. *Cells* **2021**, *10*, 629. [CrossRef] [PubMed]
173. Lubbers, E.R.; Price, M.V.; Mohler, P.J. Arrhythmogenic substrates for atrial fibrillation in obesity. *Front. Physiol.* **2018**, *9*, 1482. [CrossRef] [PubMed]
174. Niemann, B.; Rohrbach, S.; Miller, M.R.; Newby, D.E.; Fuster, V.; Kovacic, J.C. Oxidative Stress and Cardiovascular Risk: Obesity, Diabetes, Smoking, and Pollution: Part 3 of a 3-Part Series. *J. Am. Coll. Cardiol.* **2017**, *70*, 230–251. [CrossRef] [PubMed]
175. Niemann, B.; Chen, Y.; Teschner, M.; Li, L.; Silber, R.E.; Rohrbach, S. Obesity induces signs of premature cardiac aging in younger patients: The role of mitochondria. *J. Am. Coll. Cardiol.* **2011**, *57*, 577–585. [CrossRef] [PubMed]
176. Gramlich, Y.; Daiber, A.; Buschmann, K.; Oelze, M.; Vahl, C.F.; Münzel, T.; Hink, U. Oxidative stress in cardiac tissue of patients undergoing coronary artery bypass graft surgery: The effects of overweight and obesity. *Oxid. Med. Cell. Longev.* **2018**, *2018*, 6598326. [CrossRef] [PubMed]
177. Xuan, Y.; Gào, X.; Holleczek, B.; Brenner, H.; Schöttker, B. Prediction of myocardial infarction, stroke and cardiovascular mortality with urinary biomarkers of oxidative stress: Results from a large cohort study. *Int. J. Cardiol.* **2018**, *273*, 223–229. [CrossRef]
178. Godreau, A.; Lee, K.E.; Klein, B.E.K.; Shankar, A.; Tsai, M.Y.; Klein, R. Association of Oxidative Stress with Mortality: The Beaver Dam Eye Study. *Oxid. Antioxid. Med. Sci.* **2012**, *1*, 161–167. [CrossRef]

179. Carmona-Maurici, J.; Cuello, E.; Ricart-Jané, D.; Miñarro, A.; Olsina Kissler, J.J.; Baena-Fustegueras, J.A.; Peinado-Onsurbe, J.; Pardina, E. Effect of bariatric surgery in the evolution of oxidative stress depending on the presence of atheroma in patients with morbid obesity. *Surg. Obes. Relat. Dis.* **2020**, *16*, 1258–1265. [CrossRef]

180. Barteková, M.; Adameová, A.; Görbe, A.; Ferenczyová, K.; Pecháňová, O.; Lazou, A.; Dhalla, N.S.; Ferdinandy, P.; Giricz, Z. Natural and synthetic antioxidants targeting cardiac oxidative stress and redox signaling in cardiometabolic diseases. *Free Radic. Biol. Med.* **2021**, *169*, 446–477. [CrossRef]

181. Münzel, T.; Camici, G.G.; Maack, C.; Bonetti, N.R.; Fuster, V.; Kovacic, J.C. Impact of Oxidative Stress on the Heart and Vasculature: Part 2 of a 3-Part Series. *J. Am. Coll. Cardiol.* **2017**, *70*, 212–229. [CrossRef]

182. Świątkiewicz, I.; Di Somma, S.; De Fazio, L.; Mazzilli, V.; Taub, P.R. Effectiveness of intensive cardiac rehabilitation in high-risk patients with cardiovascular disease in real-world practice. *Nutrients* **2021**, *13*, 3883. [CrossRef] [PubMed]

183. Sutton, E.F.; Beyl, R.; Early, K.S.; Cefalu, W.T.; Ravussin, E.; Peterson, C.M. Early Time-Restricted Feeding Improves Insulin Sensitivity, Blood Pressure, and Oxidative Stress Even without Weight Loss in Men with Prediabetes. *Cell Metab.* **2018**, *27*, 1212–1221.e3. [CrossRef] [PubMed]

184. Cienfuegos, S.; Gabel, K.; Kalam, F.; Ezpeleta, M.; Wiseman, E.; Pavlou, V.; Lin, S.; Oliveira, M.L.; Varady, K.A. Effects of 4- and 6-h Time-Restricted Feeding on Weight and Cardiometabolic Health: A Randomized Controlled Trial in Adults with Obesity. *Cell Metab.* **2020**, *32*, 366–378.e3. [CrossRef] [PubMed]

185. Świątkiewicz, I.; Mila-Kierzenkowska, C.; Woźniak, A.; Szewczyk-Golec, K.; Nuszkiewicz, J.; Wróblewska, J.; Rajewski, P.; Eussen, S.J.P.M.; Færch, K.; Manoogian, E.N.C.; et al. Pilot clinical trial of time-restricted eating in patients with metabolic syndrome. *Nutrients* **2021**, *13*, 346. [CrossRef]

186. Varadharaj, S.; Kelly, O.J.; Khayat, R.N.; Kumar, P.S.; Ahmed, N.; Zweier, J.L. Role of Dietary Antioxidants in the Preservation of Vascular Function and the Modulation of Health and Disease. *Front. Cardiovasc. Med.* **2017**, *4*, 64. [CrossRef]

187. Garcia, M.L.; Pontes, R.B.; Nishi, E.E.; Ibuki, F.K.; Oliveira, V.; Sawaya, A.C.H.; Carvalho, P.O.; Nogueira, F.N.; Do Carmo Franco, M.; Campos, R.R.; et al. The antioxidant effects of green tea reduces blood pressure and sympathoexcitation in an experimental model of hypertension. *J. Hypertens.* **2017**, *35*, 348–354. [CrossRef]

188. Rasool, A.H.G.; Yuen, K.H.; Yusoff, K.; Wong, A.R.; Rahman, A.R.A. Dose dependent elevation of plasma tocotrienol levels and its effect on arterial compliance, plasma total antioxidant status, and lipid profile in healthy humans supplemented with tocotrienol rich vitamin E. *J. Nutr. Sci. Vitaminol.* **2006**, *52*, 473–478. [CrossRef]

189. Boshtam, M.; Rafiei, M.; Sadeghi, K.; Sarraf-Zadegan, N. Vitamin E can reduce blood pressure in mild hypertensives. *Int. J. Vitam. Nutr. Res.* **2002**, *72*, 309–314. [CrossRef]

190. Zaulkffali, A.S.; Razip, N.N.M.; Alwi, S.S.S.; Jalil, A.A.; Mutalib, M.S.A.; Gopalsamy, B.; Chang, S.K.; Zainal, Z.; Ibrahim, N.N.; Zakaria, Z.A.; et al. Vitamins D and E stimulate the PI3K-AKT signalling pathway in insulin-resistant SK-N-SH neuronal cells. *Nutrients* **2019**, *11*, 2525. [CrossRef]

191. Yusuf, S.; Dagenais, G.; Pogue, J.; Bosch, J.; Sleight, P. Vitamin E supplementation and cardiovascular events in high-risk patients. *N. Engl. J. Med.* **2000**, *342*, 154–160. [CrossRef]

192. Investigators, G.-P. Dietary Supplementation With n-3 Polyunsaturated Fatty Acids and Vitamin E After Myocardial Infarction: Results of the Gissi-Prevenzione Trial. *J. Cardiopulm. Rehabil.* **2000**, *20*, 131. [CrossRef]

193. Sergi, C.; Chiu, B.; Feulefack, J.; Shen, F.; Chiu, B. Usefulness of resveratrol supplementation in decreasing cardiometabolic risk factors comparing subjects with metabolic syndrome and healthy subjects with or without obesity: Meta-analysis using multinational, randomised, controlled trials. *Arch. Med. Sci.—Atheroscler. Dis.* **2020**, *5*, 98–111. [CrossRef] [PubMed]

194. Al Hroob, A.M.; Abukhalil, M.H.; Hussein, O.E.; Mahmoud, A.M. Pathophysiological mechanisms of diabetic cardiomyopathy and the therapeutic potential of epigallocatechin-3-gallate. *Biomed. Pharmacother.* **2019**, *109*, 2155–2172. [CrossRef] [PubMed]

195. Xie, Y.; Tang, Q.; Chen, G.; Xie, M.; Yu, S.; Zhao, J.; Chen, L. New insights into the circadian rhythm and its related diseases. *Front. Physiol.* **2019**, *10*, 682. [CrossRef]

196. de Cabo, R.; Mattson, M.P. Effects of Intermittent Fasting on Health, Aging, and Disease. *N. Engl. J. Med.* **2019**, *381*, 2541–2551. [CrossRef]

197. Kubica, A.; Obońska, K.; Kasprzak, M.; Sztuba, B.; Navarese, E.P.; Koziński, M.; Świątkiewicz, I.; Kieszkowska, M.; Ostrowska, M.; Grześk, G.; et al. Prediction of high risk of non-adherence to antiplatelet treatment. *Kardiol. Pol.* **2016**, *74*, 61–67. [CrossRef]

198. Oh, Y.S.; Jun, H.S. Effects of glucagon-like peptide-1 on oxidative stress and Nrf2 signaling. *Int. J. Mol. Sci.* **2018**, *19*, 26. [CrossRef]

199. Oeseburg, H.; De Boer, R.A.; Buikema, H.; Van Der Harst, P.; Van Gilst, W.H.; Silljé, H.H.W. Glucagon-like peptide 1 prevents reactive oxygen species-induced endothelial cell senescence through the activation of protein kinase A. *Arterioscler. Thromb. Vasc. Biol.* **2010**, *30*, 1407–1414. [CrossRef]

200. Cai, X.; She, M.; Xu, M.; Chen, H.; Li, J.; Chen, X.; Zheng, D.; Liu, J.; Chen, S.; Zhu, J.; et al. GLP-1 treatment protects endothelial cells from oxidative stress-induced autophagy and endothelial dysfunction. *Int. J. Biol. Sci.* **2018**, *14*, 1696–1708. [CrossRef]

201. del Olmo García, M.I.; Merino-Torres, J.F. GLP 1 receptor agonists, glycemic variability, oxidative stress and acute coronary syndrome. *Med. Hypotheses* **2020**, *136*, 109504. [CrossRef]

202. Ceriello, A.; Novials, A.; Canivell, S.; La Sala, L.; Pujadas, G.; Esposito, K.; Testa, R.; Bucciarelli, L.; Rondinelli, M.; Genovese, S. Simultaneous GLP-1 and insulin administration acutely enhances their vasodilatory, antiinflammatory, and antioxidant action in type 2 diabetes. *Diabetes Care* **2014**, *37*, 1938–1943. [CrossRef] [PubMed]

203. Du, Y.; Zhu, Y.J.; Zhou, Y.X.; Ding, J.; Liu, J.Y. Metformin in therapeutic applications in human diseases: Its mechanism of action and clinical study. *Mol. Biomed.* **2022**, *3*, 41. [CrossRef] [PubMed]
204. LaMoia, T.E.; Shulman, G.I. Cellular and Molecular Mechanisms of Metformin Action. *Endocr. Rev.* **2021**, *42*, 77–96. [CrossRef] [PubMed]

International Journal of **Molecular Sciences**

Review

Portrayal of NLRP3 Inflammasome in Atherosclerosis: Current Knowledge and Therapeutic Targets

Daniela Maria Tanase [1,2,†], Emilia Valasciuc [1,2,†], Evelina Maria Gosav [1,2,*], Anca Ouatu [1,2], Oana Nicoleta Buliga-Finis [1,2], Mariana Floria [1,2,*], Minela Aida Maranduca [2,3,‡] and Ionela Lacramioara Serban [3]

[1] Department of Internal Medicine, "Grigore T. Popa" University of Medicine and Pharmacy, 700115 Iasi, Romania; tanasedm@gmail.com (D.M.T.); dr.emiliavalasciuc@gmail.com (E.V.); ank_mihailescu@yahoo.com (A.O.); oana_finish@yahoo.com (O.N.B.-F.)
[2] Internal Medicine Clinic, "St. Spiridon" County Clinical Emergency Hospital Iasi, 700111 Iasi, Romania; minela.maranduca@umfiasi.ro
[3] Department of Morpho-Functional Sciences II, Discipline of Physiology, "Grigore T. Popa" University of Medicine and Pharmacy, 700115 Iasi, Romania; ionela.serban@umfiasi.ro
* Correspondence: dr.evelinagosav@gmail.com (E.M.G.); floria_mariana@yahoo.com (M.F.)
† These authors contributed equally to this work.
‡ This author contributed as second author.

Abstract: We are witnessing the globalization of a specific type of arteriosclerosis with rising prevalence, incidence and an overall cardiovascular disease burden. Currently, atherosclerosis increasingly affects the younger generation as compared to previous decades. While early preventive medicine has seen improvements, research advances in laboratory and clinical investigation promise to provide us with novel diagnosis tools. Given the physio-pathological complexity and epigenetic patterns of atherosclerosis and the discovery of new molecules involved, the therapeutic field of atherosclerosis has room for substantial growth. Thus, the scientific community is currently investigating the role of nucleotide-binding and oligomerization domain-like receptor family pyrin domain-containing 3 (NLRP3) inflammasome, a crucial component of the innate immune system in different inflammatory disorders. NLRP3 is activated by distinct factors and numerous cellular and molecular events which trigger NLRP3 inflammasome assembly with subsequent cleavage of pro-interleukin (IL)-1β and pro-IL-18 pathways via caspase-1 activation, eliciting endothelial dysfunction, promotion of oxidative stress and the inflammation process of atherosclerosis. In this review, we introduce the basic cellular and molecular mechanisms of NLRP3 inflammasome activation and its role in atherosclerosis. We also emphasize its promising therapeutic pharmaceutical potential.

Keywords: Inflammasome; atherosclerosis; NLRP3; IL-1β; IL-18; therapeutic target; NLRP3 inhibitors

Citation: Tanase, D.M.; Valasciuc, E.; Gosav, E.M.; Ouatu, A.; Buliga-Finis, O.N.; Floria, M.; Maranduca, M.A.; Serban, I.L. Portrayal of NLRP3 Inflammasome in Atherosclerosis: Current Knowledge and Therapeutic Targets. *Int. J. Mol. Sci.* **2023**, *24*, 8162. https://doi.org/10.3390/ijms24098162

Academic Editor: Yutang Wang

Received: 11 April 2023
Revised: 26 April 2023
Accepted: 1 May 2023
Published: 3 May 2023

1. Introduction

Atherosclerosis, one of the 21st century's fastest rising health emergencies, is defined as the accumulation of lipids, inflammatory cells, fibrous tissue, and calcification within vessels, especially in large arteries [1,2]. Exact data about its prevalence are hard to obtain, but its extent can be estimated by studying the multiple complications that result from and atherosclerotic disease [1,2]. The vascular pathological consequences of the macrovascular and microvascular systems, such as cardiovascular disease (CVD) and cerebrovascular events, are the most significant causes of morbidity and mortality in patients and place a substantial financial burden on the provision of equal access to treatment [3–6]. Of the four stages within the pathophysiology of atherosclerosis, comprising endothelial dysfunction succeeded by lipoprotein deposition, foam cell formation, inflammatory cell proliferation and migration, the inflammatory response appears to have the most prominent role. It is involved in both the initiation and the progression of atherogenesis [3,7]. After dyslipidemia emerged as a major risk factor for atherosclerosis which may lead to CVD, further research proved that atherosclerosis-related cardiovascular events are not entirely contingent solely on plasma lipid levels, prompting the emergence of additional risk factors underlying the development of atherosclerosis [3,6]. From the first reports about the inflammatory theory of atherosclerosis in 1999 [8] to the present date, more research is increasingly focused on systemic inflammation as a particular risk factor based on the relationship established between increased cardiovascular events and inflammatory markers, such as interleukin (IL)-6 and high-sensitivity C-reactive protein (hsCRP) [9]. Recent results from the Canakinumab Anti-inflammatory Thrombosis Outcome Study (CANTOS), in which anti-IL-1β treatment significantly reduced cardiovascular events independent of lipid levels, further support the idea of addressing inflammation as a key process in arresting atherosclerosis with a focus on inflammasomes [2,10,11].

There is increasing evidence that damage-associated molecular patterns (DAMPs), alongside hyperglycemia and hyperlipidemia, are associated with the accelerated onset of atherosclerosis via NLRP3 inflammasomes [12,13]. Furthermore, it has been established that NLRP3 inflammasome associated with the activation of interleukin-1β (IL-1β) and interleukin-18 (IL-18) amplifies vascular endothelial cell (VECs) damage, monocyte adhesion and infiltration, vascular smooth muscle cell (VSMC) proliferation, and promotes secondary plaque vulnerability [14–16].

While the role of inflammation in atherosclerosis is currently extensively explored, the precise deleterious effects on endothelial integrity and the involvement of molecules, such as NLRP3 inflammasomes, in the pathogenesis and evolution of atherosclerosis remain elusive [17,18].

In this narrative review, we aim to give an up-to-date perspective on the implications of NLRP3 inflammasome in atherosclerosis by describing their intricate physiopathological relationship, their known therapeutic pathways and the existence of newer molecules that can modulate atherosclerosis via NLRP3 inflammasome. Finally, as a new approach, we point out in parallel the most recent reviews, their different focuses and their main highlights.

2. Portrayal of NLRP3 Inflammasome

As explained afterward, chronic inflammation is considered an essential part of the underlying multifactorial pathways of atherosclerosis, along with NLRP3 inflammasome activation. Inflammasomes were first introduced by Tschopp's research in 2002 and soon after, innate immunity and cellular signal transduction became the main topics of research; inflammasomes in particular began to dominate this field of study [19,20].

Inflammasomes are intracellular protein complexes that are formed as pattern recognition receptors (PRRs) and interact either with DAMPs or with pathogen-associated molecular patterns (PAMPs) [21]. Examples of intrinsic molecules known as DAMPs that are generated in response to injury or distress include extracellular adenosine triphosphate (ATP) and cholesterol crystals (CCs). PAMPs are molecules of external origins, such as toxins generated by bacteria and viruses [22]. Although the number of other recognition receptors identified as having the ability to trigger inflammasome generation is consistently increasing, currently only nucleotide-binding oligomerization domain (NOD)-like receptor (NLR) family pyrin domain (PYD) containing 1 (NLRP1), NLRP3, NLR family caspase-recruitment domain (CARD) containing 4 (NLRC4), absent in melanoma 2 (AIM2) and pyrin are accepted as inflammasome receptors [6,14].

The NLRP3 inflammasome ubiquitously present in the cytosol of numerous cell types (monocytes, macrophages, T and B cells, fibroblasts) is the most extensively researched and prominent family member of inflammasomes [23]. Infections, cholesterol crystals, uric acid, bacteria, and a plethora of different ligands associated with underlying sterile inflammation in pathologies such as diabetes, hypertension, and atherosclerosis are just a few of the signals that cause NLPR3 to be activated [24–26]. By promoting caspase-1 activation to further break down pro-IL-1β and pro-IL-18 into mature and physiologically active forms

(IL-1 and IL-18), it functions as a molecular switch for the inflammatory pathway that initiates and propagates atherogenesis [6,21].

Nevertheless, the precise pathway through which the NLRP3 inflammasome impacts atherosclerosis is elusive; hence, comprehending the inflammasome activation pathways is pivotal for developing innovative targeted and efficient treatments [27].

2.1. Structure of NLRP3 Inflammasome

The sequentially organized process of assembling an inflammasome usually involves a sensor protein, an adaptor protein, and an effector protein [12]. The NLRP3 inflammasome is a three-domain cytosolic protein compound consisting of three domains: a C-terminal leucine-rich repeat (LRR) domain, a central nucleotide-binding and oligomerization (NACHT/NOD) domain and an N-terminal effector PYD that interfaces with apoptosis-associated speck-like protein containing a caspase recruitment domain (ASC) [6,21,28].

The bridge between NLRP3 and caspase-1 is in turn mediated by ASC, an adaptor protein with an N-terminal PYD and a C-terminal caspase recruitment domain (CARD) [16,29]. In addition, the interaction with NLRP3 and other inflammasomal proteins, as well as ASC self-association, relies on the PYD domain [16,29]. Caspase-1, also known as the IL-1β converting enzyme (ICE), originally produced as an inactive zymogen through proteolytic cleavage [30,31], plays a major part in inflammation by mediating the conversion of the proinflammatory cytokines pro-IL-1β and pro-IL-18 into their mature and metabolically active forms, i.e., IL-1 and IL-18 [6,30,31].

In response to a particular stimulus, the NLRP3 sensing protein couples with ASC using homotypic PYD-PYD domain interactions, creating a single ASC "speck" residing within the activated cell and subsequently attracting pro-caspase-1 via CARD-CARD domain interactions [22,32]. Pro-caspase-1 undergoes autoproteolytic cleavage subsequent to the assembly of NLRP3, ASC and pro-caspase-1, releasing its active p20/10 subunits, which contribute to the self-inhibition of proteolysis. Additionally, proinflammatory cytokines from the IL-1 family, including IL-1β and IL-18, are cleaved by active caspase-1, releasing their mature forms [22,31–34]. Similar to the aforementioned, mature caspase-1 has just demonstrated its contribution to the proteolytic cleavage of gasdermin D (GSDMD), taking part in the development of the oligomeric membrane pore and inflammatory cell death mediated by inflammasomes known as pyroptosis [34]. Notably, pyroptosis is considered a highly inflammatory form of lytic programmed cell death mediated by the gasdermin family of proteins. It is initiated upon intracellular danger signals and acts as a defense mechanism against infection by inducing pathological inflammation, accompanied by the activation of inflammasomes and the maturation of pro-inflammatory cytokines [34].

2.2. Mechanisms of NLRP3 Inflammasome Activation

To date, several NLRP3 inflammasome activation pathways have been established: the canonical pathway which comprises of a two-signal model involving priming (signal 1) and activation (signal 2); a non-canonical pathway that necessitates caspase-4/caspase-5, secretion of IL-1β and IL-18 which respond to a particular intracellular lipopolysaccharide (LPS)-induced infection of gram-negative bacteria; and the alternative pathway driven by toll-like receptor 2 (TLR2) or TLR4 signaling without implicating additional secondary activators [35–37]. We further described the canonical pathway in the paragraphs that follow, given that is the main culprit for underlying atherosclerosis.

Two steps are required for canonical NLRP3 inflammasome activation: initiation (signal 1) and activation (signal 2). Both constitute simultaneous defense mechanisms that tightly regulate inflammatory cells (Figure 1) [37].

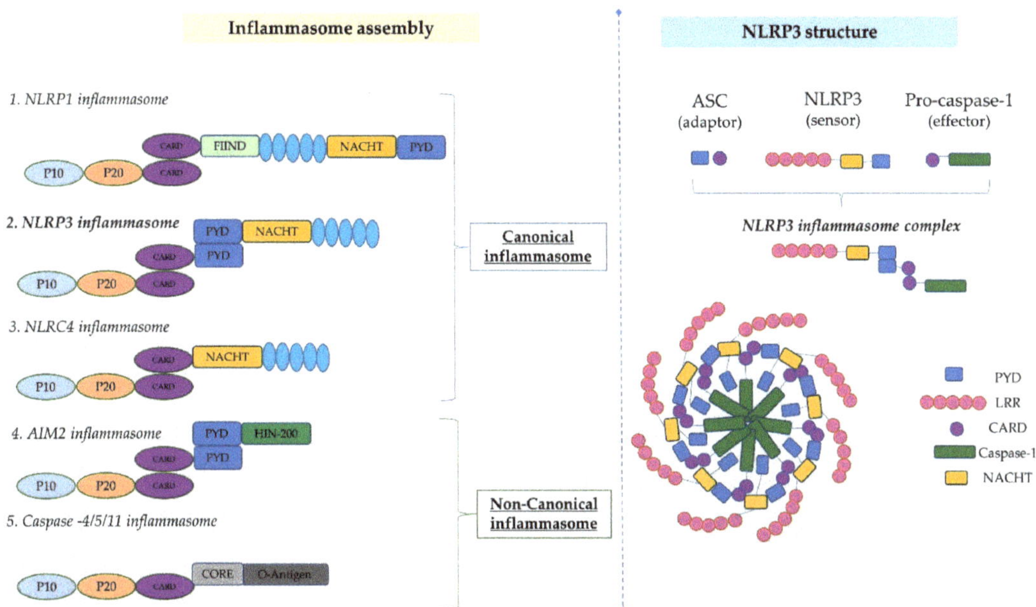

Figure 1. A diagrammatic representation of the assembly of different known inflammasomes and the NLRP3 inflammasome complex structure. Nucleotide-binding and oligomerization domain-like receptor family pyrin domain-containing 3 (NLRP3); NLR Family CARD domain-containing protein 4 (NLRC4); the interferon-inducible protein/absent in melanoma 2 (AIM2); pyrin domain (PYD); leucine-rich-repeat domain (LRR); function-to-find domain (FIIND); caspase recruitment domain (CARD); apoptosis-associated speck-like protein containing a CARD (ASC); caspase-1/ IL-1β converting enzyme (ICE); central nucleotide-binding and oligomerization domain (NACHT/NOD); hematopoietic interferon-inducible nuclear proteins with a 200-amino-acid repeat (HIN-200).

The priming process involves the recognition of PAMPs and DAMPs by PRRs (TLRs, TL-1Rs, and cytokine receptors) followed by the activation of nuclear factor-κB (NF-κB), resulting in transcriptional activation of NLRP3, pro-IL-1β, pro-IL-18 and structural protein shifts, such as ASC phosphorylation and de-ubiquitination of NLRP3 [16,38]. The activation step involves NLRP3 oligomerization and NLRP3, ASC, and pro-caspase-1 complex formation, prompting caspase-1 activation as well as IL-1β and IL-18 generation [39].

Mechanisms of NLRP3 inflammasome activation continue to be a matter of interest, despite being thoroughly explored. Ion fluxes (K^+ efflux, Ca^{2+} influx, and Cl^- efflux), mitochondrial malfunction, reactive oxygen species (ROS) accumulation, cathepsin B release from unstable lysosomes and trans-Golgi decay are several of the hypotheses that have been put forward [14,16,40].

2.2.1. Ionic Fluxes

Most of the NLRP3 activators, which include extracellular ATP, nigericin (K^+ ionophore), and particulate matter, activate inflammasome assembly via K^+ efflux; this pathway has been deemed to be the predominant route of activation [41]. Conflicting findings have emerged as the alternative NLRP3 inflammasome pathway does not require K^+ efflux, whereas the caspase-11-mediated non-canonical inflammasome pathway does [14,16,40]. More recent research has revealed that pharmacological substances, such as GB111-NH2, imiquimod and CL097, can activate NLRP3 by bypassing potassium efflux, implying that this event is presently accepted as a required but not crucial step for inflammasome activation [42].

In addition, potassium efflux triggers Ca^{2+} independent phospholipase A2, enabling IL-1β formation. Further cellular pathways, such as Ca^{2+} mobilization via calcium-sensing receptor (CaSR), Na^+ influx and Cl^- efflux via volume-regulated anion channel (VRAC) and chloride intracellular channels (CLICs), were also proposed to indirectly activate the inflammasome by modulating K+ efflux [6,16]. The intermediate phase of never-in-mitosis A-related kinase 7 (NEK7)-NLRP3 molecular complex assembly underlies the mechanism by which K^+ efflux-mediated activation occurs, being crucial for subsequent inflammasome activation [22,25,43]. Nonetheless, the precise regulatory mechanisms of these cellular processes are incompletely unraveled and still under dispute.

2.2.2. Oxidative Stress

Another crucial element in NLRP3 inflammasome activation is the generation of mitochondrial ROS (mtROS). In particular, numerous studies have shown that ROS transform mitochondrial DNA (mtDNA) generated in response to NLRP3 activators into an oxidized form, thereby promoting inflammasome activation [6,16,44]. Whereas mtDNA interacts with NLRP3 and AIM2, the oxidized mtDNA is precisely required for triggering TLR signaling and NLRP3 inflammasome activation [16,44,45].

The manner in which NADPH-oxidase (NOX), another known important ROS generator, affects NLRP3 inflammasome activation is questionable. Although it was previously stated that NLRP3 inflammasome activation was not impacted by genetic and pharmaceutical inhibition of NOX, it has been proposed that NOX4 may play a role by controlling carnitine palmitoyl transferase 1A (CPT1A) and by increasing fatty acid oxidation, which is a potent inflammasome promoter [27,46]. ROS production in relation to oxidative stress also promotes inflammasome activation by dissociation of a thioredoxin-interacting protein (TXNIP) [27,46].

2.2.3. Lysosomal Damage, Autophagy and the Trans-Golgi Network

Cathepsin release from the injured lysosome is an additional cellular driver of NLRP3 inflammasome activation. Degradation occurs if monosodium urate, CCs, asbestos, silica, β-amyloid, calcium crystals and silica particles are scavenged by macrophages but insufficiently degraded in lysosomes [40,47]. However, it is not yet known how lysosomal disruption and NLRP3 inflammasome activation are linked.

The build-up of diacylglycerol (DAG) was also linked to the potential involvement of the Golgi apparatus in NLRP3 inflammasome activation since Golgi membranes are surrounded by DAG and mitochondria-associated endoplasmic reticulum membranes (MAM) [43,48]. In response to NLRP3 activators, the protein kinase D (PKD) attraction induces NLRP3 phosphorylation, thereby facilitating inflammasome assembly [43,48]. The discovery of Nek7, alongside new developments in the role of a dispersed trans-Golgi network and mtDNA in NLRP3 inflammasome activation, constitutes important novel findings in this area [49].

It has been documented that autophagy is an alternative pathway that diminishes NLRP3 inflammasome activation through the removal of activators and intracellular components. In addition, pyrin, often referred to as the tripartite motif 20, serves as a targeted receptor that mediates precise autophagy of NLRP3 and pro-caspase-1 to trigger autodegradation [22]. Further research is warranted to unravel the activation process and the integration of stimulus-induced signaling events that activate the NLRP3 inflammasome [32].

2.2.4. Regulation of NLRP3 Inflammasome

In addition, various other regulatory mechanisms, including post-translational modification (PTM), microRNA (miRNA) and endogenous modulator (CARD proteins, pirin proteins), control NLRP3 inflammasome expression and function. PTMs involving ubiquitination, phosphorylation, nitrosylation, sumoylation, glycosylation, and acetylation may control the initial activation of NLRP3 as well as its consequent priming [16,21,50]. In the regulation of ATS-NLRP3 inflammasome activation, several miRNAs, including miR-9, miR-155, miR-30c-5p, miR-181a, miR-181b-5p and miR-20a, were revealed to be implicated [43]. However, so far, this domain has not identified as a commonly shared mechanism of NLRP3 inflammasome activation. NLRP3 stimuli have been reported to induce activation of the inflammasome and a variety of cell signaling sequences. Arguably, the most critical signaling event amongst these is still the K^+ efflux, which is required for most stimuli to activate the NLRP3 inflammasome [21]. The relevance of other events, including Ca^{2+} mobilization, Cl^- efflux, ROS, and mitochondrial dysfunction, are currently uncertain and under investigation [29,32,50].

2.3. Role of NLRP3 Inflammasome in Atherosclerosis

Localized inflammation in the vascular wall is prompted by dyslipidemia, high low-density lipoprotein (LDL) cholesterol and lipoproteins, which are all amplified in type 2 diabetes mellitus and enhance the development of atherosclerotic plaques [24]. IL-1β and IL-18, both by-products of NLRP3 inflammasome activation, appear to have a relevant contribution in the occurrence and propagation of atherosclerosis which is corroborated by abundant data obtained from the evaluation of atherosclerotic plaques in rodents and humans [26,31]. Besides the expression of adhesion molecules, such as the intercellular adhesion molecule-1 (ICAM-1) and the vascular cell adhesion molecule-1 (VCAM-1), inflammatory cytokines and chemokines, such as IL-6, IL-8, IL-1β, monocyte chemoattractant protein-1/chemokine(C-C motif) ligand 2 (MCP-1/CCL2) and matrix metalloproteinases (MMPs), elicit an inflammatory phenotype in endothelial cells and VSMC that enable macrophage build up [51]. While IL-18 receptors α/β are expressed in macrophages, endothelial cells and VSMC, IL-18 is expressed only in macrophages [40].

The discovery by Duewell et al. [52] in 2010 that low-density lipoprotein receptor (LDLR)$^{-/-}$ atheroprone mice exhibit diminished atherosclerotic lesions consequent to their lack of NLRP3, ASC or IL-1α/β in bone marrow cells yielded the first concrete evidence that the NLRP3 inflammasome contributes to the onset of atherosclerosis [48,52]. This discovery paved the path for further research into this theory, and abundant examples of NLRP3 activating stimuli related to atherosclerosis are shown (Table 1); CCs are reported among the most effective activators of the NLRP3 inflammasome that occur during all phases of ATS [50,53]. Secondary to failure of macrophages to adequately achieve CC phagocytosis, it causes lysosomal instability and cathepsin efflux, which activates the NLRP3 inflammasome [50,53]. Moreover, it has also been documented that the generation of neutrophil extracellular traps (NETs), which primes macrophages, is initiated by CCs [21,50]. Furthermore, CCs contribute to a vicious cycle by enhancing NET release and macrophage priming, further complementing IL-1β, IL-18 and NET formation and NLRP3 inflammasome activation [6,50]. Nonetheless, the underlying pathway relating NETs to NLRP3 activation has not been fully determined and requires further exploration. Since oxidized low-density-lipoproteins (oxLDLs) can generate the initiation signal, this may be sufficient to send the NLRP3 inflammasome to both the activation and the priming signals [53]. The uptake of oxLDL by macrophage scavenger receptors, such as cluster of differentiation (CD) 36, initiates the formation of a TLR4/TLR6 heterodimer, which further activates the inflammasome and triggers NF-κB signaling [36,54].

ATP-dependent NLRP3 activation revealed itself as an important player in diet-induced atherosclerotic lesions via the purinergic 2X7 receptor (P2X7R), whose deficiency has been reported to suppress the extent of atherosclerotic plaques and decrease inflammasome activation [23,27,55].

Novel experimental data using animal and cellular designs of atherosclerosis provide mechanism-based perspectives on inflammasome modulation, particularly in the matter of diabetic macrovascular dysfunction [56]. Both in vitro and in vivo, NLRP3 promoted hyperglycemia-induced endothelial inflammation [13,57,58]. OxLDLs and high mobility group box protein 1, two agonist ligands of receptor for advances glycation endproducts

(RAGE), have also been linked to NLRP3 activation which occurs in parallel with the atherosclerotic process in conjunction with hyperglycemia-induced ROS overproduction. Despite inconsistent published results, it is becoming evident that TXNIP, another redox signaling regulator, is significantly increased in response to hyperglycemia and may act as a direct ligand of the NLRP3 inflammasome [29,33,59]. Whilst advanced glycation end products(AGEs) have undisputedly contributed to diabetic atherosclerosis, it is not currently established whether the AGE/RAGE axis likewise activates NLRP3 in the atherogenic process [60–62]. Besides glucose toxicity, another potential inflammasome regulatory molecule, i.e., sterol regulatory element binding protein-1 (SREBP-1), is reported to be a key player in oxLDL-induced excessive lipid accumulation, causing foam cell formation and de novo lipid synthesis through the ROS-mediated NLRP3/IL-1β/SREBP-1 pathway [20,33,63].

Although endogenous factors that can activate inflammasomes (Figure 2) have been uncovered, additional research is required to comprehend how these signals are direct determinants of diabetes-related macrovascular dysfunction [23,64,65]. While endogenous signals that can activate inflammasome assembly have been identified, further research is needed to understand how they are directly linked to diabetes-related macrovascular injury.

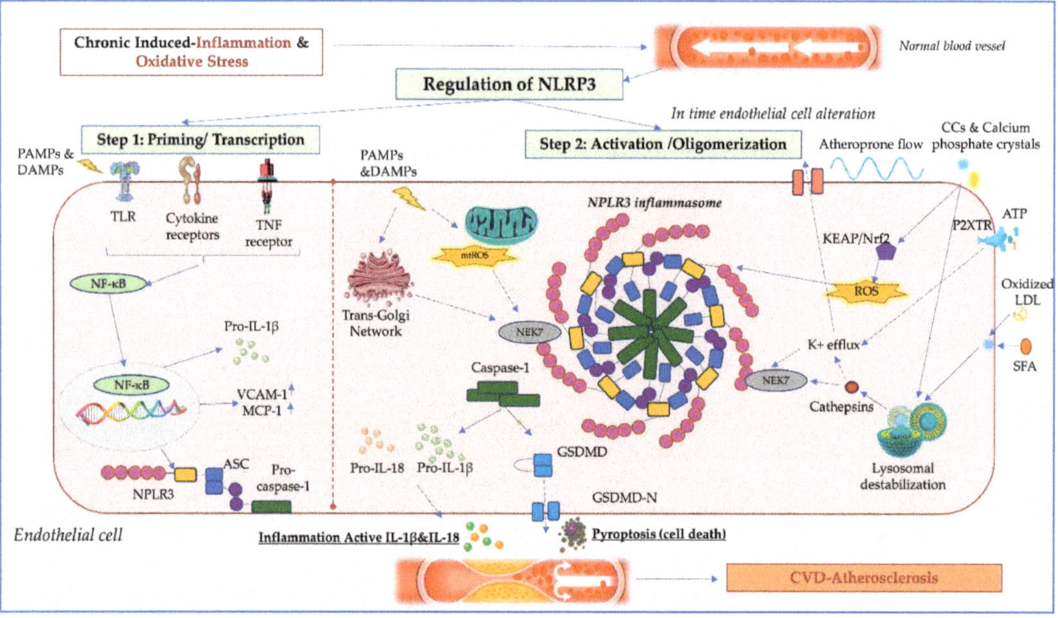

Figure 2. Two step canonical activation of NLRP3 inflammasome-driven downstream events in the arterial endothelial cells. Signal 1 (priming) or ubiquitination phosphorylation, is activated by TLRs or cytokine receptors that recognize and combine the corresponding signals to activate NF-κB at the transcriptional level. The activation signal (Signal 2) is mediated by common intracellular events, such as mitochondrial injury and ROS generation; K+ efflux; lysosome rupture and cathepsin B release; and dispersal of the trans-Golgi network. The NLRP3 inflammasome activates mature caspase-1 that cleaves pro-IL-1b and pro-IL-18 into their active forms. Active caspase-1 also cleaves GSDMD, and its cleaved N-terminus (GSDMD-N) forms subsequently with pyroptosis. ROS = radical oxygen species; nuclear factor kappa-light-chain-enhancer of activated B cells (NF-κB); apoptosis-associated speck-like protein containing a CARD (ASC); P2X purinoreceptor 7 (P2RX7); vascular cell adhesion molecule 1 (VCAM-1); monocyte chemoattractant protein 1 (MCP-1); IL-= in-terleukin-1; IL-18= interleukin-18; never-in-mitosis A-related kinase 7 (NEK7); gasdermin D (GSDMD); terminal domain (GSDMD-N; the up-arrow indicates upregulation.

Table 1. Role of NLRP3 inflammasome in atherosclerosis.

Subjects	Notable NLRP3 Effects in ATS	Refs.
Ascending aortic tissue (CABG patients)	- NLRP3 expression higher in patients with AS and correlated with the degree of coronary artery disease	[66]
Human carotid atherosclerotic plaques	- NLRP3 inflammasome and components (ASC, caspase-1, IL-1β and IL-18) higher expression in unstable atherosclerotic plaques	[67]
Atherosclerotic plaques (ischemic cerebrovascular disease, MI patients)	- NLRP3–mRNA expression higher in symptomatic AS patients	[68]
Peripheral blood monocyte (chronic heart disease and acute coronary syndrome patients)	- NLRP3 inflammasome positive correlation with coronary atherosclerosis	[69]
ApoE$^{-/-}$ mice fed with a HF and HM diet	- Increased NLRP3 expression and proinflammatory effect in hyperhomocysteinemia-induced atherosclerosis	[70]
ApoE$^{-/-}$ mice fed with a HF diet	- NLRP3 inflammasome inhibition increased plaque stability - NLRP3 inflammasome inhibition reduced the size of atherosclerotic plaques and IL-1β and IL-18 levels - CCs activate NLRP3 inflammasome	[71] [72] [52]
ApoE$^{-/-}$ mice chow diet	- NLRP3 inflammasome activation via Sirt3/FOXO3a/Parkin signaling pathway reduced atherosclerotic progression	[73]
ApoE$^{-/-}$ mice western-type diet	- Specific NLRP3 inflammasome inhibition reduced atherosclerotic plaque development	[74]
Ldlr$^{-/-}$ mice fed with PUFAs diet	- NLRP3 inflammasome inhibition reduced atherosclerosis by macrophage autophagy activation	[75]
ApoE$^{-/-}$ mice	- The oxLDLs promote direct NLRP3 inflammasome activation and indirect via ERK1/2 pathway	[76]
ApoE$^{-/-}$/caspase-1$^{-/-}$ double knockout mice	- The extent of the area of atherosclerotic plaque reduced in caspase-1 deficient mice	[77]
Macrophages incubated with oxLDLs	- NLRP3 inflammasome activation, increase in IL-1β and IL-18 levels	[52]
NLRP3-deficient THP-1 cells	- NLRP3 inhibition reduces foam cell formation of THP-1 macrophages by oxLDL uptake suppression and increasing cholesterol efflux	[73]
HAECs	- NLRP3 inflammasome is activated by nicotine which promotes pyroptosis, proinflammatory cytokines secretion and atherosclerosis	[78]
Human and mice aortic endothelial cells	- Melatonin inhibits pyroptosis through the MEG3/miR-223/NLRP3 signaling axis	[79]
HAECs	- NLRP3 inhibitor Microrna-30c-5p inhibits inflammation and pyroptosis via FOX03 pathway	[80]
VSMC	- AIM2 can stimulate caspase-1 via NLRP3 pathway and then mediates the inflammatory response by slicing GSDMD	[81]

coronary artery bypass grafting (CABG); myocardial infarction (MI); microRNA (mRNA); apolipoprotein E-deficient (ApoE$^{-/-}$); high-fat and high-methionine (HF and HM); silent information regulator 3 (Sirt3); forkhead box O3 (FOXO3); LDL receptor knock-out (Ldlr$^{-/-}$); polyunsaturated fatty acids (PUFAs); human aortic endothelial cells (HAECs); vascular smooth muscle cells (VSMC); absent in melanoma 2 (AIM2).

3. Atherosclerosis

Despite all the major advances in addressing risk factors, such as obesity, hypertension, and dyslipidemia, atherosclerotic-CVD manifested by myocardial infarction, stroke, and peripheral vascular disease, with an accelerated progression and concurrent impairment of several arterial territories, remains a primary cause of mortality with a significant negative influence on the quality of life [60,82]. Therefore, developing new strategies for atherosclerosis prevention and treatment is essential, given the rising prevalence and the severity of atherosclerosis-associated complications encumbered by its progression [83].

Endothelial dysfunction caused by oxidative stress, glyco-oxidation, and systemic inflammation amplified by hyperglycemia and dyslipidemia, allows mainly LDL and lipoprotein(a) (Lp(a)) to infiltrate vascular walls and enhance the migration of inflammatory cells through the expression of leukocyte adhesion markers, such as E-selectin, P-selectin, ICAM-1 and VCAM-1 [84,85]. LDL oxidation and an inflammatory immune response mediated by T lymphocytes and monocytes are prolonged in plaque-like lesions composed of inflammatory cells trapped in the subendothelial zone [10]. OxLDLs engulfed by monocyte-derived macrophages form foam cells that cluster in the vascular intima and cause necrotic core formation, inflammation, and phenotypic transition of VSMCs [60].

An extracellular matrix is produced in the fibrous cap as a result of VSMC migration and proliferation, collagen build-up and subsequent calcification [86]. As a result, the intima thickens and a self-perpetuating cycle of localized inflammation and apoptosis begins, causing gradual endothelial damage and the development of lipid-rich plaques with fibrous capsules, especially at the arterial emergence sites or bifurcations that are predisposed to blood flow alteration [87,88].

Whilst the early stages of endothelial dysfunction are relatively understood, it remains puzzling to comprehend how the progression and destabilization phases of atheromatous plaque take place [89].

3.1. Endothelial Dysfunction and Oxidative Stress

The majority of pathologies that are related to atherosclerosis display vascular endothelial dysfunction [90]. Local hemodynamics of blood flow constitute a regional risk factor for atherogenesis by producing injury prone areas primarily where laminar flow is altered. Endothelial cells, through their ability to modulate an optimal hemodynamic response to fluctuations in blood flow, play an instrumental role in atherogenesis [83,91,92].

A vicious loop that promotes apoptosis and increases extracellular matrix synthesis results in increased vascular permeability, activation of NOX, and worsened vasodilation due to decreased NO generation and increased ROS formation [88,93]. ROS production through the activation of several enzymes, including those in the mitochondrial respiratory chain, NOX, endothelial uncoupled nitric oxide oxidase (eNOS), cyclooxygenase, and xanthine (XO), provoke vasodilation abnormalities. This increases vascular permeability, thus encouraging the production of more adhesion molecules, such as ICAM-1, VCAM-1, and growth factors, such as vascular endothelial growth factor (VEGF), ET-1 and plasminogen activator inhibitor 1 (PAI-1), which accelerate the development of vascular sclerosis [94–96]. Increased activity of NF-κB triggers the production of proinflammatory cytokines, including IL-1β, IL-6 and TNF-α, and is secondary to reduced NO bioavailability, thus further attracting the creation of a pro-coagulant state by promoting the expression of tissue factor PAI-1 and the von Willebrand factor besides their role in maintaining the inflammatory milieu [60,86,95,97].

Under the influence of ROS production, as illustrated above, the structural integrity of the vascular endothelium is altered, resulting in the elevated expression of adhesion molecules (ICAM-1 and VCAM-1) and the adhesion of monocytes into the subendothelial space [2,10,98]. The differentiation of monocytes into macrophages attracts the release of IL-1β, IL-18, TNF-α, INF-γ, and MCP-1, leading to a vicious pathway of ROS production [88,93]. Subsequently, LDL cholesterol infiltrates into the subendothelial space in the intima, where it is deposited and modified into oxLDLs [99]. Further, the differentiation of monocytes in the subendothelial region into macrophages that release proinflammatory cytokines and ingest oxLDLs leads to the production of foam cells [97,99,100]. In addition, oxLDL causes the upregulation and release of inflammatory modulators, which further encourage the migration of monocytes, an increase in the density of scavenger macrophage receptors and the uptake of oxLDL during foam cell formation [99,101].

Furthermore, ROS potentiate various inflammatory pathways through redox modification of inflammatory mediators, such as DAMPs, transcription factors, nuclear factor erythroid 2-related factor 2 (Nrf2), NF-κB, hypoxia-inducible factor 1 (HIF-1) and activator protein 1 (AP-1), and through the formation of redox-dependent protein complexes (Nrf2- kelch-like ECH-associated protein 1 (Keap1)) [94,102]. In conjunction with causing insulin resistance, stimulation of the MAPK pathway also affects the phosphatidylinositol-3-kinase/serine-threonine kinase (PI3K/AKT) eNOS modulatory mechanism [86,98,103].

By modulating the NLRP3 inflammasome, the interaction between oxidative stress (OS) and inflammation through cytokine release is likewise documented [57,60]. As a result of a myriad of conditions, secondary dysfunctional endothelium implies a change from a quiescent to a proinflammatory phenotype and serves as the first stage in the development of an atherosclerotic lesion.

3.2. Inflammation in Atherosclerosis

As stated earlier in the previous part of this review, inflammation is a pivotal element in the pathophysiology of accelerated ATS, from the earliest stages of development to the final thrombotic outcomes; thus, it has a potential therapeutical endpoint [104]. The progression of atherosclerotic plaques is a continuum of processes involving both immune and non-immune vascular cells [10,24].

LDLs build up in the subendothelial region within the early stages of atherosclerosis, where they are transformed into oxLDLs and can activate ECs and macrophages in a proinflammatory manner, aggravate endothelial damage and attract leukocytes [10,99]. When VSMCs are subjected to modified LDLs, CCL2, CCL5, and MCP-1 are released, thus mediating monocyte recruitment [35,86,105]. In response to the local macrophage colony-stimulating factor (M-CSF), underlying subendothelial monocytes divide into macrophages, which additionally release IL-1, IL-18, TNF-α, and INF- γ [24,35]. Under the influence of TNF-α, INF-γ and TLR ligands, macrophages adjacent to the lipid core differentiate into the proinflammatory M1 phenotype that correlates with self-reinforcing inflammatory processes, development of vulnerable plaques and the progression of atherosclerosis [35,106,107].

Another facet of atherosclerosis pathogenesis is the upregulation and activation of TLR4 and NLRP3 inflammasome and the nuclear transcription factor NF-kB as an extension of the endogenous host response to cholesterol efflux signaling misregulation, which is mediated by oxLDL and CCs [33,54,105,108,109]. Furthermore, NLRP3 inflammasomes have been linked to plaque development and progression through increased IL-1β and IL-18 production, increased MCP-1 and VCAM-1 and accumulation of the vascular extracellular matrix; these effects result from the JNK-induced apoptotic pathway, ASC and pro-caspase-1 [14,31,54,110,111]. In addition, the remodeling that proceeds with plaque development, intraplaque neovascularization, matrix depletion with thinning of the fibrous cap and eventually fibrous cap rupture causing thrombosis have all been connected to inflammation [2].

Considering the association between the accelerated progression of atherosclerosis induced by chronic sterile inflammation, addressing the disorder from this perspective may be a more reliable way to limit the development of its vascular consequences. Currently, no consensus guidelines are targeting specific inflammatory pathways that can improve outcomes for patients; therefore, there is a compelling need to further attractive research alternatives and therapeutic strategies that may arise from inhibiting inflammatory pathways.

4. Therapeutic Targets

4.1. Hypoglycemic Agents

Anti-diabetic drugs have recently emerged as important novel mediators in treating diabetic vascular dysfunction and atherosclerosis by inhibiting NLRP3 inflammasome activation and reducing endothelial damage.

Glyburide, an FDA-approved sulfonylurea drug, interferes with the release of insulin from pancreatic β cells by inhibiting ATP-sensitive potassium channels (KATP) [57]. Glyburide was reported to exhibit anti-inflammatory effects mediated by the benzamide and sulfonyl group specific to NLRP3, thus blocking caspase-1 activation, IL-1β secretion and crystal-induced activation without relying on KATP channels and acting further downstream of P2X7R and upstream of inflammasome formation [112,113]. While glyburide effectively inhibits NLRP3 activation in vitro, the high doses required in vivo limit its usefulness as a treatment due to major adverse effects. Its exact mechanism of action remains not yet fully understood [21].

Since 1995, metformin, which was originally synthesized in 1922, has become broadly recognized as the first-line treatment for T2DM, thus starting a new phase in the growing burden of diabetes [114]. The main therapeutic mechanisms behind the drug are primarily the result of adenosine monophosphate-activated protein kinase (AMPK) activation, which further reduces hepatic gluconeogenesis and improves insulin resistance while improving glucose uptake in peripheral tissues [115]. Metformin is currently being investigated for new roles and pharmacological processes in light of growing interest in the role of inflammation in the etiology of both T2DM and atherosclerosis [114,116]. The main data revealed that metformin increased the production of AMPK and protein phosphatase 2A (PP2A) expression, leading to reduced expression and inhibition of NLRP3 inflammasome activation in oxLDL-stimulated macrophages [117]. These findings were additionally supported by the fact that metformin averted accelerated diabetic AS in apoE$^{-/-}$ mice by activating the Trx-1/Txn pathway [32,116,118].

The Empagliflozin Cardiovascular Outcome Event Trial in Type 2 Diabetes Patients (EMPA-REG OUTCOMES) study reported that treatment with sodium–glucose cotransporter 2 (SGLT2) inhibitors, a novel class of hypoglycemic agents, reduced the rate of all-cause cardiovascular death in patients with T2DM at high risk of cardiovascular events [119]. The mechanism of action is mediated by the proximal kidney tubules, where SGLT2 inhibitors reduce renal glucose reabsorption and increase urinary glucose excretion [32,57]. In ex vivo research employing human macrophages, empagliflozin administration, in addition to previous effects, diminished NLRP3 inflammasome activation and IL-1 release, partly by increasing β-hydroxybutyrate (BHB) and by reducing serum insulin, glucose and uric acid levels [120].

Dapagliflozin, another family member, alleviated diabetic cardiomyopathy by inhibiting NLRP3 through activation of the AMPK system and blocked the TXNIP/NLRP3 pathway [54,121]. Furthermore, dapagliflozin therapy suppressed the generation of serum NLRP3, IL-1β, IL-18 levels and ROS in the vasculature of atherosclerotic aortic lesions, thus reducing the progression of ATS, diminishing macrophage infiltration and improving lesion stability [122]. Regardless, the reported results strongly indicate compelling reasons to further research the effects of SGLT2 inhibitors as a therapeutic solution for NLRP3 inflammasome inhibition in diabetic patients [32,57].

Apart from their hypoglycemic action, dipeptidyl peptidase 4 (DPP-4) inhibitors and glucagon-like peptide receptor (GLP-1R) agonists have been demonstrated to improve inflammatory markers, oxidative stress, endothelial function and, to some extent, the atheroprotective features in patients with T2DM [123]. Saxagliptin was reported to reduce myocardial injury by inhibiting the ERK/TLR4/NLRP3 and p38/miR-146b/TLR4/NLRP3 pathways, implying that DPP4 inhibition blocks the first step involved in inflammasome activation [124]. In oxLDL-induced THP-1 cells, sitagliptin elicited a substantial downregulation of NLRP3, TLR4 and IL-1 expression and an increase in GLP-1R expression [43]. Vildagliptin, as per Qi et al., mitigates endothelial dysfunction mediated by elevated free fatty acid (FFA) levels by inhibiting the AMPK-NLRP3-HMGB1 pathway, preserves mitochondrial function and restores eNOS levels while decreasing cellular lactate dehydrogenase (LDH) production and ROS levels [125]. Anagliptin, a new DPP-4 inhibitor approved for treating T2DM, was recently reported to reverse endothelial dysfunction by SIRT1-dependent inhibition of NLRP3 inflammasome activation and suppression of NOX4-ROS-TXNIP-NLRP3 crosstalk, in this way creating scope for additional research [46,126]. In addition, glucagon-like peptide-1 receptor (GLP-1 RA) agonists provide beneficial effects on the cardiac function as well. Dulaglutide, a novel representant of the class, inhibits NLRP3 inflammasome, NOX4 and TXNIP expression in endothelial cells in a SIRT1-dependent manner, protecting against the effects of increased glucose on the NLRP3 inflammasome activation [43,46,127].

4.2. Direct and Indirect NLRP3 Inhibitors

Aiming to prevent inflammasome-mediated cell death and thereby decrease local inflammation, pharmacological suppression of NLRP3 inflammasome activation may represent a more precise and certain therapeutic approach for CVD [21]. Agents with small molecules, such as MCC950, dapansutrile (OLT1177), CY-09 or tranilast, block NLRP3 inflammasome activation by specifically targeting the NATCH domain of the NLRP3 structure while numerous other inhibitors act by impairing ATPase function, such as 3,4-methylenedioxy-nitrostyrene (MNS), Bay 11-7082, BOT-4-one, parthenolides and INF39 [128]. Furthermore, because of many other biological functions, as shown in more detail below, these agents are unlikely to function as specific NLRP3 inhibitors [21]. Nevertheless, the scarcity of in vivo studies and NLRP3-dependent models underlines the shortage of information on some compounds and the reasons why they remain unapproved as treatments.

4.2.1. MCC950

MCC950, a sulfonylurea compound that was first identified in 2001 as CP-456773 and subsequently known as cytokine release inhibitor drug 3 (CRID3), is an effective and selective NLRP3 inflammasome inhibitor that binds to a specific residue in the walker B motif of the NATCH domain of NLRP3 to prevent ASC oligomerization and NLRP3 ATP hydrolysis [129]. Since it was established as an NLRP3 inhibitor, it has been used as a pharmacological instrument to unravel the pathogenic functions of the NLRP3 inflammasome in several disease models, including T2DM, ATS, and many others [46]. Corcoran et al. comprehensively analyzed its functions in more than 100 preclinical models of inflammatory diseases [130]. At nanomolar doses, MCC950 inhibits NLRP3 inflammasome activation by suppressing IL-1β secretion but not NLRP1, NLRC4, or AIM2 inflammasomes, suggesting a capacity to modulate the immune system by only partial interleukin inhibition [35,59,131]. By indirectly influencing intraplaque macrophage composition or through processes such as macrophage proliferation, the diminished size of atherosclerotic lesions represents the probable therapeutic benefit of MCC950 in ApoE$^{-/-}$ mice [43]. In addition, through sirtuin 1 and superoxide dismutase (SOD), tissular overexpression of MCC950 exhibited antioxidant capacity [132]. Regardless of the favorable outcomes in animal models, the compound is currently withheld from clinical trials because of its liver toxicity [24,36].

4.2.2. CY-09

By interfering directly with the ATP-binding Walker A motif protein, another small-molecule inhibitor, CY-09, selectively inhibits the NLRP3 inflammasome ATPase activity [21,133]. Ex vivo and in vivo designs for cryopyrin-associated periodic syndromes (CAPS), type 2 diabetes, gout and non-alcoholic fatty liver disease (NAFLD) have established the efficacy of CY-09, although further research is required to materialize its full therapeutic value [134,135].

4.2.3. OLT1177/Dapansutrile

OLT1177, also referred to as dapanustrile, is a nitrile sulfonyl derivative that selectively inhibits the inflammasome ATPase activity, disabling NLRP3-ASC and NLRP3-caspase-1 interactions and IL-1β generation in neutrophils and monocyte-derived macrophages [136]. In a phase I clinical trial, dapanustrile, proved to be safe and tolerated in oral application in healthy adults, with both in vitro and in vivo efficacy distinguishing it from MCC950 [36,137].

4.2.4. Tranilast

Originally discovered to suppress NF-κB by disrupting the association between NF-κB and CREB-binding protein (CBP), tranilast, an analogue of a tryptophan metabolite, has been utilized to treat bronchial asthma in Japan, China and South Korea since 1982 [43]. Highlighted in current emerging research as a selective NLRP3 inhibitor and IL-1β down-regulator, in contrast to other direct inhibitors, the compound mediates only the NLRP3-ASC interplay and does not affect ATPase activity of the protein [138]. Tranilast suppresses inflammation by inhibiting NLRP3 inflammasomes but not AIM2 or NLRC4 inflammasomes. Additionally, it blocks LPS-induced pro-IL-1β and IL-6 production, despite having an in vivo potency that is 5–10 times less effective than MCC950 [21].

In addition to its anti-inflammatory properties, it further exhibited antioxidant effects demonstrated by ROS scavenging and direct suppression of xanthine oxidase activity in vitro, as well as reducing TXNIP expression and ROS generation in rats that were administered streptozotocin [36].

4.2.5. Oridonin

Oridonin (Ori), a natural constituent derived from the plant *Rabdosia rubescens*, is frequently consumed in East Asia as a dietary supplement for its anti-inflammatory, antitumor, antimicrobial and neuroprotective effects. Through the disruption of NF-κB signaling, nuclear translocation and DNA binding, it mediates the suppression of various pro-inflammatory regulators [139–141]. Recently, ori has emerged as a selective, non-reversible inhibitor of NLRP3 that suppresses IL-1β generation and NLRP3-NEK7 interaction by covalently altering the C279 cysteine residue in the NACHT domain without impacting the ATPase [142].

Concerns regarding the toxicity and the detrimental pharmacological consequences of ori exist, despite the in vivo effects observed in preclinical investigations [143]. To increase pharmacological activity and bioavailability, more studies on ori analogues are warranted.

4.2.6. 3,4-Methylenedioxy-β-Nitrostyrene (MNS)

By interfering with the synthesis of IL-1β, IL-18 and caspase-1 activation, the spleen tyrosine kinase (Syk) and Src tyrosine kinase inhibitor, 3,4-methylenedioxy-β-nitrostyrene (MNS) is likewise a reliable and selective inhibitor of the NLRP3 inflammasome [144]. Subsequent findings indicate MNS binds to the LRR and NACHT domains of the NLRP3 protein, inhibiting ATPase function but not the inflammasome AIM2 or NLRC4. However, its extreme toxicity hindered further research [47].

4.2.7. INF Analogues

INF analogues were first introduced in 2014; one of the compounds, INF4E (compound 9), irreversibly inhibits the NLRP3 protein via its Michael acceptor. They display a potential role as targets for IL-1β secretion, ATPase and caspase-1 activity inhibition [145]. In addition, INF4E inhibits ATP- and nigericin-induced cell death and inhibits pyroptosis, demonstrating novel additional cardioprotective benefits in myocardial ischemia/reperfusion injury via the ERK/Akt/GSK-3β RISK pathway [54,146]. The discovery of the analogue INF58 (compound 14), which irreversibly inhibits the NLRP3 ATPase with enhanced potency, and INF39 (compound 11), which has reduced cytotoxicity and reactivity while attenuating NEK7-NLRP3 interactions, was made possible by derivatization and complementary structural analysis [145].

With no inhibition of AIM2 or NLRC4 inflammasomes but inhibition of ASC speck production and cleavage of caspase-1, IL-1β and GSDMD proteins, INF39 is the most studied representative as a potential future NLRP3 selective inhibitor [147].

4.2.8. BAY 11-7082

BAY 11-7082 (BAY), a phenylvinylsulfone with multiple pharmacological and potential pathways, is known to reduce TNF-α, which inhibits the nuclear factor-κB (IκB) kinase (IKK) phosphorylation that in turn reduces NF-κB signaling and NLRP3 production [36,43]. By completely alkylating cysteine residues in the NLRP3-ATPase region through a Michael coupling process, BAY decreases NLRP3's ATPase function and prevents ASC oligomerization [148]. The covalent cysteine residue C191 alkylation by the compound is involved in the GSDMD pore formation, as was observed by Hu et al. [149].

4.2.9. VX-740 (Pralnacasan) and VX-765

The orally targeted caspase-1 inhibitors VX-740 (Pralnacasan) and their analogue VX-765 represent two novel compounds that covalently alter the catalytic cysteine residues in the active sites of caspase-1, thereby impairing caspase-1 cleavage and pro-IL-1β/18 processing [150]. While both rheumatoid arthritis (RA) and osteoarthritis (OA) have been successfully managed with VX-740 in phase IIa clinical trials, additional research has been suspended due to the severe hepatotoxicity uncovered with prolonged use [43]. Since VX-765 was proved to significantly downregulate pyroptosis and IL-1β expression in SMCs incubated with oxLDL and suppress AS formation in ApoE$^{-/-}$ mice, it has become the most researched caspase-1 inhibitor [151].

4.2.10. Anakinra (Kineret), Rilonacept (Arcalyst) and Canakinumab (Ilaris)

Anakinra is a synthetic IL-1Ra. Rilonacept is a dimeric fusion protein with decoy receptors containing extracellular residues of the two IL-1R subunits. Canakinumab is an anti-IL-1β monoclonal antibody that specifically binds to and neutralizes IL-1β. These are the three biological therapies currently approved by the FDA that additionally target the NLRP3 pathway [6]. Neither of the biologics previously listed has received approval for use in clinical practice as an NLRP3 inhibitor alone, despite the NLRP3-IL-1b inflammasome pathway being a potentially valuable therapeutic target for patients with atherosclerosis and other inflammatory conditions [15]. Regarding CVDs, some clinical trials with anakinra have been carried out in patients with myocardial infarction (VCU-ART and VCU-ART2; MRC-ILA-Heart Study) [152,153], but the stated outcomes are controversial due to the sparse patient samples involved [15,154].

A large-scale clinical trial, the Canakinumab Anti-inflammatory Thrombosis Outcomes Study (CANTOS), was conducted to explore the theory of inflammation interplay in ATS [11]. Although neutralizing IL-1β with the canakinumab antibody (at a dose of 150 mg every three months) improved cardiovascular events related to atherosclerosis, a major impediment in admitting this compound as a therapy are the undefined strict criteria for eligibility as the study revealed substantial side effects of the treatment [11,24,43].

4.2.11. Colchicine

In addition to its proven proficiency in treating gouty arthritis and pericarditis, colchicine, a widespread drug, has proven recently to be an effective anti-inflammatory compound by blocking NLRP3 inflammasome activity and suppressing IL-1β and IL-18 secretion [31,43].

Patients with acute coronary syndrome who were administered short-term colchicine in the Colchicine Cardiovascular Outcomes Trial (COLCOT) displayed markedly decreased IL-1 and caspase-1 production when compared to untreated patients, indicating that colchicine has potential atheroprotective effects [155]. The incidence of cardiovascular mortality, myocardial infarction, ischemic stroke and ischemia-induced coronary revascularization was decreased by 31% in the Low-Dose Colchicine (LoDoCo20) trial, which enrolled patients with chronic stable CAD after one month of colchicine treatment (0.5 mg once daily). The LoDoCo2 trial that included patients with stable chronic CAD after one month of colchicine use (0.5 mg once daily) had a 31% reduction in the incidence of cardiovascular death, myocardial infarction, ischemic stroke and ischemia-driven coronary revascularization when compared with patients receiving placebo [156]. The results reinforce the promising advantages of anti-inflammatory agents in patients with coronary artery disease and are concordant with those of the COLCOT study and the first LoDoCo study [24,155,157].

Even though CANTOS and LoDoCo2 have still not adjusted the therapeutic agenda for cardiovascular risk reduction in clinical practice [11,156], these pivotal studies constitute an important step in the practical implementation of immunomodulatory therapies targeting NLRP3 for CVD.

4.2.12. Less Known NLRP3 Inhibitors

Other NLRP3 inflammasome inhibitors are currently being studied, but they are in an early stage of research and, considering that new molecules are constantly emerging, require further evaluation to be considered as potential future therapies.

The benzoxathiol derivative BOT-4-one exerts anti-inflammatory properties through various pathways, including inhibition of the NF-κB pathway by alkylation of IKKβ and consequently decreased production of NLRP3 [149]. Although the effect on ubiquitination is uncertain, the molecule's capacity to arrest ATP-induced IL-1 release deserves further exploration in atherosclerosis [47,158]. By disrupting the transcription factors NF-κB and signal transducer and activator of transcription 1 (STAT1), methylene blue, an anti-inflammatory, antioxidant and neuroprotective substance, has lately proved its effects in reducing NLRP3, pro-IL-1β and iNOS expression [159,160].

Disulfiram, the FDA-approved treatment in alcohol dependence, additionally exhibited lysosomal protection, mitochondria-independent ROS generation and NLRP3 inhibition [161]. In particular, disulfiram has been reported to suppress GSDMD pore development, hindering IL-1 β release and associated pyroptosis [162]. Supplementary research is needed to retarget disulfiram treatment for inflammatory pathologies.

Some phenamic acid derivatives have been reported to interact with NLRP3 inflammasomes by disrupting the Cl⁻ volume-regulated anion channel (VRAC) as a complement to their extensive use as nonsteroidal anti-inflammatory drugs (NSAIDs) that inhibit cyclooxygenase (COX) [163]. The detection of pleiotropic effects in antidepressants has been enabled by acknowledgement of the importance of inflammatory routes in the pathophysiology of psychiatric diseases. Recently, fluoxetine revealed direct NLRP3 inhibitory effects and downstream IL-1β regulation in macrophage cells and retinal pigment epithelium (RPE) cells [164]. This opens potential new areas of use for fluoxetine and phenamic acid derivatives, but further studies are necessary to determine their efficacy in various conditions. Another innovative approach is ursodeoxycholic acid (UDCA), which decreased intracellular CC by dissolving in macrophages and thereby reducing IL-1β secretion [165].

Considering its entanglement in inflammatory disorders, with the direct consequence of endorsing cytokine release and mediation of endothelial pyroptosis, NLRP3 inflammasome is an attractive therapeutic target. As observed, new pharmacological approaches hamper its pathway by targeting a specific phase via different components or by directly inhibiting NLRP3 and fully stopping its effects (Table 2).

Table 2. Therapeutic pharmacological agents against NLRP3 inflammasome according to their activation phases.

Agent	Study Design	Salient Results	Ref.
		Phase I (priming)	
Bay 11-7082	Preclinical (Experimental study)	- Via NF-κB inhibition selectively blocks IKKβ kinase activity with subsequent inhibition of the NLRP3 inflammasome activation; - Inhibited nigericin-induced and MSU-induced caspase-1 activation by the NLRP3 inflammasome.	[148]
		NLRP3 oligomerization	
CY-09 (glitazone derivate)	Preclinical (Experimental study)	- It binds to the ATP-binding motif of NLRP3 NACHT domain and inhibits NLRP3 ATPase assembly and activity in macrophages; - Inhibits NLRP3 ATPase activity; - Reverses metabolic effects via NLRP3 inhibition.	[166]

Table 2. *Cont.*

Agent	Study Design	Salient Results	Ref.
MCC 950	Preclinical (Experimental study)	- Inhibits NLRP3 inflammasome activation by suppressing IL-1β secretion; - It does not affect NLRP1, NLRC4 or AIM2 inflammasomes; - Atheroprotective activity by reducing the size of the plaque; - Reversed the impaired endothelial dysfunction.	[129–131]
Dapansutrile	CT (Phase I, randomized controlled trial)	- Inhibits NLRP3-ASC band NLRP3-caspase-1 interaction.	[137]
Tranilast	Approved (Experimental study)	- Reduces ROS, TXNIP expression and directly inhibits xhantine oxidase activity in vitro; - Via binding to the NATCH domain of NLRP3, inhibits assembly and its effects.	[21,36]
		Phase II (activation)	
Ang-(1-7)	Preclinical (Experimental study)	- Anti-inflammatory and anti-senescent action through RAAS; - Inhibits IL-1 -induced iNOS expression and NF-κB activation in vascular smooth muscle cells; - Diminishes NLRP3 inflammasome/IL-1 over-activation loop.	[167]
HL2351	CT (Phase I, randomized controlled trial)	-Inhibition of IL-1 function with indirect NLRP3 inflammasome action.	[113]
GSK1070806	CT (Phase II, randomized, placebo-controlled)	- Inhibition of IL-18; - Inhibition of IL-18 did not lead to any improvements in glucose control.	[168]
Rilonacept	Approved (Phase III, double-blind, randomized-withdrawal)	- Inhibition of the IL-1 pathway; - Reduced the activation of endothelial cell NADPH oxidase.	[169]
Canakinumab	Approved (Randomized, double-blind)	- Direct blockade of IL-1 or its receptor; - Antioxidant effects.	[11]
Anakinra	Approved (Randomized, double-blind)	- Modulation of mitochondrial ROS production by activating SOD2.	[11]

Iκβ kinase β (IKKβ); monosodium urate (MSU); clinical trial (CT); the anti-interleukin-18 monoclonal antibody (GSK1070806); renin-angiotensin-aldosterone system (RAAS); superoxide dismutase 2 (SOD2).

4.3. Statins

Modern research shows that statins, as analogues of 3-hydroxy-3-methylglutaryl coenzyme A inhibitor (HMG-CoA) and typically utilized for their cholesterol-lowering ability, also have pleiotropic effects, including immunomodulatory and anti-inflammatory abilities.

Wu et al. [170] revealed that atorvastatin, the most researched member of the class, inhibits the NLRP3 inflammasome and the molecules involved in pyroptosis, such as caspase-1 and IL-1 β, by increasing the production of NEXN-antisense RNA1 (AS1) lncRNA and its related gene NEXN [170,171]. In addition, the anti-inflammatory effects translated as IL-1β downregulation are supported by inhibition of the TLR4/MyD88/NF-κB pathway demonstrated in THP-1 cells stimulated with phorbol 12-myristate 13-acetate (PMA) [27,172]. In this account, it is reasonable to hypothesize that reduced NLRP3 inflammasome signaling is at least broadly accountable for the effects of atorvastatin on the progression of atherogenic pathways and chronic inflammation [43].

Simvastatin therapy also substantially decreased expression levels of NLRP3, cathepsin B, IL-1β and IL-18 in peripheral blood monocytes in response to CC stimulation both in vitro and in vivo [32,173]. Additionally, another atheroprotective pathway emerged and is exemplified by the endothelial Kruppel-like factor 2 (Klf2)-Forkhead box P transcription factor 1 (Foxp1)–NLRP3 inflammasome network; it warrants more consideration for its potential effect in delaying T2DM-associated atherosclerosis progression [46,174]. Another class representative, rosuvastatin, displayed in vivo decreased NLRP3 activation mostly via crosstalk with the oxidative stress enzymatic players, such as SOD, glutathione peroxidase and catalase (CAT) [54,175].

When considered collectively, preclinical research findings imply that statins have the potential to be an efficient therapeutic and prevention approach for cardiovascular disease by controlling NLRP3 activation.

4.4. Natural Compounds

A vast array of novel substances identified from a variety of natural compounds that have been used for many years in conventional, traditional medicine qualify as potential NLRP3 inflammasome inhibitors that can provide effective therapies. However, the mechanisms underlying their immune modulatory function have not been thoroughly investigated.

4.4.1. Flavonoids

Many fruits, vegetables and grains contain a phytonutrient class commonly known as flavonoids; it is widely recognized for its neuroprotective, anti-inflammatory, and antioxidant effects. The ability of flavonoids to decrease ROS production and expression of inflammasome components, such as ASC, NLRP3 and active caspase 1, as well as the consistent release of IL-1β, are presumed to be molecular mechanisms by which they modulate inflammasome activity [47,176]. The most promising NLRP3 inflammasome antagonists will hereafter be outlined.

Reductions in levels of the pro-inflammatory cytokines IL-6, IL-1β and TNF-α, coupled with additional inhibition of the ERK1/2 and NF-κB pathways, are accomplished by the flavone apigenin, contained in vegetables such as celery and parsley [32,177]. Although apigenin does not affect ASC protein levels despite reduced NF-κB activation, it has recently been reported to reduce phosphorylation of two key enzymes involved in ASC phosphorylation: Syk and the protein tyrosine kinase 2 (Pyk2) [176]. Additional research is required to define the specific NLRP3 inhibitory mechanism since the compound exhibits non-selective function by interfering with IL-1β production via AIM2 inflammasome [178].

Cardamonin, a naturally occurring chalcone identified in the Alpinia katsumadai Hayata plant, has been documented to reduce nitric oxide (NO) levels via NF-κB inhibition and by increasing the AhR/Nrf2/NQO1 pathway, a recognized NLRP3 downregulator [179]. However, there is a lack of research papers on the effects of cardamonin on ATS [180].

Because of their high concentrations of flavonoids, chalcones and other phytonutrients, including isoliquiritigenin (ILG) and glycyrrhizin (GL), two extracts from the Glycyrrhiza uralensis (licorice) plant, have traditionally been utilized to treat a series of conditions [47]. Either compound inhibits NF-κB and MAPK activity by repressing the TLR4/MD-2 molecule [54]. A single study concluded that isoliquiritigenin therapy reduced NLRP3 inflammasome, caspase-1 activation and IL-1β generation in obesity through interference with both the inflammasome initiation and activation phases [181]. Albeit in a non-specific manner modulated by NF-κB luteolin present in plants such as broccoli, celery, thyme and pepper, these compounds can reduce ROS and NLRP3 generation besides other inflammatory mediators [182,183]. Further testing raised the hypothesis of luteonin inflammasome inhibitor potential, but none focused on ATS [182,184]. IL-1β secretion via NLRP3 and AIM2 inflammasome pathways is diminished by the non-selective inflammasome inhibitor quercetin, commonly present in a variety of grains, vegetables, and fruits [185]. Quercetin represses the NLRP3 inflammasome activation phase by impairing ASC-speck oligomerization, although the exact mechanism is currently unclear [47,186].

4.4.2. Phenols

Several species of the genus *Artemisia* have prompted interest in further study of their anti-inflammatory benefits due to their current use in traditional Asian remedies. As a pertinent example, the dysregulated activation of NLRP3 or AIM2 inflammasomes yields the anti-inflammatory effects of Artemisia princeps extract (APO), which suppresses the generation of ASC speck [187,188]. Additional constituents, such as Artemisinin from *Artemisia annua*, ameliorated foam cell production by impairing the inflammatory AMPK/ NF-κB-NLRP3 pathway response in macrophages [32,189]. By modulating ASC-PYD binding and inhibiting NLRP3-ASC interplay, the phenolic class representant caffeic acid phenethyl ester (CAPE) found in bee propolis might decrease NF-κB activation and hinder NLRP3 and AIM2 inflammasome activation [190].

Curcumin, a common polyphenol substance renowned for its anti-inflammatory and antioxidant properties, reduced NLRP3 expression due to IKK phosphorylation inhibition that further impaired the TLR4/MyD88, NF-κB and P2X7R inflammasome pathways [32,191], thus dismissing the hypothesis of inhibition by the antioxidant mechanism [192].

While a promising target for suppressing the AIM2 inflammasome, the terpene group compounds, such as andrographolide and parthenolide, produce only mild inhibitory effects on the NLRP3 inflammasome [47,192,193].

4.4.3. Miscellaneous

The traditional Chinese herbal naftoquinone compound Shikonin, extracted from the roots of *Lithospermum erythrorhizon*, displays numerous anti-inflammatory mechanisms by targeting NF-κB suppression and NLRP3 mediated IL-1β production; however, the effects on cardiovascular disease are still unknown [194,195]. In addition, *Salvia miltiorrhiza* Bunge contains salvianolic acid A (SAA), a phenolic molecule that hinders NLRP3 activation in aortic tissues via the NF-κB pathway and reduces early-stage ATS [196]. Moreover, 6-shogal, the most potent ginger-derived compound, displayed positive effects by reducing Akt activation, ROS production and NLRP3 inflammasome activation, thereby alleviating hyperglycemia-induced calcification of arterial smooth muscle cells [55,197]. Inhibition of ROS/TXNIP/NLRP3 crosstalk has been established in mangiferin, puerarin, and rutin administration that can further mitigate oxidative stress and inflammation in vascular endothelial cells [56,198]. As AGEs induced endothelial dysfunction, salidroside might play a pivotal role by regulating AMPK/NF-κB/NLRP3 signaling [199].

Inhibiting diabetes-related atherosclerosis by raising eNOS levels and decreasing NF-κB and TXNIP expression in response to the priming signal from NLRP3 inflammasome activation, Biejiajian (BJJ), a traditional Chinese medicine, has lately exhibited promising results [12]. The LOX-1/NLRP3 pathway recently emerged as an important mediator in inflammasome activation upon the use of astragaloside IV (AS-IV), the active component of *Astragalus membranaceus* (Fisch.) Bge., an antioxidant treatment [200,201].

The aforementioned natural elements function essentially as antioxidants. In part, this can account for their function as NLRP3 inflammasome inhibitors by reducing ROS generation. However, they also provide new insights into the complex interaction between inflammation and oxidation and suggest potential therapeutic options for vascular problems [56]. Current studies are insufficient to completely explain the therapeutic effects of these substances, and further research is necessary to better understand their inflammasome-inhibiting potential.

5. Discussion

Although inflammatory biomarkers (hsCRP, IL-6, IL-1, TNF-, MCP-1), endothelial biomarkers (ICAM-1, VCAM-1) and selectins are established in both human and experimental studies as routinely detected in retrieved blood samples, we are still bound by particular detection methods when it comes to quantifying inflammation as a target. While in preclinical studies these changes have been detected in samples of peripheral blood monocytes, macrophages, aortic endothelial cells from mouse tissues and cell cultures using a variety of staining methods and tests (e.g., Western blotting, real-time polymerase chain reaction), the assessment of the therapeutic–inflammatory response is not clearly delineated in most human clinical studies. This underlines the necessity of designing a specific test for sensing and assessing anti-inflammatory therapy that can be implemented in clinical practice in conjunction with a more extensive panel of inflammatory biomarkers.

We need to emphasize some of the contrasts between our research and some of the most important reviews that have been conducted in this topic over the last three years after thorough research. For instance, in their work, Silvis et al. [31] focused on the mechanisms and the involvement of NLRP3 inflammasome strictly in coronary artery disease and acute myocardial infarction with a brief mention of therapy options mentioning only clinical immunotherapy trials. Jiang et al. [43], in their review, discussed in detail the NLRP3 inflammasome components and their pathophysiological implications in atherosclerosis, with a focus only on direct and indirect NLRP3 inflammasome inhibitors and setting aside natural components which could prove an inexpensive source of treatment in the future. Sharma B.R. et al. [38] reviewed the mechanisms of NLRP3 regarding chronic inflammation in atherosclerosis and cancer, without any mention of treatment options, while Sharma A. et al. [132] raised awareness about NLRP3 inflammasome involvement in diabetic atherosclerosis in relation to only one therapeutic option. In contrast, Burger et al. [202] pointed out the role of NLRP3 inflammasome in plaque destabilization and only briefly covered the treatment potential of colchicine. Takahashi et al. [6] covered NLRP3 inflammasome involvement in atherosclerosis in addition to the vascular pathology of different causes (aortic aneurysm, Kawasaki disease). They discuss known clinical trials (e.g., CANTOS, COLCOT) with a concise overview of preclinical studies. Zeng et al. [203] discussed the implications and therapeutic possibilities of pyroptosis but concerning only one aspect of the myriad ramifications of the NLRP3 inflammasome in the pathophysiology of atherosclerosis; they only discuss one therapeutic product.

While pharmacological inhibitors of NLRP3 inflammasome and natural compounds have shown promise in animal models, there are limitations to their use in clinical practice. The lack of comprehensive studies on a larger scale in humans and high-quality research prevents the Food and Drug Administration (FDA) and medical experts from recommending the use of numerous molecules in the form of affordable dietary supplements that contain mainly the natural compounds presented above for the prevention and treatment of atherosclerosis. Furthermore, these agents may have limited bioavailability and can be metabolized quickly, making them less effective in vivo. For example, some inhibitors have low specificity and can also affect other inflammasomes, leading to potential off-target effects. Inhibiting only IL-1 and IL-8 may not fully address the complex inflammatory response downstream of NLRP3 activation, as its activation leads to the production of various cytokines and chemokines that contribute to inflammation, including IL-1β, IL-18, IL-6, TNF-α and IL-8. Additionally, the multiple downstream mediators inhibition treatments targeting IL-1 and IL-18 do not have NLRP3 specificity and can also inhibit the normal immune response pathways. Additionally, the long-term use of interleukin inhibitors is well known for their immune-related adverse events, including low blood pressure, nausea, vomiting, liver enzyme elevations, diarrhea and increased rates of infection. While NLRP3 is a key regulator of inflammasome activation, there is increasing evidence that other inflammasome sensors, such as NLRC4 and AIM2, also play a role in atherosclerosis. Therefore, targeting NLRP3 alone may not be sufficient to fully modulate the inflammasome response in atherosclerosis. Combining NLRP3 inhibitors with other agents that target other aspects of atherosclerosis., such as lipid metabolism or immune cell recruitment, may provide synergistic effects and enhance therapeutic efficacy. Advances in drug discovery and development may lead to identifying new, more specific and potent NLRP3 inhibitors with improved bioavailability and pharmacokinetic properties, potentially resulting in fewer adverse effects. While NLRP3 inflammasome activation is a promising therapeutic target for the treatment of atherosclerosis (Figure 3), and most of its beneficial effects are seen in experimental studies, developing other NLRP3-modifying agents is necessary to address the limitations of current agents. It is also necessary to account for NLRP3-independent mechanisms and explore potential combination therapies.

As noted, we conclude that our updated review provides scientific data not only about the pathophysiology underlying atherosclerosis, which involves the complex relationship between inflammation, hyperglycemia and oxidative stress and the role of NRLP3 in these pathways, but also encompasses a wider array of therapy information. In addition, our review outlines both preclinical and clinical findings on NLRP3 inhibitors and includes integrative reviews of other lesser-known NLRP3 inhibitors. The aim of this manuscript is to reveal the vastness of this field, which encompasses still unexplored pathways and molecules with biomarker potency and less addressed potential therapeutic targets. Hopefully, it can be of help for future researchers.

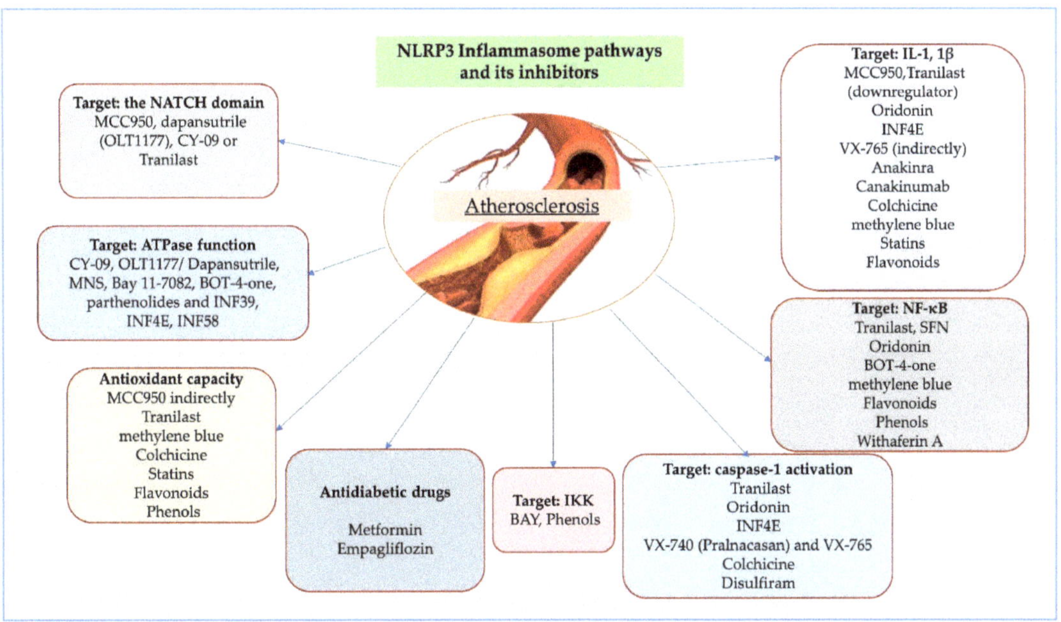

Figure 3. NLRP3 Inflammasome pathways and its inhibitors according to their mechanism of action.

6. Conclusions

Unfortunately, as the human population grows, unhealthy lifestyles are increasingly adopted, leading to a rise in the incidence and prevalence of metabolic diseases. All things considered, despite the efforts of bringing new molecules for atherosclerosis treatment and up-to-date management guidelines, atherosclerosis still holds the title of one of the most debilitating diseases when its multiple central and peripheric vascular complications are considered.

For this reason, over the last decades, the scientific community has directed its attention to the supplementary investigation of the pathogenesis behind this disease, with more concentration on inflammasome pathway research. A plethora of evidence noted the role of NLRP3 inflammasome in endothelial dysfunction and atherosclerosis. Human and preclinical studies predominantly displayed the beneficial effects of direct or indirect NLRPR signaling/assembly/activation phase inhibitors.

A better understanding through translational research could enlighten the full role and potential of inflammasomes in CVD disease. Thus, perhaps the discovery of clinically approved novel pharmacological molecular inhibitors of NLRP3, along with other adjuvant therapy agents, may bring new therapeutic strategies with a reduced burden of atherosclerosis complications.

Author Contributions: Conceptualization, D.M.T., E.V. and E.M.G.; methodology, O.N.B.-F., A.O. and M.A.M.; software, A.O.; validation, D.M.T., E.M.G. and I.L.S.; formal analysis, A.O.; investigation, M.A.M., M.F. and E.V.; resources, O.N.B.-F.; data curation, D.M.T.; writing—original draft preparation, D.M.T., E.V., E.M.G. and I.L.S.; writing—review and editing, I.L.S.; visualization, M.F.; supervision, E.V., E.M.G. and M.A.M.; project administration, D.M.T. and I.L.S.; funding acquisition, A.O. All authors have read and agreed to the published version of the manuscript.

Funding: This research received no external funding.

Institutional Review Board Statement: Not applicable.

Informed Consent Statement: Not applicable.

Data Availability Statement: Not applicable.

Conflicts of Interest: The authors declare no conflict of interest.

References

1. Libby, P.; Buring, J.E.; Badimon, L.; Hansson, G.K.; Deanfield, J.; Bittencourt, M.S.; Tokgözoğlu, L.; Lewis, E.F. Atherosclerosis. *Nat. Rev. Dis. Prim.* **2019**, *5*, 56. [CrossRef]
2. Libby, P. The Changing Landscape of Atherosclerosis. *Nature* **2021**, *592*, 524–533. [CrossRef]
3. Liberale, L.; Montecucco, F.; Schwarz, L.; Lüscher, T.F.; Camici, G.G. Inflammation and Cardiovascular Diseases: Lessons from Seminal Clinical Trials. *Cardiovasc. Res.* **2021**, *117*, 411–422. [CrossRef] [PubMed]
4. Montecucco, F.; Liberale, L.; Bonaventura, A.; Vecchiè, A.; Dallegri, F.; Carbone, F. The Role of Inflammation in Cardiovascular Outcome. *Curr. Atheroscler. Rep.* **2017**, *19*, 11. [CrossRef]
5. Liberale, L.; Holy, E.W.; Akhmedov, A.; Bonetti, N.R.; Nietlispach, F.; Matter, C.M.; Mach, F.; Montecucco, F.; Beer, J.H.; Paneni, F.; et al. Interleukin-1β Mediates Arterial Thrombus Formation via NET-Associated Tissue Factor. *JCM* **2019**, *8*, 2072. [CrossRef] [PubMed]
6. Takahashi, M. NLRP3 Inflammasome as a Key Driver of Vascular Disease. *Cardiovasc. Res.* **2022**, *118*, 372–385. [CrossRef] [PubMed]
7. Fernandez, D.M.; Rahman, A.H.; Fernandez, N.F.; Chudnovskiy, A.; Amir, E.D.; Amadori, L.; Khan, N.S.; Wong, C.K.; Shamailova, R.; Hill, C.A.; et al. Single-Cell Immune Landscape of Human Atherosclerotic Plaques. *Nat. Med.* **2019**, *25*, 1576–1588. [CrossRef] [PubMed]
8. Ross, R. Atherosclerosis—An Inflammatory Disease. *New Engl. J. Med.* **1999**, *340*, 115–126. [CrossRef] [PubMed]
9. Kritharides, L. Inflammatory Markers and Outcomes in Cardiovascular Disease. *PLoS Med.* **2009**, *6*, e1000147. [CrossRef] [PubMed]
10. Björkegren, J.L.M.; Lusis, A.J. Atherosclerosis: Recent Developments. *Cell* **2022**, *185*, 1630–1645. [CrossRef] [PubMed]
11. Ridker, P.M.; Everett, B.M.; Thuren, T.; MacFadyen, J.G.; Chang, W.H.; Ballantyne, C.; Fonseca, F.; Nicolau, J.; Koenig, W.; Anker, S.D.; et al. Antiinflammatory Therapy with Canakinumab for Atherosclerotic Disease. *N. Engl. J. Med.* **2017**, *377*, 1119–1131. [CrossRef]
12. Fu, Y.; Yuan, J.; Sang, F.; Shao, M.; Yan, S.; Li, L.; Zhu, R.; Wang, Z. Biejiajian Pill Ameliorates Diabetes-Associated Atherosclerosis through Inhibition of the NLRP3 Inflammasome. *Evid. -Based Complement. Altern. Med.* **2022**, *2022*, 9131178. [CrossRef] [PubMed]
13. Wan, Z.; Fan, Y.; Liu, X.; Xue, J.; Han, Z.; Zhu, C.; Wang, X. NLRP3 Inflammasome Promotes Diabetes-Induced Endothelial Inflammation and Atherosclerosis. *DMSO* **2019**, *12*, 1931–1942. [CrossRef] [PubMed]
14. Grebe, A.; Hoss, F.; Latz, E. NLRP3 Inflammasome and the IL-1 Pathway in Atherosclerosis. *Circ. Res.* **2018**, *122*, 1722–1740. [CrossRef] [PubMed]
15. Abbate, A.; Toldo, S.; Marchetti, C.; Kron, J.; Van Tassell, B.W.; Dinarello, C.A. Interleukin-1 and the Inflammasome as Therapeutic Targets in Cardiovascular Disease. *Circ. Res.* **2020**, *126*, 1260–1280. [CrossRef]
16. Kelley, N.; Jeltema, D.; Duan, Y.; He, Y. The NLRP3 Inflammasome: An Overview of Mechanisms of Activation and Regulation. *IJMS* **2019**, *20*, 3328. [CrossRef]
17. Fatima, S.; Ambreen, S.; Mathew, A.; Elwakiel, A.; Gupta, A.; Singh, K.; Krishnan, S.; Rana, R.; Khawaja, H.; Gupta, D.; et al. ER-Stress and Senescence Coordinately Promote Endothelial Barrier Dysfunction in Diabetes-Induced Atherosclerosis. *Nutrients* **2022**, *14*, 2786. [CrossRef]
18. Mikkelsen, R.R.; Hundahl, M.P.; Torp, C.K.; Rodríguez-Carrio, J.; Kjolby, M.; Bruun, J.M.; Kragstrup, T.W. Immunomodulatory and Immunosuppressive Therapies in Cardiovascular Disease and Type 2 Diabetes Mellitus: A Bedside-to-Bench Approach. *Eur. J. Pharmacol.* **2022**, *925*, 174998. [CrossRef]
19. Martinon, F.; Burns, K.; Tschopp, J. The Inflammasome: A Molecular Platform Triggering Activation of Inflammatory Caspases and Processing of ProIL-Beta. *Mol. Cell* **2002**, *10*, 10. [CrossRef]
20. Zhou, W.; Chen, C.; Chen, Z.; Liu, L.; Jiang, J.; Wu, Z.; Zhao, M.; Chen, Y. NLRP3: A Novel Mediator in Cardiovascular Disease. *J. Immunol. Res.* **2018**, *2018*, 5702103. [CrossRef]

21. Swanson, K.V.; Deng, M.; Ting, J.P.-Y. The NLRP3 Inflammasome: Molecular Activation and Regulation to Therapeutics. *Nat. Rev. Immunol.* **2019**, *19*, 477–489. [CrossRef]

22. Biasizzo, M.; Kopitar-Jerala, N. Interplay Between NLRP3 Inflammasome and Autophagy. *Front. Immunol.* **2020**, *11*, 591803. [CrossRef]

23. Gora, I.M.; Ciechanowska, A.; Ladyzynski, P. NLRP3 Inflammasome at the Interface of Inflammation, Endothelial Dysfunction, and Type 2 Diabetes. *Cells* **2021**, *10*, 314. [CrossRef] [PubMed]

24. Engelen, S.E.; Robinson, A.J.B.; Zurke, Y.-X.; Monaco, C. Therapeutic Strategies Targeting Inflammation and Immunity in Atherosclerosis: How to Proceed? *Nat. Rev. Cardiol.* **2022**, *19*, 522–542. [CrossRef] [PubMed]

25. Ding, S.; Xu, S.; Ma, Y.; Liu, G.; Jang, H.; Fang, J. Modulatory Mechanisms of the NLRP3 Inflammasomes in Diabetes. *Biomolecules* **2019**, *9*, 850. [CrossRef]

26. Lee, J.; Wan, J.; Lee, L.; Peng, C.; Xie, H.; Lee, C. Study of the NLRP3 Inflammasome Component Genes and Downstream Cytokines in Patients with Type 2 Diabetes Mellitus with Carotid Atherosclerosis. *Lipids Health Dis.* **2017**, *16*, 217. [CrossRef] [PubMed]

27. Parsamanesh, N.; Moossavi, M.; Bahrami, A.; Fereidouni, M.; Barreto, G.; Sahebkar, A. NLRP3 Inflammasome as a Treatment Target in Atherosclerosis: A Focus on Statin Therapy. *Int. Immunopharmacol.* **2019**, *73*, 146–155. [CrossRef] [PubMed]

28. Baldrighi, M.; Mallat, Z.; Li, X. NLRP3 Inflammasome Pathways in Atherosclerosis. *Atherosclerosis* **2017**, *267*, 127–138. [CrossRef]

29. Tong, Y.; Wang, Z.; Cai, L.; Lin, L.; Liu, J.; Cheng, J. NLRP3 Inflammasome and Its Central Role in the Cardiovascular Diseases. *Oxidative Med. Cell. Longev.* **2020**, *2020*, 4293206. [CrossRef]

30. Meyers, A.K.; Zhu, X. The NLRP3 Inflammasome: Metabolic Regulation and Contribution to Inflammaging. *Cells* **2020**, *9*, 1808. [CrossRef]

31. Silvis, M.J.M.; Demkes, E.J.; Fiolet, A.T.L.; Dekker, M.; Bosch, L.; van Hout, G.P.J.; Timmers, L.; de Kleijn, D.P.V. Immunomodulation of the NLRP3 Inflammasome in Atherosclerosis, Coronary Artery Disease, and Acute Myocardial Infarction. *J. Cardiovasc. Trans. Res.* **2021**, *14*, 23–34. [CrossRef] [PubMed]

32. Liu, Y.; Li, C.; Yin, H.; Zhang, X.; Li, Y. NLRP3 Inflammasome: A Potential Alternative Therapy Target for Atherosclerosis. *Evid. -Based Complement. Altern. Med.* **2020**, *2020*, 1561342. [CrossRef]

33. Wang, Y.; Liu, X.; Shi, H.; Yu, Y.; Yu, Y.; Li, M.; Chen, R. NLRP3 Inflammasome, an Immune-inflammatory Target in Pathogenesis and Treatment of Cardiovascular Diseases. *Clin. Transl. Med.* **2020**, *10*, 91–106. [CrossRef] [PubMed]

34. Zeng, X.; Liu, D.; Huo, X.; Wu, Y.; Liu, C.; Sun, Q. Pyroptosis in NLRP3 Inflammasome-Related Atherosclerosis. *CST* **2022**, *6*, 79–88. [CrossRef]

35. Kong, P.; Cui, Z.-Y.; Huang, X.-F.; Zhang, D.-D.; Guo, R.-J.; Han, M. Inflammation and Atherosclerosis: Signaling Pathways and Therapeutic Intervention. *Sig. Transduct. Target* **2022**, *7*, 131. [CrossRef] [PubMed]

36. González-Moro, A.; Valencia, I.; Shamoon, L.; Sánchez-Ferrer, C.F.; Peiró, C.; de la Cuesta, F. NLRP3 Inflammasome in Vascular Disease: A Recurrent Villain to Combat Pharmacologically. *Antioxidants* **2022**, *11*, 269. [CrossRef]

37. Huang, Y.; Xu, W.; Zhou, R. NLRP3 Inflammasome Activation and Cell Death. *Cell Mol. Immunol.* **2021**, *18*, 2114–2127. [CrossRef]

38. Sharma, B.R.; Kanneganti, T.-D. NLRP3 Inflammasome in Cancer and Metabolic Diseases. *Nat. Immunol.* **2021**, *22*, 550–559. [CrossRef]

39. Moasses Ghafary, S.; Soriano-Teruel, P.M.; Lotfollahzadeh, S.; Sancho, M.; Serrano-Candelas, E.; Karami, F.; Barigye, S.J.; Fernández-Pérez, I.; Gozalbes, R.; Nikkhah, M.; et al. Identification of NLRP3PYD Homo-Oligomerization Inhibitors with Anti-Inflammatory Activity. *IJMS* **2022**, *23*, 1651. [CrossRef]

40. Hoseini, Z.; Sepahvand, F.; Rashidi, B.; Sahebkar, A.; Masoudifar, A.; Mirzaei, H. NLRP3 Inflammasome: Its Regulation and Involvement in Atherosclerosis. *J. Cell Physiol.* **2018**, *233*, 2116–2132. [CrossRef]

41. Zheng, Y.; Xu, L.; Dong, N.; Li, F. NLRP3 Inflammasome: The Rising Star in Cardiovascular Diseases. *Front. Cardiovasc. Med.* **2022**, *9*, 927061. [CrossRef]

42. Groß, C.J.; Mishra, R.; Schneider, K.S.; Médard, G.; Wettmarshausen, J.; Dittlein, D.C.; Shi, H.; Gorka, O.; Koenig, P.-A.; Fromm, S.; et al. K + Efflux-Independent NLRP3 Inflammasome Activation by Small Molecules Targeting Mitochondria. *Immunity* **2016**, *45*, 761–773. [CrossRef] [PubMed]

43. Jiang, C.; Xie, S.; Yang, G.; Wang, N. Spotlight on NLRP3 Inflammasome: Role in Pathogenesis and Therapies of Atherosclerosis. *JIR* **2021**, *14*, 7143–7172. [CrossRef] [PubMed]

44. Zhong, Z.; Liang, S.; Sanchez-Lopez, E.; He, F.; Shalapour, S.; Lin, X.; Wong, J.; Ding, S.; Seki, E.; Schnabl, B.; et al. New Mitochondrial DNA Synthesis Enables NLRP3 Inflammasome Activation. *Nature* **2018**, *560*, 198–203. [CrossRef] [PubMed]

45. Shimada, K.; Crother, T.R.; Karlin, J.; Dagvadorj, J.; Chiba, N.; Chen, S.; Ramanujan, V.K.; Wolf, A.J.; Vergnes, L.; Ojcius, D.M.; et al. Oxidized Mitochondrial DNA Activates the NLRP3 Inflammasome during Apoptosis. *Immunity* **2012**, *36*, 401–414. [CrossRef] [PubMed]

46. Bai, B.; Yang, Y.; Wang, Q.; Li, M.; Tian, C.; Liu, Y.; Aung, L.H.H.; Li, P.; Yu, T.; Chu, X. NLRP3 Inflammasome in Endothelial Dysfunction. *Cell Death Dis.* **2020**, *11*, 776. [CrossRef]

47. Blevins, H.M.; Xu, Y.; Biby, S.; Zhang, S. The NLRP3 Inflammasome Pathway: A Review of Mechanisms and Inhibitors for the Treatment of Inflammatory Diseases. *Front. Aging Neurosci.* **2022**, *14*, 879021. [CrossRef]

48. Chen, J.; Chen, Z.J. PtdIns4P on Dispersed Trans-Golgi Network Mediates NLRP3 Inflammasome Activation. *Nature* **2018**, *564*, 71–76. [CrossRef]

49. Sharif, H.; Wang, L.; Wang, W.L.; Magupalli, V.G.; Andreeva, L.; Qiao, Q.; Hauenstein, A.V.; Wu, Z.; Núñez, G.; Mao, Y.; et al. Structural Mechanism for NEK7-Licensed Activation of NLRP3 Inflammasome. *Nature* **2019**, *570*, 338–343. [CrossRef]
50. Paik, S.; Kim, J.K.; Silwal, P.; Sasakawa, C.; Jo, E.-K. An Update on the Regulatory Mechanisms of NLRP3 Inflammasome Activation. *Cell Mol. Immunol.* **2021**, *18*, 1141–1160. [CrossRef]
51. Chan, A.H.; Schroder, K. Inflammasome Signaling and Regulation of Interleukin-1 Family Cytokines. *J. Exp. Med.* **2020**, *217*, e20190314. [CrossRef]
52. Duewell, P.; Kono, H.; Rayner, K.J.; Sirois, C.M.; Vladimer, G.; Bauernfeind, F.G.; Abela, G.S.; Franchi, L.; Nuñez, G.; Schnurr, M.; et al. NLRP3 Inflammasomes Are Required for Atherogenesis and Activated by Cholesterol Crystals. *Nature* **2010**, *464*, 1357–1361. [CrossRef] [PubMed]
53. Jin, Y.; Fu, J. Novel Insights Into the NLRP3 Inflammasome in Atherosclerosis. *JAHA* **2019**, *8*, e012219. [CrossRef]
54. Pellegrini, C.; Martelli, A.; Antonioli, L.; Fornai, M.; Blandizzi, C.; Calderone, V. NLRP3 Inflammasome in Cardiovascular Diseases: Pathophysiological and Pharmacological Implications. *Med. Res. Rev.* **2021**, *41*, 1890–1926. [CrossRef] [PubMed]
55. Ho, S.-C.; Chang, Y.-H. Comparison of Inhibitory Capacities of 6-, 8- and 10-Gingerols/Shogaols on the Canonical NLRP3 Inflammasome-Mediated IL-1β Secretion. *Molecules* **2018**, *23*, 466. [CrossRef] [PubMed]
56. Bai, Y.; Mu, Q.; Bao, X.; Zuo, J.; Fang, X.; Hua, J.; Zhang, D.; Jiang, G.; Li, P.; Gao, S.; et al. Targeting NLRP3 Inflammasome in the Treatment Of Diabetes and Diabetic Complications: Role of Natural Compounds from Herbal Medicine. *Aging Dis.* **2021**, *12*, 1587. [CrossRef] [PubMed]
57. Sharma, A.; Tate, M.; Mathew, G.; Vince, J.E.; Ritchie, R.H.; de Haan, J.B. Oxidative Stress and NLRP3-Inflammasome Activity as Significant Drivers of Diabetic Cardiovascular Complications: Therapeutic Implications. *Front. Physiol.* **2018**, *9*, 114. [CrossRef]
58. Menini, S.; Iacobini, C.; Vitale, M.; Pugliese, G. The Inflammasome in Chronic Complications of Diabetes and Related Metabolic Disorders. *Cells* **2020**, *9*, 1812. [CrossRef]
59. Jorquera, G.; Russell, J.; Monsalves-Álvarez, M.; Cruz, G.; Valladares-Ide, D.; Basualto-Alarcón, C.; Barrientos, G.; Estrada, M.; Llanos, P. NLRP3 Inflammasome: Potential Role in Obesity Related Low-Grade Inflammation and Insulin Resistance in Skeletal Muscle. *IJMS* **2021**, *22*, 3254. [CrossRef]
60. Hasheminasabgorji, E.; Jha, J.C. Dyslipidemia, Diabetes and Atherosclerosis: Role of Inflammation and ROS-Redox-Sensitive Factors. *Biomedicines* **2021**, *9*, 1602. [CrossRef]
61. Pickering, R.J.; Rosado, C.J.; Sharma, A.; Buksh, S.; Tate, M.; de Haan, J.B. Recent Novel Approaches to Limit Oxidative Stress and Inflammation in Diabetic Complications. *Clin. Transl. Immunol.* **2018**, *7*, e1016. [CrossRef] [PubMed]
62. Poznyak, A.; Grechko, A.V.; Poggio, P.; Myasoedova, V.A.; Alfieri, V.; Orekhov, A.N. The Diabetes Mellitus–Atherosclerosis Connection: The Role of Lipid and Glucose Metabolism and Chronic Inflammation. *IJMS* **2020**, *21*, 1835. [CrossRef] [PubMed]
63. Varghese, J.F.; Patel, R.; Yadav, U.C.S. Sterol Regulatory Element Binding Protein (SREBP) -1 Mediates Oxidized Low-Density Lipoprotein (OxLDL) Induced Macrophage Foam Cell Formation through NLRP3 Inflammasome Activation. *Cell Signal.* **2019**, *53*, 316–326. [CrossRef] [PubMed]
64. Sun, Y.; Ding, S. NLRP3 Inflammasome in Diabetic Cardiomyopathy and Exercise Intervention. *IJMS* **2021**, *22*, 13228. [CrossRef]
65. Zhao, H.; Liu, H.; Yang, Y.; Wang, H. The Role of H2S Regulating NLRP3 Inflammasome in Diabetes. *IJMS* **2022**, *23*, 4818. [CrossRef] [PubMed]
66. Zheng, F.; Xing, S.; Gong, Z.; Xing, Q. NLRP3 Inflammasomes Show High Expression in Aorta of Patients with Atherosclerosis. *Heart Lung Circ.* **2013**, *22*, 746–750. [CrossRef]
67. Shi, X.; Xie, W.-L.; Kong, W.-W.; Chen, D.; Qu, P. Expression of the NLRP3 Inflammasome in Carotid Atherosclerosis. *J. Stroke Cerebrovasc. Dis.* **2015**, *24*, 2455–2466. [CrossRef]
68. Paramel Varghese, G.; Folkersen, L.; Strawbridge, R.J.; Halvorsen, B.; Yndestad, A.; Ranheim, T.; Krohg-Sørensen, K.; Skjelland, M.; Espevik, T.; Aukrust, P.; et al. NLRP3 Inflammasome Expression and Activation in Human Atherosclerosis. *JAHA* **2016**, *5*, e003031. [CrossRef]
69. Afrasyab, A.; Qu, P.; Zhao, Y.; Peng, K.; Wang, H.; Lou, D.; Niu, N.; Yuan, D. Correlation of NLRP3 with Severity and Prognosis of Coronary Atherosclerosis in Acute Coronary Syndrome Patients. *Heart Vessel.* **2016**, *31*, 1218–1229. [CrossRef]
70. Wang, R.; Wang, Y.; Mu, N.; Lou, X.; Li, W.; Chen, Y.; Fan, D.; Tan, H. Activation of NLRP3 Inflammasomes Contributes to Hyperhomocysteinemia-Aggravated Inflammation and Atherosclerosis in ApoE-Deficient Mice. *Lab. Investig.* **2017**, *97*, 922–934. [CrossRef]
71. Zheng, F.; Xing, S.; Gong, Z.; Mu, W.; Xing, Q. Silence of NLRP3 Suppresses Atherosclerosis and Stabilizes Plaques in Apolipoprotein E-Deficient Mice. *Mediat. Inflamm.* **2014**, *2014*, 507208. [CrossRef]
72. Abderrazak, A.; Couchie, D.; Darweesh Mahmood, D.F.; Elhage, R.; Vindis, C.; Laffargue, M.; Matéo, V.; Büchele, B.; Ayala, M.R.; Gaafary, M.E.; et al. Response to Letter Regarding Article, "Anti-Inflammatory and Antiatherogenic Effects of the Inflammasome NLRP3 Inhibitor Arglabin in ApoE2. *Ki Mice Fed a High-Fat Diet.*" *Circulation* **2015**, *132*, e250–e251. [CrossRef] [PubMed]
73. Ma, S.; Chen, J.; Feng, J.; Zhang, R.; Fan, M.; Han, D.; Li, X.; Li, C.; Ren, J.; Wang, Y.; et al. Melatonin Ameliorates the Progression of Atherosclerosis via Mitophagy Activation and NLRP3 Inflammasome Inhibition. *Oxidative Med. Cell. Longev.* **2018**, *2018*, 9286458. [CrossRef]
74. van der Heijden, T.; Kritikou, E.; Venema, W.; van Duijn, J.; van Santbrink, P.J.; Slütter, B.; Foks, A.C.; Bot, I.; Kuiper, J. NLRP3 Inflammasome Inhibition by MCC950 Reduces Atherosclerotic Lesion Development in Apolipoprotein E–Deficient Mice—Brief Report. *ATVB* **2017**, *37*, 1457–1461. [CrossRef] [PubMed]

75. Shen, L.; Yang, Y.; Ou, T.; Key, C.-C.C.; Tong, S.H.; Sequeira, R.C.; Nelson, J.M.; Nie, Y.; Wang, Z.; Boudyguina, E.; et al. Dietary PUFAs Attenuate NLRP3 Inflammasome Activation via Enhancing Macrophage Autophagy. *J. Lipid Res.* **2017**, *58*, 1808–1821. [CrossRef]

76. Yin, R.; Zhu, X.; Wang, J.; Yang, S.; Ma, A.; Xiao, Q.; Song, J.; Pan, X. MicroRNA-155 Promotes the Ox-LDL-Induced Activation of NLRP3 Inflammasomes via the ERK1/2 Pathway in THP-1 Macrophages and Aggravates Atherosclerosis in ApoE−/− Mice. *Ann. Palliat. Med.* **2019**, *8*, 676–689. [CrossRef] [PubMed]

77. Gage, J.; Hasu, M.; Thabet, M.; Whitman, S.C. Caspase-1 Deficiency Decreases Atherosclerosis in Apolipoprotein E-Null Mice. *Can. J. Cardiol.* **2012**, *28*, 222–229. [CrossRef]

78. Wu, X.; Zhang, H.; Qi, W.; Zhang, Y.; Li, J.; Li, Z.; Lin, Y.; Bai, X.; Liu, X.; Chen, X.; et al. Nicotine Promotes Atherosclerosis via ROS-NLRP3-Mediated Endothelial Cell Pyroptosis. *Cell Death Dis.* **2018**, *9*, 171. [CrossRef]

79. Zhang, Y.; Liu, X.; Bai, X.; Lin, Y.; Li, Z.; Fu, J.; Li, M.; Zhao, T.; Yang, H.; Xu, R.; et al. Melatonin Prevents Endothelial Cell Pyroptosis via Regulation of Long Noncoding RNA MEG3/MiR-223/NLRP3 Axis. *J. Pineal Res.* **2018**, *64*, e12449. [CrossRef]

80. Li, P.; Zhong, X.; Li, J.; Liu, H.; Ma, X.; He, R.; Zhao, Y. MicroRNA-30c-5p Inhibits NLRP3 Inflammasome-Mediated Endothelial Cell Pyroptosis through FOXO3 down-Regulation in Atherosclerosis. *Biochem. Biophys. Res. Commun.* **2018**, *503*, 2833–2840. [CrossRef]

81. Man, S.M.; Zhu, Q.; Zhu, L.; Liu, Z.; Karki, R.; Malik, A.; Sharma, D.; Li, L.; Malireddi, R.K.S.; Gurung, P.; et al. Critical Role for the DNA Sensor AIM2 in Stem Cell Proliferation and Cancer. *Cell* **2015**, *162*, 45–58. [CrossRef] [PubMed]

82. Virani, S.S.; Alonso, A.; Benjamin, E.J.; Bittencourt, M.S.; Callaway, C.W.; Carson, A.P.; Chamberlain, A.M.; Chang, A.R.; Cheng, S.; Delling, F.N.; et al. Heart Disease and Stroke Statistics—2020 Update: A Report From the American Heart Association. *Circulation* **2020**, *141*, e139–e596. [CrossRef] [PubMed]

83. Kotlyarov, S.; Kotlyarova, A. Molecular Pharmacology of Inflammation Resolution in Atherosclerosis. *IJMS* **2022**, *23*, 4808. [CrossRef]

84. Xu, S.; Ilyas, I.; Little, P.J.; Li, H.; Kamato, D.; Zheng, X.; Luo, S.; Li, Z.; Liu, P.; Han, J.; et al. Endothelial Dysfunction in Atherosclerotic Cardiovascular Diseases and Beyond: From Mechanism to Pharmacotherapies. *Pharm. Rev.* **2021**, *73*, 924–967. [CrossRef] [PubMed]

85. Jebari-Benslaiman, S.; Galicia-García, U.; Larrea-Sebal, A.; Olaetxea, J.R.; Alloza, I.; Vandenbroeck, K.; Benito-Vicente, A.; Martín, C. Pathophysiology of Atherosclerosis. *IJMS* **2022**, *23*, 3346. [CrossRef] [PubMed]

86. Charlton, A.; Garzarella, J.; Jandeleit-Dahm, K.A.M.; Jha, J.C. Oxidative Stress and Inflammation in Renal and Cardiovascular Complications of Diabetes. *Biology* **2021**, *10*, 18. [CrossRef]

87. Haas, A.V.; McDonnell, M.E. Pathogenesis of Cardiovascular Disease in Diabetes. *Endocrinol. Metab. Clin. North. Am.* **2018**, *47*, 51–63. [CrossRef] [PubMed]

88. Higashi, Y. Roles of Oxidative Stress and Inflammation in Vascular Endothelial Dysfunction-Related Disease. *Antioxidants* **2022**, *11*, 1958. [CrossRef]

89. Li, J.; Fu, X.; Yang, R.; Zhang, W. Atherosclerosis Vascular Endothelial Secretion Dysfunction and Smooth Muscle Cell Proliferation. *J. Healthc. Eng.* **2022**, *2022*, 9271879. [CrossRef]

90. Petrie, J.R.; Guzik, T.J.; Touyz, R.M. Diabetes, Hypertension, and Cardiovascular Disease: Clinical Insights and Vascular Mechanisms. *Can. J. Cardiol.* **2018**, *34*, 575–584. [CrossRef]

91. Souilhol, C.; Serbanovic-Canic, J.; Fragiadaki, M.; Chico, T.J.; Ridger, V.; Roddie, H.; Evans, P.C. Endothelial Responses to Shear Stress in Atherosclerosis: A Novel Role for Developmental Genes. *Nat. Rev. Cardiol.* **2020**, *17*, 52–63. [CrossRef]

92. Campinho, P.; Vilfan, A.; Vermot, J. Blood Flow Forces in Shaping the Vascular System: A Focus on Endothelial Cell Behavior. *Front. Physiol.* **2020**, *11*, 552. [CrossRef]

93. Shaito, A.; Aramouni, K.; Assaf, R.; Parenti, A.; Orekhov, A.; Yazbi, A.E.; Pintus, G.; Eid, A.H. Oxidative Stress-Induced Endothelial Dysfunction in Cardiovascular Diseases. *Front. Biosci. (Landmark Ed)* **2022**, *27*, 0105. [CrossRef] [PubMed]

94. Katakami, N. Mechanism of Development of Atherosclerosis and Cardiovascular Disease in Diabetes Mellitus. *J. Atheroscler. Thromb.* **2018**, *25*, 27–39. [CrossRef] [PubMed]

95. Janjusevic, M.; Fluca, A.L.; Gagno, G.; Pierri, A.; Padoan, L.; Sorrentino, A.; Beltrami, A.P.; Sinagra, G.; Aleksova, A. Old and Novel Therapeutic Approaches in the Management of Hyperglycemia, an Important Risk Factor for Atherosclerosis. *IJMS* **2022**, *23*, 2336. [CrossRef]

96. Poznyak, A.V.; Grechko, A.V.; Orekhova, V.A.; Chegodaev, Y.S.; Wu, W.-K.; Orekhov, A.N. Oxidative Stress and Antioxidants in Atherosclerosis Development and Treatment. *Biology* **2020**, *9*, 60. [CrossRef] [PubMed]

97. Yang, X.; Li, Y.; Li, Y.; Ren, X.; Zhang, X.; Hu, D.; Gao, Y.; Xing, Y.; Shang, H. Oxidative Stress-Mediated Atherosclerosis: Mechanisms and Therapies. *Front. Physiol.* **2017**, *8*, 600. [CrossRef]

98. Beverly, J.K.; Budoff, M.J. Atherosclerosis: Pathophysiology of Insulin Resistance, Hyperglycemia, Hyperlipidemia, and Inflammation. *J. Diabetes* **2020**, *12*, 102–104. [CrossRef]

99. Poznyak, A.V.; Nikiforov, N.G.; Markin, A.M.; Kashirskikh, D.A.; Myasoedova, V.A.; Gerasimova, E.V.; Orekhov, A.N. Overview of OxLDL and Its Impact on Cardiovascular Health: Focus on Atherosclerosis. *Front. Pharmacol.* **2021**, *11*, 613780. [CrossRef]

100. Wang, D.; Yang, Y.; Lei, Y.; Tzvetkov, N.T.; Liu, X.; Yeung, A.W.K.; Xu, S.; Atanasov, A.G. Targeting Foam Cell Formation in Atherosclerosis: Therapeutic Potential of Natural Products. *Pharm. Rev.* **2019**, *71*, 596–670. [CrossRef]

101. Jiang, H.; Zhou, Y.; Nabavi, S.M.; Sahebkar, A.; Little, P.J.; Xu, S.; Weng, J.; Ge, J. Mechanisms of Oxidized LDL-Mediated Endothelial Dysfunction and Its Consequences for the Development of Atherosclerosis. *Front. Cardiovasc. Med.* **2022**, *9*, 925923. [CrossRef]

102. Huang, Z.; Wu, M.; Zeng, L.; Wang, D. The Beneficial Role of Nrf2 in the Endothelial Dysfunction of Atherosclerosis. *Cardiol. Res. Pract.* **2022**, *2022*, 4287711. [CrossRef] [PubMed]

103. Feng, Z.; Wang, C.; Jin, Y.; Meng, Q.; Wu, J.; Sun, H. Kaempferol-Induced GPER Upregulation Attenuates Atherosclerosis via the PI3K/AKT/Nrf2 Pathway. *Pharm. Biol.* **2021**, *59*, 1104–1114. [CrossRef] [PubMed]

104. Samarghandian, S.; Azimi-Nezhad, M.; Farkhondeh, T. Immunomodulatory and Antioxidant Effects of Saffron Aqueous Extract (Crocus Sativus L.) on Streptozotocin-Induced Diabetes in Rats. *Indian Heart J.* **2017**, *69*, 151–159. [CrossRef] [PubMed]

105. Shah, P.K.; Lecis, D. Inflammation in Atherosclerotic Cardiovascular Disease. *F1000Res* **2019**, *8*, 1402. [CrossRef]

106. Drareni, K.; Gautier, J.-F.; Venteclef, N.; Alzaid, F. Transcriptional Control of Macrophage Polarisation in Type 2 Diabetes. *Semin. Immunopathol.* **2019**, *41*, 515–529. [CrossRef] [PubMed]

107. Lin, P.; Ji, H.-H.; Li, Y.-J.; Guo, S.-D. Macrophage Plasticity and Atherosclerosis Therapy. *Front. Mol. Biosci.* **2021**, *8*, 679797. [CrossRef] [PubMed]

108. Niyonzima, N.; Bakke, S.S.; Gregersen, I.; Holm, S.; Sandanger, Ø.; Orrem, H.L.; Sporsheim, B.; Ryan, L.; Kong, X.Y.; Dahl, T.B.; et al. Cholesterol Crystals Use Complement to Increase NLRP3 Signaling Pathways in Coronary and Carotid Atherosclerosis. *EBioMedicine* **2020**, *60*, 102985. [CrossRef]

109. Qi, H.-M.; Cao, Q.; Liu, Q. TLR4 Regulates Vascular Smooth Muscle Cell Proliferation in Hypertension. *Am. J. Transl. Res.* **2021**, *13*, 314–325.

110. Poznyak, A.V.; Melnichenko, A.A.; Wetzker, R.; Gerasimova, E.V.; Orekhov, A.N. NLPR3 Inflammasomes and Their Significance for Atherosclerosis. *Biomedicines* **2020**, *8*, 205. [CrossRef] [PubMed]

111. Baragetti, A.; Catapano, A.L.; Magni, P. Multifactorial Activation of NLRP3 Inflammasome: Relevance for a Precision Approach to Atherosclerotic Cardiovascular Risk and Disease. *IJMS* **2020**, *21*, 4459. [CrossRef]

112. Hill, J.R.; Coll, R.C.; Sue, N.; Reid, J.C.; Dou, J.; Holley, C.L.; Pelingon, R.; Dickinson, J.B.; Biden, T.J.; Schroder, K.; et al. Sulfonylureas as Concomitant Insulin Secretagogues and NLRP3 Inflammasome Inhibitors. *ChemMedChem* **2017**, *12*, 1449–1457. [CrossRef]

113. Ngo, L.; Oh, J.; Kim, A.; Back, H.; Kang, W.; Chae, J.; Yun, H.; Lee, H. Development of a Pharmacokinetic Model Describing Neonatal Fc Receptor-Mediated Recycling of HL2351, a Novel Hybrid Fc-Fused Interleukin-1 Receptor Antagonist, to Optimize Dosage Regimen. *CPT Pharmacomet. Syst. Pharmacol.* **2020**, *9*, 584–595. [CrossRef]

114. Yang, F.; Qin, Y.; Wang, Y.; Meng, S.; Xian, H.; Che, H.; Lv, J.; Li, Y.; Yu, Y.; Bai, Y.; et al. Metformin Inhibits the NLRP3 Inflammasome via AMPK/MTOR-Dependent Effects in Diabetic Cardiomyopathy. *Int. J. Biol. Sci.* **2019**, *15*, 1010–1019. [CrossRef] [PubMed]

115. Xian, H.; Liu, Y.; Rundberg Nilsson, A.; Gatchalian, R.; Crother, T.R.; Tourtellotte, W.G.; Zhang, Y.; Aleman-Muench, G.R.; Lewis, G.; Chen, W.; et al. Metformin Inhibition of Mitochondrial ATP and DNA Synthesis Abrogates NLRP3 Inflammasome Activation and Pulmonary Inflammation. *Immunity* **2021**, *54*, 1463–1477.e11. [CrossRef] [PubMed]

116. Feng, X.; Chen, W.; Ni, X.; Little, P.J.; Xu, S.; Tang, L.; Weng, J. Metformin, Macrophage Dysfunction and Atherosclerosis. *Front. Immunol.* **2021**, *12*, 682853. [CrossRef]

117. Zhang, L.; Lu, L.; Zhong, X.; Yue, Y.; Hong, Y.; Li, Y.; Li, Y. Metformin Reduced NLRP3 Inflammasome Activity in Ox-LDL Stimulated Macrophages through Adenosine Monophosphate Activated Protein Kinase and Protein Phosphatase 2A. *Eur. J. Pharmacol.* **2019**, *852*, 99–106. [CrossRef]

118. Tang, G.; Duan, F.; Li, W.; Wang, Y.; Zeng, C.; Hu, J.; Li, H.; Zhang, X.; Chen, Y.; Tan, H. Metformin Inhibited Nod-like Receptor Protein 3 Inflammasomes Activation and Suppressed Diabetes-Accelerated Atherosclerosis in ApoE−/− Mice. *Biomed. Pharmacother.* **2019**, *119*, 109410. [CrossRef] [PubMed]

119. Zinman, B.; Wanner, C.; Lachin, J.M.; Fitchett, D.; Bluhmki, E.; Hantel, S.; Mattheus, M.; Devins, T.; Johansen, O.E.; Woerle, H.J.; et al. Empagliflozin, Cardiovascular Outcomes, and Mortality in Type 2 Diabetes. *N. Engl. J. Med.* **2015**, *373*, 2117–2128. [CrossRef]

120. Kim, S.R.; Lee, S.-G.; Kim, S.H.; Kim, J.H.; Choi, E.; Cho, W.; Rim, J.H.; Hwang, I.; Lee, C.J.; Lee, M.; et al. SGLT2 Inhibition Modulates NLRP3 Inflammasome Activity via Ketones and Insulin in Diabetes with Cardiovascular Disease. *Nat. Commun.* **2020**, *11*, 2127. [CrossRef]

121. Chen, H.; Tran, D.; Yang, H.-C.; Nylander, S.; Birnbaum, Y.; Ye, Y. Dapagliflozin and Ticagrelor Have Additive Effects on the Attenuation of the Activation of the NLRP3 Inflammasome and the Progression of Diabetic Cardiomyopathy: An AMPK–MTOR Interplay. *Cardiovasc. Drugs* **2020**, *34*, 443–461. [CrossRef]

122. Liu, Z.; Ma, X.; Ilyas, I.; Zheng, X.; Luo, S.; Little, P.J.; Kamato, D.; Sahebkar, A.; Wu, W.; Weng, J.; et al. Impact of Sodium Glucose Cotransporter 2 (SGLT2) Inhibitors on Atherosclerosis: From Pharmacology to Pre-Clinical and Clinical Therapeutics. *Theranostics* **2021**, *11*, 4502–4515. [CrossRef] [PubMed]

123. Razavi, M.; Wei, Y.-Y.; Rao, X.-Q.; Zhong, J.-X. DPP-4 Inhibitors and GLP-1RAs: Cardiovascular Safety and Benefits. *Mil. Med. Res.* **2022**, *9*, 45. [CrossRef]

124. Ye, Y.; Bajaj, M.; Yang, H.-C.; Perez-Polo, J.R.; Birnbaum, Y. SGLT-2 Inhibition with Dapagliflozin Reduces the Activation of the Nlrp3/ASC Inflammasome and Attenuates the Development of Diabetic Cardiomyopathy in Mice with Type 2 Diabetes. Further Augmentation of the Effects with Saxagliptin, a DPP4 Inhibitor. *Cardiovasc. Drugs* **2017**, *31*, 119–132. [CrossRef] [PubMed]

125. Qi, Y.; Du, X.; Yao, X.; Zhao, Y. Vildagliptin Inhibits High Free Fatty Acid (FFA)-Induced NLRP3 Inflammasome Activation in Endothelial Cells. *Artif. Cell. Nanomed. Biotechnol.* **2019**, *47*, 1067–1074. [CrossRef] [PubMed]

126. Jiang, T.; Jiang, D.; Zhang, L.; Ding, M.; Zhou, H. Anagliptin Ameliorates High Glucose- Induced Endothelial Dysfunction via Suppression of NLRP3 Inflammasome Activation Mediated by SIRT1. *Mol. Immunol.* **2019**, *107*, 54–60. [CrossRef]

127. Luo, X.; Hu, Y.; He, S.; Ye, Q.; Lv, Z.; Liu, J.; Chen, X. Dulaglutide Inhibits High Glucose- Induced Endothelial Dysfunction and NLRP3 Inflammasome Activation. *Arch. Biochem. Biophys.* **2019**, *671*, 203–209. [CrossRef]

128. Yang, Y.; Wang, H.; Kouadir, M.; Song, H.; Shi, F. Recent Advances in the Mechanisms of NLRP3 Inflammasome Activation and Its Inhibitors. *Cell Death Dis.* **2019**, *10*, 128. [CrossRef]

129. Coll, R.C.; Hill, J.R.; Day, C.J.; Zamoshnikova, A.; Boucher, D.; Massey, N.L.; Chitty, J.L.; Fraser, J.A.; Jennings, M.P.; Robertson, A.A.B.; et al. MCC950 Directly Targets the NLRP3 ATP-Hydrolysis Motif for Inflammasome Inhibition. *Nat. Chem. Biol.* **2019**, *15*, 556–559. [CrossRef]

130. Corcoran, S.E.; Halai, R.; Cooper, M.A. Pharmacological Inhibition of the Nod-Like Receptor Family Pyrin Domain Containing 3 Inflammasome with MCC950. *Pharm. Rev.* **2021**, *73*, 968–1000. [CrossRef]

131. Mekni, N.; De Rosa, M.; Cipollina, C.; Gulotta, M.R.; De Simone, G.; Lombino, J.; Padova, A.; Perricone, U. In Silico Insights towards the Identification of NLRP3 Druggable Hot Spots. *IJMS* **2019**, *20*, 4974. [CrossRef]

132. Sharma, A.; Choi, J.S.Y.; Stefanovic, N.; Al-Sharea, A.; Simpson, D.S.; Mukhamedova, N.; Jandeleit-Dahm, K.; Murphy, A.J.; Sviridov, D.; Vince, J.E.; et al. Specific NLRP3 Inhibition Protects Against Diabetes-Associated Atherosclerosis. *Diabetes* **2021**, *70*, 772–787. [CrossRef] [PubMed]

133. Wang, Z.; Zhang, S.; Xiao, Y.; Zhang, W.; Wu, S.; Qin, T.; Yue, Y.; Qian, W.; Li, L. NLRP3 Inflammasome and Inflammatory Diseases. *Oxidative Med. Cell. Longev.* **2020**, *2020*, 4063562. [CrossRef]

134. Fan, Y.; Xue, G.; Chen, Q.; Lu, Y.; Dong, R.; Yuan, H. CY-09 Inhibits NLRP3 Inflammasome Activation to Relieve Pain via TRPA1. *Comput. Math. Methods Med.* **2021**, *2021*, 9806690. [CrossRef] [PubMed]

135. Yu, L.; Hong, W.; Lu, S.; Li, Y.; Guan, Y.; Weng, X.; Feng, Z. The NLRP3 Inflammasome in Non-Alcoholic Fatty Liver Disease and Steatohepatitis: Therapeutic Targets and Treatment. *Front. Pharmacol.* **2022**, *13*, 780496. [CrossRef]

136. Lonnemann, N.; Hosseini, S.; Marchetti, C.; Skouras, D.B.; Stefanoni, D.; D'Alessandro, A.; Dinarello, C.A.; Korte, M. The NLRP3 Inflammasome Inhibitor OLT1177 Rescues Cognitive Impairment in a Mouse Model of Alzheimer's Disease. *Proc. Natl. Acad. Sci. USA* **2020**, *117*, 32145–32154. [CrossRef] [PubMed]

137. Marchetti, C.; Swartzwelter, B.; Gamboni, F.; Neff, C.P.; Richter, K.; Azam, T.; Carta, S.; Tengesdal, I.; Nemkov, T.; D'Alessandro, A.; et al. OLT1177, a β-Sulfonyl Nitrile Compound, Safe in Humans, Inhibits the NLRP3 Inflammasome and Reverses the Metabolic Cost of Inflammation. *Proc. Natl. Acad. Sci. USA* **2018**, *115*, E1530–E1539. [CrossRef]

138. Huang, Y.; Jiang, H.; Chen, Y.; Wang, X.; Yang, Y.; Tao, J.; Deng, X.; Liang, G.; Zhang, H.; Jiang, W.; et al. Tranilast Directly Targets NLRP 3 to Treat Inflammasome-driven Diseases. *EMBO Mol. Med.* **2018**, *10*, e8689. [CrossRef]

139. Cummins, C.B.; Wang, X.; Sommerhalder, C.; Bohanon, F.J.; Nunez Lopez, O.; Tie, H.-Y.; Rontoyanni, V.G.; Zhou, J.; Radhakrishnan, R.S. Natural Compound Oridonin Inhibits Endotoxin-Induced Inflammatory Response of Activated Hepatic Stellate Cells. *BioMed Res. Int.* **2018**, *2018*, 6137420. [CrossRef]

140. Xu, J.; Wold, E.; Ding, Y.; Shen, Q.; Zhou, J. Therapeutic Potential of Oridonin and Its Analogs: From Anticancer and Antiinflammation to Neuroprotection. *Molecules* **2018**, *23*, 474. [CrossRef]

141. Xu, M.; Wan, C.; Huang, S.; Wang, H.; Fan, D.; Wu, H.-M.; Wu, Q.; Ma, Z.; Deng, W.; Tang, Q.-Z. Oridonin Protects against Cardiac Hypertrophy by Promoting P21-Related Autophagy. *Cell Death Dis.* **2019**, *10*, 403. [CrossRef]

142. He, H.; Jiang, H.; Chen, Y.; Ye, J.; Wang, A.; Wang, C.; Liu, Q.; Liang, G.; Deng, X.; Jiang, W.; et al. Oridonin Is a Covalent NLRP3 Inhibitor with Strong Anti-Inflammasome Activity. *Nat. Commun.* **2018**, *9*, 2550. [CrossRef] [PubMed]

143. Li, X.; Zhang, C.-T.; Ma, W.; Xie, X.; Huang, Q. Oridonin: A Review of Its Pharmacology, Pharmacokinetics and Toxicity. *Front. Pharmacol.* **2021**, *12*, 645824. [CrossRef]

144. Zheng, J.; Jiang, Z.; Song, Y.; Huang, S.; Du, Y.; Yang, X.; Xiao, Y.; Ma, Z.; Xu, D.; Li, J. 3,4-Methylenedioxy-β-Nitrostyrene Alleviates Dextran Sulfate Sodium–Induced Mouse Colitis by Inhibiting the NLRP3 Inflammasome. *Front. Pharmacol.* **2022**, *13*, 866228. [CrossRef] [PubMed]

145. Cocco, M.; Pellegrini, C.; Martínez-Banaclocha, H.; Giorgis, M.; Marini, E.; Costale, A.; Miglio, G.; Fornai, M.; Antonioli, L.; López-Castejón, G.; et al. Development of an Acrylate Derivative Targeting the NLRP3 Inflammasome for the Treatment of Inflammatory Bowel Disease. *J. Med. Chem.* **2017**, *60*, 3656–3671. [CrossRef] [PubMed]

146. Mastrocola, R.; Penna, C.; Tullio, F.; Femminò, S.; Nigro, D.; Chiazza, F.; Serpe, L.; Collotta, D.; Alloatti, G.; Cocco, M.; et al. Pharmacological Inhibition of NLRP3 Inflammasome Attenuates Myocardial Ischemia/Reperfusion Injury by Activation of RISK and Mitochondrial Pathways. *Oxidative Med. Cell. Longev.* **2016**, *2016*, 5271251. [CrossRef]

147. Shi, Y.; Lv, Q.; Zheng, M.; Sun, H.; Shi, F. NLRP3 Inflammasome Inhibitor INF39 Attenuated NLRP3 Assembly in Macrophages. *Int. Immunopharmacol.* **2021**, *92*, 107358. [CrossRef]

148. Juliana, C.; Fernandes-Alnemri, T.; Wu, J.; Datta, P.; Solorzano, L.; Yu, J.-W.; Meng, R.; Quong, A.A.; Latz, E.; Scott, C.P.; et al. Anti-Inflammatory Compounds Parthenolide and Bay 11-7082 Are Direct Inhibitors of the Inflammasome. *J. Biol. Chem.* **2010**, *285*, 9792–9802. [CrossRef]

149. Hu, J.J.; Liu, X.; Zhao, J.; Xia, S.; Ruan, J.; Luo, X.; Kim, J.; Lieberman, J.; Wu, H. Identification of Pyroptosis Inhibitors That Target a Reactive Cysteine in Gasdermin D. *Nat. Immunol.* **2018**, *7*, A132.

150. Zahid, A.; Li, B.; Kombe, A.J.K.; Jin, T.; Tao, J. Pharmacological Inhibitors of the NLRP3 Inflammasome. *Front. Immunol.* **2019**, *10*, 2538. [CrossRef]

151. Li, Y.; Niu, X.; Xu, H.; Li, Q.; Meng, L.; He, M.; Zhang, J.; Zhang, Z.; Zhang, Z. VX-765 Attenuates Atherosclerosis in ApoE Deficient Mice by Modulating VSMCs Pyroptosis. *Exp. Cell Res.* **2020**, *389*, 111847. [CrossRef]

152. Abbate, A.; Kontos, M.C.; Grizzard, J.D.; Biondi-Zoccai, G.G.L.; Van Tassell, B.W.; Robati, R.; Roach, L.M.; Arena, R.A.; Roberts, C.S.; Varma, A.; et al. Interleukin-1 Blockade With Anakinra to Prevent Adverse Cardiac Remodeling After Acute Myocardial Infarction (Virginia Commonwealth University Anakinra Remodeling Trial [VCU-ART] Pilot Study). *Am. J. Cardiol.* **2010**, *105*, 1371–1377.e1. [CrossRef] [PubMed]

153. Abbate, A.; Van Tassell, B.W.; Biondi-Zoccai, G.; Kontos, M.C.; Grizzard, J.D.; Spillman, D.W.; Oddi, C.; Roberts, C.S.; Melchior, R.D.; Mueller, G.H.; et al. Effects of Interleukin-1 Blockade With Anakinra on Adverse Cardiac Remodeling and Heart Failure After Acute Myocardial Infarction [from the Virginia Commonwealth University-Anakinra Remodeling Trial (2) (VCU-ART2) Pilot Study]. *Am. J. Cardiol.* **2013**, *111*, 1394–1400. [CrossRef] [PubMed]

154. Toldo, S.; Abbate, A. The NLRP3 Inflammasome in Acute Myocardial Infarction. *Nat. Rev. Cardiol.* **2018**, *15*, 203–214. [CrossRef] [PubMed]

155. Tardif, J.-C.; Kouz, S.; Waters, D.D.; Bertrand, O.F.; Diaz, R.; Maggioni, A.P.; Pinto, F.J.; Ibrahim, R.; Gamra, H.; Kiwan, G.S.; et al. Efficacy and Safety of Low-Dose Colchicine after Myocardial Infarction. *N. Engl. J. Med.* **2019**, *381*, 2497–2505. [CrossRef]

156. Nidorf, S.M.; Fiolet, A.T.L.; Eikelboom, J.W.; Schut, A.; Opstal, T.S.J.; Bax, W.A.; Budgeon, C.A.; Tijssen, J.G.P.; Mosterd, A.; Cornel, J.H.; et al. The Effect of Low-Dose Colchicine in Patients with Stable Coronary Artery Disease: The LoDoCo2 Trial Rationale, Design, and Baseline Characteristics. *Am. Heart J.* **2019**, *218*, 46–56. [CrossRef]

157. Nidorf, S.M.; Eikelboom, J.W.; Budgeon, C.A.; Thompson, P.L. Low-Dose Colchicine for Secondary Prevention of Cardiovascular Disease. *J. Am. Coll. Cardiol.* **2013**, *61*, 404–410. [CrossRef]

158. Lee, H.G.; Cho, N.; Jeong, A.J.; Li, Y.-C.; Rhie, S.-J.; Choi, J.S.; Lee, K.-H.; Kim, Y.; Kim, Y.-N.; Kim, M.-H.; et al. Immunomodulatory Activities of the Benzoxathiole Derivative BOT-4-One Ameliorate Pathogenic Skin Inflammation in Mice. *J. Investig. Dermatol.* **2016**, *136*, 107–116. [CrossRef]

159. Ahn, H.; Kang, S.G.; Yoon, S.; Ko, H.-J.; Kim, P.-H.; Hong, E.-J.; An, B.-S.; Lee, E.; Lee, G.-S. Methylene Blue Inhibits NLRP3, NLRC4, AIM2, and Non-Canonical Inflammasome Activation. *Sci. Rep.* **2017**, *7*, 12409. [CrossRef]

160. Hao, J.; Zhang, H.; Yu, J.; Chen, X.; Yang, L. Methylene Blue Attenuates Diabetic Retinopathy by Inhibiting NLRP3 Inflammasome Activation in STZ-Induced Diabetic Rats. *Ocul. Immunol. Inflamm.* **2019**, *27*, 836–843. [CrossRef]

161. Deng, W.; Yang, Z.; Yue, H.; Ou, Y.; Hu, W.; Sun, P. Disulfiram Suppresses NLRP3 Inflammasome Activation to Treat Peritoneal and Gouty Inflammation. *Free Radic. Biol. Med.* **2020**, *152*, 8–17. [CrossRef]

162. Hu, J.J.; Liu, X.; Xia, S.; Zhang, Z.; Zhang, Y.; Zhao, J.; Ruan, J.; Luo, X.; Lou, X.; Bai, Y.; et al. FDA-Approved Disulfiram Inhibits Pyroptosis by Blocking Gasdermin D Pore Formation. *Nat. Immunol.* **2020**, *21*, 736–745. [CrossRef]

163. Daniels, M.J.D.; Rivers-Auty, J.; Schilling, T.; Spencer, N.G.; Watremez, W.; Fasolino, V.; Booth, S.J.; White, C.S.; Baldwin, A.G.; Freeman, S.; et al. Fenamate NSAIDs Inhibit the NLRP3 Inflammasome and Protect against Alzheimer's Disease in Rodent Models. *Nat. Commun.* **2016**, *7*, 12504. [CrossRef] [PubMed]

164. Ambati, M.; Apicella, I.; Wang, S.; Narendran, S.; Leung, H.; Pereira, F.; Nagasaka, Y.; Huang, P.; Varshney, A.; Baker, K.L.; et al. Identification of Fluoxetine as a Direct NLRP3 Inhibitor to Treat Atrophic Macular Degeneration. *Proc. Natl. Acad. Sci. USA* **2021**, *118*, e2102975118. [CrossRef] [PubMed]

165. Holtmann, T.M.; Inzaugarat, M.E.; Knorr, J.; Geisler, L.; Schulz, M.; Bieghs, V.; Frissen, M.; Feldstein, A.E.; Tacke, F.; Trautwein, C.; et al. Bile Acids Activate NLRP3 Inflammasome, Promoting Murine Liver Inflammation or Fibrosis in a Cell Type-Specific Manner. *Cells* **2021**, *10*, 2618. [CrossRef] [PubMed]

166. Jiang, H.; He, H.; Chen, Y.; Huang, W.; Cheng, J.; Ye, J.; Wang, A.; Tao, J.; Wang, C.; Liu, Q.; et al. Identification of a Selective and Direct NLRP3 Inhibitor to Treat Inflammatory Disorders. *J. Exp. Med.* **2017**, *214*, 3219–3238. [CrossRef]

167. Romero, A.; San Hipólito-Luengo, Á.; Villalobos, L.A.; Vallejo, S.; Valencia, I.; Michalska, P.; Pajuelo-Lozano, N.; Sánchez-Pérez, I.; León, R.; Bartha, J.L.; et al. The Angiotensin-(1-7)/Mas Receptor Axis Protects from Endothelial Cell Senescence via Klotho and Nrf2 Activation. *Aging Cell* **2019**, *18*, e12913. [CrossRef]

168. McKie, E.A.; Reid, J.L.; Mistry, P.C.; DeWall, S.L.; Abberley, L.; Ambery, P.D.; Gil-Extremera, B. A Study to Investigate the Efficacy and Safety of an Anti-Interleukin-18 Monoclonal Antibody in the Treatment of Type 2 Diabetes Mellitus. *PLoS ONE* **2016**, *11*, e0150018. [CrossRef]

169. Klein, A.L.; Imazio, M.; Cremer, P.; Brucato, A.; Abbate, A.; Fang, F.; Insalaco, A.; LeWinter, M.; Lewis, B.S.; Lin, D.; et al. Phase 3 Trial of Interleukin-1 Trap Rilonacept in Recurrent Pericarditis. *N. Engl. J. Med.* **2021**, *384*, 31–41. [CrossRef]

170. Wu, L.-M.; Wu, S.-G.; Chen, F.; Wu, Q.; Wu, C.-M.; Kang, C.-M.; He, X.; Zhang, R.-Y.; Lu, Z.-F.; Li, X.-H.; et al. Atorvastatin Inhibits Pyroptosis through the LncRNA NEXN-AS1/NEXN Pathway in Human Vascular Endothelial Cells. *Atherosclerosis* **2020**, *293*, 26–34. [CrossRef]

171. Peng, S.; Xu, L.-W.; Che, X.-Y.; Xiao, Q.-Q.; Pu, J.; Shao, Q.; He, B. Atorvastatin Inhibits Inflammatory Response, Attenuates Lipid Deposition, and Improves the Stability of Vulnerable Atherosclerotic Plaques by Modulating Autophagy. *Front. Pharmacol.* **2018**, *9*, 438. [CrossRef]

172. Kong, F.; Ye, B.; Lin, L.; Cai, X.; Huang, W.; Huang, Z. Atorvastatin Suppresses NLRP3 Inflammasome Activation via TLR4/MyD88/NF-KB Signaling in PMA-Stimulated THP-1 Monocytes. *Biomed. Pharmacother.* **2016**, *82*, 167–172. [CrossRef]

173. Zhu, J.; Wu, S.; Hu, S.; Li, H.; Li, M.; Geng, X.; Wang, H. NLRP3 Inflammasome Expression in Peripheral Blood Monocytes of Coronary Heart Disease Patients and Its Modulation by Rosuvastatin. *Mol. Med. Rep.* **2019**, *20*, 1826–1836. [CrossRef] [PubMed]

174. Zhuang, T.; Liu, J.; Chen, X.; Zhang, L.; Pi, J.; Sun, H.; Li, L.; Bauer, R.; Wang, H.; Yu, Z.; et al. Endothelial Foxp1 Suppresses Atherosclerosis via Modulation of Nlrp3 Inflammasome Activation. *Circ. Res.* **2019**, *125*, 590–605. [CrossRef] [PubMed]

175. Yu, Y.; Jin, L.; Zhuang, Y.; Hu, Y.; Cang, J.; Guo, K. Cardioprotective Effect of Rosuvastatin against Isoproterenol-Induced Myocardial Infarction Injury in Rats. *Int. J. Mol. Med.* **2018**, *41*, 3509–3516. [CrossRef] [PubMed]

176. Lim, H.; Min, D.S.; Park, H.; Kim, H.P. Flavonoids Interfere with NLRP3 Inflammasome Activation. *Toxicol. Appl. Pharmacol.* **2018**, *355*, 93–102. [CrossRef] [PubMed]

177. Yamagata, K.; Hashiguchi, K.; Yamamoto, H.; Tagami, M. Dietary Apigenin Reduces Induction of LOX-1 and NLRP3 Expression, Leukocyte Adhesion, and Acetylated Low-Density Lipoprotein Uptake in Human Endothelial Cells Exposed to Trimethylamine-N-Oxide. *J. Cardiovasc. Pharmacol.* **2019**, *74*, 558–565. [CrossRef]

178. Salehi, B.; Venditti, A.; Sharifi-Rad, M.; Kręgiel, D.; Sharifi-Rad, J.; Durazzo, A.; Lucarini, M.; Santini, A.; Souto, E.; Novellino, E.; et al. The Therapeutic Potential of Apigenin. *IJMS* **2019**, *20*, 1305. [CrossRef]

179. Wang, K.; Lv, Q.; Miao, Y.; Qiao, S.; Dai, Y.; Wei, Z. Cardamonin, a Natural Flavone, Alleviates Inflammatory Bowel Disease by the Inhibition of NLRP3 Inflammasome Activation via an AhR/Nrf2/NQO1 Pathway. *Biochem. Pharmacol.* **2018**, *155*, 494–509. [CrossRef]

180. Liao, N.-C.; Shih, Y.-L.; Chou, J.-S.; Chen, K.-W.; Chen, Y.-L.; Lee, M.-H.; Peng, S.-F.; Leu, S.-J.; Chung, J.-G. Cardamonin Induces Cell Cycle Arrest, Apoptosis and Alters Apoptosis Associated Gene Expression in WEHI-3 Mouse Leukemia Cells. *Am. J. Chin. Med.* **2019**, *47*, 635–656. [CrossRef]

181. Honda, H.; Nagai, Y.; Matsunaga, T.; Okamoto, N.; Watanabe, Y.; Tsuneyama, K.; Hayashi, H.; Fujii, I.; Ikutani, M.; Hirai, Y.; et al. Isoliquiritigenin Is a Potent Inhibitor of NLRP3 Inflammasome Activation and Diet-induced Adipose Tissue Inflammation. *J. Leukoc. Biol.* **2014**, *96*, 1087–1100. [CrossRef]

182. Imran, M.; Rauf, A.; Abu-Izneid, T.; Nadeem, M.; Shariati, M.A.; Khan, I.A.; Imran, A.; Orhan, I.E.; Rizwan, M.; Atif, M.; et al. Luteolin, a Flavonoid, as an Anticancer Agent: A Review. *Biomed. Pharmacother.* **2019**, *112*, 108612. [CrossRef] [PubMed]

183. Zhang, B.-C.; Li, Z.; Xu, W.; Xiang, C.-H.; Ma, Y.-F. Luteolin Alleviates NLRP3 Inflammasome Activation and Directs Macrophage Polarization in Lipopolysaccharide-Stimulated RAW264.7 Cells. *Am. J. Transl. Res.* **2018**, *10*, 265–273. [PubMed]

184. Li, B.; Du, P.; Du, Y.; Zhao, D.; Cai, Y.; Yang, Q.; Guo, Z. Luteolin Alleviates Inflammation and Modulates Gut Microbiota in Ulcerative Colitis Rats. *Life Sci.* **2021**, *269*, 119008. [CrossRef] [PubMed]

185. Domiciano, T.P.; Wakita, D.; Jones, H.D.; Crother, T.R.; Verri, W.A.; Arditi, M.; Shimada, K. Quercetin Inhibits Inflammasome Activation by Interfering with ASC Oligomerization and Prevents Interleukin-1 Mediated Mouse Vasculitis. *Sci. Rep.* **2017**, *7*, 41539. [CrossRef] [PubMed]

186. Jiang, W.; Huang, Y.; Han, N.; He, F.; Li, M.; Bian, Z.; Liu, J.; Sun, T.; Zhu, L. Quercetin Suppresses NLRP3 Inflammasome Activation and Attenuates Histopathology in a Rat Model of Spinal Cord Injury. *Spinal Cord* **2016**, *54*, 592–596. [CrossRef]

187. Chen, P.; Bai, Q.; Wu, Y.; Zeng, Q.; Song, X.; Guo, Y.; Zhou, P.; Wang, Y.; Liao, X.; Wang, Q.; et al. The Essential Oil of Artemisia Argyi H.Lév. and Vaniot Attenuates NLRP3 Inflammasome Activation in THP-1 Cells. *Front. Pharmacol.* **2021**, *12*, 712907. [CrossRef]

188. Kwak, S.-B.; Koppula, S.; In, E.-J.; Sun, X.; Kim, Y.-K.; Kim, M.-K.; Lee, K.-H.; Kang, T.-B. *Artemisia* Extract Suppresses NLRP3 and AIM2 Inflammasome Activation by Inhibition of ASC Phosphorylation. *Mediat. Inflamm.* **2018**, *2018*, 6054069. [CrossRef]

189. Jiang, Y.; Du, H.; Liu, X.; Fu, X.; Li, X.; Cao, Q. Artemisinin Alleviates Atherosclerotic Lesion by Reducing Macrophage Inflammation via Regulation of AMPK/NF-KB/NLRP3 Inflammasomes Pathway. *J. Drug Target.* **2020**, *28*, 70–79. [CrossRef]

190. Lee, H.E.; Yang, G.; Kim, N.D.; Jeong, S.; Jung, Y.; Choi, J.Y.; Park, H.H.; Lee, J.Y. Targeting ASC in NLRP3 Inflammasome by Caffeic Acid Phenethyl Ester: A Novel Strategy to Treat Acute Gout. *Sci. Rep.* **2016**, *6*, 38622. [CrossRef]

191. Kong, F.; Ye, B.; Cao, J.; Cai, X.; Lin, L.; Huang, S.; Huang, W.; Huang, Z. Curcumin Represses NLRP3 Inflammasome Activation via TLR4/MyD88/NF-KB and P2X7R Signaling in PMA-Induced Macrophages. *Front. Pharmacol.* **2016**, *7*, 369. [CrossRef]

192. Rahban, M.; Habibi-Rezaei, M.; Mazaheri, M.; Saso, L.; Moosavi-Movahedi, A.A. Anti-Viral Potential and Modulation of Nrf2 by Curcumin: Pharmacological Implications. *Antioxidants* **2020**, *9*, 1228. [CrossRef]

193. Gao, J.; Peng, S.; Shan, X.; Deng, G.; Shen, L.; Sun, J.; Jiang, C.; Yang, X.; Chang, Z.; Sun, X.; et al. Inhibition of AIM2 Inflammasome-Mediated Pyroptosis by Andrographolide Contributes to Amelioration of Radiation-Induced Lung Inflammation and Fibrosis. *Cell Death Dis.* **2019**, *10*, 957. [CrossRef] [PubMed]

194. Liebman, S.E.; Le, T.H. Eat Your Broccoli: Oxidative Stress, NRF2, and Sulforaphane in Chronic Kidney Disease. *Nutrients* **2021**, *13*, 266. [CrossRef] [PubMed]

195. Wang, F.; Yao, X.; Zhang, Y.; Tang, J. Synthesis, Biological Function and Evaluation of Shikonin in Cancer Therapy. *Fitoterapia* **2019**, *134*, 329–339. [CrossRef] [PubMed]

196. Ma, Q.; Yang, Q.; Chen, J.; Yu, C.; Zhang, L.; Zhou, W.; Chen, M. Salvianolic Acid A Ameliorates Early-Stage Atherosclerosis Development by Inhibiting NLRP3 Inflammasome Activation in Zucker Diabetic Fatty Rats. *Molecules* **2020**, *25*, 1089. [CrossRef] [PubMed]

197. Chen, T.-C.; Yen, C.-K.; Lu, Y.-C.; Shi, C.-S.; Hsieh, R.-Z.; Chang, S.-F.; Chen, C.-N. The Antagonism of 6-Shogaol in High-Glucose-Activated NLRP3 Inflammasome and Consequent Calcification of Human Artery Smooth Muscle Cells. *Cell Biosci.* **2020**, *10*, 5. [CrossRef]

198. Lian, D.; Yuan, H.; Yin, X.; Wu, Y.; He, R.; Huang, Y.; Chen, Y. Puerarin Inhibits Hyperglycemia-Induced Inter-Endothelial Junction through Suppressing Endothelial Nlrp3 Inflammasome Activation via ROS-Dependent Oxidative Pathway. *Phytomedicine* **2019**, *55*, 310–319. [CrossRef]

199. Hu, R.; Wang, M.; Ni, S.; Wang, M.; Liu, L.; You, H.; Wu, X.; Wang, Y.; Lu, L.; Wei, L. Salidroside Ameliorates Endothelial Inflammation and Oxidative Stress by Regulating the AMPK/NF-KB/NLRP3 Signaling Pathway in AGEs-Induced HUVECs. *Eur. J. Pharmacol.* **2020**, *867*, 172797. [CrossRef]

200. Feng, H.; Zhu, X.; Tang, Y.; Fu, S.; Kong, B.; Liu, X. Astragaloside IV Ameliorates Diabetic Nephropathy in *Db/Db* Mice by Inhibiting NLRP3 Inflammasome-mediated Inflammation. *Int. J. Mol. Med.* **2021**, *48*, 164. [CrossRef]

201. Qian, W.; Cai, X.; Qian, Q.; Zhuang, Q.; Yang, W.; Zhang, X.; Zhao, L. Astragaloside IV Protects Endothelial Progenitor Cells from the Damage of Ox-LDL via the LOX-1/NLRP3 Inflammasome Pathway. *DDDT* **2019**, *13*, 2579–2589. [CrossRef]

202. Burger, F.; Baptista, D.; Roth, A.; Da Silva, R.F.; Montecucco, F.; Mach, F.; Brandt, K.J.; Miteva, K. NLRP3 Inflammasome Activation Controls Vascular Smooth Muscle Cells Phenotypic Switch in Atherosclerosis. *IJMS* **2021**, *23*, 340. [CrossRef] [PubMed]

203. Zeng, W.; Wu, D.; Sun, Y.; Suo, Y.; Yu, Q.; Zeng, M.; Gao, Q.; Yu, B.; Jiang, X.; Wang, Y. The Selective NLRP3 Inhibitor MCC950 Hinders Atherosclerosis Development by Attenuating Inflammation and Pyroptosis in Macrophages. *Sci. Rep.* **2021**, *11*, 19305. [CrossRef] [PubMed]

International Journal of
Molecular Sciences

Review

A Receptor Story: Insulin Resistance Pathophysiology and Physiologic Insulin Resensitization's Role as a Treatment Modality

Stanley T. Lewis [1], Frank Greenway [2], Tori R. Tucker [3], Michael Alexander [4], Levonika K. Jackson [5], Scott A. Hepford [5], Brian Loveridge [5] and Jonathan R. T. Lakey [4,6,*]

[1] Eselle Health, La Jolla, CA 92037, USA; stanleytlewis2@gmail.com
[2] Clinical Trials Unit, Pennington Biomedical Research Center, Louisiana State University, Baton Rouge, LA 77808, USA; frank.greenway@pbrc.edu
[3] Department of Developmental and Cell Biology, University of California Irvine, Irvine, CA 92617, USA; trtucker@uci.edu
[4] Department of Surgery, University of California Irvine, Orange, CA 92686, USA; michalex858@gmail.com
[5] Well Cell Global, Medical and Scientific Advisory Board, Houston, TX 77079, USA; levy@wellcellsupport.com (L.K.J.); scott@wellcellglobal.com (S.A.H.); brian@diabetesrelief.com (B.L.)
[6] Department of Biomedical Engineering, University of California Irvine, Irvine, CA 92868, USA
[*] Correspondence: jlakey@uci.edu

Citation: Lewis, S.T.; Greenway, F.; Tucker, T.R.; Alexander, M.; Jackson, L.K.; Hepford, S.A.; Loveridge, B.; Lakey, J.R.T. A Receptor Story: Insulin Resistance Pathophysiology and Physiologic Insulin Resensitization's Role as a Treatment Modality. *Int. J. Mol. Sci.* **2023**, *24*, 10927. https://doi.org/10.3390/ijms241310927

Academic Editors: Yutang Wang and Dianna Magliano

Received: 10 June 2023
Revised: 23 June 2023
Accepted: 25 June 2023
Published: 30 June 2023

Abstract: Physiologic insulin secretion consists of an oscillating pattern of secretion followed by distinct trough periods that stimulate ligand and receptor activation. Apart from the large postprandial bolus release of insulin, β cells also secrete small amounts of insulin every 4–8 min independent of a meal. Insulin resistance is associated with a disruption in the normal cyclical pattern of insulin secretion. In the case of type-2 diabetes, β-cell mass is reduced due to apoptosis and β cells secrete insulin asynchronously. When ligand/receptors are constantly exposed to insulin, a negative feedback loop down regulates insulin receptor availability to insulin, creating a relative hyperinsulinemia. The relative excess of insulin leads to insulin resistance (IR) due to decreased receptor availability. Over time, progressive insulin resistance compromises carbohydrate metabolism, and may progress to type-2 diabetes (T2D). In this review, we discuss insulin resistance pathophysiology and the use of dynamic exogenous insulin administration in a manner consistent with more normal insulin secretion periodicity to reverse insulin resistance. Administration of insulin in such a physiologic manner appears to improve insulin sensitivity, lower HgbA1c, and, in some instances, has been associated with the reversal of end-organ damage that leads to complications of diabetes. This review outlines the rationale for how the physiologic secretion of insulin orchestrates glucose metabolism, and how mimicking this secretion profile may serve to improve glycemic control, reduce cellular inflammation, and potentially improve outcomes in patients with diabetes.

Keywords: T2D; diabetes; carbohydrate metabolism; insulin resistance; physiologic insulin resensitization (PIR)

1. Introduction

Hormones that are released in an oscillating pattern (e.g., insulin) are physiologically designed to have receptors on the surface of effector cells that bind to their respective ligands. The cyclical release of these hormones plays a crucial role in maintaining homeostasis and regulating various physiological processes. In some cases, the ligand is brought into the cell where it separates from its receptor, and the receptor then migrates back to the cell's surface, ready to bind with another ligand molecule. In the case of insulin, the receptor is activated as a tyrosine protein kinase, initiating a cascade of intracellular signaling events. The receptor is internalized as it undergoes an auto-phosphorylation sequence [1]. The reconfiguration of the receptor to its active extracellular configuration takes approximately

4 min after the ligand receptor complex formation, allowing for the subsequent binding and activation of additional ligand molecules [2]. This intricate process of ligand–receptor interaction, internalization, and reconfiguration plays a vital role in the precise control and regulation of insulin signaling.

For nearly 40 years, scientific research has established that pancreatic β cells secrete insulin in a dynamic biphasic manner by responding to a square wave increase in blood glucose concentrations. This periodic release pattern allows for optimal cellular carbohydrate metabolism, striking a delicate balance between providing sufficient insulin for glucose utilization, while avoiding overexposure of the insulin receptor and the consequent reverse feedback loop [3,4]. Intriguingly, insulin is released every 4–8 min or more commonly, 5–6 min, independent of the ingestion of food. The initial phase of insulin secretion involves the rapid release of insulin by β cells, which lasts for about 10 min. This rapid release of insulin by β cells is facilitated by the storage of insulin in miniature membrane-bound secretory granules that are primed to fuse with plasma [5]. The physiological release of insulin is vital for maintaining glucose homeostasis, and disruptions in this process during the initial phase have been linked to the development of type-2 diabetes (T2D) [6]. Following the initial phase, a second phase ensues, characterized by a plateau in insulin release that can last for 2–3 h. During the physiological release of insulin, the discrete oscillatory secretions and distinct trough periods stimulate ligand and receptor activation. The ability of insulin release to exhibit such oscillatory behavior can be attributed to the unique arrangement of β cells within the islets of Langerhans, which are cell clusters residing in the pancreas. Within these clusters, β cells maintain close contact with one another, forming intricate networks of autonomic nerves that enable the coordination of dynamic, cyclical patterns of insulin secretion [7]. Thus, the periodic pattern of insulin secretion is most likely a result of intrinsic β-cell mechanisms that become modified by exogenous stimulus such as hormones and neuronal inputs.

2. Pancreatic Neuronal Network

Under normal physiological conditions, insulin oscillations are mediated by a complex pancreatic neuronal network that connects cells within the islets of Langerhans [8,9]. The brain and its neuronal network have a large influence on glucose homeostasis via the autonomic nerves that regulate the endocrine function of the pancreas. From tracing studies, neuronal networks have been mapped to pancreatic islets, which are innervated by neuronal circuits that come from the hypothalamus [10]. Within the pancreatic islets, both sympathetic and parasympathetic nerves of the autonomic nervous system can be found, indicating their involvement in modulating insulin release [11,12]. Furthermore, tracing studies also revealed that vagal afferent axons can be found within the pancreas [13]. Serotonin is a stimulating molecule for vagal afferent neurons and within the pancreas, β cells are known to produce serotonin to communicate with neighboring cells within the islet. It was recently thought that serotonin is a signaling molecule β cells use to communicate with the brain by vagal afferent neurons [10].

Within the neural network, an individual β cell is connected to other β cells, which mediates insulin secretion. β cells can be characterized as either "Leader cells" or "Follower cells" based on their functional characteristics [14]. Leader β cells possess pacemaker properties, allowing them to respond to glucose and synchronize communication patterns with follower cells. In response to glucose stimulation, leader β cells trigger a calcium (Ca^{2+}) influx [15], which diffuses to the follower β cells. The influx of intracellular Ca^{2+} levels leads to β-cell depolarization, initiating electrical activity and a subsequent wave of action potentials. The precise coordination of β cell-to-β cell communication, facilitated by the interplay between Ca^{2+} influx and action potentials, is closely associated with the oscillatory pattern of insulin secretion [16].

3. Pancreatic Inflammation

An insult or injury (i.e., autoimmunity, obesity, toxin, trauma, stress etc.) causes inflammation in the pancreas that disrupts normal pancreatic insulin rhythmicity [17]. Type-one diabetes (T1D) is an autoimmune condition that results in islet inflammation caused by the infiltration of immune cells, leading to the progressive destruction of β cells. Individuals first diagnosed with T1D usually require small doses of insulin that eventually increases as the body begins to produce less insulin as β-cell mass decreases. Upon initial diagnosis of T1D, patients experience the honeymoon phase as β cells try to recover or compensate for the decline in β-cell mass [18]. The honeymoon phase is characterized as a partial remission that occurs shortly after a T1D patient begins insulin therapy. During this period, the individual's diabetes may improve, and the patient may require less insulin than the first few days after diagnosis. The honeymoon phase can last approximately 3–12 months. Eventually, T1D patients will no longer experience normal physiological insulin secretion and will instead require lifelong exogenous insulin therapy via insulin injections or a continuous subcutaneous insulin pump.

In contrast, T2D continues to produce insulin, but in an uncoordinated pattern without a true peak and trough waveform due to the inflammation and breakdown in the pancreatic neuronal network. Lang et al. found that healthy humans have intermittent insulin secretions and trough periods where T2D subjects had lost this native physiologic architecture [4]. It is thought that the disruption of insulin secretion may be a result of inflammation in the pancreas due to conditions such as obesity, toxins, trauma, etc. In early studies, it was found that insulin oscillates for approximately 15 min in fasting humans that do not have diabetes [19–21]. With an improvement in technology to measure insulin pulses, more recent studies have determined that the in vivo oscillation of insulin during fasting is closer to 5 min. For those who have T2D, Lang et al. revealed that insulin oscillations were shorter and highly irregular with a mean oscillation period of 8.8 min compared to control subjects who had a period of 10.7 min. Hunter et al. showed that control subjects had a higher glucose clearance and insulin sensitivity than those with T2D [22]. T2D subjects had greater insulin resistance and hepatic glucose output. In another study by Peiris et al., the number of insulin pulsations observed were correlated with a decrease in glucose clearance in diabetic subjects. These results supported the idea that abnormal insulin secretion and impaired insulin action are linked. This idea was driven by the hypothesis that a decline in insulin sensitivity is driven by insulin resistance (IR) [23].

4. Impaired Physiologic Pattern of Insulin

As β cells cease to secrete insulin in the periodic synchronous pattern, insulin receptors undergo downregulation or internalization when they are continuously exposed to insulin. This continuous exposure disrupts the normal dynamics of insulin signaling and leads to a decrease in receptor responsiveness. A study demonstrated that subjecting healthy individuals to 20 h of constant insulin exposure at a steady glucose level resulted in a reduction in insulin action [24]. This prolonged exposure to insulin led to a decrease in insulin sensitivity and compromised insulin response [25].

It is important to recognize that hyperinsulinemia, or elevated levels of insulin, is relative to the normal insulin levels observed during intervals between periodic secretions (i.e., troughs). During routine fasted assessments, the absolute insulin level may fall within the broad normal range, making it challenging to detect impaired insulin function. This is why insulin impairment can often go undetected for an extended period, as conventional measurements may not capture the underlying abnormalities in insulin signaling and receptor dynamics.

Understanding the intricate relationship between insulin secretion, receptor responsiveness, and the consequences of continuous insulin exposure is critical in elucidating the mechanisms underlying IR and related metabolic disorders.

5. Impacts of Hyperinsulinemia

Hyperinsulinemia occurs when the fasting concentration of insulin in the blood remains higher than normal levels for prolonged periods. Various factors can contribute to hyperinsulinemia, including incretin hormones, obesity, diet, visceral fat accumulation, as well as genetic and environmental factors (Figure 1) [26]. In rare cases, hyperinsulinemia can be attributed to specific conditions such as insulinoma, a tumor in β cells that leads to excessive insulin production, or nesidioblastosis, characterized by an abnormal increase in β-cell mass within the pancreas. One of the diseases most often associated with hyperinsulinemia is T2D, caused by progressive IR. In some cases of T2D, the body tries to compensate for IR by increasing the amount of insulin secreted by β cells in order to decrease blood glucose and maintain homeostasis. However, hyperinsulinemia itself can further induce insulin resistance over time because as described previously, constant exposure to insulin triggers a negative feedback loop, which causes insulin receptors to become less responsive to insulin and downregulate [27]. One study showed that when cells in vitro were chronically exposed to insulin, they had diminished insulin receptor tyrosine and serine autophosphorylation [28]. When chronic insulin exposure is removed, normal insulin receptor function can be achieved only if exposed cells maintain a molecular memory prior to chronic insulin exposure. However, if cells are exposed to persistent chronic levels of insulin for long periods of time, the molecular memory of the cell becomes reprogrammed in such a way that causes IR. Thus, these cells do not recover their normal insulin response even after chronic insulin exposure is removed [29].

6. Inadequate Receptor Function

Over time, inadequate insulin receptor function and insulin resistance can contribute to the development of various diseases, including diabetes and other metabolic disorders. It is well-established that unchecked IR is the underlying pathophysiologic precursor to the development of T2D. As described earlier, T2D can result from inadequate receptor function that is often linked to obesity, but there are genetic predispositions that can result in receptor dysfunction. However, it is worth noting that genetic factors can also play a role in receptor dysfunction, leading to impaired insulin signaling. Two notable examples of genetic involvement are the *IRS1* and *IRS2* insulin receptor substrate genes. These genes encode peptides that hold significant importance in insulin-signaling pathways. Studies have demonstrated that specific polymorphisms within these genes are associated with a decrease in insulin sensitivity, ultimately predisposing individuals to T2D [30,31]. These genetic variations can disrupt the normal functioning of insulin receptors, hindering their ability to effectively respond to insulin molecules.

One metabolic disorder that can result from an increase in IR is lipotoxicity. Free fatty acids (FFA) are an increased risk factor for those who have IR [32]. FFA in the plasma is regulated by insulin. Under normal physiological conditions, FFA increases during fasting and decreases after meal consumption. In obese patients, FFAs are often high and can contribute to insulin resistance and T2D [33]. Additionally, lipodystrophy and adipose tissue dysfunction can also contribute to elevated FFAs in the plasma. The two FFAs that are primarily responsible for insulin resistance and a decrease in insulin sensitivity are diacyglycerol (DAG) and ceramide. However, physical exercise can reduce the amount of DAG and ceramide found in skeletal muscle, which can improve insulin sensitivity [34]. In addition, a recent meta-analysis showed that intermittent fasting may also improve insulin sensitivity [35]. Two causes for FFA elevation are often associated with hyperlipidemia and inflammation. It has also been shown that increased IR is associated with an increase in cholesterol synthesis and a decrease in the uptake of cholesterol [36]. When IR increases, there is a shift in cholesterol metabolism. Often, patients with IR have decreased high-density lipoprotein (HDL) cholesterol, increased low-density lipoprotein (LDL) cholesterol, and increased triglycerides levels [36–38]. Disruption to lipid metabolism has mostly been associated with T2D. However, lipid metabolism disorders have also been seen in those with T1D. Lipid disorders can result from poor glycemic control in those who have either T1D or T2D [39]. Disruptions to lipid metabolism have been associated with an increased risk of developing coronary artery disease (CAD) and various other cardiovascular diseases [40,41]. More broadly, insulin resistance has been associated with metabolic diseases that masquerade as neurologic disorders. Diseases such as Parkinson's disease and Alzheimer's disease are associated with insulin resistance (Figure 1) [42].

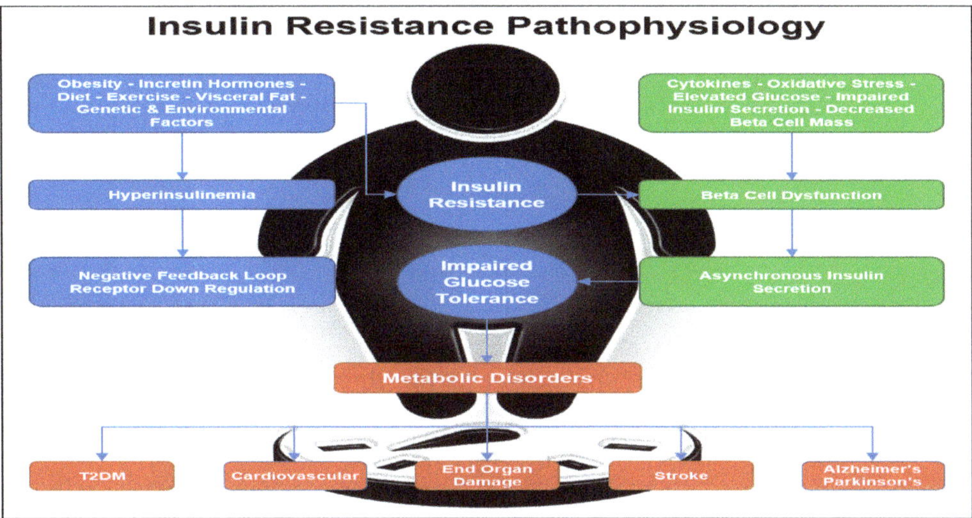

Figure 1. Insulin Resistance Pathophysiology. Insulin resistance can be initiated by obesity with an increase in visceral fat, inactivity, genetics, diet, and environmental factors. Insulin resistance leads to hyperinsulinemia with down regulation of the insulin receptors and a further increase in insulin resistance. Diabetes is a progressive disease with beta-cell dysfunction and loss of insulin pulses. Diabetes leads to diabetes complications, cardiovascular disease, metabolic syndrome, and is associated with neurodegenerative diseases [26,40–42].

7. Physiologic Insulin Administration

One approach to counter IR would be to reintroduce precision physiologic insulin delivery. Currently, patients with T1D and T2D receive a continuous delivery of insulin either by insulin injections or a continuous subcutaneous insulin pump. IR often results from insulin receptors being constantly exposed to levels of insulin with no trough periods to allow the insulin receptor adequate time to reset. Counter to the goal of reducing serum glucose, the constant exposure to insulin can serve to create tolerance. Decreased sensitivity of receptors results in a decreased uptake of serum glucose and subsequent hyperglycemia. Often, type-2 diabetics are prescribed medications such as metformin or thiazolidinediones (i.e., rosiglitazone and pioglitazone) to combat abnormal insulin secretion caused by IR [43,44]. Although these medications are classified as insulin sensitizing, they do not restore or prevent abnormal insulin-secretion patterns.

In 2012, Matveyenko et al. reported that insulin delivered in small intermittent boluses via the portal vein is more efficacious than receiving a constant insulin infusion [45]. In this study, rats received insulin continuously and in a more physiologic pattern of intermittent boluses. Rats that received a constant delivery of insulin had a delay and impairment in the activation of hepatic insulin receptors, which resulted in an impaired activation of downstream signaling. It is speculated that hepatic IR in diabetes is likely a result of the impaired physiologic insulin secretion experienced.

Physiological insulin resensitization (PIR) provides for insulin to be delivered peripherally in a dynamic pattern of physiologic boluses that mimics the healthy pancreas. This is achieved by inserting an intravenous access connected to a precision intravenous infusion pump that can be programmed to dynamically deliver insulin to mimic normal glucose metabolism. The physiologic administration of insulin consists of periodic cycling of up to 3 IU of regular fast-acting insulin infused intermittently between 4–8 min (usually 5–6 min), based on the body's utilization for 2 to 4 h. These infusions are given based on the degree of insulin resistance, and precision dosing is undertaken to mimic hormonal signaling patterns in a concentration and frequency dependent on individual patient condition. During the infusion process, oral glucose is given to patients to simulate a meal and keep blood glucose levels in a prescribed range. Patients are observed during the process until glucose levels are stable after the physiologic insulin infusion is administered [46]. The rationale for the peripheral administration of IV insulin in a rhythmic pattern is to replace impaired physiological signals that are critical to cellular glucose metabolism.

8. Insulin Receptor Upregulation

Physiological insulin secretion exhibits a greater efficacy in upregulating insulin receptors compared to constant exposure to insulin. This effect is particularly notable in the liver, which experiences more insulin exposure than any other organ in the body due to the direct flow of insulin from the pancreas through the portal vein, which then spreads throughout the rest of the body [47]. Prolonged exposure to insulin leads to the downregulation of insulin receptors, ultimately contributing to IR [48]. The goal of PIR aims to address this issue by administering insulin in a manner that mimics the native pattern of hormone secretion. This is achieved by inducing metabolic activity to consume carbohydrates rather than fat. Glucose can be used and stored effectively when the body has not just enough insulin, but the right timing of insulin. However, when there is a lack of insulin or an over exposure of insulin over time causing a reverse feedback loop to downregulate the insulin receptor, a decline in carbohydrate metabolism occurs, and the body switches to fat metabolism in a process called ketosis. Ketosis triggers the production of ketones and, in severe cases, the accumulation of ketones can result in a life-threatening condition called diabetic ketoacidosis (DKA) [49]. Additionally, reducing excessive fat metabolism, through PIR, may counter the harmful effects of free radicals and oxidative stress that occur in the setting of elevated IL-6 levels found in T2D [50].

9. Cellular Glucose Uptake and Adenosine Triphosphate Production

Restoring the physiologic cadence of insulin improves IR and allows for an increase in carbohydrate metabolism and a decrease in fat metabolism by restoring insulin's ability to suppress lipolysis. As described previously, insulin binds to its receptor on cells, which are then endocytosed into the cell where glucose is further metabolized into "cellular energy". Inside the cell, glucose undergoes a series of enzymatic reactions, primarily through the process of glycolysis. Initially, glucose is converted into pyruvate, which then enters the mitochondria to undergo additional processing into acetyl-CoA. This conversion occurs within the tricarboxylic acid (TCA) cycle, where acetyl-CoA participates in a series of chemical reactions. Along this cycle, the reducing agent NADH is produced, which is subsequently shuttled through complex I and II of the electron transport chain.

As electrons are transported, free energy is released, which actively pumps protons across the inner membrane of the mitochondria. As protons are pumped across the membrane, a proton gradient is formed, which creates an electrochemical gradient that fuels the production of adenosine triphosphate (ATP), commonly known as cellular energy (Figure 2) [51]. Thus, carbohydrate metabolism is essential for cells to produce cellular energy in the form of ATP.

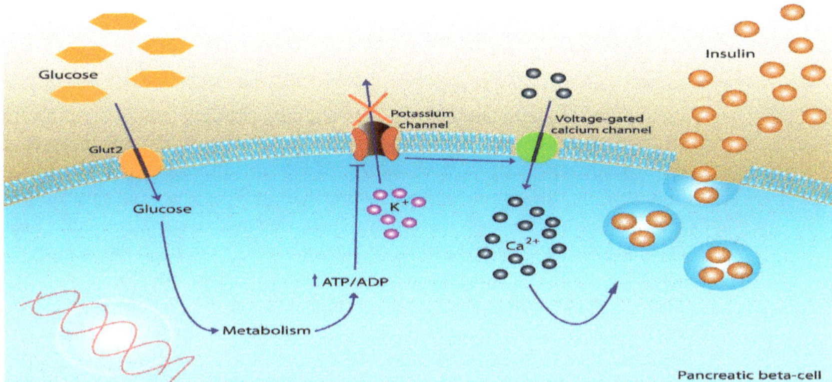

Figure 2. Cellular glucose uptake and ATP production [51,52]. Glucose is brought into the pancreatic beta cell by the GLUT-2 transporter and converted to glucose-6-phosphate by glucokinase initiating the glycolytic pathway and resulting in the generation of pyruvate. Pyruvate is transported into the mitochondria where ATP is generated through the electron transport chain. ATP depolarizes the ATP-sensitive potassium channel causing depolarization of the voltage dependent calcium channel, which causes prepackaged insulin granules to fuse with the cell membrane and release insulin into the blood stream (Image used under license from: https://support.shutterstock.com/s/article/Can-I-use-Images-on-my-website?language=en_US#: ~:text=Shutterstock%20Photos%20Are%20Royalty%2DFree,multiple%20ways%20on%20multiple% 20applications, accessed on 20 June 2023).

10. Cellular Restoration

Cell growth and proliferation are highly dependent upon carbohydrate metabolism. Glucose metabolism provides cellular energy (ATP), but it also supplies metabolites that are critical for the cellular synthesis of nucleic acids, proteins, and lipids [53,54]. One study showed that glucose metabolism is involved in the healing process of tendons [55]; a second study showed that glucose is important in human gingival fibroblast maintenance after periodontal surgery [56]. In the case of diabetes, blood glucose control and metabolism are important for wound healing. Furthermore, prolonged hyperglycemia in diabetes causes tissue to become energy-deficient, which over time can over result in a decline in wound healing [57]. For instance, the process of angiogenesis is required to promote tissue wound healing. Hyperglycemia decreases angiogenesis in endothelial cells, and it affects the stability of hypoxia-inducible factor 1-alpha (HIF-1a) to target genes such as the vascular endothelial growth factor (*VEGF*) to promote wound healing and tissue repair [58]. Often, diabetics with poor glucose control have wounds, most commonly foot ulcers, that go unnoticed, take longer to heal, become infected, or never adequately heal. Foot ulcers and their poor healing are often due to underlying neuropathy and a poorly functioning microcirculation. Nerve and microvascular tissue clearly benefit from optimized glucose ATP production and the related decrease in inflammation based on the body utilizing glucose as its primary fuel source instead of fat metabolism [50].

Glucose metabolism is essential for the development and maintenance of several tissues, including skeletal muscle, vasculature, liver, heart, and adipose tissue. The retina is one example of a tissue that is affected by the dysregulation of glucose metabolism. Diabetic retinopathy often results 10–20 years after the onset of diabetes due to glucose metabolism failure. Retinopathy is a disease of the microvascular system of the retina when the retina experiences hypoxia. Blood passes via the optic artery and perfuses the retina. In the retina, arteries radiate outward to form a capillary network that supplies blood vessels and oxygen. When high blood glucose levels persist, this can weaken and damage blood vessels within the retina. Damaged blood vessels starve the retina of oxygen, which leads to abnormal blood vessel outgrowth to compensate for oxygen deprivation. HIF-1a has a role in the stability of retina homeostasis, but intense fluctuations in carbohydrate metabolism can cause HIF-1a to fluctuate, which results in abnormal blood vessel outgrowth and progression of retinopathy [59].

Neuronal tissue appears to be the most sensitive tissue type affected when there is a decrease in energy from carbohydrate metabolism. Neuropathy in diabetes can result from glucose metabolism failure. Symptoms often include numbness, pain, muscle weakness, or tingling in the hands and feet. Hyperglycemia can induce mitochondria dysfunction/overload of glucose and increase reactive oxygen species (ROS) [59]. The accumulation of ROS results in induced cellular apoptosis and a decrease in nerve blood flow. Increased oxidative stress upregulates the polyADP-ribose polymerase (PARP) pathway, which increases inflammation and causes neuronal dysfunction. Neurovascular dysfunction from hyperglycemia affects five main pathways, including: the polyol pathway, the advanced glycation end-product (AGE) pathway, the protein kinase C pathway, the PARP pathway, and the hexosamine pathway [60]. One of the consequences of neuropathy is a decrease in wound detection and healing. To combat neuropathy, cell therapies have been used such as growth factors to promote neuron survival, function, and repair. Schratzberger et al. demonstrated that growth factors such as *VEGF, FGF2, NGF, BDNF,* and *IGF1* have neurotropic properties in treating diabetic neuropathy [61].

11. Discussion

Perhaps one of the most perplexing challenges in managing patients with diabetes and other metabolic disorders is IR. In patients with diabetes, the conventional approach to controlling serum blood glucose involves subcutaneous insulin administration or the use of oral medications that increase insulin production. However, this continuous exposure of insulin receptors to their ligand can lead to the development of tolerance. Continuous exposure to high levels of insulin causes insulin receptors to internalize, rendering them unavailable for engagement with insulin molecules. This internalization process contributes to the development of IR. Paradoxically, the administration of higher doses of insulin to reduce serum blood glucose levels can further worsen IR. This is due to a negative feedback loop, where higher levels of insulin in the bloodstream are associated with decreased availability of receptors, exacerbating the IR phenomenon.

In the normal physiologic state, characterized by insulin sensitivity, insulin is secreted from pancreatic β cells in a specific cadence, with regular intervals of insulin secretion and troughs occurring approximately every 4–8 min. This intermittent exposure of peripheral insulin receptors to insulin plays a crucial role in maintaining the sensitivity of these receptors, preventing receptor tolerance and preserving insulin sensitivity. However, in patients who have developed diabetes or other IR phenotypes, the natural cadence of insulin secretion is disrupted. To address this, treatment strategies have been developed to mimic the physiologic intermittent secretion of insulin. These treatments, known as PIR, aim to restore the normal pattern of insulin exposure to peripheral insulin receptors. By administering insulin in a manner that resembles the natural insulin secretion pattern, PIR treatments have been associated with improvements in fasting serum glucose levels and HbA1c.

Beyond these assessments that correspond to improvements in diabetes management, improvements in diabetic complications have also been observed. The rationale for improvements in conditions such as diabetic neuropathy, retinopathy, nephropathy, etc., may be related to improvements in glucose utilization and signaling associated with the physiologic administration of insulin. That is, insulin administration that mimics normal physiology appears to temporarily restore signaling that results in enhanced energy production from improvements in carbohydrate metabolism. While empiric assessments of insulin receptor availability and function are lacking, the apparent improvements in insulin sensitivity may be related to receptor function. The phenotype of IR appears to be a story of receptor availability mediated by the intermittent administration of insulin. Mimicry of this cadence of peaks and troughs appears to relieve IR and facilitate the improvement of complications associated with diabetes.

12. Conclusions

Insulin receptor modulation is a multifaceted process influenced and mediated, in part, by the intermittent peaks and troughs of the ligand. However, dysfunction in the autonomic nerves within the pancreas can disrupt the finely tuned oscillatory secretion of insulin, leading to abnormal insulin patterns and compromising glucose regulation. This disruption of the characteristic higher and lower concentrations observed in normal physiologic pancreatic β-cell secretion patterns is closely associated with insulin resistance, the onset of diabetes, and the subsequent development of complications and perhaps other disorders.

In contrast to conventional standard medicinal interventions, there is emerging evidence supporting the benefits of intravenous insulin administration that mimics the healthy pancreatic insulin secretion pattern. By restoring this physiological cadence, improvements in insulin resistance, restoration of metabolic function, enhanced cellular repair mechanisms, and a reduced risk of long-term complications have been observed.

Author Contributions: Conceptualization, S.T.L., F.G., S.A.H., B.L. and J.R.T.L. writing—original draft preparation, S.T.L., F.G., T.R.T., M.A., B.L. and J.R.T.L.: writing—reviewing and editing, S.T.L., F.G., T.R.T., L.K.J., S.A.H., B.L. and J.R.T.L.; funding acquisition, F.G. All authors have read and agreed to the published version of the manuscript.

Funding: This manuscript was supported in part by 1 U54 GM104940 from the National Institute of General Medical Sciences of the National Institutes of Health, which funds the Louisiana Clinical and Translational Science Center. The content is solely the responsibility of the authors and does not necessarily represent the official views of the National Institutes of Health. The authors also acknowledge the support and assistance of the Department of Surgery at the University of California Irvine.

Institutional Review Board Statement: Not applicable.

Informed Consent Statement: Not applicable.

Data Availability Statement: Not applicable.

Conflicts of Interest: Stanley T. Lewis, Scott A. Hepford, and Brian Loveridge are partial owners of Well Cell Global. Levonika K. Jackson and Brian Loveridge are employees of an affiliate of Well Cell Global.

References

1. De Meyts, P. *The Insulin Receptor and Its Signal Transduction Network*; Kenneth, R.F., Ed.; MDText.com, Inc.: Dartmouth, MA, USA, 2000; Endotext. Available online: https://www.ncbi.nlm.nih.gov/books/NBK378978/ (accessed on 15 September 2022).
2. Chen, Y.; Huang, L.; Qi, X.; Chen, C. Insulin Receptor Trafficking: Consequences for Insulin Sensitivity and Diabetes. *Int. J. Mol. Sci.* **2019**, *20*, 5007. [CrossRef]
3. Curry, D.L.; Bennett, L.L.; Grodsky, G.M. Dynamics of Insulin Secretion by the Perfused Rat Pancreas. *Endocrinology* **1968**, *83*, 572–584. [CrossRef]
4. Lang, D.A.; Matthews, D.R.; Peto, J.; Turner, R.C. Cyclic Oscillations of Basal Plasma Glucose and Insulin Concentrations in Human Beings. *N. Engl. J. Med.* **1979**, *301*, 1023–1027. [CrossRef]
5. Germanos, M.; Gao, A.; Taper, M.; Yau, B.; Kebede, M. Inside the Insulin Secretory Granule. *Metabolites* **2021**, *11*, 515. [CrossRef] [PubMed]
6. Cerasi, E.; Luft, R. "What Is Inherited—What Is Added" Hypothesis for the Pathogenesis of Diabetes Mellitus. *Diabetes* **1967**, *16*, 615–627. [CrossRef] [PubMed]
7. Ahrén, B. Autonomic Regulation of islet Hormone Secretion—Implications for Health and Disease. *Diabetologia* **2000**, *43*, 393–410. [CrossRef] [PubMed]
8. Stanley, S.; Moheet, A.; Seaquist, E.R. Central Mechanisms of Glucose Sensing and Counterregulation in Defense of Hypoglycemia. *Endocr. Rev.* **2019**, *40*, 768–788. [CrossRef]
9. Rosario, W.; Singh, I.; Wautlet, A.; Patterson, C.; Flak, J.; Becker, T.C.; Ali, A.; Tamarina, N.; Philipson, L.H.; Enquist, L.W.; et al. The Brain–to–Pancreatic Islet Neuronal Map Reveals Differential Glucose Regulation from Distinct Hypothalamic Regions. *Diabetes* **2016**, *65*, 2711–2723. [CrossRef]

10. Makhmutova, M.; Weitz, J.; Tamayo, A.; Pereira, E.; Boulina, M.; Almaça, J.; Rodriguez-Diaz, R.; Caicedo, A. Pancreatic β-Cells Communicate with Vagal Sensory Neurons. *Gastroenterology* **2021**, *160*, 875–888.e11. [CrossRef]
11. Taborsky, G.J. Islets Have a Lot of Nerve! Or Do They? *Cell Metab.* **2011**, *14*, 5–6. [CrossRef]
12. Rodriguez-Diaz, R.; Abdulreda, M.H.; Formoso, A.L.; Gans, I.; Ricordi, C.; Berggren, P.; Caicedo, A. Innervation Patterns of Autonomic Axons in the Human Endocrine Pancreas. *Cell Metab.* **2011**, *14*, 45–54. [CrossRef] [PubMed]
13. Carobi, C. Capsaicin-sensitive vagal afferent neurons innervating the rat pancreas. *Neurosci. Lett.* **1987**, *77*, 5–9. [CrossRef] [PubMed]
14. Langlois, A.; Dumond, A.; Vion, J.; Pinget, M.; Bouzakri, K. Crosstalk Communications Between Islets Cells and Insulin Target Tissue: The Hidden Face of Iceberg. *Front. Endocrinol.* **2022**, *13*, 836344. [CrossRef]
15. Longo, E.A.; Tornheim, K.; Deeney, J.T.; Varnum, B.A.; Tillotson, D.; Prentki, M.; E Corkey, B. Oscillations in cytosolic free Ca^{2+}, oxygen consumption, and insulin secretion in glucose-stimulated rat pancreatic islets. *J. Biol. Chem.* **1991**, *266*, 9314–9319. [CrossRef] [PubMed]
16. Jacobson, D.A.; Philipson, L.H. Action potentials and insulin secretion: New insights into the role of Kv channels. *Diabetes Obes. Metab.* **2007**, *9* (Suppl. S2), 89–98. [CrossRef] [PubMed]
17. Gál, E.; Dolenšek, J.; Stožer, A.; Czakó, L.; Ébert, A.; Venglovecz, V. Mechanisms of Post-Pancreatitis Diabetes Mellitus and Cystic Fibrosis-Related Diabetes: A Review of Preclinical Studies. *Front. Endocrinol.* **2021**, *12*, 715043. [CrossRef]
18. Sokołowska, M.; Chobot, A.; Jarosz-Chobot, P. The honeymoon phase—What we know today about the factors that can modulate the remission period in type 1 diabetes. *Pediatr. Endocrinol. Diabetes Metab.* **2016**, *22*, 66–70. [CrossRef]
19. Ellulu, M.S.; Patimah, I.; KhazáAi, H.; Rahmat, A.; Abed, Y. Obesity and inflammation: The linking mechanism and the complications. *Arch. Med. Sci.* **2017**, *13*, 851–863. [CrossRef]
20. Horii, T.; Fujita, Y.; Ishibashi, C.; Fukui, K.; Eguchi, H.; Kozawa, J.; Shimomura, I. Islet inflammation is associated with pancreatic fatty infiltration and hyperglycemia in type 2 diabetes. *BMJ Open Diabetes Res. Care* **2020**, *8*, e001508. [CrossRef]
21. Chen, Y.; Zhang, P.; Lv, S.; Su, X.; Du, Y.; Xu, C.; Jin, Z. Ectopic fat deposition and its related abnormalities of lipid metabolism followed by nonalcoholic fatty pancreas. *Endosc. Ultrasound* **2022**, *11*, 407–413. [CrossRef]
22. Hunter, S.J.; Atkinson, A.B.; Ennis, C.N.; Sheridan, B.; Bell, P.M. Association Between Insulin Secretory Pulse Frequency and Peripheral Insulin Action in NIDDM and Normal Subjects. *Diabetes* **1996**, *45*, 683–686. [CrossRef] [PubMed]
23. Peiris, A.N.; I Stagner, J.; Vogel, R.L.; Nakagawa, A.; Samols, E. Body fat distribution and peripheral insulin sensitivity in healthy men: Role of insulin pulsatility. *J. Clin. Endocrinol. Metab.* **1992**, *75*, 290–294. [CrossRef] [PubMed]
24. Satin, L.S.; Butler, P.C.; Ha, J.; Sherman, A.S. Pulsatile insulin secretion, impaired glucose tolerance and type 2 diabetes. *Mol. Asp. Med.* **2015**, *42*, 61–77. [CrossRef]
25. Meier, J.J.; Veldhuis, J.D.; Butler, P.C. Pulsatile Insulin Secretion Dictates Systemic Insulin Delivery by Regulating Hepatic Insulin Extraction in Humans. *Diabetes* **2005**, *54*, 1649–1656. [CrossRef]
26. Thomas, D.D.; E Corkey, B.E.; Istfan, N.W.; Apovian, C.M. Hyperinsulinemia: An Early Indicator of Metabolic Dysfunction. *J. Endocr. Soc.* **2019**, *3*, 1727–1747. [CrossRef] [PubMed]
27. Garvey, W.T.; Olefsky, J.M.; Marshall, S. Insulin receptor down-regulation is linked to an insulin-induced postreceptor defect in the glucose transport system in rat adipocytes. *J. Clin. Investig.* **1985**, *76*, 22–30. [CrossRef] [PubMed]
28. Eliel, L.P.; Smith, W.O.; Chanes, R.; Hawrylko, J. Magnesium Metabolism in Hyperparathyroidism and Osteolytic Disease. *Ann. N. Y. Acad. Sci.* **1969**, *162*, 810–830. [CrossRef]
29. Laedtke, T.; Pørksen, N.; Schmitz, O.; Kjems, L.; Veldhuis, J.; Kao, P.C.; Butler, P.C. Overnight inhibition of insulin secretion restores pulsatility and proinsulin/insulin ratio in type 2 diabetes. *Am. J. Physiol. Metab.* **2000**, *279*, E520–E528. [CrossRef]
30. Clausen, J.; Hansen, T.; Rbaek, C.B.O.; Echwald, S.; Urhammer, S.; Rasmussen, S.; Andersen, C.; Hansen, L.; Almind, K.; Winther, K.; et al. Insulin resistance: Interactions between obesity and a common variant of insulin receptor substrate-1. *Lancet* **1995**, *346*, 397–402. [CrossRef]
31. Le Fur, S.; Le Stunff, C.; Bougnères, P. Increased Insulin Resistance in Obese Children Who Have Both 972 IRS-1 and 1057 IRS-2 Polymorphisms. *Diabetes* **2002**, *51* (Suppl. S3), S304–S307. [CrossRef]
32. Boden, G. Free fatty acids and insulin secretion in humans. *Curr. Diabetes Rep.* **2005**, *5*, 167–170. [CrossRef]
33. Paolisso, G.; Tataranni, P.A.; Foley, J.E.; Bogardus, C.; Howard, B.V.; Ravussin, E. A high concentration of fasting plasma non-esterified fatty acids is a risk factor for the development of NIDDM. *Diabetologia* **1995**, *38*, 1213–1217. [CrossRef]
34. Schenk, S.; Horowitz, J.F. Acute exercise increases triglyceride synthesis in skeletal muscle and prevents fatty acid–induced insulin resistance. *J. Clin. Investig.* **2007**, *117*, 1690–1698. [CrossRef] [PubMed]
35. Yuan, X.; Wang, J.; Yang, S.; Gao, M.; Cao, L.; Li, X.; Hong, D.; Tian, S.; Sun, C. Effect of Intermittent Fasting Diet on Glucose and Lipid Metabolism and Insulin Resistance in Patients with Impaired Glucose and Lipid Metabolism: A Systematic Review and Meta-Analysis. *Int. J. Endocrinol.* **2022**, *2022*, 6999907. [CrossRef] [PubMed]
36. Hoenig, M.R.; Sellke, F.W. Insulin resistance is associated with increased cholesterol synthesis, decreased cholesterol absorption and enhanced lipid response to statin therapy. *Atherosclerosis* **2010**, *211*, 260–265. [CrossRef] [PubMed]
37. Howard, B.V. Insulin Resistance and Lipid Metabolism. *Am. J. Cardiol.* **1999**, *84*, 28J–32J. [CrossRef]
38. Pihlajamäki, J.; Gylling, H.; Miettinen, T.A.; Laakso, M. Insulin resistance is associated with increased cholesterol synthesis and decreased cholesterol absorption in normoglycemic men. *J. Lipid Res.* **2004**, *45*, 507–512. [CrossRef] [PubMed]
39. Vergès, B. Lipid disorders in type 1 diabetes. *Diabetes Metab.* **2009**, *35*, 353–360. [CrossRef]

40. Hsu, C.-C.; Chang, H.-Y.; Huang, M.; Hwang, S.; Yang, Y.; Chang, C.; Chang, C.-J.; Li, Y.; Shin, S. Association between Insulin Resistance and Development of Microalbuminuria in Type 2 Diabetes: A prospective cohort study. *Diabetes Care* **2011**, *34*, 982–987. [CrossRef]
41. Ormazabal, V.; Nair, S.; Elfeky, O.; Aguayo, C.; Salomon, C.; Zuñiga, F.A. Association between insulin resistance and the development of cardiovascular disease. *Cardiovasc. Diabetol.* **2018**, *17*, 122. [CrossRef]
42. Athauda, D.; Foltynie, T. Insulin resistance and Parkinson's disease: A new target for disease modification? *Prog. Neurobiol.* **2016**, *145–146*, 98–120. [CrossRef]
43. Lupi, R.; Bugliani, M.; Del Guerrera, S.; Del Prato, S.; Marchetti, P.; Boggi, U.; Filipponi, F.; Mosca, F. Transcription factors of beta-cell differentiation and maturation in isolated human islets: Effects of high glucose, high free fatty acids and type 2 diabetes. *Nutr. Metab. Cardiovasc. Dis.* **2006**, *16*, e7–e8. [CrossRef] [PubMed]
44. Patanè, G.; Piro, S.; Rabuazzo, A.M.; Anello, M.; Vigneri, R.; Purrello, F. Metformin restores insulin secretion altered by chronic exposure to free fatty acids or high glucose: A direct metformin effect on pancreatic beta-cells. *Diabetes* **2000**, *49*, 735–740. [CrossRef] [PubMed]
45. Matveyenko, A.V.; Liuwantara, D.; Gurlo, T.; Kirakossian, D.; Man, C.D.; Cobelli, C.; White, M.F.; Copps, K.D.; Volpi, E.; Fujita, S.; et al. Pulsatile Portal Vein Insulin Delivery Enhances Hepatic Insulin Action and Signaling. *Diabetes* **2012**, *61*, 2269–2279. [CrossRef] [PubMed]
46. Greenway, F.; Loveridge, B.; Grimes, R.M.; Tucker, T.R.; Alexander, M.; Hepford, S.A.; Fontenot, J.; Nobles-James, C.; Wilson, C.; Starr, A.M.; et al. Physiologic Insulin Resensitization as a Treatment Modality for Insulin Resistance Pathophysiology. *Int. J. Mol. Sci.* **2022**, *23*, 1884. [CrossRef]
47. Najjar, S.M.; Perdomo, G. Hepatic Insulin Clearance: Mechanism and Physiology. *Physiology* **2019**, *34*, 198–215. [CrossRef]
48. Kitabchi, A.E.; Wall, B.M. Diabetic ketoacidosis. *Med Clin. N. Am.* **1995**, *79*, 9–37. [CrossRef]
49. Kaufman, B.A.; Li, C.; Soleimanpour, S.A. Mitochondrial regulation of β-cell function: Maintaining the momentum for insulin release. *Mol. Asp. Med.* **2015**, *42*, 91–104. [CrossRef]
50. Qu, D.; Liu, J.; Lau, C.W.; Huang, Y. IL-6 in diabetes and cardiovascular complications. *Br. J. Pharmacol.* **2014**, *171*, 3595–3603. [CrossRef]
51. Kwak, S.H.; Park, K.S.; Lee, K.; Lee, H.K. Mitochondrial metabolism and diabetes. *J. Diabetes Investig.* **2010**, *1*, 161–169. [CrossRef]
52. Fauci, A.S.; Kaspar, D.L.; Braunwald, E.; Hauser, S.L.; Longo, D.L.; Jameson, J.L.; Loscalzo, J. *Harrison's Principles of Internal Medicine*, 17th ed. Available online: http://www.accessmedicine.com (accessed on 6 May 2023).
53. Lunt, S.Y.; Vander Heiden, M.G. Aerobic Glycolysis: Meeting the Metabolic Requirements of Cell Proliferation. *Annu. Rev. Cell Dev. Biol.* **2011**, *27*, 441–464. [CrossRef] [PubMed]
54. Hernandez, J.M.; Fedele, M.J.; Farrell, P.A. Time course evaluation of protein synthesis and glucose uptake after acute resistance exercise in rats. *J. Appl. Physiol.* **2000**, *88*, 1142–1149. [CrossRef] [PubMed]
55. Izumi, S.; Otsuru, S.; Adachi, N.; Akabudike, N.; Enomoto-Iwamoto, M. Control of glucose metabolism is important in tenogenic differentiation of progenitors derived from human injured tendons. *PloS ONE* **2019**, *14*, e0213912. [CrossRef]
56. Li, R.; Kato, H.; Taguchi, Y.; Umeda, M. Intracellular glucose starvation affects gingival homeostasis and autophagy. *Sci. Rep.* **2022**, *12*, 1230. [CrossRef] [PubMed]
57. Spampinato, S.F.; Caruso, G.I.; De Pasquale, R.; Sortino, M.A.; Merlo, S. The Treatment of Impaired Wound Healing in Diabetes: Looking among Old Drugs. *Pharmaceuticals* **2020**, *13*, 60. [CrossRef]
58. Semenza, G.L.; Yang, Y.; Fu, Q.; Wang, X.; Liu, Y.; Zeng, Q.; Li, Y.; Gao, S.; Bao, L.; Liu, S.; et al. HIF-1: Mediator of physiological and pathophysiological responses to hypoxia. *J. Appl. Physiol.* **2000**, *88*, 1474–1480. [CrossRef] [PubMed]
59. Gunton, J.E. Hypoxia-inducible factors and diabetes. *J. Clin. Investig.* **2020**, *130*, 5063–5073. [CrossRef]
60. Han, J.W.; Sin, M.Y.; Yoon, Y.-S. Cell Therapy for Diabetic Neuropathy Using Adult Stem or Progenitor Cells. *Diabetes Metab. J.* **2013**, *37*, 91–105. [CrossRef]
61. Schratzberger, P.; Walter, D.H.; Rittig, K.; Bahlmann, F.H.; Pola, R.; Curry, C.; Silver, M.; Krainin, J.G.; Weinberg, D.H.; Ropper, A.H.; et al. Reversal of experimental diabetic neuropathy by VEGF gene transfer. *J. Clin. Investig.* **2001**, *107*, 1083–1092. [CrossRef]

International Journal of
Molecular Sciences

Review

Smooth Muscle Heterogeneity and Plasticity in Health and Aortic Aneurysmal Disease

Yunwen Hu, Zhaohua Cai * and Ben He *

Department of Cardiology, Shanghai Chest Hospital, Shanghai Jiao Tong University School of Medicine, Shanghai 200030, China
* Correspondence: caizhaohua@shchest.org (Z.C.); heben@shchest.org (B.H.)

Abstract: Vascular smooth muscle cells (VSMCs) are the predominant cell type in the medial layer of the aorta, which plays a critical role in the maintenance of aortic wall integrity. VSMCs have been suggested to have contractile and synthetic phenotypes and undergo phenotypic switching to contribute to the deteriorating aortic wall structure. Recently, the unprecedented heterogeneity and diversity of VSMCs and their complex relationship to aortic aneurysms (AAs) have been revealed by high-resolution research methods, such as lineage tracing and single-cell RNA sequencing. The aortic wall consists of VSMCs from different embryonic origins that respond unevenly to genetic defects that directly or indirectly regulate VSMC contractile phenotype. This difference predisposes to hereditary AAs in the aortic root and ascending aorta. Several VSMC phenotypes with different functions, for example, secreting VSMCs, proliferative VSMCs, mesenchymal stem cell-like VSMCs, immune-related VSMCs, proinflammatory VSMCs, senescent VSMCs, and stressed VSMCs are identified in non-hereditary AAs. The transformation of VSMCs into different phenotypes is an adaptive response to deleterious stimuli but can also trigger pathological remodeling that exacerbates the pathogenesis and development of AAs. This review is intended to contribute to the understanding of VSMC diversity in health and aneurysmal diseases. Papers that give an update on VSMC phenotype diversity in health and aneurysmal disease are summarized and recent insights on the role of VSMCs in AAs are discussed.

Keywords: abdominal aortic aneurysm; thoracic aortic aneurysm; VSMC; phenotypic switching

Citation: Hu, Y.; Cai, Z.; He, B. Smooth Muscle Heterogeneity and Plasticity in Health and Aortic Aneurysmal Disease. *Int. J. Mol. Sci.* **2023**, *24*, 11701. https://doi.org/10.3390/ijms241411701

Academic Editors: Yutang Wang and Dianna Magliano

Received: 25 June 2023
Revised: 16 July 2023
Accepted: 18 July 2023
Published: 20 July 2023

1. Introduction

The aorta is a large blood vessel consisting of three layers: intima, media, and adventitia, extending from the thorax to the abdomen. The aortic wall is eternally exposed to infinite variations of hemodynamics and neurohumoral regulators and its pathological degeneration is believed to generate aortic aneurysms (AAs) [1]. Characterized by permanent local dilation $\geq 50\%$, AAs occur in multiple aortic segments and are insidious, progressive, and fatal if untreated. Heritable genetic variations confer high AA risk. Alternatively, a causative gene mutation is always absent in patients with non-hereditary AAs [2–5].

As the major cellular component of aortic walls, vascular smooth muscle cells (VSMCs) ought to be dynamic to adapt to the fluctuating microenvironment and to maintain the intact structure and functionality of the aortic wall [6]. Unlike terminally differentiated cells, VSMCs can change their morphological and functional characteristics under certain conditions. VSMCs isolated from porcine aortas in sparse culture were observed to switch from a spindle to a polymorphic shape and exhibit logarithmic growth responses to mitogens [7]. Years of research have led to the contractile-synthetic binary phenotype theory of VSMCs. It is believed that VSMCs have 2 distinct but interchangeable phenotypes. The contractile VSMCs, which contribute to the maintenance of vascular tone, express high levels of contraction-related genes, including *Acta2*, *Myh11*, *Tagln*, and *Cnn1*. In contrast, synthetic VSMCs are characterized by reduced expression of contraction-related genes expression and increased cell proliferation, migration, inflammation, and extracellular matrix (ECM) synthesis [1,7–16]. According to the theory, healthy aortas contain only contractile VSMCs, and in pathological conditions, contractile VSMCs are transformed into synthetic VSMCs. Evidence that VSMCs are much more sophisticated and versatile is accumulating with recent advances in lineage tracing and single-cell RNA sequencing (scRNA-seq), challenging the binary phenotype theory [17,18].

2. VSMC Heterogeneity in Normal Aorta

The unique embryonic context is a source of VSMC heterogeneity. As reviewed by Majesky, fate mapping technology in developing embryos has revealed a mosaic distribution of VSMCs from different precursor sources in the aorta, which includes VSMCs derived from the secondary heart fields (SHF), cardiac neural crest (CNC), somite, and splanchnic mesoderm [19]. For instance, VSMCs of the aortic root are mainly of SHF origin, those of the ascending aorta and aortic arch originate mainly from the CNC, and VSMCs composing the descending aorta are derived from mesoderm [20–22]. Even the same aortic segment can consist of VSMCs of different embryonic origins. SHF-derived VSMCs were previously thought to be restricted in the aortic root, forming a suture with CNC-derived VSMCs at the transition from the aortic root and the ascending aorta. However, lineage tracing shows that SHF-derived VSMCs extend distally to the innominate artery orifice. The ascending aorta actually contains both SHF-derived and CNC-derived VSMCs, with the SHF-derived VSMCs wrapping around the outside of the CNC-derived VSMCs in a sleeve-shaped form (Figure 1A) [20,23].

According to classical theory, lineage heterogeneous VSMCs have a common developmental fate that converts to a contractile phenotype, but the scRNA-seq results reveal richer VSMC diversity than previously thought. Despite being the predominant cell type of the aorta, VSMCs have received much less attention than immune cells and endothelial cells in studies to obtain single-cell ATLAS of the aortic wall. Therefore, most of the published results simply identify VSMCs based on canonical VSMC markers without doing sub-clustering [24,25]. Nevertheless, limited data still provide a glimpse into the heterogeneity of VSMCs in normal aorta walls from a transcriptome perspective. In one study, a well-performing classifier based on the VSMC transcriptome obtained by scRNA-seq was constructed to discriminate the regional identity of VSMCs (AUC = 0.9967), suggesting that aortic segmental identity is preserved in individual VSMCs [26]. Even within aortic segments, VSMCs are heterogeneous and have significant intercellular variation in genes associated with contraction, inflammation, local adhesion, migration, and proliferation [26]. In addition, sc-RNAseq revealed six VSMC subpopulations distributed throughout the aorta, and no segment-specific subpopulation was identified [27].

Notably, Lina and her team innovatively combined lineage tracing, fluorescence-activated cell sorting (FACS), and scRNA-seq to identify and isolate a small subset of Sca1-expressing VSMCs from normal mouse thoracic aortas [26]. Sca1$^+$ VSMCs express reduced contractile signatures and show enrichment in genes associated with migration, proliferation, synthesis, and secretion compared to Sca1$^-$ VSMCs. Given that Sca1 has been widely used as a biomarker of stem cells and cell stemness, it is possible that Sca1$^+$ VSMCs represent a de-differentiated state [26,28]. Although Sca1$^+$ VSMCs are rare, they are widely distributed throughout the aorta. Furthermore, Sca1 expression was detected in all VSMC subgroups with different contractile identities. The above results demonstrate a ubiquitous distribution of Sca1$^+$ VSMCs [27]. Mature VSMC can express detectable levels of Sca1 in two different contexts: those in or about to undergo phenotypic transformation and those newly differentiated or recruited from vascular stem cell pools [29–32]. Because scRNA-seq can only capture transient transcriptomic data, we cannot determine whether Sca1 expression in mature VSMCs is transient. It is also difficult to label and follow the fate of Sca1$^+$ VSMCs due to the scarcity of these cells. Nonetheless, these results are significant as it indicates that VSMCs in healthy aortic walls are not static.

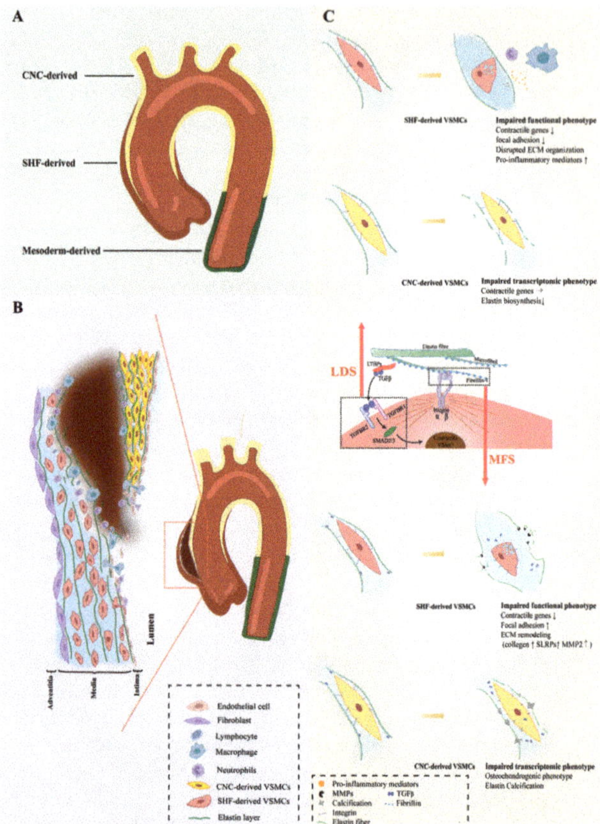

Figure 1. VSMC phenotypic diversity in hereditary AAs. The aortic wall consists of VSMCs from different embryonic origins that respond unevenly to genetic defects that directly or indirectly regulate VSMC contractile phenotype, which results in heterogeneous susceptibility to aortic diseases. (**A**) provides an overview of the origin of VSMCs in the aortic wall. The ascending aorta is composed of VSMCs derived from SHF and CNC. The former wraps outside of the latter in a "sleeve-shaped" form, However, mesoderm-derived VSMCs predominate in the abdominal aorta. (**B**) depicts a plausible hypothesis that the ascending aorta is susceptible to AAs and ADs when genetic defects are present. VSMCs with distinct embryonic backgrounds show different responses to signaling perturbations. Compared to CNC-derived VSMCs, SHF-derived VSMCs are more prone to lose their contractile phenotype and cause pathological ECM remodeling. This leads to a reduction in the strength of the aortic wall, making it susceptible to swelling. Moreover, the turbulent blood flow can easily cause tearing along the interface between the 2 VSMCs, leading to the occurrence of ADs in the presence of damaged endothelium. (**C**) shows the dysfunctional phenotypes of VSMCs from different origins with genetic defects. Fibrillin-1 is the primary protein found in the microfibril extensions of elastic lamellae, which are attached to focal adhesions on the surface of VSMCs. It is also associated with TGF-β binding proteins (LTBP), which are involved in the TGF-β signaling pathway. Genetic defects in fibrillin-1 (MFS) and defects in the TGF-β signaling pathway (LDS) affect the phenotype of both VSMCs, with SHF-derived VSMCs being more susceptible to loss of the contractile phenotype. The direction of arrows indicates the trend of gene expression levels: upward represent upregulation, horizontal indicate stable expression levels, and downward indicate downregulation. Vascular smooth muscle cells, VSMCs; Secondary heart field, SHF; Cardiac neural crest, CNC; Extracellular matrix, ECM; Aortic aneurysms, AAs; Aortic dissections, ADs; Loeys-Dietz syndrome, LDS; Marfan syndrome, MFS.

The heterogeneity of VSMCs may be a physiological adaptation. Hemodynamics vary in different segments of the aorta. Blood velocity in the ascending aorta and aortic arch is faster than in the abdominal aorta, causing greater vessel wall impact [33,34]. Due to the anatomical curvature, the blood flow in the ascending aorta, especially in the aortic arch, changes drastically, resulting in vortex flow and increased demands on the strength and self-healing ability of the vascular wall [35–37]. Vim-expressing VSMCs enriched for genes associated with shear stress, atherogenesis, WNT, and MAPK pathways populate the thoracic aorta and aortic arch, representing a group of flow-shock-adapted VSMCs that contribute to higher elastin/collagen content [27]. Interestingly, the regional distribution of VSMCs from different embryologic backgrounds is consistent with hemodynamics, with a progressive increase in the proportion of CNC-derived VSMCs along the ascending aorta to the aortic arch suggesting that the composition of the VSMCs of distinct lineage influences the intensity of the arterial wall. Indeed, VSMCs of different embryonic origins are phenotypically and functionally distinct, with different proliferation and protein expression patterns in response to cytokines [38]. Stimulation with wound repair factor transforming growth factor-beta 1 (TGF-β1) or platelet-derived growth factor BB (PDGF-BB) significantly enhanced the biosynthesis of ECM components such as Col1A1 in CNC-derived VSMCs and promoted cell proliferation by increasing c-myb expression, but there was no effect on VSMCs of mesodermal origin, suggesting that embryonic lineage is a major source of aortic VSMC heterogeneity [39–41].

3. VSMC Phenotypic Diversity in Hereditary AAs

The susceptibility to aortopathy varies between aortic segments. The ascending aorta is less prone to atherosclerosis than the abdominal aorta, and even thoracoabdominal aortic graft exchange does not alter atherosclerotic propensity, suggesting that the intrinsic nature of the arterial wall influences vulnerability to arterial disease [42–44]. Segment-specific distribution of heterogeneous VSMCs likely contributes significantly to aortic disease characteristics. Further support for this view is the fact that the aortic arch, which is composed of CNC-derived VSMCs, is the segment of the thoracic aorta that is more susceptible to calcification [45,46]. As VSMCs of different embryonic origins have heterogeneous patterns in response to environmental cues, genetic mutations that can affect intracellular signaling could also have heterogeneous effects on them. Patients with genetic defects in focal adhesion mediators or components of the TGF-β signaling cascade are predisposed to thoracic aortic aneurysms (TAAs) with manifestations of syndromic TAAs such as Marfan syndrome (MFS), Loeys–Dietz syndrome (LDS), or TAAs alone [47–55]. Importantly, almost all AAs in LDS and MFS patients occur in the aortic root and ascending arteries, and dissections often arise in the ascending aorta (Figure 1B), suggesting that heterogeneous lineage plasticity of VSMCs determines the pathological characteristics of hereditary aneurysms [56,57].

3.1. VSMC Phenotypic Diversity in LDS

LDS is an aneurysm-predisposing disease caused by defects in the canonical positive regulators of TGF-β, which regulate the differentiation and maturation of VSMCs [58,59]. Heterozygous loss-of-function mutations in genes encoding multiple components of the TGF-β signaling pathway, including the ligands (TGFB2/3), transforming growth factor-β receptor types I and II (TGFBR1 and TGFBR2) and downstream effectors (SMAD2/3), cause abnormalities in the function of VSMCs (Figure 1C).

Approximately 20–25% of patients with LDS have mutations in the *TGFBR1* genes [60]. The embryonic background of the VSMCs determines the effect of this genetic defect. SHF-derived VSMCs, but not CNC-derived VSMCs, exhibited impaired Smad2/3 activation, increased angiotensin II type 1 receptor (Agtr1a), and TGF-β ligands expression in the LDS model carrying heterozygous Tgfbr1 inactivation [61]. A human-induced pluripotent stem cell (hiPSC) model derived from an LDS patient family with the *TGFBR1*^{A230T} variant also detected disruption of SMAD3 and AKT activation and significantly reduced contractile transcript and protein levels in SHF-derived VSMCs, but not in CNC-derived

VSMCs [62]. Single-cell transcriptomic data revealed molecular similarities between SHF-derived VSMCs with the *TGFBR1^{A230T}* variant and SHF-derived VSMCs with a loss-of-function SMAD3 mutation. Additionally, pharmacological activation of the intracellular SMAD2/3 signaling cascade can rescue contractile gene expression and contractile function in TGFBR1 mutant SHF-derived VSMCs [62]. The above results indicate that TGFBR1 maintains the contractile phenotype of SHF-derived VSMCs through the activation of SMAD3. Smad3 interacts with multiple VSMC-specific promoters to facilitate a contractile phenotype [63]. The deficiency of SMAD3 in SHF-derived VSMCs induced from hiPSC significantly impaired canonical TGF-β signaling, decreased VSMC key regulators and markers, and increased collagen expression, suggesting a transition towards a less contractile phenotype [64]. Disruption of pathways related to ECM organization and down-regulation of key focal adhesion components, including integrins and anchoring protein, were also found in *Smad3$^{-/-}$* VSMCs derived from SHF [65]. Furthermore, selective knockdown of Smad3 in VSMC resulted in an up-regulation of pro-pathogenic factors such as thrombospondin-1, angiotensin-converting enzyme, and pro-inflammatory mediators in VSMC subsets located in the aortic root (mainly derived from SHF), leading to greater dilation and histologic abnormalities [65]. Reduced expression levels of contractile genes and dissociation from ECM components of SHF-derived VSMCs explain the reduction in vascular wall tone and local dilation in the aortic root [66]. CNC-derived VSMCs, however, were less affected by impaired Tgfbr1-Smad3 pathway and can always maintain the contractile phenotype by upregulating p-Smad2, suggesting a bypass might be exploited [61,62,64]. The inconsistency of phenotypic changes of CNC and SHF-VSMCs results in the decreased intensity of the interface between them in the ascending aorta, making it susceptible to type A aortic dissection (AD) [67]. Remarkably, although CNC-derived VSMCs retained significant VSMC markers in the presence of Tgfbr1-Smad3 dysregulation, elastin expression was significantly decreased, indicating phenotypic changes at the transcriptome level.

Approximately 55–60% of LDS patients carry *Tgfbr2* mutations [60]. The effect of Tgfbr2-mediated signaling activation on VSMC gene expression depends on the location of VSMCs in the aorta [68]. Smooth muscle specific *Tgfbr2* deficiency significantly enhances the progression of ascending but not abdominal aortic pathology, including increased intramural hemorrhage, medial thinning, and epicardial thickening, and damage was more pronounced outside of the media layer than inside [69,70]. Although rigorous experimental evidence such as lineage-stratified sc-RNAseq and lineage tracing is required, it is speculated that VSMCs of different embryonic origins respond heterogeneously to Tgfbr2 deficiency. Whether *Tgfbr2* exerts a facilitative or inhibitory effect on the differentiation of neural crest cells (NCCs) into VSMCs remains unclear. TGF-β signaling is involved in NCCs migration and differentiation and germline knockout of *Tgfbr2* in VSMCs leads to extensive and lethal cardiovascular malformation in mouse embryos [68,71–73]. *Tgfbr2* deficiency was found to hinder NCC-to-VSMC differentiation, resulting in a deficiency of CNC-derived VSMCs in the aorta [73,74]. However, recent studies have shown that *Tgfbr2* deficiency does not impede the VSMC fate of NCCs and may even lead to premature differentiation of NCCs into VSMCs [68,75,76]. Smad2 is confirmed as an essential regulator in progenitor-specific VSMC development and physiological differences between CNC- and mesoderm-derived VSMCs [72,77,78]. Although in vitro studies found Smad2 functionally intact in *Tgfbr2*-deficient VSMCs, animal models demonstrated reduced p-Smad2 levels [70,77,79]. Embryonic *Smad2* knockdown in NCCs reduces the number of CNC-derived VSMCs and produces a damaged, fragmented elastin lamina in the media of the ascending aorta [70,77,80]. Strikingly, postnatal knockdown of Tgfbr2 in VSMCs, while reducing Tgfb signaling, has little effect on contractile gene expression levels and contractile function in VSMCs, but rapidly induces elastin fragmentation and ascending thoracic aorta disruption (ATAD) [70,76]. Although conditional knockout of *Tgfbr2* at 11 months of age failed to induce TAAD, the ascending aorta developed pathological changes similar to those observed in elastin-deficient (*ELN$^{+/-}$*) mice [81,82]. Given that the deposition of the elastin membrane occurs only during early life and *TGFBR2* mutation was associated with a significantly higher risk of aortic events with childhood onset [83]. CNC-derived VSMCs are speculated to involve in the assembly and deposition of elastin lamina during embryonic development and are primarily responsible for the maintenance of elastin lamina integrity in adulthood [84,85]. Since the elastin lamina is a key component in maintaining the elasticity of the vessel wall, disruption of this structure impairs the absorption and buffering of kinetic energy from the blood flow by the aortic wall, resulting in a greater and more direct kinetic energy impact on the aorta. This provides a plausible explanation for the higher probability of developing ATAD and for the smaller mean arterial diameter at the time of dissection in LDS patients with Tgfbr2 mutations compared to LDS patients with Tgfbr1 mutations [86–88].

3.2. VSMC Phenotypic Diversity in MFS

FBN1 gene mutations predispose Marfan syndrome (MFS) patients to TAA at a young age and premature death from catastrophic aneurysm rupture or dissection [56]. The *FBN1* gene does not directly regulate the contraction-related phenotype of VSMC, but its coding product fibrillin-1 is an important component of the ECM and acts to mediate the attachment of VSMCs to elastin laminae [89]. Fibrillin-1, the encoded product of FBN1, is an important component of the ECM that mediates the attachment of VSMCs to elastin laminae, the loss of which would disrupt mechanotransduction and result in detached VSMCs with altered morphology, reduced contractile signatures, and upregulation of several ECM elements and elastolysis mediators indicating cellular plasticity. (Figure 1C) [90].

MFS VSMCs derived from iPSCs showed a lineage-specific protein profile. Compared to CNC-derived VSMCs, SHF-derived VSMCs have a proteome with less VSMC identity (*TAGLN*), but increased levels of focal adhesion components (integrin αV, fibronectin) and ECM remodeling molecules (collagen type 1, MMP2) [91]. As a novel protein marker associated with MFS aneurysms, uPARAP, an endocytic receptor responsible for receptor-mediated internalization of collagen for degradation, was specifically downregulated in SHF-derived VSMCs [91,92]. Recent work has shown that uPARAP also acts as an endocytic receptor for thrombospondin-1 (TSP-1), which has been reported to promote AAs by increasing aortic inflammation and TGF-β activity [93–97]. Single-cell transcriptomics provides evidence for lineage-specific VSMC plasticity in MFS aortic aneurysms. Albert and colleagues identified a group of modulated VSMCs (modVSMCs) in aortic aneurysm tissue from $Fbn1^{C1041G/+}$ mice with a transcriptome similar to that of VSMC-lineage-derived fibromyocytes in atheroma [98]. The combination of lineage tracing and scRNA-seq uncovered that SHF-derived VSMCs and CNC-derived VSMCs contribute equally to modVSMCs [99]. However, a lineage-stratified perspective reveals heterogeneous transcriptomes of VSMCs from different embryologic origins. SHF-derived VSMCs increased biosynthesis of collagens and small leucine-rich proteoglycans (SLRPs), contributing to the excessive deposition of ECM in MFS aortic aneurysms [98,99]. In contrast, CNC-derived VSMCs showed an osteochondrogenic phenotype, which is relevant for the medial calcification of the aorta [45,100]. In particular, few studies suggested that MFS aortic aneurysms have a hypercontractile phenotype with significant cytoskeletal density, upregulation of focal adhesions, and some contractile regulators presumably resulting from overactivation of TGF-β signaling [101,102]. Because the current scRNA-seq analyses focus on subdividing VSMCs with downregulated contractile markers, the hypercontractile VSMCs are likely to be buried in the normal contractile VSMC subgroup. Taking into account the lineage-specific response pattern of VSMCs to TGF-β, hypercontractile VSMCs, if they exist, are more possible to be derived from CNC-derived VSMCs.

4. VSMC Phenotypic Diversity in Non-Hereditary AAs

Apart from a few that can be explained by causative genetic mutations, most sporadic patients do not have genetic defects and these AAs are called non-hereditary AAs [103–105]. The etiology of non-hereditary AAs is still enigmatic and previous studies have shown that risk factors for non-hereditary AAs, such as hypertension, aging, and hyperlipidemia, transform contractile VSMCs into synthetic VSMCs [105–110]. However, the definition of synthetic VSMCs is ambiguous and recent studies demonstrate that they are actually conglomerates of several VSMC phenotypes with different functions (Figure 2 and Table 1).

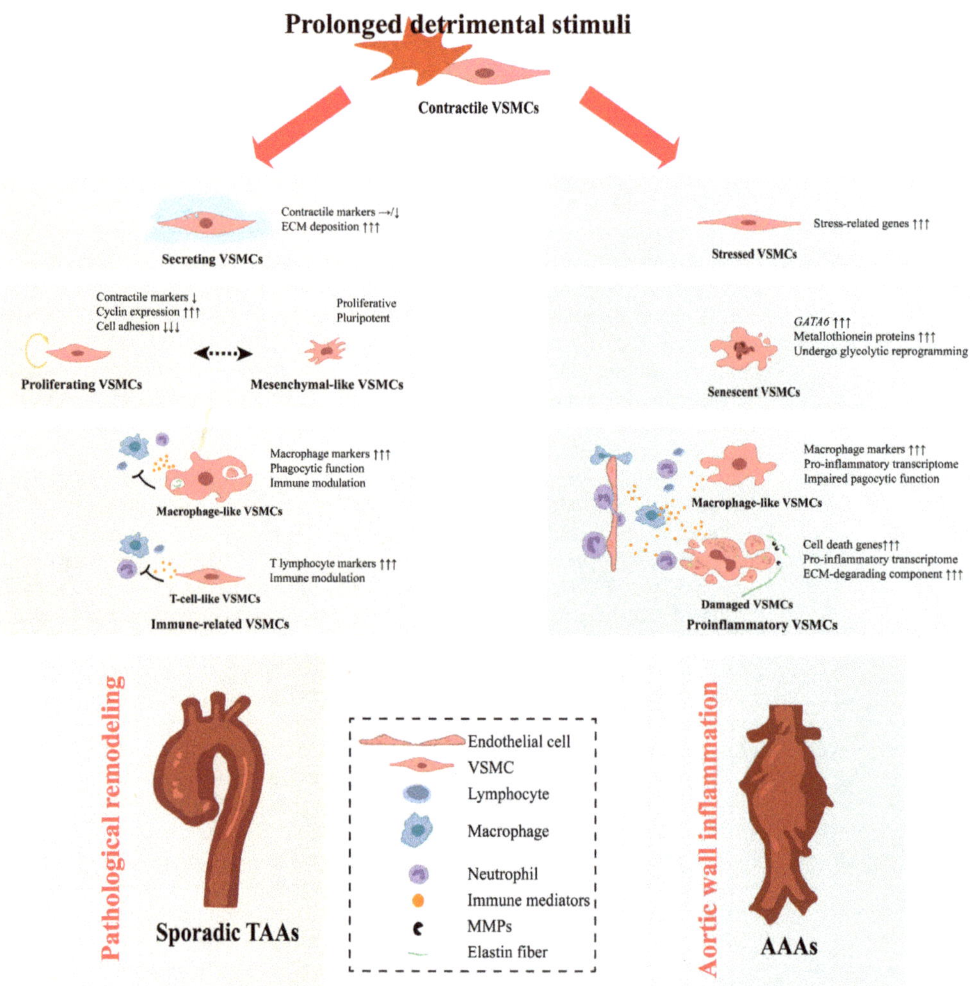

Figure 2. VSMC phenotypic diversity in non-hereditary AAs. VSMCs transform into different phenotypes and show a high degree of plasticity when exposed to detrimental stimuli. In the thoracic aorta, secreting VSMCs cause deposition of ECM components. Proliferative VSMCs undergo active proliferation and immune-related VSMCs will exert immune clearance and immunomodulatory effects. In the long term, these non-contractile phenotypes can result in pathological remodeling and, ultimately, to TAAs. In the abdominal aorta, stressed VSMCs are standing at the at the crossroads of necrosis/apoptosis or compensation. Pro-inflammatory VSMCs maintain and exacerbate vessel wall inflammation, eventually exhausting proliferative VSMCs and triggering a senescent phenotype, which leads to VSMC depletion and disruption of aortic wall integrity, triggering AAAs. Arrows indicates the gene expression changes: the number of arrows indicates the magnitude of expression change, and the direction of arrows indicates the trend, upward represent upregulation, horizontal indicate stable expression levels, and downward indicate downregulation1.

Table 1. Studies of VSMC phenotypes in non-hereditary AAs.

VSMC Phenotypes	Technique	Tissue	Annotation	Marker Genes	Characteristics
Secreting VSMC [111]	scRNA-seq	3 normal controls; 8 aneurysmal aorta samples from ATAA patients;	Fibromyocyte	*ACTA2; MYL9; COL1A2; COL8A1*	Share properties of fibroblasts and muscle cells; Upregulate collagen, proteoglycan genes.
Secreting VSMC [112]	scRNA-seq	3 normal controls; 8 aneurysmal aorta samples from ATAA patients;	Secreting VSMCs	*ACTA2; MYH11; COL1A1; COL1A2; FABP4*	Significantly increase their percentage in AA lesions.
Secreting VSMC [113]	scRNA-seq	3 normal controls; 8 aneurysmal aorta samples from ATAA patients;	Myofibroblast	*ACTA2; TAGLN; DCN; LUM; VCAN; CLU; FN1; LTBP2; CTGF; AEBP1*	Share characteristics of VSMCs and fibroblasts; Characterized by ECM organization, proliferation and migration.
Secreting VSMC [114]	scRNA-seq	WT and *Yap1$^{f/f}$; Myh11-CreER^{T2}* mice; Challenged by HFD and AngII	ECM-producing VSMCs	Not discussed	Not discussed
Proliferating VSMC [111]	scRNA-seq	3 normal controls; 8 aneurysmal aorta samples from ATAA patients;	Proliferating VSMC1	*TPM2; MAP1B; MYH11; CCND1; CALD1; MYH10; TPM4; FGF1; SPARC; FTH1*	Upregulate cyclin gene expression; Downregulate cell–cell junction; Express high level of synthetic VSMC marker genes; Express low levels of collagen and proteoglycan genes.
			Proliferating VSMC2	*GAS6; IGFBP2; MGP; FTH1; SPARC; FGF1; TPM4; MYH10; CCND1; TPM2*	Same as above
Proliferating VSMC [115]	Lineage tracing	*Apoe$^{-/-}$ (Apoe$^{-/-}$; Myh11CreERT2^{T2};mT/mG$^{f/f}$)* mice; TGFβR2^{iSMC} (*Myh11CreER^{T2};mT/mG$^{f/f}$;Tgfbr2$^{f/f}$*) mice; Challenged by HCHFD;	MSC-like VSMCs	*CD105; CD73; CD90; CD44; Sca-1*	Monoclonal expansion of lineage-traced VSMCs was found in 9/10 samples.
Proliferating VSMC [116]	Lineage tracing	*Myh11-CreER^{t2}/Rosa26-Confetti* mice; Challenged by AngII and anti-TGF;	/	/	Monochromatic patches of VSMCs were found in the medial layer of 5 out of the 6 animals analyzed.
Mesenchymal VSMC [112]	scRNA-seq	3 normal controls; 8 aneurysmal aorta samples from ATAA patients;	Msenchymal-like VSMCs	*ACTA2; MYH11; CD34; PDGFRA*	Play a poor role in the communication network with immune cells.
Immune-related VSMC [112]	scRNA-seq	3 normal controls; 8 aneurysmal aorta samples from ATAA patients;	Macrophage-like VSMCs	*ACTA2; MYH11; CD14; CD68*	Communicate with all immune cells and sent maximum number of the signaling pathways.

Table 1. *Cont.*

VSMC Phenotypes	Technique	Tissue	Annotation	Marker Genes	Characteristics
			T-cell-like VSMCs	*ACTA2; MYH11; CD3D; CD3G*	Act as a signal sender in the communication network with immune cells.
Immune-related VSMC [111]	scRNA-seq	3 normal controls; 8 aneurysmal aorta samples from ATAA patients;	Inflammatory1	*CXCR4; CCL4; CXCL12; TNF; CCL20; IFNG; XCL1; XCL2; LTB; CCL5; TNFSF9; IL32*	Show a T lymphocyte–like gene expression profile.
			Inflammatory2	*C1QA; C1QB; IL1RN; IL1B; IL18; CCL3; CCL4L2; CCL3L1; CXCL1; CXCL3; CXCL8; CXCL16*	Characterized by expression of macrophage markers.
			Inflammatory3	*IFIT1; IFI6; CXCL9; CXCL10; CCL28; IL15*	Expressed interferon-induced gene.
Immune-related VSMC [117]	scRNA-seq	4 healthy controls; 5 diseased aortas from ATAD patients;	Monocyte-like VSMCs	*CD93; THBD*	Characterized by expression of monocyte markers.
Immune-related VSMC [115]	Lineage tracing	$Apoe^{-/-}$ ($Apoe^{-/-}$; $Myh11CreERT2^{T2};mT/mG^{f/f}$) mice; $TGF\beta R2^{iSMC}$ ($Myh11CreER^{T2};mT/mG^{f/f};Tgfbr2^{f/f}$) mice; Challenged by HCHFD;	Macrophage-like cells	*Mac2*	Characterized by expression of macrophage markers.
Proinflammatory VSMC [118]	scRNA-seq	C57BL/6J mice; Challenged by elastase;	Inflammatory-like SMCs	*Mac2; Sparcl1; Igfbp5; Sncg; Thbs1; Notch3; Cd68; Cxcl1; Cxcl2; Il1r1*	Displayed a pro-inflammatory phenotype; Increased at late stage of AAA development.
Senescent VSMC [117]	scRNA-seq	4 healthy controls; 5 diseased aortas from ATAD patients;	VSMC8	*MT1G; MT1M*	Express metallothionein superfamily genes; Have minimum differentiation potential.
Senescent VSMC [118]	scRNA-seq	C57BL/6J mice; Challenged by elastase;	Dedifferentiated SMCs	*Ifrd1; Klf4; gata6; Mt1; Mt2; Hk2;*	Show the lowest expression of contractile markers; Have altered metabolic profile to glycolysis.
Stressed VSMC [111]	scRNA-seq	3 normal controls; 8 aneurysmal aorta samples from ATAA patients;	Stressed VSMCs	*FOS; ATF3; JUN; HSPB8; ACTC1; BRD2; KLF10; DUSP1;*	Share several features with contractile VSMCs, except the activation of stress response genes.
Stressed VSMC [118]	scRNA-seq	C57BL/6J mice; Challenged by elastase	Proliferative VSMCs	*Fos; Jun; Klf2; Klf4; gata6; Atf3*	Upregulate expression of genes associated with stress; Gain an inflammation-activated profile.

Vascular smooth muscle cell (VSMC), aortic aneurysm (AA), ascending thoracic aortic aneurysm (ATAA), single-cell RNA sequencing (scRNA-seq), high fat diet (HFD), mesenchymal stem cell (MSC), high cholesterol high fat diet (HCHFD), Angiotensin II (AngII), acute thoracic aortic dissection (ATAD).

4.1. VSMC Diversity in TAAs

4.1.1. Secreting VSMCs

Although investigators annotate VSMCs differently, several studies have found a subcluster of VSMCs positioned between VSMCs and fibroblasts in both UMAP and tSNE projections [111–113,119]. It maintains or moderately reduces the expression of contraction-related genes, but upregulates genes enriched in cell adhesion and ECM remodeling [15,16]. This subcluster is referred to as fibromyocytes or secreting VSMCs due to their properties of both contractile VSMCs and fibroblasts. Unlike "synthetic" VSMCs, secreting VSMCs do not activate cyclin and pro-inflammatory gene expression. Although rare in healthy aortic walls, secreting VSMCs significantly increased their proportion in aneurysmal aorta, further reducing VSMC identity markers such as *ACTA2* and *MYH11* and expressing ECM genes such as *SERPINE1* and *FN1* [119]. *VCAN*, a gene encoding Versican, a chondroitin sulfate proteoglycan composing the ECM, was identified as a marker of secreting VSMCs. Trajectory analysis of VCAN$^+$ VSMCs further supports the phenotypic transition from contractile VSMCs to fibroblasts [113,120]. It should be noted that the transition to secreting VSMCs is common in several aortic diseases. In atheroma, VSMC-traced scRNA-seq confirmed that the fibromyocytes are derived from contractile VSMCs, and they can be clustered with mod-VSMCs in MFS [98,121,122]. In addition, secreting VSMCs in hypertension-induced TAAs also have similar transcriptomes to modVSMCs in MFS [123]. Fibromyocytes contribute to the formation of the fibrous cap and are regarded as a protective factor in atherosclerosis [76,77] A protecting role of secreting VSMCs has also been suggested in TAAs, as they are associated with an increase in aortic wall thickness [114]. Interestingly, GO analysis showed that genes enriched in cardiac muscle cell development were highly expressed in secreting VSMCs which implies that secreting VSMCs may originate from a group of VSMCs regressed to an embryonic state and that HDAC9 and TGF-β pathways are involved in this change [119].

4.1.2. Proliferating VSMCs

VSMC proliferation is a hallmark of occlusive aortic diseases such as restenosis after angioplasty or stenting [124]. Upon injury to the vascular wall, the quiescent, differentiated VSMCs have been reported to be activated to a proliferative phenotype at low frequencies and such oligoclonal contribution of VSMCs is a feature of obstructive arterial disease [125–128]. Fate mapping analyses in TAA models induced by disruption of the TGF-β pathway also revealed oligoclonal expanding of the VSMCs [115,116]. Enhanced VSMC proliferation was demonstrated in sporadic TAAs [111,112,114]. In human TAA samples, sc-RNAseq identified a subpopulation of VSMCs characterized by high levels of cyclin expression and low levels of cell adhesion, which is highly suggestive of ongoing cell proliferation and was annotated as proliferating VSMCs [111]. Despite expressing some canonical synthetic marker genes such as *MGP*, *TPM4*, and *MYH10*, the proliferating VSMCs do not upregulate ECM and inflammatory genes, which is different from synthetic VSMCs [111]. In particular, the moderately reduced expression of contraction-related genes strongly suggests that this phenotype is activated from quiescent, mature VSMCs [111]. These results indicate that some mature VSMCs in the aortic wall have proliferative potential and can be activated to a proliferating state in TAAs. However, the identity of contractile VSMCs seems to be lost to varying degrees during the initiation of cell proliferation. Re-analysis of the sc-RNAseq data in reference [112] revealed the presence of mesenchymal-like VSMCs with a characteristic expression of ACTA2, MYH11, CD34, and PDGFRA and with a minor role in the communication network [112]. Fate mapping analyses have observed mesenchymal cells (osteoblasts, chondrocytes, adipocytes, and macrophages) derived from clone-expanded VSMCs in the adventitia of mice with TAAs, suggesting the pluripotency of proliferating VSMCs [115,116]. The aforementioned results suggest that the proliferating VSMCs represent a group of partially dedifferentiated VSMCs. However, it is still uncertain whether the proliferating VSMCs are the result of phenotypic transformation or clonal selection.

4.1.3. Immune-Related VSMCs

In the context of atherosclerosis, VSMCs have been shown to have the potential to switch to monocyte/macrophage lineages, implicating them in the modulation of inflammation in the vascular wall [128–131]. Although medial degeneration has been confirmed as the most common pathological manifestation of TAAs, aortitis, and atherosclerosis remain the main histopathological substrates in 1/3 of TAA patients [132,133]. The prevalence of inflammation-related histologic changes underscores the pivotal role of the inflammatory response in the pathophysiology of TAAs. VSMCs that exhibit similar characteristics of monocyte/macrophage and T-lymphocyte are identified in TAAs and referred to as immune-related VSMCs [111,112,117]. Additionally, an intermediate state, which expresses interferon-induced genes, has the potency to switch to T-cell-like VSMCs or macrophage-like VSMCs, suggesting that VSMCs are activated by inflammation [111]. Bioinformatics analysis revealed that immune-related VSMCs exert immunomodulatory functions through rich and strong interactions with immune cells [112]. T-cell-like VSMCs recruit immune cells mainly through the CXCL12-CXCR4 pair. Macrophage-like VSMCs in TAAs can communicate with all infiltrating immune cells through PTN and galectin, which are related to the suppressive immune microenvironment in malignancies [134–136]. The immunosuppressive function of macrophage-like VSMCs may explain why the inflammatory response in Ang II-induced TAAs is not as strong as in AAAs. Notably, macrophage-like VSMCs were detected in the adventitia and pseudolumen, accumulating in thrombotic/hemorrhagic regions and upregulating phagocytic and lysosomal pathways, which suggests that macrophage-like VSMCs have vigorous phagocytic activity [116]. In combination with the promotion of macrophage-mediated efferocytosis for the prevention and treatment of vascular disease, macrophage-like VSMCs in TAAs are hypothesized to play an immune clearance role by removing cells and ECM from injured sites in preparation for clones of proliferative VSMCs and ECM newly synthesized by secreting VSMCs to replace the injured tissue [137,138]. In addition, macrophage-like VSMCs, unlike terminally differentiated peripheral blood-derived macrophages, were recently found to be plastic and capable of changing their lineage characteristics toward fibroblasts and pericytes in atheroma, supporting a role for them in maintaining vessel wall stability by exerting a repair mechanism [131].

4.2. VSMC Diversity in AAAs

4.2.1. Proinflammatory VSMCs

AAA is considered to be a chronic inflammatory disease. Pro-inflammatory VSMCs defined by their function to promote aortic wall inflammation are identified. They have a proinflammatory transcriptome characterized by the expression of proinflammatory cytokines (IL-1ß, IL-6, CCL2, CCL5, and TNFα), chemokines (CXCL10 and ICAM1), and high expression of ECM-degrading components such as MMP2/9. Aortic wall inflammation can damage VSMCs, activate the expression of inflammation-related genes through innate immune pathways, and further promote vascular wall inflammation, forming a vicious cycle [139–141]. VSMCs expressing macrophage markers are also present in AAAs. [27,118]. In contrast to their immunomodulatory counterparts in TAAs, macrophage marker-expressing VSMCs in AAAs play an active role in inflammation and ECM degradation. Studies have shown that not only do these cells have high expression of pro-inflammatory cytokines and chemokines, but they also have impaired phagocytosis and efferocytosis and are unable to remove apoptotic VSMCs in a timely manner, perpetuating the pro-inflammatory effects of injured VSMCs [142–144]. Mechanically, disturbances in intracellular calcium ion balance drive the conversion of VSMCs into pro-inflammatory macrophages-like phenotype. Downregulation of the sarcoplasmic reticulum calcium-pumping ATPase was found in macrophage-like VSMCs in AAAs suggesting impaired intracellular calcium homeostasis [118]. Consistently, activation of P2Y2 receptors and TRPV4 channels on VSMCs by Panx1-mediated ATP release from endothelial cells destabilized intracellular Ca^{2+} homeostasis, stimulated proinflammatory cytokine secretion, and facilitated increased MMP2 activity [145]. Notably, antibody-dependent adaptive immune responses are involved. For example, IgG immune complex can polarize VSMCs from a contractile phenotype to a pro-inflammatory phenotype expressing M1 macrophage markers through FcγR in a TLR4-dependent manner [144,146].

4.2.2. Senescent VSMCs

Chronic inflammation of the vessel wall is bound to cause the death of VSMCs, and the depletion of contractile VSMCs is a hallmark of AAAs. The scRNA-seq analyses failed to identify any proliferating VSMCs in the AAA specimens, indicating the failure of the VSMCs to activate into a proliferative phenotype. Nevertheless, a group of VSMCs that specifically express high levels of GATA6 and undergo glycolytic reprogramming has been identified [118]. GATA6 plays a multifaceted role in regulating VSMC phenotype. GATA6 has been implicated in the differentiation of VSMCs, as studies have shown that specific knockout of GATA6 in VSMCs can activate the expression of synthetic markers [147,148]. Recently, however, it has been shown that GATA6 overexpression promotes TGF-β and MAPK signaling, resulting in increased proliferative and migratory capacity in VSMCs [149]. Notably, GATA6 is involved in the aging of mesenchymal stem cells by regulating the Hedgehog and FOXP1 signaling pathways [150]. Based on the increased expression of age-related metallothionein proteins, this VSMC subgroup is speculated to be a subset of senescent dedifferentiated VSMCs [151]. It is reasonable to speculate that senescence and loss of stemness of VSMCs are critical in the pathogenesis of AAAs, given the therapeutic efficacy of mesenchymal cell intervention for AAAs in preclinical studies [152]. It is also notable that He et al. identified a group of sparse VSMCs in patients with ATAD. These VSMCs simultaneously decreased the levels of VSMC markers, cell adhesion, the calcium-mediated signaling pathway, the cGMP metabolism, as well as the translation and RNA metabolism process. Due to the extremely low differentiation potential and the high expression of metallothionein, we speculate that these are also senescent VSMCs, although they are not annotated by the author [117].

4.2.3. Stressed VSMCs

Stress events such as oxidative stress, mitochondrial dysfunction, and unfolded protein response have been shown to alter the VSMC phenotype toward pro-inflammation and proliferation [153–156]. A group of VSMCs with transcriptomic features similar to contractile VSMCs, except a characteristically high expression of stress-related genes (*FOS, ATF3, JUN*, and *HSPB8*), were identified in TAAs and annotated as stressed VSMCs [111]. VSMCs with contractile VSMC transcriptome and high *FOS, Jun, Klf2*, and *ATF3* expression were also identified in elastase-induced AAAs [118]. Due to the relatively high expression of genes associated with proliferation, they were annotated as proliferative VSMCs by the other. However, the actual proliferative capacity of this group of cells is uncertain because they express a high level of mitogen-activated protein kinase phosphatase-1 (Dusp1), which inhibits proliferation by suppressing MAPK activity. In addition, upon elastase challenge, this group of cells exhibits an inflammation-activated profile, with significantly upregulated expression levels of pro-inflammatory cytokines (Cxcl2 and Ccl2). Therefore, it would be more appropriate to annotate them as stressed VSMCs [118]. Utilization of the transcription-dependent adaptive response implies that cells are subjected to prolonged stimuli, suggesting that they are at the crossroads of cell fate: necrosis/apoptosis or adaptive compensation. Given the significant reduction of medial VSMCs and the presence of a senescent phenotype in AAAs, stressed VSMCs in the abdominal aorta tend to go to extinction compared to their counterparts in the thoracic aorta, but the reasons for this remain to be elucidated.

5. Role of VSMC Phenotypic Diversity in AAs

In clinical samples from patients with AAs, changes in VSMC phenotype affect hemostasis of the arterial wall, and VSMC plasticity has been correlated with disease severity. Such studies, however, represent only the final stage of the disease. Thus, whether

VSMC heterogeneity and plasticity play a causative or compensatory role in AAs remains controversial. Animal models are tools for studying VSMC pathogenesis and phenotypic disruption of VSMCs can be inferred to cause hereditary aortic aneurysms from genetically engineered models [62,66,157]. VSMCs of different lineages respond differently to the same genetic mutation, and SHF-derived VSMCs are more phenotypically vulnerable than CNC-derived VSMCs. A recent study reported that the spatial specificity of TAAs has been linked to the downregulation of LRP and TGF-β signaling in SHF-derived VSMCs in response to Ang II stimulation [158]. Epigenetic mechanisms may account for the phenomenon that VSMCs of different lineages respond heterogeneously to the same signaling dysregulation. Aneurysmal lesions of MFS are characterized by dysregulated miRNA networks and gene expression profiles [159]. Epigenetic regulatory networks also have an impact on the response of VSMCs to pathogenic signals such as Ang II and TGF-β [160–162]. However, studies to date have mainly focused on epigenetic dysregulation in pathological states, and few have compared differences between VSMCs in different parts of the aorta in physiological or pathological states from an epigenetic perspective [163].

In contrast to hereditary AA, sporadic AA is a multifactorial disease that poses difficulties in the construction of animal models. At present, no animal model perfectly mimics human non-genetic AAs [164–166]. Results from non-hereditary models of AA support to some extent, the compensatory role of VSMC phenotypic transformation. To induce AA in WT mice, a minimum of two risk factors as well as sufficient challenge time are required. Challenging the aorta with a single risk factor such as Ang II infusion, hypertension, or hypercholesterolemia alone can elicit phenotypic changes of VSMCs but is not insufficient to induce AAs [105–110]. Extreme agents such as calcium chloride and elastase, which directly destroy the aortic wall, do induce AAs [167–169]. However, such "one-hit" models failed to reproduce the progression of human AAs, indicating that a stable adaptive state can be established [118,165]. Several VSMC subsets are thought to participate in compensatory responses to maintain aortic wall integrity. Collagen and proteoglycan produced by secreting VSMCs strengthen the aortic wall and prevent it from rupturing. The immune-related VSMCs suppress the immune response and assist in the removal of damaged cells from the aortic wall. The proliferative and MSC-like VSMCs are able to replicate and replace the damaged VSMCs that have been depleted by the immune cells. Even in inherited AA caused by genetic mutations that damage the "elastin-VSMC contractile unit" represented by MFS, the compensatory effect of the VSMC was observed [53,66,170]. In MFS aortic aneurysms, compensatory manifestations of impaired aortic wall force generation include the transition of contractile VSMCs to modVSMCs with increased ECM production capabilities and hypercontractile VSMCs and thickened aortic walls.

Despite the potential protective effects of the phenotypic alteration of VSMCs, the development and progression of sporadic AAs are indicative of maladaptation. We propose that the sustained presence of noxious stimuli causes a prolonged phenotypic transformation of VSMCs, which causes pathological remodeling of the aortic wall and eventually leads to non-hereditary AAs. Proteoglycan and collagen accumulation prevent aortic dissection and rupture, however, overaccumulation of proteoglycans and collagen results in stiffening of the aortic wall which disrupts the phenotypic change of VSMCs [171,172]. Elevated ECM stiffness has been reported to polarize contractile VSMCs toward proinflammatory phenotypes through mechanotransduction [173–176]. Chronic vascular wall inflammation then leads to persistent programmed death of VSMCs, ultimately exhausting the proliferative potential of VSMCs. Ideally, the transition to proliferative VSMCs is reversible and proliferative VSMCs can re-differentiate into mature, contractile VSMCs. Nevertheless, if the stimuli are prolonged, epigenetic changes can occur, making the change irreversible [16,177,178]. In addition, the precise microenvironment that regulates VSMC phenotype is perturbed in the pathologically remodeled aortic wall, resulting in disrupted VSMC differentiation. For example, an aberrant equilibrium between ECM stiffness and growth factors in aneurysmal tissue will impair the differentiation of MSC-like cells toward contractile VSMCs [179,180]. Moreover, high levels of TGF-β activation in AAs are thought to over-activate nonclassical TGF-β signaling in VSMCs, which breaks the normal physiological balance and results in AAs [181]. Chronic activations of mTOR signaling would elevate the expression and activity of MMPs derived from VSMCs, resulting in irreversible destruction of the elastin membrane and VSMC microcalcifications [100,182,183].

6. Conclusions and Future Perspectives

The arterial wall is composed of VSMCs of distinct origins that exhibit different transcriptomic profiles and dynamic phenotypes to adapt to changing environments. Their unique distribution and lineage-specific biological behaviors in the presence of AA-causing gene mutations determine the segmental vulnerability of hereditary AAs. VSMCs are highly plastic and can transform into cellular phenotypes with distinct functions in non-hereditary AAs. Reversible phenotypic transformation might play a beneficial role in maintaining the integrity of the aortic wall. However, chronic exposure to risk factors leads to irreversible phenotypic changes in VSMCs and to pathological remodeling of the aortic wall, which contributes to the pathogenesis and development of non-hereditary aortic aneurysms.

Lineage tracing and sc-RNAseq have allowed detailed phenotyping and revealed a high degree of diversity of VSMCs in AAs. Specific patterns of modulating key pathways that regulate contractile phenotypic transformation are observed in VSMCs from different embryological backgrounds. Heterogeneous responses can be induced by genetic defects or environmental cues, as long as they can affect any part of the pathway. Thus, part of the intrinsic nature of aortic segments can be attributed to the lineage-specific VSMC response to perturbation. For instance, the unique composition of VSMCs derived from the splanchnic mesoderm contributes, to some extent, to the inflammatory tendency of the abdominal aorta. However, little is currently known about VSMCs other than those derived from SHF and CNC, and the importance of lineage stratification needs to be emphasized. In addition, comprehensive and in-depth studies are needed to depict the molecular regulatory profiles specific to different lineages of VSMCs.

VSMCs can switch to a variety of functionally distinct phenotypes. How VSMC plasticity contributes to the pathogenesis and progression of AAs remains to be fully elucidated. It is important to note that the transcriptome level heterogeneity obtained by sc-RNAseq might be a reflection of transient cell states rather than stable phenotypes. Our conclusions are mainly based on the results of sc-RNAseq, but technical limitations must be considered to fully evaluate our results. For example, heterogeneity at the transcriptome level may reflect only transient cellular states and not necessarily stable cellular phenotypes. To demonstrate the existence of specific dysfunctional phenotypes, the next step will be to attempt to isolate and induce stable VSMC subsets revealed by sc-RNAseq and summarize their ultra-structural features, which are currently not available. Furthermore, sc-RNAseq can only capture the transcriptome profile of specific developmental stages. Comparison and evaluation of VSMC phenotypes at different stages of AA development will help us to understand the pattern of VSMC phenotypic changes and their impact on aortic wall integrity which we believe that will put an end to the controversy about the causal/successive relationship between VSMC plasticity and the development of AAs and provide a theoretical basis for VSMC phenotype modulation as a potential pharmacological intervention.

Author Contributions: Y.H. and Z.C. did the majority of the literature searches and created the initial draft, including its organization. Z.C. and B.H. revised and edited the manuscript. All authors have read and agreed to the published version of the manuscript.

Funding: This work was supported by the National Natural Science Foundation of China (Grant Numbers: 82130012 and 81830010) and the Nurture projects for basic research of Shanghai Chest Hospital (Grant Number: 2022YNJCQ03).

Institutional Review Board Statement: Not applicable.

Informed Consent Statement: Not applicable.

Data Availability Statement: The authors confirm that this is a review of the literature and there is no primary data associated with this manuscript.

References

1. Shen, Y.H.; LeMaire, S.A.; Webb, N.R.; Cassis, L.A.; Daugherty, A.; Lu, H.S. Aortic Aneurysms and Dissections Series. *Arterioscler. Thromb. Vasc. Biol.* **2020**, *40*, e37–e46. [CrossRef] [PubMed]
2. Pinard, A.; Jones, G.T.; Milewicz, D.M. Genetics of Thoracic and Abdominal Aortic Diseases. *Circ. Res.* **2019**, *124*, 588–606. [CrossRef] [PubMed]
3. Arnaud, P.; Hanna, N.; Benarroch, L.; Aubart, M.; Bal, L.; Bouvagnet, P.; Busa, T.; Dulac, Y.; Dupuis-Girod, S.; Edouard, T.; et al. Genetic diversity and pathogenic variants as possible predictors of severity in a French sample of nonsyndromic heritable thoracic aortic aneurysms and dissections (nshTAAD). *Genet Med.* **2019**, *21*, 2015–2024. [CrossRef] [PubMed]
4. Bradley, D.; Badger, S.; McFarland, M.; Hughes, A. Abdominal Aortic Aneurysm Genetic Associations: Mostly False? A Systematic Review and Meta-analysis. *Eur. J. Vasc. Endovasc. Surg.* **2015**, *51*, 64–75. [CrossRef] [PubMed]
5. Dong, C.X.; Malecki, C.; Robertson, E.; Hambly, B.; Jeremy, R. Molecular Mechanisms in Genetic Aortopathy–Signaling Pathways and Potential Interventions. *Int. J. Mol. Sci.* **2023**, *24*, 1795. [CrossRef]
6. Davis, E.C. Smooth muscle cell to elastic lamina connections in developing mouse aorta. Role in aortic medial organization. *Lab. Investig.* **1993**, *68*, 89–99.
7. Chamley-Campbell, J.H.; Campbell, G.R. What controls smooth muscle phenotype? *Atherosclerosis* **1981**, *40*, 347–357. [CrossRef]
8. Pyun, J.-H.; Ahn, B.-Y.; Vuong, T.A.; Kim, S.W.; Jo, Y.; Jeon, J.; Baek, S.H.; Kim, J.; Park, S.; Bae, G.-U.; et al. Inducible Prmt1 ablation in adult vascular smooth muscle leads to contractile dysfunction and aortic dissection. *Exp. Mol. Med.* **2021**, *53*, 1569–1579. [CrossRef]
9. Chiarini, A.; Onorati, F.; Marconi, M.; Pasquali, A.; Patuzzo, C.; Malashicheva, A.; Irtyega, O.; Faggian, G.; Pignatti, P.F.; Trabetti, E.; et al. Studies on sporadic non-syndromic thoracic aortic aneurysms: II. Alterations of extra-cellular matrix components and focal adhesion proteins. *Eur. J. Prev. Cardiol.* **2018**, *25*, 51–58. [CrossRef]
10. Chiarini, A.; Onorati, F.; Marconi, M.; Pasquali, A.; Patuzzo, C.; Malashicheva, A.; Irtyega, O.; Faggian, G.; Pignatti, P.F.; Trabetti, E.; et al. Studies on sporadic non-syndromic thoracic aortic aneurysms: 1. Deregulation of Jagged/Notch 1 homeostasis and selection of synthetic/secretor phenotype smooth muscle cells. *Eur. J. Prev. Cardiol.* **2018**, *25*, 42–50. [CrossRef]
11. Bochaton-Piallat, M.-L.; Bäck, M. Novel concepts for the role of smooth muscle cells in vascular disease: Towards a new smooth muscle cell classification. *Cardiovasc. Res.* **2018**, *114*, 477–480. [CrossRef]
12. Stenmark, K.R.; Frid, M.G.; Graham, B.B.; Tuder, R.M. Dynamic and diverse changes in the functional properties of vascular smooth muscle cells in pulmonary hypertension. *Cardiovasc. Res.* **2018**, *114*, 551–564. [CrossRef]
13. Frid, M.G.; Dempsey, E.C.; Durmowicz, A.G.; Stenmark, K.R. Smooth Muscle Cell Heterogeneity in Pulmonary and Systemic Vessels. Importance in vascular disease. *Arterioscler. Thromb. Vasc. Biol.* **1997**, *17*, 1203–1209. [CrossRef]
14. Allahverdian, S.; Chaabane, C.; Boukais, K.; Francis, G.A.; Bochaton-Piallat, M.-L. Smooth muscle cell fate and plasticity in atherosclerosis. *Cardiovasc. Res.* **2018**, *114*, 540–550. [CrossRef]
15. Gomez, D.; Owens, G.K. Smooth muscle cell phenotypic switching in atherosclerosis. *Cardiovasc. Res.* **2012**, *95*, 156–164. [CrossRef]
16. Campbell, J.H.; Campbell, G.R. Smooth Muscle Phenotypic Modulation—A Personal Experience. *Arterioscler. Thromb. Vasc. Biol.* **2012**, *32*, 1784–1789. [CrossRef]
17. Jauhiainen, S.; Kiema, M.; Hedman, M.; Laakkonen, J.P. Large Vessel Cell Heterogeneity and Plasticity: Focus in Aortic Aneurysms. *Arterioscler. Thromb. Vasc. Biol.* **2022**, *42*, 811–818. [CrossRef]
18. Liu, M.; Gomez, D. Smooth Muscle Cell Phenotypic Diversity. *Arterioscler. Thromb. Vasc. Biol.* **2019**, *39*, 1715–1723. [CrossRef]
19. Majesky, M.W. Developmental Basis of Vascular Smooth Muscle Diversity. *Arterioscler. Thromb. Vasc. Biol.* **2007**, *27*, 1248–1258. [CrossRef]
20. Waldo, K.L.; Hutson, M.R.; Ward, C.C.; Zdanowicz, M.; Stadt, H.A.; Kumiski, D.; Abu-Issa, R.; Kirby, M.L. Secondary heart field contributes myocardium and smooth muscle to the arterial pole of the developing heart. *Dev. Biol.* **2005**, *281*, 78–90. [CrossRef]
21. Jiang, X.; Rowitch, D.H.; Soriano, P.; McMahon, A.P.; Sucov, H.M. Fate of the mammalian cardiac neural crest. *Development* **2000**, *127*, 1607–1616. [CrossRef] [PubMed]
22. Wasteson, P.; Johansson, B.R.; Jukkola, T.; Breuer, S.; Akyürek, L.M.; Partanen, J.; Lindahl, P. Developmental origin of smooth muscle cells in the descending aorta in mice. *Development* **2008**, *135*, 1823–1832. [CrossRef] [PubMed]
23. Sawada, H.; Rateri, D.L.; Moorleghen, J.J.; Majesky, M.W.; Daugherty, A. Smooth Muscle Cells Derived From Second Heart Field and Cardiac Neural Crest Reside in Spatially Distinct Domains in the Media of the Ascending Aorta—Brief Report. *Arterioscler. Thromb. Vasc. Biol.* **2017**, *37*, 1722–1726. [CrossRef] [PubMed]
24. Winkels, H.; Ehinger, E.; Vassallo, M.; Buscher, K.; Dinh, H.Q.; Kobiyama, K.; Hamers, A.A.; Cochain, C.; Vafadarnejad, E.; Saliba, A.-E.; et al. Atlas of the Immune Cell Repertoire in Mouse Atherosclerosis Defined by Single-Cell RNA-Sequencing and Mass Cytometry. *Circ. Res.* **2018**, *122*, 1675–1688. [CrossRef]

25. Kalluri, A.S.; Vellarikkal, S.K.; Edelman, E.R.; Nguyen, L.; Subramanian, A.; Ellinor, P.T.; Regev, A.; Kathiresan, S.; Gupta, R.M. Single-Cell Analysis of the Normal Mouse Aorta Reveals Functionally Distinct Endothelial Cell Populations. *Circulation* **2019**, *140*, 147–163. [CrossRef]
26. Dobnikar, L.; Taylor, A.L.; Chappell, J.; Oldach, P.; Harman, J.L.; Oerton, E.; Dzierzak, E.; Bennett, M.R.; Spivakov, M.; Jørgensen, H.F. Disease-relevant transcriptional signatures identified in individual smooth muscle cells from healthy mouse vessels. *Nat. Commun.* **2018**, *9*, 4567. [CrossRef]
27. Yu, L.; Zhang, J.; Gao, A.; Zhang, M.; Wang, Z.; Yu, F.; Guo, X.; Su, G.; Zhang, Y.; Zhang, C. An intersegmental single-cell profile reveals aortic heterogeneity and identifies a novel Malat1+ vascular smooth muscle subtype involved in abdominal aortic aneurysm formation. *Signal Transduct. Target. Ther.* **2022**, *7*, 125. [CrossRef]
28. Upadhyay, G.; Yin, Y.; Yuan, H.; Li, X.; Derynck, R.; Glazer, R.I. Stem cell antigen-1 enhances tumorigenicity by disruption of growth differentiation factor-10 (GDF10)–dependent TGF-β signaling. *Proc. Natl. Acad. Sci. USA* **2011**, *108*, 7820–7825. [CrossRef]
29. Hu, Y.; Zhang, Z.; Torsney, E.; Afzal, A.R.; Davison, F.; Metzler, B.; Xu, Q. Abundant progenitor cells in the adventitia contribute to atherosclerosis of vein grafts in ApoE-deficient mice. *J. Clin. Investig.* **2004**, *113*, 1258–1265. [CrossRef]
30. Wang, G.; Jacquet, L.; Karamariti, E.; Xu, Q. Origin and differentiation of vascular smooth muscle cells. *J. Physiol.* **2015**, *593*, 3013–3030. [CrossRef]
31. Sajeesh, S.; Broekelman, T.; Mecham, R.P.; Ramamurthi, A. Stem cell derived extracellular vesicles for vascular elastic matrix regenerative repair. *Acta Biomater.* **2020**, *113*, 267–278. [CrossRef]
32. Sainz, J.; Zen, A.A.H.; Caligiuri, G.; Demerens, C.; Urbain, D.; Lemitre, M.; Lafont, A. Isolation of "Side Population" Progenitor Cells From Healthy Arteries of Adult Mice. *Arterioscler. Thromb. Vasc. Biol.* **2006**, *26*, 281–286. [CrossRef]
33. Sugiura, T.; Freis, E.D. Pressure Pulse in Small Arteries. *Circ. Res.* **1962**, *11*, 838–842. [CrossRef]
34. Leuprecht, A.; Perktold, K.; Kozerke, S.; Boesiger, P. Combined CFD and MRI study of blood flow in a human ascending aorta model. *Biorheology* **2002**, *39*, 425–429.
35. Vincent, P.E.; Plata, A.M.; Hunt, A.A.E.; Weinberg, P.D.; Sherwin, S.J. Blood flow in the rabbit aortic arch and descending thoracic aorta. *J. R. Soc. Interface* **2011**, *8*, 1708–1719. [CrossRef]
36. Stein, P.D.; Sabbah, H.N. Turbulent blood flow in the ascending aorta of humans with normal and diseased aortic valves. *Circ. Res.* **1976**, *39*, 58–65. [CrossRef]
37. Chandran, K.B. Flow Dynamics in the Human Aorta. *J. Biomech. Eng.* **1993**, *115*, 611–616. [CrossRef]
38. Cheung, C.; Bernardo, A.S.; Trotter, M.W.B.; Pedersen, R.A.; Sinha, S. Generation of human vascular smooth muscle subtypes provides insight into embryological origin–dependent disease susceptibility. *Nat. Biotechnol.* **2012**, *30*, 165–173. [CrossRef]
39. Gadson, P.F., Jr.; Dalton, M.L.; Patterson, E.; Svoboda, D.D.; Hutchinson, L.; Schram, D.; Rosenquist, T.H. Differential Response of Mesoderm- and Neural Crest-Derived Smooth Muscle to TGF-β1: Regulation of c-myb and α1 (I) Procollagen Genes. *Exp. Cell Res.* **1997**, *230*, 169–180. [CrossRef]
40. Topouzis, S.; Majesky, M.W. Smooth Muscle Lineage Diversity in the Chick Embryo. Two types of aortic smooth muscle cell differ in growth and receptor-mediated transcriptional responses to transforming growth factor-beta. *Dev. Biol.* **1996**, *178*, 430–445. [CrossRef]
41. Thieszen, S.L.; Dalton, M.; Gadson, P.F.; Patterson, E.; Rosenquist, T.H. Embryonic Lineage of Vascular Smooth Muscle Cells Determines Responses to Collagen Matrices and Integrin Receptor Expression. *Exp. Cell Res.* **1996**, *227*, 135–145. [CrossRef] [PubMed]
42. Oyama, N.; Gona, P.; Salton, C.J.; Chuang, M.L.; Jhaveri, R.R.; Blease, S.J.; Manning, A.R.; Lahiri, M.; Botnar, R.M.; Levy, D.; et al. Differential Impact of Age, Sex, and Hypertension on Aortic Atherosclerosis: The Framingham Heart Study. *Arterioscler. Thromb. Vasc. Biol.* **2008**, *28*, 155–159. [CrossRef] [PubMed]
43. Neisius, U.; Gona, P.N.; Oyama-Manabe, N.; Chuang, M.L.; O'donnell, C.J.; Manning, W.J.; Tsao, C.W. Relation of MRI Aortic Wall Area and Plaque to Incident Cardiovascular Events: The Framingham Heart Study. *Radiology* **2022**, *304*, 542–550. [CrossRef] [PubMed]
44. Haimovici, H.; Maier, N. Fate of aortic homografts in canine atherosclerosis. 3. study of fresh abdominal and thoracic aortic implants into thoracic aorta: Role of tissue susceptibility in atherogenesis. *Arch. Surg.* **1964**, *89*, 961–969. [CrossRef]
45. Leroux-Berger, M.; Queguiner, I.; Maciel, T.T.; Ho, A.; Relaix, F.; Kempf, H. Pathologic calcification of adult vascular smooth muscle cells differs on their crest or mesodermal embryonic origin. *J. Bone Miner. Res.* **2011**, *26*, 1543–1553. [CrossRef]
46. Muyor, K.; Laget, J.; Cortijo, I.; Duranton, F.; Jover, B.; Argilés, À.; Gayrard, N. Vascular calcification in different arterial beds in ex vivo ring culture and in vivo rat model. *Sci. Rep.* **2022**, *12*, 11861. [CrossRef]
47. Li, Y.; Fang, M.; Yang, J.; Yu, C.; Kuang, J.; Sun, T.; Fan, R. Analysis of the contribution of 129 candidate genes to thoracic aortic aneurysm or dissection of a mixed cohort of sporadic and familial cases in South China. *Am. J. Transl. Res.* **2021**, *13*, 4281–4295.
48. Isselbacher, E.M.; Cardenas, C.L.; Lindsay, M.E. Hereditary Influence in Thoracic Aortic Aneurysm and Dissection. *Circulation* **2016**, *133*, 2516–2528. [CrossRef]
49. Renard, M.; Francis, C.; Ghosh, R.; Scott, A.F.; Witmer, P.D.; Adès, L.C.; Andelfinger, G.U.; Arnaud, P.; Boileau, C.; Callewaert, B.L.; et al. Clinical Validity of Genes for Heritable Thoracic Aortic Aneurysm and Dissection. *J. Am. Coll. Cardiol.* **2018**, *72*, 605–615. [CrossRef]
50. Sakai, L.Y.; Keene, D.R. Fibrillin protein pleiotropy: Acromelic dysplasias. *Matrix Biol.* **2019**, *80*, 6–13. [CrossRef]

51. Prakash, S.K.; LeMaire, S.A.; Guo, D.-C.; Russell, L.; Regalado, E.S.; Golabbakhsh, H.; Johnson, R.J.; Safi, H.J.; Estrera, A.L.; Coselli, J.S.; et al. Rare Copy Number Variants Disrupt Genes Regulating Vascular Smooth Muscle Cell Adhesion and Contractility in Sporadic Thoracic Aortic Aneurysms and Dissections. *Am. J. Hum. Genet.* **2010**, *87*, 743–756. [CrossRef]
52. Guo, J.; Cai, L.; Jia, L.; Li, X.; Xi, X.; Zheng, S.; Liu, X.; Piao, C.; Liu, T.; Sun, Z.; et al. Wide mutation spectrum and frequent variant Ala27Thr of FBN1 identified in a large cohort of Chinese patients with sporadic TAAD. *Sci. Rep.* **2015**, *5*, 13115. [CrossRef]
53. Li, Y.; Gao, S.; Han, Y.; Song, L.; Kong, Y.; Jiao, Y.; Huang, S.; Du, J.; Li, Y. Variants of Focal Adhesion Scaffold Genes Cause Thoracic Aortic Aneurysm. *Circ. Res.* **2021**, *128*, 8–23. [CrossRef]
54. Liu, C.-L.; Liu, X.; Zhang, Y.; Liu, J.; Yang, C.; Luo, S.; Liu, T.; Wang, Y.; Lindholt, J.S.; Diederichsen, A.; et al. Eosinophils Protect Mice From Angiotensin-II Perfusion–Induced Abdominal Aortic Aneurysm. *Circ. Res.* **2021**, *128*, 188–202. [CrossRef]
55. Benjamins, J.W.; Yeung, M.W.; van de Vegte, Y.J.; Said, M.A.; van der Linden, T.; Ties, D.; Juarez-Orozco, L.E.; Verweij, N.; van der Harst, P. Genomic insights in ascending aortic size and distensibility. *Ebiomedicine* **2022**, *75*, 103783. [CrossRef]
56. Milewicz, D.M.; Braverman, A.C.; De Backer, J.; Morris, S.A.; Boileau, C.; Maumenee, I.H.; Jondeau, G.; Evangelista, A.; Pyeritz, R.E. Marfan syndrome. *Nat. Rev. Dis. Primers* **2021**, *7*, 64. [CrossRef]
57. Van Laer, L.; Dietz, H.; Loeys, B. Loeys-Dietz syndrome. *Adv. Exp. Med. Biol.* **2014**, *802*, 95–105. [CrossRef]
58. Gouda, P.; Kay, R.; Habib, M.; Aziz, A.; Aziza, E.; Welsh, R. Clinical features and complications of Loeys-Dietz syndrome: A systematic review. *Int. J. Cardiol.* **2022**, *362*, 158–167. [CrossRef]
59. Schoenhoff, F.S.; Alejo, D.E.; Black, J.H.; Crawford, T.C.; Dietz, H.C.; Grimm, J.C.; Magruder, J.T.; Patel, N.D.; Vricella, L.A.; Young, A.; et al. Management of the aortic arch in patients with Loeys–Dietz syndrome. *J. Thorac. Cardiovasc. Surg.* **2020**, *160*, 1166–1175. [CrossRef]
60. Meester, J.A.N.; Verstraeten, A.; Schepers, D.; Alaerts, M.; Van Laer, L.; Loeys, B.L. Differences in manifestations of Marfan syndrome, Ehlers-Danlos syndrome, and Loeys-Dietz syndrome. *Ann. Cardiothorac. Surg.* **2017**, *6*, 582–594. [CrossRef]
61. MacFarlane, E.G.; Parker, S.J.; Shin, J.Y.; Ziegler, S.G.; Creamer, T.J.; Bagirzadeh, R.; Bedja, D.; Chen, Y.; Calderon, J.F.; Weissler, K.; et al. Lineage-specific events underlie aortic root aneurysm pathogenesis in Loeys-Dietz syndrome. *J. Clin. Investig.* **2019**, *129*, 659–675. [CrossRef] [PubMed]
62. Zhou, D.; Feng, H.; Yang, Y.; Huang, T.; Qiu, P.; Zhang, C.; Olsen, T.R.; Zhang, J.; Chen, Y.E.; Mizrak, D.; et al. hiPSC Modeling of Lineage-Specific Smooth Muscle Cell Defects Caused by *TGFBR1* ^{A230T} Variant, and Its Therapeutic Implications for Loeys-Dietz Syndrome. *Circulation* **2021**, *144*, 1145–1159. [CrossRef] [PubMed]
63. Mack, C.P. Signaling Mechanisms That Regulate Smooth Muscle Cell Differentiation. *Arterioscler. Thromb. Vasc. Biol.* **2011**, *31*, 1495–1505. [CrossRef]
64. Gong, J.; Zhou, D.; Jiang, L.; Qiu, P.; Milewicz, D.M.; Chen, Y.E.; Yang, B. In Vitro Lineage-Specific Differentiation of Vascular Smooth Muscle Cells in Response to SMAD3 Deficiency: Implications for SMAD3-Related Thoracic Aortic Aneurysm. *Arterioscler. Thromb. Vasc. Biol.* **2020**, *40*, 1651–1663. [CrossRef] [PubMed]
65. Bramel, E.E.; Creamer, T.J.; Saqib, M.; Nunez, W.A.C.; Bagirzadeh, R.; Roker, L.A.; Goff, L.A.; MacFarlane, E.G. Postnatal Smad3 Inactivation in Murine Smooth Muscle Cells Elicits a Temporally and Regionally Distinct Transcriptional Response. *Front. Cardiovasc. Med.* **2022**, *9*, 826495. [CrossRef]
66. Milewicz, D.M.; Trybus, K.M.; Guo, D.-C.; Sweeney, H.L.; Regalado, E.; Kamm, K.; Stull, J.T. Altered Smooth Muscle Cell Force Generation as a Driver of Thoracic Aortic Aneurysms and Dissections. *Arterioscler. Thromb. Vasc. Biol.* **2017**, *37*, 26–34. [CrossRef]
67. Seike, Y.; Matsuda, H.; Ishibashi-Ueda, H.; Morisaki, H.; Morisaki, T.; Minatoya, K.; Ogino, H. Surgical Outcome and Histological Differences between Individuals with TGFBR1 and TGFBR2 Mutations in Loeys-Dietz Syndrome. *Ann. Thorac. Cardiovasc. Surg.* **2021**, *27*, 56–63. [CrossRef]
68. Jaffe, M.; Sesti, C.; Washington, I.M.; Du, L.; Dronadula, N.; Chin, M.T.; Stolz, D.B.; Davis, E.C.; Dichek, D.A. Transforming Growth Factor-β Signaling in Myogenic Cells Regulates Vascular Morphogenesis, Differentiation, and Matrix Synthesis. *Arterioscler. Thromb. Vasc. Biol.* **2012**, *32*, e1–e11. [CrossRef]
69. Angelov, S.N.; Hu, J.H.; Wei, H.; Airhart, N.; Shi, M.; Dichek, D.A. TGF-β (Transforming Growth Factor-β) Signaling Protects the Thoracic and Abdominal Aorta From Angiotensin II-Induced Pathology by Distinct Mechanisms. *Arterioscler. Thromb. Vasc. Biol.* **2017**, *37*, 2102–2113. [CrossRef]
70. Li, W.; Li, Q.; Jiao, Y.; Qin, L.; Ali, R.; Zhou, J.; Ferruzzi, J.; Kim, R.W.; Geirsson, A.; Dietz, H.C.; et al. Tgfbr2 disruption in postnatal smooth muscle impairs aortic wall homeostasis. *J. Clin. Investig.* **2014**, *124*, 755–767. [CrossRef]
71. Gittenbergerdegroot, A.; Azhar, M.; Molin, D. Transforming Growth Factor β–SMAD2 Signaling and Aortic Arch Development. *Trends Cardiovasc. Med.* **2006**, *16*, 1–6. [CrossRef] [PubMed]
72. Molin, D.G.; Poelmann, R.E.; DeRuiter, M.C.; Azhar, M.; Doetschman, T.; Groot, A.C.G.-D. Transforming Growth Factor β–SMAD2 Signaling Regulates Aortic Arch Innervation and Development. *Circ. Res.* **2004**, *95*, 1109–1117. [CrossRef] [PubMed]
73. Langlois, D.; Hneino, M.; Bouazza, L.; Parlakian, A.; Sasaki, T.; Bricca, G.; Li, J.Y. Conditional inactivation of TGF-β type II receptor in smooth muscle cells and epicardium causes lethal aortic and cardiac defects. *Transgenic Res.* **2010**, *19*, 1069–1082. [CrossRef] [PubMed]
74. Wurdak, H.; Ittner, L.M.; Lang, K.S.; Leveen, P.; Suter, U.; Fischer, J.A.; Karlsson, S.; Born, W.; Sommer, L. Inactivation of TGFβ signaling in neural crest stem cells leads to multiple defects reminiscent of DiGeorge syndrome. *Genes Dev.* **2005**, *19*, 530–535. [CrossRef]

75. Choudhary, B.; Ito, Y.; Makita, T.; Sasaki, T.; Chai, Y.; Sucov, H.M. Cardiovascular malformations with normal smooth muscle differentiation in neural crest-specific type II TGFβ receptor (Tgfbr2) mutant mice. *Dev. Biol.* **2006**, *289*, 420–429. [CrossRef]

76. Zhu, J.; Angelov, S.; Yildirim, I.A.; Wei, H.; Hu, J.H.; Majesky, M.W.; Brozovich, F.V.; Kim, F.; Dichek, D.A. Loss of Transforming Growth Factor Beta Signaling in Aortic Smooth Muscle Cells Causes Endothelial Dysfunction and Aortic Hypercontractility. *Arterioscler. Thromb. Vasc. Biol.* **2021**, *41*, 1956–1971. [CrossRef]

77. Xie, W.-B.; Li, Z.; Shi, N.; Guo, X.; Tang, J.; Ju, W.; Han, J.; Liu, T.; Bottinger, E.P.; Chai, Y.; et al. Smad2 and Myocardin-Related Transcription Factor B Cooperatively Regulate Vascular Smooth Muscle Differentiation From Neural Crest Cells. *Circ. Res.* **2013**, *113*, e76–e86. [CrossRef]

78. Huang, W.-Y.; Xie, W.; Guo, X.; Li, F.; Jose, P.A.; Chen, S.-Y.; Jiang, H.; Chen, Y.; Yu, T.; Zhao, X.; et al. Smad2 and PEA3 cooperatively regulate transcription of response gene to complement 32 in TGF-β-induced smooth muscle cell differentiation of neural crest cells. *Am. J. Physiol. Cell Physiol.* **2011**, *301*, C499–C506. [CrossRef]

79. Inamoto, S.; Kwartler, C.S.; Lafont, A.L.; Liang, Y.Y.; Fadulu, V.T.; Duraisamy, S.; Willing, M.; Estrera, A.; Safi, H.; Hannibal, M.C.; et al. TGFBR2 mutations alter smooth muscle cell phenotype and predispose to thoracic aortic aneurysms and dissections. *Cardiovasc. Res.* **2010**, *88*, 520–529. [CrossRef]

80. Ferruzzi, J.; Murtada, S.-I.; Li, G.; Jiao, Y.; Uman, S.; Ting, M.Y.L.; Tellides, G.; Humphrey, J.D. Pharmacologically Improved Contractility Protects against Aortic Dissection in Mice with Disrupted Transforming Growth Factor-β Signaling Despite Compromised Extracellular Matrix Properties. *Arterioscler. Thromb. Vasc. Biol.* **2016**, *36*, 919–927. [CrossRef]

81. Lee, C.Y.; Angelov, S.N.; Zhu, J.; Bi, L.; Sanford, N.; Yildirim, I.A.; Dichek, D.A. Blockade of TGF-β (Transforming Growth Factor Beta) Signaling by Deletion of *Tgfbr2* in Smooth Muscle Cells of 11-Month-Old Mice Alters Aortic Structure and Causes Vasomotor Dysfunction—Brief Report. *Arterioscler. Thromb. Vasc. Biol.* **2022**, *42*, 764–771. [CrossRef]

82. Krishnamurthy, V.K.; Evans, A.N.; Wansapura, J.P.; Osinska, H.; Maddy, K.E.; Biechler, S.V.; Narmoneva, D.A.; Goodwin, R.L.; Hinton, R.B. Asymmetric Cell–Matrix and Biomechanical Abnormalities in Elastin Insufficiency Induced Aortopathy. *Ann. Biomed. Eng.* **2014**, *42*, 2014–2028. [CrossRef]

83. Regalado, E.S.; Morris, S.A.; Braverman, A.C.; Hostetler, E.M.; De Backer, J.; Li, R.; Pyeritz, R.E.; Yetman, A.T.; Cervi, E.; Shalhub, S.; et al. Comparative Risks of Initial Aortic Events Associated with Genetic Thoracic Aortic Disease. *J. Am. Coll. Cardiol.* **2022**, *80*, 857–869. [CrossRef]

84. Safar, M.E.; Boudier, H.S. Vascular Development, Pulse Pressure, and the Mechanisms of Hypertension. *Hypertension* **2005**, *46*, 205–209. [CrossRef]

85. Fhayli, W.; Boyer, M.; Ghandour, Z.; Jacob, M.; Andrieu, J.; Starcher, B.; Estève, E.; Faury, G. Chronic administration of minoxidil protects elastic fibers and stimulates their neosynthesis with improvement of the aorta mechanics in mice. *Cell. Signal.* **2019**, *62*, 109333. [CrossRef]

86. Pannu, H.; Fadulu, V.T.; Chang, J.; Lafont, A.; Hasham, S.N.; Sparks, E.; Giampietro, P.F.; Zaleski, C.; Estrera, A.L.; Safi, H.J.; et al. Mutations in Transforming Growth Factor-β Receptor Type II Cause Familial Thoracic Aortic Aneurysms and Dissections. *Circulation* **2005**, *112*, 513–520. [CrossRef]

87. Tran-Fadulu, V.; Pannu, H.; Kim, D.H.; Vick, G.W., 3rd; Lonsford, C.M.; Lafont, A.L.; Boccalandro, C.; Smart, S.; Peterson, K.L.; Hain, J.Z.; et al. Analysis of multigenerational families with thoracic aortic aneurysms and dissections due to TGFBR1 or TGFBR2 mutations. *J. Med. Genet.* **2009**, *46*, 607–613. [CrossRef]

88. Jondeau, G.; Ropers, J.; Regalado, E.; Braverman, A.; Evangelista, A.; Teixedo, G.; De Backer, J.; Muiño-Mosquera, L.; Naudion, S.; Zordan, C.; et al. International Registry of Patients Carrying *TGFBR1* or *TGFBR2* Mutations: Results of the MAC (Montalcino Aortic Consortium). *Circ. Cardiovasc. Genet.* **2016**, *9*, 548–558. [CrossRef]

89. Robertson, I.; Jensen, S.; Handford, P. TB domain proteins: Evolutionary insights into the multifaceted roles of fibrillins and LTBPs. *Biochem. J.* **2011**, *433*, 263–276. [CrossRef]

90. Bunton, T.E.; Biery, N.J.; Myers, L.; Gayraud, B.; Ramirez, F.; Dietz, H.C. Phenotypic Alteration of Vascular Smooth Muscle Cells Precedes Elastolysis in a Mouse Model of Marfan Syndrome. *Circ. Res.* **2001**, *88*, 37–43. [CrossRef]

91. Iosef, C.; Pedroza, A.J.; Cui, J.Z.; Dalal, A.R.; Arakawa, M.; Tashima, Y.; Koyano, T.K.; Burdon, G.; Churovich, S.M.P.; Orrick, J.O.; et al. Quantitative proteomics reveal lineage-specific protein profiles in iPSC-derived Marfan syndrome smooth muscle cells. *Sci. Rep.* **2020**, *10*, 20392. [CrossRef] [PubMed]

92. Elango, J.; Hou, C.; Bao, B.; Wang, S.; de Val, J.E.M.S.; Wenhui, W. The Molecular Interaction of Collagen with Cell Receptors for Biological Function. *Polymers* **2022**, *14*, 876. [CrossRef] [PubMed]

93. Murphy-Ullrich, J.E.; Suto, M.J. Thrombospondin-1 regulation of latent TGF-β activation: A therapeutic target for fibrotic disease. *Matrix Biol.* **2018**, *68–69*, 28–43. [CrossRef] [PubMed]

94. Yamashiro, Y.; Thang, B.Q.; Shin, S.J.; Lino, C.A.; Nakamura, T.; Kim, J.; Sugiyama, K.; Tokunaga, C.; Sakamoto, H.; Osaka, M.; et al. Role of Thrombospondin-1 in Mechanotransduction and Development of Thoracic Aortic Aneurysm in Mouse and Humans. *Circ. Res.* **2018**, *123*, 660–672. [CrossRef]

95. Liu, Z.; Morgan, S.; Ren, J.; Wang, Q.; Annis, D.S.; Mosher, D.F.; Zhang, J.; Sorenson, C.M.; Sheibani, N.; Liu, B.; et al. Thrombospondin-1 (TSP1) Contributes to the Development of Vascular Inflammation by Regulating Monocytic Cell Motility in Mouse Models of Abdominal Aortic Aneurysm. *Circ. Res.* **2015**, *117*, 129–141. [CrossRef]

96. Li, H.; Xu, H.; Wen, H.; Wang, H.; Zhao, R.; Sun, Y.; Bai, C.; Ping, J.; Song, L.; Luo, M.; et al. Lysyl hydroxylase 1 (LH1) deficiency promotes angiotensin II (Ang II)-induced dissecting abdominal aortic aneurysm. *Theranostics* **2021**, *11*, 9587–9604. [CrossRef]

97. Yang, H.; Zhou, T.; Sorenson, C.M.; Sheibani, N.; Liu, B. Myeloid-Derived TSP1 (Thrombospondin-1) Contributes to Abdominal Aortic Aneurysm Through Suppressing Tissue Inhibitor of Metalloproteinases-1. *Arterioscler. Thromb. Vasc. Biol.* **2020**, *40*, e350–e366. [CrossRef]
98. Pedroza, A.J.; Tashima, Y.; Shad, R.; Cheng, P.; Wirka, R.; Churovich, S.; Nakamura, K.; Yokoyama, N.; Cui, J.Z.; Iosef, C.; et al. Single-Cell Transcriptomic Profiling of Vascular Smooth Muscle Cell Phenotype Modulation in Marfan Syndrome Aortic Aneurysm. *Arterioscler. Thromb. Vasc. Biol.* **2020**, *40*, 2195–2211. [CrossRef]
99. Pedroza, A.J.; Dalal, A.R.; Shad, R.; Yokoyama, N.; Nakamura, K.; Cheng, P.; Wirka, R.C.; Mitchel, O.; Baiocchi, M.; Hiesinger, W.; et al. Embryologic Origin Influences Smooth Muscle Cell Phenotypic Modulation Signatures in Murine Marfan Syndrome Aortic Aneurysm. *Arterioscler. Thromb. Vasc. Biol.* **2022**, *42*, 1154–1168. [CrossRef]
100. Wanga, S.; Hibender, S.; Ridwan, Y.; van Roomen, C.; Vos, M.; van der Made, I.; van Vliet, N.; Franken, R.; van Riel, L.A.; Groenink, M.; et al. Aortic microcalcification is associated with elastin fragmentation in Marfan syndrome. *J. Pathol.* **2017**, *243*, 294–306. [CrossRef]
101. Crosas-Molist, E.; Meirelles, T.; López-Luque, J.; Serra-Peinado, C.; Selva, J.; Caja, L.; Del Blanco, D.G.; Uriarte, J.J.; Bertran, E.; Mendizábal, Y.; et al. Vascular Smooth Muscle Cell Phenotypic Changes in Patients with Marfan Syndrome. *Arterioscler. Thromb. Vasc. Biol.* **2015**, *35*, 960–972. [CrossRef]
102. Dawson, A.; Li, Y.; Li, Y.; Ren, P.; Vasquez, H.G.; Zhang, C.; Rebello, K.R.; Ageedi, W.; Azares, A.R.; Mattar, A.B.; et al. Single-Cell Analysis of Aneurysmal Aortic Tissue in Patients with Marfan Syndrome Reveals Dysfunctional TGF-β Signaling. *Genes* **2021**, *13*, 95. [CrossRef]
103. Pyeritz, R.E. Heritable thoracic aortic disorders. *Curr. Opin. Cardiol.* **2014**, *29*, 97–102. [CrossRef]
104. Robertson, E.N.; Hambly, B.D.; Jeremy, R.W. Thoracic aortic dissection and heritability: Forensic implications. *Forensic Sci. Med. Pathol.* **2016**, *12*, 366–368. [CrossRef]
105. Lu, H.; Daugherty, A. Aortic Aneurysms. *Arterioscler. Thromb. Vasc. Biol.* **2017**, *37*, e59–e65. [CrossRef]
106. Kan, H.; Zhang, K.; Mao, A.; Geng, L.; Gao, M.; Feng, L.; You, Q.; Ma, X. Single-cell transcriptome analysis reveals cellular heterogeneity in the ascending aortas of normal and high-fat diet-fed mice. *Exp. Mol. Med.* **2021**, *53*, 1379–1389. [CrossRef]
107. Zhang, K.; Kan, H.; Mao, A.; Geng, L.; Ma, X. Single-cell analysis of salt-induced hypertensive mouse aortae reveals cellular heterogeneity and state changes. *Exp. Mol. Med.* **2021**, *53*, 1866–1876. [CrossRef]
108. Salmon, M.; Johnston, W.F.; Woo, A.; Pope, N.H.; Su, G.; Upchurch, G.R., Jr.; Owens, G.K.; Ailawadi, G. KLF4 Regulates Abdominal Aortic Aneurysm Morphology and Deletion Attenuates Aneurysm Formation. *Circulation* **2013**, *128*, S163–S174. [CrossRef]
109. Shankman, L.S.; Gomez, D.; Cherepanova, O.A.; Salmon, M.; Alencar, G.F.; Haskins, R.M.; Swiatlowska, P.; Newman, A.A.C.; Greene, E.S.; Straub, A.C.; et al. KLF4-dependent phenotypic modulation of smooth muscle cells has a key role in atherosclerotic plaque pathogenesis. *Nat. Med.* **2015**, *21*, 628–637. [CrossRef]
110. Cheng, J.; Gu, W.; Lan, T.; Deng, J.; Ni, Z.; Zhang, Z.; Hu, Y.; Sun, X.; Yang, Y.; Xu, Q. Single-cell RNA sequencing reveals cell type- and artery type-specific vascular remodelling in male spontaneously hypertensive rats. *Cardiovasc. Res.* **2021**, *117*, 1202–1216. [CrossRef]
111. Li, Y.; Ren, P.; Dawson, A.; Vasquez, H.G.; Ageedi, W.; Zhang, C.; Luo, W.; Chen, R.; Li, Y.; Kim, S.; et al. Single-Cell Transcriptome Analysis Reveals Dynamic Cell Populations and Differential Gene Expression Patterns in Control and Aneurysmal Human Aortic Tissue. *Circulation* **2020**, *142*, 1374–1388. [CrossRef] [PubMed]
112. Cao, G.; Lu, Z.; Gu, R.; Xuan, X.; Zhang, R.; Hu, J.; Dong, H. Deciphering the Intercellular Communication Between Immune Cells and Altered Vascular Smooth Muscle Cell Phenotypes in Aortic Aneurysm From Single-Cell Transcriptome Data. *Front. Cardiovasc. Med.* **2022**, *9*, 936287. [CrossRef]
113. Song, W.; Qin, L.; Chen, Y.; Chen, J.; Wei, L. Single-cell transcriptome analysis identifies Versican(+) myofibroblast as a hallmark for thoracic aortic aneurysm marked by activation of PI3K-AKT signaling pathway. *Biochem. Biophys. Res. Commun.* **2022**, *643*, 175–185. [CrossRef] [PubMed]
114. Zhang, C.; Li, Y.; Chakraborty, A.; Li, Y.; Rebello, K.R.; Ren, P.; Luo, W.; Zhang, L.; Lu, H.S.; Cassis, L.A.; et al. Aortic Stress Activates an Adaptive Program in Thoracic Aortic Smooth Muscle Cells That Maintains Aortic Strength and Protects against Aneurysm and Dissection in Mice. *Arterioscler. Thromb. Vasc. Biol.* **2023**, *43*, 234–252. [CrossRef] [PubMed]
115. Chen, P.-Y.; Qin, L.; Li, G.; Malagon-Lopez, J.; Wang, Z.; Bergaya, S.; Gujja, S.; Caulk, A.W.; Murtada, S.-I.; Zhang, X.; et al. Smooth Muscle Cell Reprogramming in Aortic Aneurysms. *Cell Stem Cell* **2020**, *26*, 542–557.e511. [CrossRef] [PubMed]
116. Clément, M.; Chappell, J.; Raffort, J.; Lareyre, F.; Vandestienne, M.; Taylor, A.L.; Finigan, A.; Harrison, J.; Bennett, M.R.; Bruneval, P.; et al. Vascular Smooth Muscle Cell Plasticity and Autophagy in Dissecting Aortic Aneurysms. *Arterioscler. Thromb. Vasc. Biol.* **2019**, *39*, 1149–1159. [CrossRef]
117. He, Y.-B.; Jin, H.-Z.; Zhao, J.-L.; Wang, C.; Ma, W.-R.; Xing, J.; Zhang, X.-B.; Zhang, Y.-Y.; Dai, H.-D.; Zhao, N.-S.; et al. Single-cell transcriptomic analysis reveals differential cell subpopulations and distinct phenotype transition in normal and dissected ascending aorta. *Mol. Med.* **2022**, *28*, 158. [CrossRef]
118. Zhao, G.; Lu, H.; Chang, Z.; Zhao, Y.; Zhu, T.; Chang, L.; Guo, Y.; Garcia-Barrio, M.T.; Chen, Y.E.; Zhang, J. Single-cell RNA sequencing reveals the cellular heterogeneity of aneurysmal infrarenal abdominal aorta. *Cardiovasc. Res.* **2020**, *117*, 1402–1416. [CrossRef]

119. Chou, E.L.; Chaffin, M.; Simonson, B.; Pirruccello, J.P.; Akkad, A.-D.; Nekoui, M.; Cardenas, C.L.L.; Bedi, K.C.; Nash, C.; Juric, D.; et al. Aortic Cellular Diversity and Quantitative Genome-Wide Association Study Trait Prioritization Through Single-Nuclear RNA Sequencing of the Aneurysmal Human Aorta. *Arterioscler. Thromb. Vasc. Biol.* **2022**, *42*, 1355–1374. [CrossRef]
120. Wight, T.N.; Kang, I.; Evanko, S.P.; Harten, I.A.; Chang, M.Y.; Pearce, O.M.T.; Allen, C.E.; Frevert, C.W. Versican—A Critical Extracellular Matrix Regulator of Immunity and Inflammation. *Front. Immunol.* **2020**, *11*, 512. [CrossRef]
121. Wirka, R.C.; Wagh, D.; Paik, D.T.; Pjanic, M.; Nguyen, T.; Miller, C.L.; Kundu, R.; Nagao, M.; Coller, J.; Koyano, T.K.; et al. Atheroprotective roles of smooth muscle cell phenotypic modulation and the TCF21 disease gene as revealed by single-cell analysis. *Nat. Med.* **2019**, *25*, 1280–1289. [CrossRef]
122. Pan, H.; Xue, C.; Auerbach, B.J.; Fan, J.; Bashore, A.C.; Cui, J.; Yang, D.Y.; Trignano, S.B.; Liu, W.; Shi, J.; et al. Single-Cell Genomics Reveals a Novel Cell State During Smooth Muscle Cell Phenotypic Switching and Potential Therapeutic Targets for Atherosclerosis in Mouse and Human. *Circulation* **2020**, *142*, 2060–2075. [CrossRef]
123. Pedroza, A.J.; Shad, R.; Dalal, A.R.; Yokoyama, N.; Nakamura, K.; Hiesinger, W.; Fischbein, M.P. Acute Induced Pressure Overload Rapidly Incites Thoracic Aortic Aneurysmal Smooth Muscle Cell Phenotype. *Hypertension* **2022**, *79*, e86–e89. [CrossRef]
124. Ma, Q.; Yang, Q.; Xu, J.; Zhang, X.; Kim, D.; Liu, Z.; Da, Q.; Mao, X.; Zhou, Y.; Cai, Y.; et al. ATIC-Associated De Novo Purine Synthesis Is Critically Involved in Proliferative Arterial Disease. *Circulation* **2022**, *146*, 1444–1460. [CrossRef]
125. Majesky, M.W.; Horita, H.; Ostriker, A.; Lu, S.; Regan, J.N.; Bagchi, A.; Dong, X.R.; Poczobutt, J.; Nemenoff, R.A.; Weiser-Evans, M.C.; et al. Differentiated Smooth Muscle Cells Generate a Subpopulation of Resident Vascular Progenitor Cells in the Adventitia Regulated by Klf4. *Circ. Res.* **2017**, *120*, 296–311. [CrossRef]
126. Shikatani, E.A.; Chandy, M.; Besla, R.; Li, C.C.; Momen, A.; El-Mounayri, O.; Robbins, C.S.; Husain, M. c-Myb Regulates Proliferation and Differentiation of Adventitial Sca1 [+] Vascular Smooth Muscle Cell Progenitors by Transactivation of Myocardin. *Arterioscler. Thromb. Vasc. Biol.* **2016**, *36*, 1367–1376. [CrossRef]
127. Worssam, M.D.; Lambert, J.; Oc, S.; Taylor, J.C.K.; Taylor, A.L.; Dobnikar, L.; Chappell, J.; Harman, J.L.; Figg, N.L.; Finigan, A.; et al. Cellular mechanisms of oligoclonal vascular smooth muscle cell expansion in cardiovascular disease. *Cardiovasc. Res.* **2022**, *119*, 1279–1294. [CrossRef]
128. Chappell, J.; Harman, J.L.; Narasimhan, V.M.; Yu, H.; Foote, K.; Simons, B.D.; Bennett, M.R.; Jørgensen, H.F. Extensive Proliferation of a Subset of Differentiated, yet Plastic, Medial Vascular Smooth Muscle Cells Contributes to Neointimal Formation in Mouse Injury and Atherosclerosis Models. *Circ. Res.* **2016**, *119*, 1313–1323. [CrossRef]
129. Feil, S.; Fehrenbacher, B.; Lukowski, R.; Essmann, F.; Schulze-Osthoff, K.; Schaller, M.; Feil, R. Transdifferentiation of Vascular Smooth Muscle Cells to Macrophage-Like Cells During Atherogenesis. *Circ. Res.* **2014**, *115*, 662–667. [CrossRef]
130. Harman, J.L.; Jørgensen, H.F. The role of smooth muscle cells in plaque stability: Therapeutic targeting potential. *Br. J. Pharmacol.* **2019**, *176*, 3741–3753. [CrossRef]
131. Li, Y.; Zhu, H.; Zhang, Q.; Han, X.; Zhang, Z.; Shen, L.; Wang, L.; Lui, K.O.; He, B.; Zhou, B. Smooth muscle-derived macrophage-like cells contribute to multiple cell lineages in the atherosclerotic plaque. *Cell Discov.* **2021**, *7*, 111. [CrossRef] [PubMed]
132. Leone, O.; Corsini, A.; Pacini, D.; Corti, B.; Lorenzini, M.; Laus, V.; Foà, A.; Reggiani, M.L.B.; Di Marco, L.; Rapezzi, C. The complex interplay among atherosclerosis, inflammation, and degeneration in ascending thoracic aortic aneurysms. *J. Thorac. Cardiovasc. Surg.* **2019**, *160*, 1434–1443.e6. [CrossRef] [PubMed]
133. Nesi, G.; Anichini, C.; Tozzini, S.; Boddi, V.; Calamai, G.; Gori, F. Pathology of the thoracic aorta: A morphologic review of 338 surgical specimens over a 7-year period. *Cardiovasc. Pathol.* **2009**, *18*, 134–139. [CrossRef] [PubMed]
134. Negedu, M.N.; Duckworth, C.A.; Yu, L.-G. Galectin-2 in Health and Diseases. *Int. J. Mol. Sci.* **2022**, *24*, 341. [CrossRef]
135. Minas, T.Z.; Candia, J.; Dorsey, T.H.; Baker, F.; Tang, W.; Kiely, M.; Smith, C.J.; Zhang, A.L.; Jordan, S.V.; Obadi, O.M.; et al. Serum proteomics links suppression of tumor immunity to ancestry and lethal prostate cancer. *Nat. Commun.* **2022**, *13*, 1759. [CrossRef]
136. Huang, Y.; Wang, H.-C.; Zhao, J.; Wu, M.-H.; Shih, T.-C. Immunosuppressive Roles of Galectin-1 in the Tumor Microenvironment. *Biomolecules* **2021**, *11*, 1398. [CrossRef]
137. Zhang, J.; Zhao, X.; Guo, Y.; Liu, Z.; Wei, S.; Yuan, Q.; Shang, H.; Sang, W.; Cui, S.; Xu, T.; et al. Macrophage ALDH2 (Aldehyde Dehydrogenase 2) Stabilizing Rac2 Is Required for Efferocytosis Internalization and Reduction of Atherosclerosis Development. *Arterioscler. Thromb. Vasc. Biol.* **2022**, *42*, 700–716. [CrossRef]
138. Kojima, Y.; Werner, N.; Ye, J.; Nanda, V.; Tsao, N.; Wang, Y.; Flores, A.M.; Miller, C.L.; Weissman, I.; Deng, H.; et al. Proefferocytic Therapy Promotes Transforming Growth Factor-β Signaling and Prevents Aneurysm Formation. *Circulation* **2018**, *137*, 750–753. [CrossRef]
139. López-Otín, C.; Kroemer, G. Hallmarks of Health. *Cell* **2021**, *184*, 33–63. [CrossRef]
140. Luo, W.; Wang, Y.; Zhang, L.; Ren, P.; Zhang, C.; Li, Y.; Azares, A.R.; Zhang, M.; Guo, J.; Ghaghada, K.B.; et al. Critical Role of Cytosolic DNA and Its Sensing Adaptor STING in Aortic Degeneration, Dissection, and Rupture. *Circulation* **2020**, *141*, 42–66. [CrossRef]
141. Saito, T.; Hasegawa, Y.; Ishigaki, Y.; Yamada, T.; Gao, J.; Imai, J.; Uno, K.; Kaneko, K.; Ogihara, T.; Shimosawa, T.; et al. Importance of endothelial NF-κB signalling in vascular remodelling and aortic aneurysm formation. *Cardiovasc. Res.* **2013**, *97*, 106–114. [CrossRef]
142. Rong, J.X.; Shapiro, M.; Trogan, E.; Fisher, E.A. Transdifferentiation of mouse aortic smooth muscle cells to a macrophage-like state after cholesterol loading. *Proc. Natl. Acad. Sci. USA* **2003**, *100*, 13531–13536. [CrossRef]

143. Vengrenyuk, Y.; Nishi, H.; Long, X.; Ouimet, M.; Savji, N.; Martinez, F.O.; Cassella, C.P.; Moore, K.J.; Ramsey, S.A.; Miano, J.M.; et al. Cholesterol Loading Reprograms the MicroRNA-143/145–Myocardin Axis to Convert Aortic Smooth Muscle Cells to a Dysfunctional Macrophage-Like Phenotype. *Arterioscler. Thromb. Vasc. Biol.* **2015**, *35*, 535–546. [CrossRef]

144. Lopez-Sanz, L.; Bernal, S.; Jimenez-Castilla, L.; Prieto, I.; La Manna, S.; Gomez-Lopez, S.; Blanco-Colio, L.M.; Egido, J.; Martin-Ventura, J.L.; Gomez-Guerrero, C. Fcγ receptor activation mediates vascular inflammation and abdominal aortic aneurysm development. *Clin. Transl. Med.* **2021**, *11*, e463. [CrossRef]

145. Filiberto, A.C.; Spinosa, M.D.; Elder, C.T.; Su, G.; Leroy, V.; Ladd, Z.; Lu, G.; Mehaffey, J.H.; Salmon, M.D.; Hawkins, R.B.; et al. Endothelial pannexin-1 channels modulate macrophage and smooth muscle cell activation in abdominal aortic aneurysm formation. *Nat. Commun.* **2022**, *13*, 1521. [CrossRef]

146. Shao, F.; Miao, Y.; Zhang, Y.; Han, L.; Ma, X.; Deng, J.; Jiang, C.; Kong, W.; Xu, Q.; Feng, J.; et al. B cell-derived anti-beta 2 glycoprotein I antibody contributes to hyperhomocysteinaemia-aggravated abdominal aortic aneurysm. *Cardiovasc. Res.* **2020**, *116*, 1897–1909. [CrossRef]

147. Kurz, J.; Weiss, A.-C.; Lüdtke, T.H.-W.; Deuper, L.; Trowe, M.-O.; Thiesler, H.; Hildebrandt, H.; Heineke, J.; Duncan, S.A.; Kispert, A. GATA6 is a crucial factor for *Myocd* expression in the visceral smooth muscle cell differentiation program of the murine ureter. *Development* **2022**, *149*, dev200522. [CrossRef]

148. Zhuang, T.; Liu, J.; Chen, X.; Pi, J.; Kuang, Y.; Wang, Y.; Tomlinson, B.; Chan, P.; Zhang, Q.; Li, Y.; et al. Cell-Specific Effects of GATA (GATA Zinc Finger Transcription Factor Family)-6 in Vascular Smooth Muscle and Endothelial Cells on Vascular Injury Neointimal Formation. *Arterioscler. Thromb. Vasc. Biol.* **2019**, *39*, 888–901. [CrossRef]

149. Alajbegovic, A.; Daoud, F.; Ali, N.; Kawka, K.; Holmberg, J.; Albinsson, S. Transcription factor GATA6 promotes migration of human coronary artery smooth muscle cells in vitro. *Front. Physiol.* **2022**, *13*, 1054819. [CrossRef]

150. Jiao, H.; Walczak, B.E.; Lee, M.-S.; Lemieux, M.E.; Li, W.-J. GATA6 regulates aging of human mesenchymal stem/stromal cells. *Stem Cells* **2021**, *39*, 62–77. [CrossRef]

151. Pabis, K.; Chiari, Y.; Sala, C.; Straka, E.; Giacconi, R.; Provinciali, M.; Li, X.; Brown-Borg, H.; Nowikovsky, K.; Valencak, T.G.; et al. Elevated metallothionein expression in long-lived species mediates the influence of cadmium accumulation on aging. *Geroscience* **2021**, *43*, 1975–1993. [CrossRef] [PubMed]

152. Li, X.; Wen, H.; Lv, J.; Luan, B.; Meng, J.; Gong, S.; Wen, J.; Xin, S. Therapeutic efficacy of mesenchymal stem cells for abdominal aortic aneurysm: A meta-analysis of preclinical studies. *Stem Cell Res. Ther.* **2022**, *13*, 81. [CrossRef] [PubMed]

153. Galluzzi, L.; Bravo-San Pedro, J.M.; Kepp, O.; Kroemer, G. Regulated cell death and adaptive stress responses. *Cell. Mol. Life Sci.* **2016**, *73*, 2405–2410. [CrossRef] [PubMed]

154. Chattopadhyay, A.; Kwartler, C.S.; Kaw, K.; Li, Y.; Kaw, A.; Chen, J.; LeMaire, S.A.; Shen, Y.H.; Milewicz, D.M. Cholesterol-Induced Phenotypic Modulation of Smooth Muscle Cells to Macrophage/Fibroblast–like Cells Is Driven by an Unfolded Protein Response. *Arterioscler. Thromb. Vasc. Biol.* **2021**, *41*, 302–316. [CrossRef] [PubMed]

155. Hamczyk, M.R.; Villa-Bellosta, R.; Quesada, V.; Gonzalo, P.; Vidak, S.; Nevado, R.M.; Andrés-Manzano, M.J.; Misteli, T.; López-Otín, C.; Andrés, V. Progerin accelerates atherosclerosis by inducing endoplasmic reticulum stress in vascular smooth muscle cells. *EMBO Mol. Med.* **2019**, *11*, e9736. [CrossRef]

156. Oller, J.; Gabandé-Rodríguez, E.; Ruiz-Rodríguez, M.J.; Desdín-Micó, G.; Aranda, J.F.; Rodrigues-Diez, R.; Ballesteros-Martínez, C.; Blanco, E.M.; Roldan-Montero, R.; Acuña, P.; et al. Extracellular Tuning of Mitochondrial Respiration Leads to Aortic Aneurysm. *Circulation* **2021**, *143*, 2091–2109. [CrossRef]

157. Asano, K.; Cantalupo, A.; Sedes, L.; Ramirez, F. Pathophysiology and Therapeutics of Thoracic Aortic Aneurysm in Marfan Syndrome. *Biomolecules* **2022**, *12*, 128. [CrossRef]

158. Sawada, H.; Katsumata, Y.; Higashi, H.; Zhang, C.; Li, Y.; Morgan, S.; Lee, L.H.; Singh, S.A.; Chen, J.Z.; Franklin, M.K.; et al. Second Heart Field–Derived Cells Contribute to Angiotensin II–Mediated Ascending Aortopathies. *Circulation* **2022**, *145*, 987–1001. [CrossRef]

159. D'amico, F.; Doldo, E.; Pisano, C.; Scioli, M.G.; Centofanti, F.; Proietti, G.; Falconi, M.; Sangiuolo, F.; Ferlosio, A.; Ruvolo, G.; et al. Specific miRNA and Gene Deregulation Characterize the Increased Angiogenic Remodeling of Thoracic Aneurysmatic Aortopathy in Marfan Syndrome. *Int. J. Mol. Sci.* **2020**, *21*, 6886. [CrossRef]

160. Si, X.; Chen, Q.; Zhang, J.; Zhou, W.; Chen, L.; Chen, J.; Deng, N.; Li, W.; Liu, D.; Wang, L.; et al. MicroRNA-23b prevents aortic aneurysm formation by inhibiting smooth muscle cell phenotypic switching via FoxO4 suppression. *Life Sci.* **2022**, *288*, 119092. [CrossRef]

161. Yang, K.; Ren, J.; Li, X.; Wang, Z.; Xue, L.; Cui, S.; Sang, W.; Xu, T.; Zhang, J.; Yu, J.; et al. Prevention of aortic dissection and aneurysm via an ALDH2-mediated switch in vascular smooth muscle cell phenotype. *Eur. Heart J.* **2020**, *41*, 2442–2453. [CrossRef]

162. Lino Cardenas, C.L.; Kessinger, C.W.; Cheng, Y.; Macdonald, C.; MacGillivray, T.; Ghoshhajra, B.; Huleihel, L.; Nuri, S.; Yeri, A.S.; Jaffer, F.A.; et al. An HDAC9-MALAT1-BRG1 complex mediates smooth muscle dysfunction in thoracic aortic aneurysm. *Nat. Commun.* **2018**, *9*, 1009. [CrossRef]

163. Zalewski, D.P.; Ruszel, K.P.; Stępniewski, A.; Gałkowski, D.; Feldo, M.; Kocki, J.; Bogucka-Kocka, A. miRNA Regulatory Networks Associated with Peripheral Vascular Diseases. *J. Clin. Med.* **2022**, *11*, 3470. [CrossRef]

164. Krishna, S.M.; Morton, S.K.; Li, J.; Golledge, J. Risk Factors and Mouse Models of Abdominal Aortic Aneurysm Rupture. *Int. J. Mol. Sci.* **2020**, *21*, 7250. [CrossRef]

165. Lu, G.; Su, G.; Davis, J.P.; Schaheen, B.; Downs, E.; Roy, R.J.; Ailawadi, G.; Upchurch, G.R. A novel chronic advanced stage abdominal aortic aneurysm murine model. *J. Vasc. Surg.* **2017**, *66*, 232–242.e4. [CrossRef]
166. Golledge, J.; Krishna, S.M.; Wang, Y. Mouse models for abdominal aortic aneurysm. *Br. J. Pharmacol.* **2020**, *179*, 792–810. [CrossRef]
167. Sulé-Suso, J.; Forster, A.; Zholobenko, V.; Stone, N.; El Haj, A. Effects of CaCl$_2$ and MgCl$_2$ on Fourier Transform Infrared Spectra of Lung Cancer Cells. *Appl. Spectrosc.* **2004**, *58*, 61–67. [CrossRef]
168. Sundararaman, S.S.; van der Vorst, E.P.C. Calcium-Sensing Receptor (CaSR), Its Impact on Inflammation and the Consequences on Cardiovascular Health. *Int. J. Mol. Sci.* **2021**, *22*, 2478. [CrossRef]
169. Wagenseil, J.E.; Mecham, R.P.; Tellides, G.; Staiculescu, M.C.; Cocciolone, A.J.; Procknow, J.D.; Kim, J.; Hawes, J.Z.; Johnson, E.O.; Murshed, M.; et al. Vascular Extracellular Matrix and Arterial Mechanics. *Physiol. Rev.* **2009**, *89*, 957–989. [CrossRef]
170. Nolasco, P.; Fernandes, C.G.; Ribeiro-Silva, J.C.; Oliveira, P.V.; Sacrini, M.; de Brito, I.V.; DE Bessa, T.; Pereira, L.V.; Tanaka, L.Y.; Alencar, A.; et al. Impaired vascular smooth muscle cell force-generating capacity and phenotypic deregulation in Marfan Syndrome mice. *Biochim. Biophys. Acta (BBA) Mol. Basis Dis.* **2019**, *1866*, 165587. [CrossRef]
171. Wong, L.; Kumar, A.; Gabela-Zuniga, B.; Chua, J.; Singh, G.; Happe, C.L.; Engler, A.J.; Fan, Y.; McCloskey, K.E. Substrate stiffness directs diverging vascular fates. *Acta Biomater.* **2019**, *96*, 321–329. [CrossRef] [PubMed]
172. Schnellmann, R.; Ntekoumes, D.; Choudhury, M.I.; Sun, S.; Wei, Z.; Gerecht, S. Stiffening Matrix Induces Age-Mediated Microvascular Phenotype Through Increased Cell Contractility and Destabilization of Adherens Junctions. *Adv. Sci.* **2022**, *9*, e2201483. [CrossRef] [PubMed]
173. Talwar, S.; Kant, A.; Xu, T.; Shenoy, V.B.; Assoian, R.K. Mechanosensitive smooth muscle cell phenotypic plasticity emerging from a null state and the balance between Rac and Rho. *Cell Rep.* **2021**, *35*, 109019. [CrossRef] [PubMed]
174. Wang, J.; Xie, S.-A.; Li, N.; Zhang, T.; Yao, W.; Zhao, H.; Pang, W.; Han, L.; Liu, J.; Zhou, J. Matrix stiffness exacerbates the proinflammatory responses of vascular smooth muscle cell through the DDR1-DNMT1 mechanotransduction axis. *Bioact. Mater.* **2022**, *17*, 406–424. [CrossRef]
175. Qian, W.; Hadi, T.; Silvestro, M.; Ma, X.; Rivera, C.F.; Bajpai, A.; Li, R.; Zhang, Z.; Qu, H.; Tellaoui, R.S.; et al. Microskeletal stiffness promotes aortic aneurysm by sustaining pathological vascular smooth muscle cell mechanosensation via Piezo1. *Nat. Commun.* **2022**, *13*, 512. [CrossRef]
176. Pasta, S.; Agnese, V.; Gallo, A.; Cosentino, F.; Di Giuseppe, M.; Gentile, G.; Raffa, G.M.; Maalouf, J.F.; Michelena, H.I.; Bellavia, D.; et al. Shear Stress and Aortic Strain Associations with Biomarkers of Ascending Thoracic Aortic Aneurysm. *Ann. Thorac. Surg.* **2020**, *110*, 1595–1604. [CrossRef]
177. Gomez, D.; Swiatlowska, P.; Owens, G.K. Epigenetic Control of Smooth Muscle Cell Identity and Lineage Memory. *Arterioscler. Thromb. Vasc. Biol.* **2015**, *35*, 2508–2516. [CrossRef]
178. Gurung, R.; Choong, A.M.; Woo, C.C.; Foo, R.; Sorokin, V. Genetic and Epigenetic Mechanisms Underlying Vascular Smooth Muscle Cell Phenotypic Modulation in Abdominal Aortic Aneurysm. *Int. J. Mol. Sci.* **2020**, *21*, 6334. [CrossRef]
179. Floren, M.; Bonani, W.; Dharmarajan, A.; Motta, A.; Migliaresi, C.; Tan, W. Human mesenchymal stem cells cultured on silk hydrogels with variable stiffness and growth factor differentiate into mature smooth muscle cell phenotype. *Acta Biomater.* **2016**, *31*, 156–166. [CrossRef]
180. Walters, B.; Turner, P.A.; Rolauffs, B.; Hart, M.L.; Stegemann, J.P. Controlled Growth Factor Delivery and Cyclic Stretch Induces a Smooth Muscle Cell-like Phenotype in Adipose-Derived Stem Cells. *Cells* **2021**, *10*, 3123. [CrossRef]
181. Tingting, T.; Wenjing, F.; Qian, Z.; Hengquan, W.; Simin, Z.; Zhisheng, J.; Shunlin, Q. The TGF-β pathway plays a key role in aortic aneurysms. *Clin. Chim. Acta* **2020**, *501*, 222–228. [CrossRef]
182. Li, G.; Wang, M.; Caulk, A.W.; Cilfone, N.A.; Gujja, S.; Qin, L.; Chen, P.-Y.; Chen, Z.; Yousef, S.; Jiao, Y.; et al. Chronic mTOR activation induces a degradative smooth muscle cell phenotype. *J. Clin. Investig.* **2020**, *130*, 1233–1251. [CrossRef]
183. Zhou, Z.; Liu, Y.; Gao, S.; Zhou, M.; Qi, F.; Ding, N.; Zhang, J.; Li, R.; Wang, J.; Shi, J.; et al. Excessive DNA damage mediates ECM degradation via the RBBP8/NOTCH1 pathway in sporadic aortic dissection. *Biochim. Biophys. Acta (BBA) Mol. Basis Dis.* **2022**, *1868*, 166303. [CrossRef]

International Journal of
Molecular Sciences

Review

Angiotensin Regulation of Vascular Homeostasis: Exploring the Role of ROS and RAS Blockers

Nikolaos Koumallos [1,*], Evangelia Sigala [1], Theodoros Milas [1], Nikolaos G. Baikoussis [1],
Dimitrios Aragiannis [2], Skevos Sideris [2] and Konstantinos Tsioufis [2]

[1] Cardiothoracic Department, Hippokration Hospital of Athens, 11527 Athens, Greece;
evitasig@hotmail.com (E.S.); theo_milas@yahoo.gr (T.M.); nikolaos.baikoussis@gmail.com (N.G.B.)

[2] Cardiology Department, Hippokration Hospital of Athens, 11527 Athens, Greece;
arajohn@hotmail.com (D.A.); skevos1@otenet.gr (S.S.); ktsioufis@hippocratio.gr (K.T.)

* Correspondence: koumallosn@yahoo.gr

Abstract: Extensive research has been conducted to elucidate and substantiate the crucial role of the Renin-Angiotensin System (RAS) in the pathogenesis of hypertension, cardiovascular disorders, and renal diseases. Furthermore, the role of oxidative stress in maintaining vascular balance has been well established. It has been observed that many of the cellular effects induced by Angiotensin II (Ang II) are facilitated by reactive oxygen species (ROS) produced by nicotinamide adenine dinucleotide phosphate (NADPH) oxidase. In this paper, we present a comprehensive overview of the role of ROS in the physiology of human blood vessels, specifically focusing on its interaction with RAS. Moreover, we delve into the mechanisms by which clinical interventions targeting RAS influence redox signaling in the vascular wall.

Keywords: redox signaling; oxidative stress; RAS; ACE inhibitors; ARBs

Citation: Koumallos, N.; Sigala, E.; Milas, T.; Baikoussis, N.G.; Aragiannis, D.; Sideris, S.; Tsioufis, K. Angiotensin Regulation of Vascular Homeostasis: Exploring the Role of ROS and RAS Blockers. *Int. J. Mol. Sci.* **2023**, *24*, 12111. https://doi.org/10.3390/ijms241512111

Academic Editors: Yutang Wang and Dianna Magliano

Received: 3 July 2023
Revised: 20 July 2023
Accepted: 21 July 2023
Published: 28 July 2023

1. Introduction

Reactive oxygen species (ROS) are considered highly reactive molecules that, depending on the amount produced, can affect cells and vascular functions [1]. Their precise regulation of production in the endothelium is crucial for controlling various cell functions in the vascular system [2–4]. Otherwise, excessive production of ROS (including superoxide O_2-, hydroxyl radical $-OH$, and peroxynitrite anion $ONOO-$) can cause the disruption of redox homeostasis, damage the immune system response, and favour the manifestation of vascular diseases [5,6]. This situation can occur in various circumstances where there is an imbalance between ROS generation and antioxidant defence mechanisms, such as chronic inflammation, ischemia-reperfusion injury, aging, and metabolic disorders. Another important factor is impaired endothelial function (since endothelium is the main source of ROS), which disrupts the production of protective blood agents and contributes to cytotoxic effects, cell death, and, inevitably, oxidative stress [7,8]. Interestingly, it is worth noting that ROS have beneficial effects due to their cytotoxic properties and can serve as a defence mechanism against infections [9].

As for the overproduction of ROS in the development of cardiovascular disease, damage to the cardiovascular system caused by oxidative stress favours diseases such as hypertension, atherosclerosis, and heart failure [10,11]. The renin-angiotensin system (RAS) interacts with ROS and contributes to these diseases. Angiotensin II (Ang II), as a key component of the RAS, plays an important role in this interaction as it can stimulate ROS production through various mechanisms, such as activating nicotinamide adenine dinucleotide phosphate (NADPH) oxidase through its binding to the AT1 receptor (AT1R) [10]. This overproduction of ROS over Ang II can in turn affect the activity of components of the RAS, such as increase in the expression of the Angiotensin I to ACE and the production of renin. In situations where excessive ROS production plays a role in the pathogenesis of disease progression, therapeutic interventions may be beneficial [12]. Modulators of this complex interaction are angiotensin-converting enzyme inhibitors (ACEIs) and angiotensin receptor blockers (ARBs).

In this review, we aim to explore the intricate relationship between ROS and the angiotensin system. We will delve into the role of ACEIs and ARBs in modulating the interplay. By exploring the physiological aspects of ROS in the vasculature and their interaction with the RAS, we can improve our understanding of the mechanisms that control redox signaling within the vascular wall. This investigation will provide valuable insights into the therapeutic implications of modulating RAS for the treatment of vascular diseases and optimization of vascular homeostasis.

2. Discussion

2.1. Physiology of the RAS in Vascular Regulation

Understanding the physiology of RAS is vital for comprehending the interplay between ROS and RAS in vascular regulation. Increased levels of Ang II, the main effector of the RAS molecule, play a crucial role in cardiovascular disease progression when its production or signaling becomes dysregulated [10,11]. The physiological intricacies of the RAS have been thoroughly examined and comprehensively expounded upon in previous scientific investigations [10]. However, in the context of our present research on the influence of ROS in the progression of cardiovascular diseases, it is imperative to revisit and restate the fundamental principles of RAS physiology. By revisiting this well-established foundation, our objective is to establish a cohesive link between the RAS and the role of ROS in the development and manifestation of cardiovascular disorders.

The release of renin (an enzyme produced by conditions of reduced perfusion such as low blood pressure, low blood volume, or sympathetic stimulation), will activate a sequential processing of glycoprotein angiotensinogen (AGT), leading to the production of the decapeptide Ang I [12]. Subsequently, Ang I is further cleaved by ACE, a membrane-bound metalloprotease, and converts to Ang II. Once Ang II is formed it binds G-protein-coupled receptors (AT1-4R) [13–15]. AT1Rs are the prime mediators of Ang II leading to vasoconstriction, aldosterone and vasopressin secretion, and sodium and water retention. Because of this, AT1Rs play a crucial role in cardiovascular regulation, inflammation, fibrosis, endothelial dysfunction, and organ damage like nephrosclerosis [16,17] (Figure 1).

On the other hand, the effects of AT2-4Rs are not fully understood, but they act differently than AT1Rs. It is believed that AT2Rs, when activated, counteract the short- and long-term effects of AT1Rs (Figure 1) leading to beneficial antiproliferative and vasodilatory effects [18]. Other potential effects of AT2Rs include the regulation of cell growth and processing of the neuronal tissue [19]. AT3Rs were recently discovered and their exact role has not been elucidated. AT4Rs play a protective role in thrombotic effects as a fibrinolysis buffer by controlling the production of plasminogen activator inhibitor-1 (PAI-1). The nature of AT4Rs has been extensively investigated in the past with insulin-regulated aminopeptidase (IRAP) being the prominent candidate [20]. Angiotensin IV not only inhibits the peptidase activity of this enzyme but also facilitates IRAP translocation to the cell surface and enhances insulin-mediated glucose uptake [20]. In addition to the impacts mediated by G-protein-coupled receptors, recent discoveries of alternative enzyme systems and novel effector peptides have broadened our conventional understanding [21]. Among these systems are the receptors for prorenin and the G-protein-coupled receptor MAS [22]. The identification of these systems enhances our understanding of the intricate nature of the RAS by revealing that prorenin can activate renin and subsequently trigger protein kinases ERK1 and ERK2 [23,24]. The activation of these receptors introduces additional proliferative and metabolic effects that are not dependent on Ang II [25]. For blood pressure regulation, Ang II fast response to vascular changes is important. G protein-depended pathways enhance smooth muscle cell contraction [26] and trigger molecules that will contract smooth muscle cells. Moreover, when Ang II activates the G protein, it triggers the activation of phospholipase C, which leads to the production of inositol-1,4,5-trisphosphate

(IP3) and diacylglycerol. These molecules then initiate the release of calcium ions (Ca^{2+}) into the cytoplasm. The binding of Ca^{2+} to calmodulin activates myosin, a protein involved in muscle contraction, and enhances its interaction with actin. As a result, smooth muscle cells contract, contributing to the modulation of blood pressure.

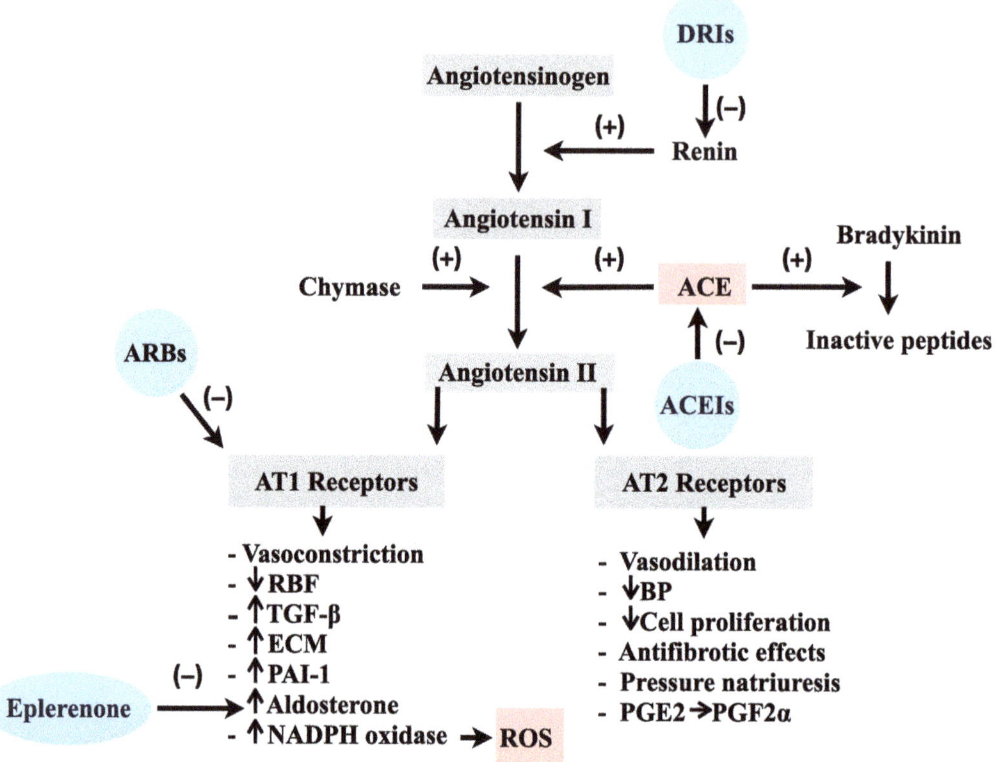

Figure 1. Interactions of the Renin-Angiotensin System, Kinin System, and Vascular Redox State: Implications for ACEIs and ARBs. ACE: Angiotensin-converting enzyme; AT1: Angiotensin type 1; AT2: Angiotensin type 2; ACEIs: ACE inhibitors; ARBs: AT1 receptor blockers; BP: Blood pressure; DRIs: Direct renin inhibitors; ECM: Extracellular matrix; NADPH: Nicotinamide adenine dinucleotide phosphate; PAI-1: Plasminogen activator inhibitor-1; PG: prostaglandin; RBF: Renal blood flow; ROS: Reactive oxygen species; TGF-β: Transforming growth factor-β.

The provided section of physiology elucidates the mechanism behind blood pressure regulation, which are mainly regulated by the rapid action of Ang II. By activating smooth muscle cell contraction through the G protein-dependent pathway, Ang II triggers a cascade of events involving phospholipase C, inositol-1,4,5-trisphosphate (IP3), diacylglycerol, and calcium ions (Ca^{2+}). This series of molecular interactions ultimately leads to the contraction of smooth muscle cells. Understanding these intricate processes helps us grasp the physiological basis of blood pressure adjustments in response to changes in posture.

2.2. Pathological Implications of the RAS and the Involvement of ROS

As established thus far, the pathological manifestations of the RAS (abnormal cellular proliferation, inflammation, disrupted vascular balance) are closely related to the excessive presence and prolonged exposure of Ang II [27]. ROS, such as $O_2^{\bullet-}$ (superoxide) and H_2O_2 (hydrogen peroxide), contribute to these pathological processes through

Ang II signaling [28–31]. In the cardiovascular system, Ang II contributes significantly to hypertension through its central, vascular, or renal effects. The multimeric enzyme of NADPH oxidase-derived ROS production is a deleterious equation in the development of Ang II-induced hypertension. NAPDH oxidase induces Ang II-induced oxidative stress, due to increased enzymatic activity. This occurs because of the rapid translocation and phosphorylation of cytosolic subunits of the small GTPase rac1 and p47phox to the cytochrome complex, via protein kinase C (PKC) [32,33]. Thereafter, PKC activates Janus kinase (JAK), which transduces and activates the JAK/STAT signaling pathway. This sequence promotes Ang II multiplication processes. Other components in revealing Ang II adverse effects are the early growth responsive genes and redox-sensitive proteins (c-Src, epidermal growth factor EGFR) [34–36]. Specifically, EGFR activates the Ras/Raf/ERK cascade, which subsequently upregulates c-Fos. Together with c-Jun, activated by c-Src via JNK, c-Fos forms the transcription factor known as activator protein-1 (AP-q1) [33].

Elevated levels of Ang II can have additional consequences on cellular viability and potentially induce DNA damage. The increased presence of $O_2^{\bullet-}$ and H_2O_2 activates additional redox-sensitive proteins, including p38/MAPK, which in turn stimulates the pro-survival factor Akt [37,38]. In a cascade of subsequent reactions, Akt inhibits various pro-apoptotic proteins. Another essential mediator of the RAS is aldosterone, which plays a crucial role in maintaining sodium and potassium balance, thus influencing extracellular volume. Aldosterone also exhibits potent pro-fibrotic effects [39]. The release of aldosterone is triggered by Ang II, and emerging evidence suggests that aldosterone may contribute to and exacerbate the detrimental effects of Ang II. Through activation of mineralocorticoid receptors, aldosterone promotes endothelial dysfunction and thrombosis, reduces vascular compliance and baroreceptor function, and induces fibrosis in both myocardial and vascular tissues [40]. The resulting increase in blood pressure and circulating volume, caused by the effects of Ang II and aldosterone on their target organs, establishes a negative feedback loop that suppresses renin release. The maintenance of this feedback inhibition critically relies on the Ang II-mediated activation of the AT1R [41].

Understanding the pathological effects of the RAS and the involvement of ROS is crucial for several reasons. It provides insights into the molecular mechanisms underlying cardiovascular diseases associated with RAS dysregulation, aiding in the advancement of precision therapeutic interventions. The identification of specific proteins and pathways involved opens opportunities for drug development, enabling the design of medications that selectively target these molecules or their associated receptors. This knowledge also contributes to personalized medicine approaches, considering individual variability in RAS responses. Moreover, it drives research advancements by uncovering novel signaling pathways and potential biomarkers, enhancing our understanding of cardiovascular disease pathophysiology. Overall, comprehending the pathological effects of RAS and ROS has significant implications for improving patient outcomes and advancing cardiovascular research.

2.3. The Application of ACEIs and ARBs in the Treatment of Cardiovascular Disorders

For a considerable period, the ACE/Ang II/AT1R pathway was recognized as the primary mechanism through which the RAS influences cardiovascular processes. Various categories of antihypertensive medications, including ACEIs, ARBs, β blockers (BBs), direct renin inhibitors (DRIs), and mineralocorticoid receptor antagonists (MRAs), have been employed to provide cardiorenal protection. ACEIs and ARBs, along with BBs, DRIs, and MRAs, act by interfering with the signaling pathways within the RAS. They are considered the first-line treatment options for managing hypertension and other cardiovascular disorders such as heart failure [42–44]. Furthermore, ACEIs and ARBs have a significant impact on the cardiovascular system and offer a protective effect against the occurrence and progression of kidney disease [45].

In addition to the well-established protection that ACEIs offer to the cardiovascular and renal systems by effectively controlling arterial pressure thresholds, they have also been associated with additional beneficial effects on the endothelium [46]. This has been confirmed in research studies conducted on patients with coronary artery disease or hypercholesterolemia, regardless of the effect on blood pressure reduction. These studies demonstrate that ACEIs and ARBs can additionally improve endothelial and vascular function in these patients. Furthermore, large studies have shown that ACEIs and ARBs are effective as monotherapy in managing other conditions such as left ventricular hypertrophy, systolic dysfunction, heart failure, and myocardial infarction [47]. Additionally, trials like the HOPE study, ONTARGET, and TRANSCEND have reported the extended benefits of ACEIs and ARBs to patients with an increased susceptibility to adverse outcomes but without left ventricular dysfunction [48]. For instance, the HOPE trial showed that ramipril significantly reduced the incidence of death, myocardial infarction, and stroke in high-risk patients [49]. Similarly, the EUROPA trial revealed a relative risk reduction in cardiovascular events with perindopril treatment in patients with stable coronary heart disease. However, the PEACE trial, which focused on stable coronary artery disease patients with preserved ventricular function, did not find a therapeutic benefit of ACEIs when added to conventional therapy, potentially because of the minimal occurrence of significant outcome events in that patient group [47,50].

The findings presented in this section contribute to our understanding of the significant benefits offered by ACEIs and ARBs in the realm of cardiovascular disease. These medications not only effectively lower blood pressure but also exhibit additional cardioprotective and reno-protective effects. The evidence highlights their ability to improve endothelial function, enhance vascular structure, and ultimately lead to improved cardiovascular outcomes. ACEIs and ARBs are essential primary therapies for high blood pressure, cardiac dysfunction, heart attack, and chronic kidney disease. Their therapeutic impact extends to high-risk patients without left ventricular dysfunction, further underscoring their clinical relevance. The comprehensive knowledge gained from large-scale randomized clinical trials supports the utilization of ACEIs and ARBs in the management of cardiovascular conditions, promoting improved patient outcomes and cardiovascular health.

2.4. Enhancing Endothelial Function and Mitigating Oxidative Stress: Exploring the Effects of ACEIs and ARBs

Several studies have demonstrated the favourable effects of ACEIs and ARBs on endothelial function and their ability to control oxidative stress levels in populations with cardiovascular diseases. As it has already been cleared, Ang II excessive presence is thought to play a pivotal role in the increased generation of $O_2^{\bullet-}$ and the impairment of endothelial function in blood vessels. This effect is achieved through the activation of NADPH oxidases, triggered by the stimulation of AT1 receptors [51]. Additionally, studies conducted in laboratory settings have revealed that AT1 receptors can be upregulated by low-density lipoproteins [52,53]. By augmenting the activity of SOD3 and blocking NADPH oxidase activation, ACEIs and ARBs provide an indirect mitigation of oxidative stress [46,54]. In relation to ACEIs, clinical trials have demonstrated favourable outcomes in terms of improving endothelial function in individuals with hypertension or cardiovascular history. Yet, there remains a scarcity of research investigating the impact of ACEI treatment regarding the decrease in C-Reactive Protein (CRP) levels, highlighting a need for further exploration in this area [55–60] (Table 1). Additionally, while not the primary focus of this review, it is worth mentioning that research conducted in experimental studies indicates that ACEIs and ARBs may offer advantages in preventing or slowing down cognitive decline and dementia [61].

Table 1. Randomized controlled trials assessing the influence of ACEIs and ARBs on serum CRP levels and endothelial function.

Author	Agents	Population	Period	Outcome
Walczak-Galezewska et al. [54]	Ramipril vs. Nebivolol	Hypertension	12 weeks	Ramipril demonstrated a decrease in CRP levels in comparison to nebivolol.
Ridker et al. [55]	HCTZ/Valsartan vs. Valsartan	Hypertension	6 weeks	CRP levels were reduced solely through treatment with valsartan.
Ghiadoni et al. [56]	Lisopril	Ventricular hypertrophy	3 years	Endothelial function showed improvement with lisinopril treatment when compared to the initial baseline.
Ghiadoni et al. [57]	Candesartan	Hypertension	1 year	Endothelial function demonstrated improvement with candesartan treatment when compared to the initial baseline.
Schiffrin et al. [58]	Losartan	Hypertension	1 year	Endothelial function showed improvement with losartan treatment in comparison to the initial baseline.
Schmieder et al. [59]	Telmisaran	Hypertension and DM	9 weeks	Endothelial function demonstrated improvement with telmisartan treatment in comparison to the initial baseline.

CRP: C-Reactive Protein; HCTZ: Hydrochlorothiazide.

In patients with coronary artery disease, it has been observed that ramipril (ACEI) and losartan (ARB) improve endothelial function by increasing the availability of nitric oxide (NO) through the mitigation of oxidative stress within the arterial wall [62]. Additionally, ARBs offer vascular protective effects [63]. Losartan promotes the phosphorylation of endothelial NO synthase (eNOS) and suppresses endothelial apoptosis induced by Tissue Necrosis Factor-α (TNF-α) through the activation of the VEGFR2/PI3K/Akt pathway [61]. Additionally, in diabetic rats, losartan restored glomerular NO production by increasing GCH1 protein expression and tetrahydrobiopterin (BH4) levels [64]. Furthermore, valsartan and irbesartan (ARBs) exert effects that counteract the development of atherosclerosis. The promotion of eNOS Ser117-phosphorylation increases eNOS mRNA stability and this leads to the reduced NADPH oxidase expression and also to the augmented vascular BH4 and restored eNOS uncoupling [65,66]. More recent studies have revealed significant suppression of NADPH oxidase p22(phox) expression in the aortic wall of patients with thoracic aortic aneurysm following ARB treatment, which the authors attribute to the pleiotropic effects of ARBs on vascular metabolism [67].

Some of the positive effects of ACEIs may be attributed to their impact on an ACE signaling cascade, resulting in improved endothelial function that appears to be independent of their effects on vasoactive substances [68]. Another mechanism suggested to beneficially affect the endothelium is the introduction of ACEIs as a treatment, with potential ACE signaling cascade involvement. This effect seems unrelated to the impact of ACEIs on vasoactive substances [69]. Moreover, when ACEI binds to the cell surface ectoenzyme ACE, it triggers a cascade that ultimately increases ACE and cyclooxygenase-2 expression. Firstly, ACEI binding activates casein kinase 2, leading to serine residue phosphorylation at the molecule's C-terminal end [69]. Secondly, ACE-associated c-Jun N-terminal kinase is activated, possibly through mitogen-activated protein kinase 7 activation [70]. This cascade eventually leads to an increase in the expression of ACE and cyclooxygenase-2 (through the accumulation of phosphorylated c-Jun in the nucleus, enhancing the DNA-binding activity of activator protein-1) [71]. Additionally, the elevated expression of cyclooxygenase-2 benefits endothelial function by promoting the production of prostacyclin, a vasodilator and antiplatelet agent, by endothelial cells [72].

Fleming et al. [71] have elucidated another aspect of ACE's outside-in signaling role. According to their study, when an ACE inhibitor interacts with ACE, it triggers a series of signaling events that affect the synthesis of multiple proteins. This implies that ACEIs exert their beneficial effects through the activation of a unique signaling cascade mediated by ACE, which goes beyond mere alterations in Ang II and bradykinin levels. Another important signaling of ACEIs is the signaling of kinin B1 and B2 receptors (B1R, B2R). B2R functions and ACEIs can act similarly as allosteric agonists. B1R and B2R signaling enhancement help ACEIs in promoting the production of NO, a major contributor to many cardiovascular conditions' treatments. Additionally, ACEIs have been shown to decrease CD40L levels, a protein predominantly present in activated T cells, as evidenced by well-established research [73,74]. Furthermore, recent studies have demonstrated that ACEIs also elevate adiponectin levels, a hormone responsible for regulating diverse metabolic processes [75,76].

Several studies have explored the shared effects of ACEIs in terms of their antihypertensive and anti-inflammatory actions. In one particular study, the ACEI captopril, which contains a sulfhydryl group, was compared to enalapril, an ACEI without a sulfhydryl group [77]. The findings suggested that ACEIs with a sulfhydryl group, such as captopril, possess the ability to protect the vascular endothelium against damages induced by L-methionine. The beneficial effects of captopril were associated with the attenuation in decreased activity of Paraoxonase-1 and NO levels. Another study investigated the role of the sulfhydryl group by comparing zofenopril and lisinopril in rats with myocardial infarction-induced heart failure. The results aligned with the previous study, indicating that the presence of a sulfhydryl group may offer a potential advantage in improving endothelial dysfunction through increased NO activity released from the endothelium into the vessel wall [78].

The concept of achieving more comprehensive inhibition of RAS and overcoming the "ACE escape" phenomenon led to the development of ARBs. ARBs effectively block the harmful effects of Ang II at the AT1 receptor, but they may diminish the beneficial effects of kinins. Interestingly, it was discovered that increased levels of kinins, previously considered an undesirable consequence of ACEIs, promote vasodilation and offer benefits. ARB therapy can activate the AT2 receptor, which potentially results in favourable anti-inflammatory, antithrombotic, and antiproliferative effects. Furthermore, losartan blockade of AT1R increases Ang II metabolism to angiotensin IV, and thus increases AT4R activation [79]. Some studies have reported relatively lower rates of myocardial infarction with non-ARB antihypertensive treatments compared to ARB-based treatments, leading to the "ARB-myocardial infarction paradox". It has been suggested that this paradox arises from the unopposed activation of the AT2 receptor [79,80]. Although these receptors are generally associated with favourable effects, studies in certain animal models have shown that they can have hypertrophic and pro-inflammatory effects [79]. For instance, mice lacking the AT2 receptor are protected against cardiac hypertrophy, while overexpression of the AT2 receptor in isolated human cardiomyocytes is linked to hypertrophy [81]. Moreover, AT2 receptor activation has been found to stimulate the production of matrix metalloproteinase-1, an enzyme involved in degrading the fibrous cap of atherosclerotic plaques [82]. Therefore, AT2 receptor activation could contribute to plaque instability and the formation of blood clots [80].

Furthermore, ACEIs and ARBs contribute to the progression of atherosclerosis and oxidative stress by decreasing asymmetric dimethylarginine (ADMA) levels through the stimulation of dimethylarginine dimethylaminohydrolase (DDAH) activity [76,83]. It is well established that decreased ADMA levels enhance the coupling of eNOS. Nevertheless, the precise mechanism underlying the influence of RAS inhibitors on ADMA metabolism remains uncertain. Ang II fosters the generation of ROS via vascular NADPH oxidase [84–87].

The inactivation of DDAH by ROS suggests that ACEIs and ARBs may improve ADMA metabolism by reducing oxidative stress. Indeed, certain studies have shown that these drugs can reduce serum markers of oxidative stress [88,89]. For example, a study where the impacts of two ACE inhibitors were compared [90] found that zofenopril, which includes reduced sulfhydryl groups and displays direct antioxidant characteristics, and enalapril, which lacks -SH groups and does not demonstrate antioxidant activity. Zofenopril was more effective in reducing ADMA concentration. However, other studies [91] have not observed changes in serum lipid peroxidation products in patients treated with ACEIs, indicating that additional mechanisms should be considered.

The decrease in ADMA levels resulting from RAS blockade could be attributed to the reduction in blood pressure, as shear stress increases ADMA production by endothelial cells [92]. Therefore, lowering blood pressure through RAS blockade could lead to a decrease in ADMA [92,93]. Some studies have reported a simultaneous decrease in both ADMA and blood pressure, supporting this possibility [93]. However, other studies have shown that ACEIs or ARBs decrease ADMA levels without affecting blood pressure [89]. Additionally, the observation that only perindopril, but not bisoprolol, reduces ADMA in hypertensive patients despite a similar decrease in blood pressure suggests that the effect on blood pressure may not be the primary factor. Most studies involving ACEIs or ARBs have not reported changes in renal function, indicating that the reduction in ADMA concentration is unlikely to be solely due to improved renal excretion. In fact, in patients with type 2 diabetes, treatment with perindopril reduced plasma ADMA levels but had no effect on urinary ADMA levels [91].

To conclude this section, it is evident that ACEIs and ARBs have shown potential in modulating ADMA metabolism and improving endothelial function. These drugs may reduce ADMA levels through various mechanisms, including activation of DDAH and attenuation of oxidative stress. However, further research is required to fully elucidate the precise mechanisms involved. Understanding the impact of ACEIs and ARBs on ADMA metabolism contributes to our knowledge of their potential benefits in cardiovascular health and warrants continued investigation in this field.

3. Clinical Implications

The findings discussed in this paper have significant clinical implications for the use of ACEIs and ARBs in the management of cardiovascular diseases. The beneficial effects of ACEIs and ARBs extend beyond their traditional role as blood pressure-lowering agents. Firstly, ACEIs and ARBs have demonstrated their efficacy in improving endothelial function, which plays a crucial role in maintaining cardiovascular health. These drugs have been shown to increase the bioavailability of NO by reducing oxidative stress and enhancing NO production. Improved endothelial function is associated with vasodilation, anti-aggregatory effects, and reduced inflammation, ultimately leading to better cardiovascular outcomes. Therefore, ACEIs and ARBs should be considered as first-line therapies in patients with hypertension, coronary artery disease, and other cardiovascular disorders characterized by endothelial dysfunction.

Secondly, ACEIs and ARBs have shown potential in modulating ADMA metabolism. Elevated levels of ADMA, an endogenous inhibitor of NOS, have been linked to endothelial dysfunction and increased cardiovascular risk. ACEIs and ARBs may reduce ADMA levels through various mechanisms, including activation of DDAH and attenuation of oxidative stress. By improving ADMA metabolism and restoring NOS activity, these drugs can further enhance endothelial function and mitigate the progression of cardiovascular diseases.

Moreover, the pleiotropic effects of ACEIs and ARBs go beyond their role in blood pressure control. These drugs exhibit antioxidant properties, inhibit inflammatory pathways, and have potential anti-atherosclerotic effects. They may also modulate the RAS beyond blocking the effects of Ang II, leading to the activation of beneficial pathways such as the AT2 receptor. These broader effects contribute to their cardioprotective actions and suggest their potential utility in a range of cardiovascular disorders, including heart failure, myocardial infarction, and chronic kidney disease.

4. Limitations

While ACEIs and ARBs have demonstrated significant clinical benefits, it is important to acknowledge certain limitations associated with their use. These limitations should be taken into consideration when interpreting the findings and implementing these therapies in clinical practice.

Firstly, individual patient variability and heterogeneity in response to ACEIs and ARBs may impact their effectiveness. The response to these drugs can vary based on factors such as genetic variations, underlying comorbidities, and concomitant medications. Some patients may experience limited or suboptimal response to ACEIs or ARBs, necessitating the consideration of alternative treatment strategies or combination therapies. Secondly, although ACEIs and ARBs have shown promising effects on endothelial function, ADMA metabolism, and other cardiovascular parameters, the precise mechanisms underlying these effects are not fully understood. The complex interplay between RAS, oxidative stress, inflammatory pathways, and other factors involved in cardiovascular pathophysiology requires further investigation. A deeper understanding of these mechanisms could potentially help refine the use of ACEIs and ARBs and identify patient subgroups who would benefit the most from these therapies. Thirdly, like any medication, ACEIs and ARBs are associated with potential adverse effects. Common side effects include hypotension, hyperkalaemia, and renal dysfunction. Monitoring of blood pressure, electrolyte levels, and renal function is important during treatment. Additionally, individual patient characteristics and preferences should be considered when selecting the most appropriate therapy, as some patients may be more prone to specific side effects or have contraindications for ACEIs or ARBs.

Furthermore, the evidence supporting the clinical use of ACEIs and ARBs is primarily based on observational studies, randomized controlled trials, and meta-analyses. While these studies provide valuable insights, they also have inherent limitations, such as the potential for selection bias, confounding factors, and limited generalizability to diverse patient populations. Future well-designed clinical trials are needed to further validate the findings and establish the optimal use of these therapies in specific patient groups. Recognizing these limitations is crucial for clinicians to make informed treatment decisions and to individualize therapy for optimal patient outcomes. Further research is needed to address these limitations and expand our knowledge on the use of ACEIs and ARBs in diverse patient populations and clinical scenarios.

5. Conclusions

In summary, this paper emphasizes the paramount significance of the RAS in the genesis and advancement of hypertension, cardiovascular diseases, and renal disorders. The RAS has been subject to extensive scrutiny and its role in these conditions has been firmly established. Additionally, this study sheds light on the vital influence of ROS in maintaining the balance within the blood vessels and their intricate interactions with the RAS, impacting the course of diseases. Notably, therapeutic interventions aimed at the RAS, such as ACEIs and ARBs, have demonstrated beneficial effects on endothelial function and oxidative stress levels. Furthermore, ongoing research is delving into the underlying mechanisms governing the interplay between the RAS and ROS, offering new insights and potential innovative therapeutic avenues. Gaining a comprehensive understanding of these complex interactions is pivotal for the development of targeted and efficacious treatments for cardiovascular conditions.

Author Contributions: Conceptualization, N.K. and K.T.; methodology, N.K., E.S. and T.M, investigation, N.K. and T.M., data curation, N.G.B. and D.A.; Writing—original draft preparation, N.K. and E.S.; recourses, T.M.; writing—review and editing N.K., E.S. and S.S.; visualization, D.A.; supervision, K.T.; project administration, K.T. Authorship is limited to those who have contributed substantially to the work reported. All authors have read and agreed to the published version of the manuscript.

Funding: This research received no external funding.

Institutional Review Board Statement: Not applicable.

Informed Consent Statement: Not applicable.

Conflicts of Interest: The authors declare no conflict of interest.

References

1. Antoniades, C.; Tousoulis, D.; Tentolouris, C.; Toutouzas, P.; Stefanadis, C. Oxidative stress, antioxidant vitamins, and atherosclerosis. From basic research to clinical practice. *Herz* **2003**, *28*, 628–638. [CrossRef] [PubMed]
2. Fujiwara, N.; Yamashita, S.; Okamoto, M.; Cooley, M.A.; Ozaki, K.; Everett, E.T.; Suzuki, M. Perfluorooctanoic acid-induced cell death via the dual roles of ROS-MAPK/ERK signaling in ameloblast-lineage cells. *Ecotoxicol. Environ. Saf.* **2023**, *260*, 115089. [CrossRef]
3. Lee, Y.-P.; Lin, C.-R.; Chen, S.-S.; Chen, R.-J.; Wu, Y.-H.; Chen, Y.-H.; Huang, B.-M. Combination treatment of cordycepin and radiation induces MA-10 mouse Leydig tumor cell death via ROS accumulation and DNA damage. *Am. J. Cancer Res.* **2023**, *13*, 1329–1346. [PubMed]
4. Wang, B.; Wang, Y.; Zhang, J.; Hu, C.; Jiang, J.; Li, Y.; Peng, Z. ROS-induced lipid peroxidation modulates cell death outcome: Mechanisms behind apoptosis, autophagy, and ferroptosis. *Arch. Toxicol.* **2023**, *97*, 1439–1451. [CrossRef] [PubMed]
5. Cheng, C.K.; Ding, H.; Jiang, M.; Yin, H.; Gollasch, M.; Huang, Y. Perivascular adipose tissue: Fine-tuner of vascular redox status and inflammation. *Redox Biol.* **2023**, *62*, 102683. [CrossRef]
6. Hernández-Reséndiz, S.; Muñoz-Vega, M.; Contreras, W.E.; Crespo-Avilan, G.E.; Rodriguez-Montesinos, J.; Arias-Carrión, O.; Pérez-Méndez, O.; Boisvert, W.A.; Preissner, K.T.; Cabrera-Fuentes, H.A. Responses of Endothelial Cells Towards Ischemic Conditioning Following Acute Myocardial Infarction. *Cond. Med.* **2018**, *1*, 247–258.
7. Da Silva, F.C.; de Araújo, B.J.; Cordeiro, C.S.; Arruda, V.M.; Faria, B.Q.; Guerra, J.F.D.C.; De Araújo, T.G.; Fürstenau, C.R. Endothelial dysfunction due to the inhibition of the synthesis of nitric oxide: Proposal and characterization of an in vitro cellular model. *Front. Physiol.* **2022**, *13*, 978378. [CrossRef]
8. Shaito, A.; Aramouni, K.; Assaf, R.; Parenti, A.; Orekhov, A.; Yazbi, A.E.; Pintus, G.; Eid, A.H. Oxidative Stress-Induced Endothelial Dysfunction in Cardiovascular Diseases. *Front. Biosci.* **2022**, *27*, 105. [CrossRef]
9. Hänze, J.; Weissmann, N.; Grimminger, F.; Seeger, W.; Rose, F. Cellular and molecular mechanisms of hypoxia-inducible factor driven vascular remodeling. *Thromb. Haemost.* **2007**, *97*, 774–787.
10. Paul, M.; Poyan Mehr, A.; Kreutz, R. Physiology of local renin-angiotensin systems. *Physiol. Rev.* **2006**, *86*, 747–803. [CrossRef]
11. Kasiakogias, A.; Rosei, E.A.; Camafort, M.; Ehret, G.; Faconti, L.; Ferreira, J.P.; Brguljan, J.; Januszewicz, A.; Kahan, T.; Manolis, A.; et al. Hypertension and heart failure with preserved ejection fraction: Position paper by the European Society of Hypertension. *J. Hypertens.* **2021**, *39*, 1522–1545. [CrossRef] [PubMed]
12. McFarlane, S.I.; Kumar, A.; Sowers, J.R. Mechanisms by which angiotensin-converting enzyme inhibitors prevent diabetes and cardiovascular disease. *Am. J. Cardiol.* **2003**, *91*, 30H–37H. [CrossRef] [PubMed]
13. Lévy, B.I. Can angiotensin II type 2 receptors have deleterious effects in cardiovascular disease? Implications for therapeutic blockade of the renin-angiotensin system. *Circulation* **2004**, *109*, 8–13. [CrossRef] [PubMed]
14. Zhang, H.; Luginina, A.; Mishin, A.; Baidya, M.; Shukla, A.K.; Cherezov, V. Structural insights into ligand recognition and activation of angiotensin receptors. *Trends Pharmacol. Sci.* **2021**, *42*, 577–587. [CrossRef]
15. Ziaja, M.; Urbanek, K.A.; Kowalska, K.; Piastowska-Ciesielska, A.W. Angiotensin II and Angiotensin Receptors 1 and 2—Multifunctional System in Cells Biology, What Do We Know? *Cells* **2021**, *10*, 381. [CrossRef]
16. Duprez, D.A. Role of the renin–angiotensin–aldosterone system in vascular remodeling and inflammation: A clinical review. *J. Hypertens.* **2006**, *24*, 983–991. [CrossRef]
17. Ferrario, C.M.; Strawn, W.B. Role of the renin-angiotensin-aldosterone system and proinflammatory mediators in cardiovascular disease. *Am. J. Cardiol.* **2006**, *98*, 121–128. [CrossRef]
18. Stoll, M.; Steckelings, U.M.; Paul, M.; Bottari, S.P.; Metzger, R.; Unger, T. The angiotensin AT2-receptor mediates inhibition of cell proliferation in coronary endothelial cells. *J. Clin. Investig.* **1995**, *95*, 651–657. [CrossRef]
19. Steckelings, U.M.; Kaschina, E.; Unger, T. The AT2 receptor—A matter of love and hate. *Peptides* **2005**, *26*, 1401–1409. [CrossRef]
20. Albiston, A.L.; McDowall, S.G.; Matsacos, D.; Sim, P.; Clune, E.; Mustafa, T.; Lee, J.; Mendelsohn, F.A.O.; Simpson, R.J.; Connolly, L.M.; et al. Evidence that the angiotensin IV (AT4) receptor is the enzyme insulin-regulated aminopeptidase. *J. Biol. Chem.* **2001**, *276*, 48623–48626. [CrossRef]
21. Silva, I.M.S.; Assersen, K.B.; Willadsen, N.N.; Jepsen, J.; Artuc, M.; Steckelings, U.M. The role of the renin-angiotensin system in skin physiology and pathophysiology. *Exp. Dermatol.* **2020**, *29*, 891–901. [CrossRef] [PubMed]
22. Nguyen, G.; Danser, A.H.J. Prorenin and (pro)renin receptor: A review of available data from in vitro studies and experimental models in rodents. *Exp. Physiol.* **2008**, *93*, 557–563. [CrossRef]
23. Santos, R.A.S.; Simoes e Silva, A.C.; Maric, C.; Silva, D.M.R.; Machado, R.P.; de Buhr, I.; Heringer-Walther, S.; Pinheiro, S.V.B.; Lopes, M.T.; Bader, M.; et al. Angiotensin-(1–7) is an endogenous ligand for the G protein-coupled receptor Mas. *Proc. Natl. Acad. Sci. USA* **2003**, *100*, 8258–8263. [CrossRef] [PubMed]

24. Pimenta, E.; Oparil, S. Role of aliskiren in cardio-renal protection and use in hypertensives with multiple risk factors. *Ther. Clin. Risk Manag.* **2009**, *5*, 459–464. [CrossRef] [PubMed]

25. Zhang, L.; Ju, Q.; Sun, J.; Huang, L.; Wu, S.; Wang, S.; Li, Y.; Guan, Z.; Zhu, Q.; Xu, Y. Discovery of Novel Dual Extracellular Regulated Protein Kinases (ERK) and Phosphoinositide 3-Kinase (PI3K) Inhibitors as a Promising Strategy for Cancer Therapy. *Molecules* **2020**, *25*, 5693. [CrossRef] [PubMed]

26. Nakamura, F.; Tsukamoto, I.; Inoue, S.; Hashimoto, K.; Akagi, M. Cyclic compressive loading activates angiotensin II type 1 receptor in articular chondrocytes and stimulates hypertrophic differentiation through a G-protein-dependent pathway. *FEBS Open Bio* **2018**, *8*, 962–973. [CrossRef]

27. Touyz, R.M. Reactive oxygen species and angiotensin II signaling in vascular cells: Implications in cardiovascular disease. *Braz. J. Med. Biol. Res.* **2004**, *37*, 1263–1273. [CrossRef] [PubMed]

28. Branicky, R.; Wang, Y.; Khaki, A.; Liu, J.-L.; Kramer-Drauberg, M.; Hekimi, S. Stimulation of RAS-dependent ROS signaling extends longevity by modulating a developmental program of global gene expression. *Sci. Adv.* **2022**, *8*, eadc9851. [CrossRef]

29. Mohamed, H.R.H. Acute Oral Administration of Cerium Oxide Nanoparticles Suppresses Lead Acetate–Induced Genotoxicity, Inflammation, and ROS Generation in Mice Renal and Cardiac Tissues. *Biol. Trace Elem. Res.* **2022**, *200*, 3284–3293. [CrossRef]

30. Barp, C.G.; Bonaventura, D.; Assreuy, J. NO, ROS, RAS, and PVAT: More Than a Soup of Letters. *Front. Physiol.* **2021**, *12*, 640021. [CrossRef]

31. Griendling, K.K.; Minieri, C.A.; Ollerenshaw, J.D.; Alexander, R.W. Angiotensin II stimulates NADH and NADPH oxidase activity in cultured vascular smooth muscle cells. *Circ. Res.* **1994**, *74*, 1141–1148. [CrossRef] [PubMed]

32. Wassmann, S.; Laufs, U.; Bäumer, A.T.; Müller, K.; Konkol, C.; Sauer, H.; Böhm, M.; Nickenig, G. Inhibition of geranylgeranylation reduces angiotensin II-mediated free radical production in vascular smooth muscle cells: Involvement of angiotensin AT1 receptor expression and Rac1 GTPase. *Mol. Pharmacol.* **2001**, *59*, 646–654. [CrossRef] [PubMed]

33. Mehta, P.K.; Griendling, K.K. Angiotensin II cell signaling: Physiological and pathological effects in the cardiovascular system. *Am. J. Physiol. Cell Physiol.* **2007**, *292*, C82–C97. [CrossRef]

34. Berk, B.C.; Corson, M.A. Angiotensin II signal transduction in vascular smooth muscle: Role of tyrosine kinases. *Circ. Res.* **1997**, *80*, 607–616. [CrossRef]

35. Ren, Y.; Xie, W.; Yang, S.; Jiang, Y.; Wu, D.; Zhang, H.; Sheng, S. Angiotensin-converting enzyme 2 inhibits inflammation and apoptosis in high glucose-stimulated microvascular endothelial cell damage by regulating the JAK2/STAT3 signaling pathway. *Bioengineered* **2022**, *13*, 10802–10810. [CrossRef] [PubMed]

36. Jung, J.-H.; Choi, J.-W.; Lee, M.-K.; Choi, Y.-H.; Nam, T.-J. Effect of Cyclophilin from *Pyropia Yezoensis* on the Proliferation of Intestinal Epithelial Cells by Epidermal Growth Factor Receptor/Ras Signaling Pathway. *Mar. Drugs* **2019**, *17*, 297. [CrossRef] [PubMed]

37. Kim, D.-H.; Chung, J.-K. Akt: Versatile mediator of cell survival and beyond. *Korean Soc. Biochem. Mol. Biol.-BMB Rep.* **2002**, *35*, 106–115. [CrossRef]

38. Azouz, A.A.; Abdel-Rahman, D.M.; Messiha, B.A.S. Balancing renal Ang-II/Ang-(1-7) by xanthenone; an ACE2 activator; contributes to the attenuation of Ang-II/p38 MAPK/NF-κB p65 and Bax/caspase-3 pathways in amphotericin B-induced nephrotoxicity in rats. *Toxicol. Mech. Methods* **2023**, *33*, 452–462. [CrossRef]

39. Johar, S.; Cave, A.C.; Narayanapanicker, A.; Grieve, D.J.; Shah, A.M. Aldosterone mediates angiotensin II-induced interstitial cardiac fibrosis via a Nox2-containing NADPH oxidase. *FASEB J.* **2006**, *20*, 1546–1548. [CrossRef]

40. Struthers, A.D.; MacDonald, T.M. Review of aldosterone- and angiotensin II-induced target organ damage and prevention. *Cardiovasc. Res.* **2004**, *61*, 663–670. [CrossRef]

41. Dehe, L.; Mousa, S.A.; Aboryag, N.; Shaqura, M.; Beyer, A.; Schäfer, M.; Treskatsch, S. Identification of Mineralocorticoid Receptors, Aldosterone, and Its Processing Enzyme CYP11B2 on Parasympathetic and Sympathetic Neurons in Rat Intracardiac Ganglia. *Front. Neuroanat.* **2021**, *15*, 802359. [CrossRef] [PubMed]

42. Jia, Q.; Han, W.; Shi, S.; Hu, Y. The effects of ACEI/ARB, aldosterone receptor antagonists and statins on preventing recurrence of atrial fibrillation: A protocol for systematic review and network meta-analysis. *Medicine* **2021**, *100*, e24280. [CrossRef] [PubMed]

43. Eberhardt, R.T.; Kevak, R.M.; Kang, P.M.; Frishman, W.H. Angiotensin II receptor blockade: An innovative approach to cardiovascular pharmacotherapy. *J. Clin. Pharmacol.* **1993**, *33*, 1023–1038. [CrossRef] [PubMed]

44. Papadopoulos, D.P.; Votteas, V. Role of perindopril in the prevention of stroke. *Recent Pat. Cardiovasc. Drug Discov.* **2006**, *1*, 283–289. [CrossRef]

45. Mogensen, U.M.; Gong, J.; Jhund, P.S.; Shen, L.; Køber, L.; Desai, A.S.; Lefkowitz, M.P.; Packer, M.; Rouleau, J.L.; Solomon, S.D.; et al. Effect of sacubitril/valsartan on recurrent events in the Prospective comparison of ARNI with ACEI to Determine Impact on Global Mortality and morbidity in Heart Failure trial (PARADIGM-HF). *Eur. J. Heart Fail.* **2018**, *20*, 760–768. [CrossRef]

46. Hornig, B.; Landmesser, U.; Kohler, C.; Ahlersmann, D.; Spiekermann, S.; Christoph, A.; Tatge, H.; Drexler, H. Comparative effect of ace inhibition and angiotensin II type 1 receptor antagonism on bioavailability of nitric oxide in patients with coronary artery disease: Role of superoxide dismutase. *Circulation* **2001**, *103*, 799–805. [CrossRef]

47. Yusuf, S.; Sleight, P.; Pogue, J.; Bosch, J.; Davies, R.; Dagenais, G.; Heart Outcomes Prevention Evaluation Study Investigators. Effects of an angiotensin-converting-enzyme inhibitor, ramipril, on cardiovascular events in high-risk patients. *N. Engl. J. Med.* **2000**, *342*, 145–153. [CrossRef]

48. Schmieder, R.E.; Martus, P.; Klingbeil, A. Reversal of left ventricular hypertrophy in essential hypertension. A meta-analysis of randomized double-blind studies. *JAMA* **1996**, *275*, 1507–1513. [CrossRef] [PubMed]

49. Fox, K.M. The EURopean Trial on Reduction of Cardiac Events with Perindopril in Stable Coronary Artery Disease Investigators. Efficacy of perindopril in reduction of cardiovascular events among patients with stable coronary artery disease: Randomised, double-blind, placebo-controlled, multicentre trial (the EUROPA study). *Lancet* **2003**, *362*, 782–788. [CrossRef] [PubMed]

50. Braunwald, E.; Domanski, M.J.; Fowler, S.E.; Geller, N.L.; Gersh, B.J.; Hsia, J.; Pfeffer, M.A.; Rice, M.M.; Rosenberg, Y.D.; Rouleau, J.L.; et al. Angiotensin-converting-enzyme inhibition in stable coronary artery disease. *N. Engl. J. Med.* **2004**, *351*, 2058–2068. [CrossRef] [PubMed]

51. Griendling, K.K.; Sorescu, D.; Ushio-Fukai, M. NAD(P)H oxidase: Role in cardiovascular biology and disease. *Circ. Res.* **2000**, *86*, 494–501. [CrossRef]

52. Warnholtz, A.; Nickenig, G.; Schulz, E.; Macharzina, R.; Bräsen, J.H.; Skatchkov, M.; Heitzer, T.; Stasch, J.P.; Griendling, K.K.; Harrison, D.G.; et al. Increased NADH-oxidase-mediated superoxide production in the early stages of atherosclerosis: Evidence for involvement of the renin-angiotensin system. *Circulation* **1999**, *99*, 2027–2033. [CrossRef] [PubMed]

53. Cernecka, H.; Doka, G.; Srankova, J.; Pivackova, L.; Malikova, E.; Galkova, K.; Kyselovic, J.; Krenek, P.; Klimas, J. Ramipril restores PPARβ/δ and PPARγ expressions and reduces cardiac NADPH oxidase but fails to restore cardiac function and accompanied myosin heavy chain ratio shift in severe anthracycline-induced cardiomyopathy in rat. *Eur. J. Pharmacol.* **2016**, *791*, 244–253. [CrossRef] [PubMed]

54. Petersen, S.V.; Poulsen, N.B.; Linneberg Matthiesen, C.; Vilhardt, F. Novel and Converging Ways of NOX2 and SOD3 in Trafficking and Redox Signaling in Macrophages. *Antioxidants* **2021**, *10*, 172. [CrossRef] [PubMed]

55. Walczak-Gałęzewska, M.; Szulińska, M.; Miller-Kasprzak, E.; Pupek-Musialik, D.; Bogdański, P. The effect of nebivolol and ramipril on selected biochemical parameters, arterial stiffness, and circadian profile of blood pressure in young men with primary hypertension: A 12-week prospective randomized, open-label study trial. *Medicine* **2018**, *97*, e11717. [CrossRef]

56. Ridker, P.M.; Danielson, E.; Rifai, N.; Glynn, R.J.; Val-MARC Investigators. Valsartan, blood pressure reduction, and C-reactive protein: Primary report of the Val-MARC trial. *Hypertension* **2006**, *48*, 73–79. [CrossRef]

57. Ghiadoni, L.; Versari, D.; Magagna, A.; Kardasz, I.; Plantinga, Y.; Giannarelli, C.; Taddei, S.; Salvetti, A. Ramipril dose-dependently increases nitric oxide availability in the radial artery of essential hypertension patients. *J. Hypertens.* **2007**, *25*, 361–366. [CrossRef]

58. Ghiadoni, L.; Virdis, A.; Magagna, A.; Taddei, S.; Salvetti, A. Effect of the angiotensin II type 1 receptor blocker candesartan on endothelial function in patients with essential hypertension. *Hypertension* **2000**, *35 Pt 2*, 501–506. [CrossRef] [PubMed]

59. Schiffrin, E.L.; Deng, L.Y. Comparison of effects of angiotensin I– converting enzyme inhibition and β-blockade for 2 years on function of small arteries from hypertensive patients. *Hypertension* **1995**, *25 Pt 2*, 699–703. [CrossRef]

60. Schmieder, R.E.; Delles, C.; Mimran, A.; Fauvel, J.P.; Ruilope, L.M. Impact of telmisartan versus ramipril on renal endothelial function in patients with hypertension and type 2 diabetes. *Diabetes Care* **2007**, *30*, 1351–1356. [CrossRef]

61. Goh, K.L.; Bhaskaran, K.; Minassian, C.; Evans, S.J.; Smeeth, L.; Douglas, I.J. Angiotensin receptor blockers and risk of dementia: Cohort study in UK Clinical Practice Research Datalink. *Br. J. Clin. Pharmacol.* **2015**, *79*, 337–350. [CrossRef] [PubMed]

62. Chao, Y.; Ye, P.; Zhu, L.; Kong, X.; Qu, X.; Zhang, J.; Luo, J.; Yang, H.; Chen, S. Low shear stress induces endothelial reactive oxygen species via the AT1R/eNOS/NO pathway. *J. Cell. Physiol.* **2018**, *233*, 1384–1395. [CrossRef] [PubMed]

63. Watanabe, T.; Suzuki, J.; Yamawaki, H.; Sharma, V.K.; Sheu, S.-S.; Berk, B.C. Losartan metabolite EXP3179 activates Akt and endothelial nitric oxide synthase via vascular endothelial growth factor receptor-2 in endothelial cells: Angiotensin II type 1 receptor-independent effects of EXP3179. *Circulation* **2005**, *112*, 1798–1805. [CrossRef] [PubMed]

64. Satoh, M.; Fujimoto, S.; Arakawa, S.; Yada, T.; Namikoshi, T.; Haruna, Y.; Horike, H.; Sasaki, T.; Kashihara, N. Angiotensin II type 1 receptor blocker ameliorates uncoupled endothelial nitric oxide synthase in rats with experimental diabetic nephropathy. *Nephrol. Dial. Transplant.* **2008**, *23*, 3806–3813. [CrossRef] [PubMed]

65. Imanishi, T.; Tsujioka, H.; Ikejima, H.; Kuroi, A.; Takarada, S.; Kitabata, H.; Tanimoto, T.; Muragaki, Y.; Mochizuki, S.; Goto, M.; et al. Renin inhibitor aliskiren improves impaired nitric oxide bioavailability and protects against atherosclerotic changes. *Hypertension* **2008**, *52*, 563–572. [CrossRef] [PubMed]

66. Nussberger, J.; Aubert, J.-F.; Bouzourene, K.; Pellegrin, M.; Hayoz, D.; Mazzolai, L. Renin inhibition by aliskiren prevents atherosclerosis progression: Comparison with irbesartan, atenolol, and amlodipine. *Hypertension* **2008**, *51*, 1306–1311. [CrossRef]

67. Honjo, T.; Yamaoka-Tojo, M.; Inoue, N. Pleiotropic effects of ARB in vascular metabolism--focusing on atherosclerosis-based cardiovascular disease. *Curr. Vasc. Pharmacol.* **2011**, *9*, 145–152. [CrossRef]

68. Fleming, I. Signaling by the angiotensin-converting enzyme. *Circ. Res.* **2006**, *98*, 887–896. [CrossRef]

69. Horne, M.C.; Hell, J.W. Angiotensin II signalling kicks out p27[Kip1]: Casein kinase 2 augmentation of Cav1.2 L-type Ca^{2+} channel activity in immature ventricular cardiomyocytes. *J. Physiol.* **2017**, *595*, 4131–4132. [CrossRef]

70. Kashihara, T.; Nakada, T.; Kojima, K.; Takeshita, T.; Yamada, M. Angiotensin II activates Ca$_V$1.2 Ca^{2+} channels through β-arrestin2 and casein kinase 2 in mouse immature cardiomyocytes. *J. Physiol.* **2017**, *595*, 4207–4225. [CrossRef]

71. Fleming, I.; Kohlstedt, K.; Busse, R. New fACEs to the renin-angiotensin system. *Physiology* **2005**, *20*, 91–95. [CrossRef] [PubMed]

72. Li, X.; Hu, H.; Wang, Y.; Xue, M.; Li, X.; Cheng, W.; Xuan, Y.; Yin, J.; Yang, N.; Yan, S. Valsartan Upregulates Kir2.1 in Rats Suffering from Myocardial Infarction via Casein Kinase 2. *Cardiovasc. Drugs Ther.* **2015**, *29*, 209–218. [CrossRef] [PubMed]

73. Daub, S.; Lutgens, E.; Münzel, T.; Daiber, A. CD40/CD40L and Related Signaling Pathways in Cardiovascular Health and Disease—The Pros and Cons for Cardioprotection. *Int. J. Mol. Sci.* **2020**, *21*, 8533. [CrossRef] [PubMed]

74. Senchenkova, E.Y.; Russell, J.; Vital, S.A.; Yildirim, A.; Orr, A.W.; Granger, D.N.; Gavins, F.N.E. A critical role for both CD40 and VLA5 in angiotensin II–mediated thrombosis and inflammation. *FASEB J.* **2018**, *32*, 3448–3456. [CrossRef] [PubMed]
75. Tousoulis, D.; Antoniades, C.; Nikolopoulou, A.; Koniari, K.; Vasiliadou, C.; Marinou, K.; Koumallos, N.; Papageorgiou, N.; Stefanadi, E.; Siasos, G.; et al. Interaction between cytokines and sCD40L in patients with stable and unstable coronary syndromes. *Eur. J. Clin. Investig.* **2007**, *37*, 623–628. [CrossRef]
76. Antoniades, C.; Antonopoulos, A.S.; Tousoulis, D.; Stefanadis, C. Adiponectin: From obesity to cardiovascular disease. *Obes. Rev.* **2009**, *10*, 269–279. [CrossRef]
77. Liu, Y.-H.; Liu, L.-Y.; Wu, J.-X.; Chen, S.-X.; Sun, Y.-X. Comparison of captopril and enalapril to study the role of the sulfhydryl-group in improvement of endothelial dysfunction with ACE inhibitors in high dieted methionine mice. *J. Cardiovasc. Pharmacol.* **2006**, *47*, 82–88. [CrossRef]
78. Buikema, H.; Monnink, S.H.J.; Tio, R.A.; Crijns, H.J.G.M.; de Zeeuw, D.; van Gilst, W.H. Comparison of zofenopril and lisinopril to study the role of the sulfhydryl-group in improvement of endothelial dysfunction with ACE-inhibitors in experimental heart failure. *Br. J. Pharmacol.* **2000**, *130*, 1999–2007. [CrossRef]
79. Julius, S.; Kjeldsen, S.E.; Weber, M.; Brunner, H.R.; Ekman, S.; Hansson, L.; Hua, T.; Laragh, J.; McInnes, G.T.; Mitchell, L.; et al. Outcomes in hypertensive patients at high cardiovascular risk treated with regimens based on valsartan or amlodipine: The VALUE randomised trial. *Lancet* **2004**, *363*, 2022–2031. [CrossRef]
80. Strauss, M.H.; Hall, A.S. Angiotensin receptor blockers may increase risk of myocardial infarction: Unraveling the ARB-MI paradox. *Circulation* **2006**, *114*, 838–854. [CrossRef]
81. Senbonmatsu, T.; Ichihara, S.; Price, E., Jr.; Gaffney, F.A.; Inagami, T. Evidence for angiotensin II type 2 receptor-mediated cardiac myocyte enlargement during in vivo pressure overload. *J. Clin. Investig.* **2000**, *106*, R25–R29. [CrossRef] [PubMed]
82. D'Amore, A.; Black, M.J.; Thomas, W.G. The angiotensin II type 2 receptor causes constitutive growth of cardiomyocytes and does not antagonize angiotensin II type 1 receptor-mediated hypertrophy. *Hypertension* **2005**, *46*, 1347–1354. [CrossRef] [PubMed]
83. Trocha, M.; Szuba, A.; Merwid-Lad, A.; Sozański, T. Effect of selected drugs on plasma asymmetric dimethylarginine (ADMA) levels. *Pharmazie* **2010**, *65*, 562–571. [PubMed]
84. Kalinowski, L.; Malinski, T. Endothelial NADH/NADPH-dependent enzymatic sources of superoxide production: Relationship to endothelial dysfunction. *Acta Biochim. Pol.* **2004**, *51*, 459–469. [CrossRef]
85. Wang, C.; Luo, Z.; Carter, G.; Wellstein, A.; Jose, P.A.; Tomlinson, J.; Leiper, J.; Welch, W.J.; Wilcox, C.S.; Wang, D. NRF2 prevents hypertension, increased ADMA, microvascular oxidative stress, and dysfunction in mice with two weeks of ANG II infusion. *Am. J. Physiol. Regul. Integr. Comp. Physiol.* **2018**, *314*, R399–R406. [CrossRef]
86. Wang, H.; Jiang, H.; Liu, H.; Zhang, X.; Ran, G.; He, H.; Liu, X. Modeling Disease Progression: Angiotensin II Indirectly Inhibits Nitric Oxide Production via ADMA Accumulation in Spontaneously Hypertensive Rats. *Front. Physiol.* **2016**, *7*, 555. [CrossRef]
87. Shahin, N.N.; Abdelkader, N.F.; Safar, M.M. A Novel Role of Irbesartan in Gastroprotection against Indomethacin-Induced Gastric Injury in Rats: Targeting DDAH/ADMA and EGFR/ERK Signaling. *Sci. Rep.* **2018**, *8*, 4280. [CrossRef]
88. Fu, Y.-F.; Xiong, Y.; Guo, Z. A reduction of endogenous asymmetric dimethylarginine contributes to the effect of captopril on endothelial dysfunction induced by homocysteine in rats. *Eur. J. Pharmacol.* **2005**, *508*, 167–175. [CrossRef]
89. Ito, A.; Egashira, K.; Narishige, T.; Muramatsu, K.; Takeshita, A. Renin-angiotensin system is involved in the mechanism of increased serum asymmetric dimethylarginine in essential hypertension. *Jpn. Circ. J.* **2001**, *65*, 775–778. [CrossRef]
90. Napoli, C.; Sica, V.; de Nigris, F.; Pignalosa, O.; Condorelli, M.; Ignarro, L.J.; Liguori, A. Sulfhydryl angiotensin-converting enzyme inhibition induces sustained reduction of systemic oxidative stress and improves the nitric oxide pathway in patients with essential hypertension. *Am. Heart J.* **2004**, *148*, 172. [CrossRef]
91. Ito, A.; Egashira, K.; Narishige, T.; Muramatsu, K.; Takeshita, A. Angiotensin-converting enzyme activity is involved in the mechanism of increased endogenous nitric oxide synthase inhibitor in patients with type 2 diabetes mellitus. *Circ. J.* **2002**, *66*, 811–815. [CrossRef] [PubMed]
92. Sibal, L.; Agarwal, S.C.; Home, P.D.; Boger, R.H. The Role of Asymmetric Dimethylarginine (ADMA) in Endothelial Dysfunction and Cardiovascular Disease. *Curr. Cardiol. Rev.* **2010**, *6*, 82–90. [CrossRef] [PubMed]
93. Chen, J.-W.; Hsu, N.-W.; Wu, T.-C.; Lin, S.-J.; Chang, M.-S. Long-term angiotensin-converting enzyme inhibition reduces plasma asymmetric dimethylarginine and improves endothelial nitric oxide bioavailability and coronary microvascular function in patients with syndrome X. *Am. J. Cardiol.* **2002**, *90*, 974–982. [CrossRef] [PubMed]

International Journal of
Molecular Sciences

Review

Dysfunctional and Dysregulated Nitric Oxide Synthases in Cardiovascular Disease: Mechanisms and Therapeutic Potential

Roman Roy [1], Joshua Wilcox [2], Andrew J. Webb [3] and Kevin O'Gallagher [1,4,*]

[1] Cardiovascular Department, King's College Hospital NHS Foundation Trust, London SE5 9RS, UK; roman.roy@kcl.ac.uk

[2] Cardiovascular Department, Guy's and St. Thomas' NHS Foundation Trust, London SE1 7EH, UK; joshua.wilcox@kcl.ac.uk

[3] Department of Clinical Pharmacology, British Heart Foundation Centre, School of Cardiovascular and Metabolic Medicine and Sciences, King's College London, London SE1 7EH, UK; andrew.1.webb@kcl.ac.uk

[4] British Heart Foundation Centre of Research Excellence, School of Cardiovascular and Metabolic Medicine & Sciences, Faculty of Life Sciences & Medicine, King's College London, London SE5 9NU, UK

* Correspondence: kevin.o'gallagher@kcl.ac.uk; Tel.: +44-020-7448-5332

Abstract: Nitric oxide (NO) plays an important and diverse signalling role in the cardiovascular system, contributing to the regulation of vascular tone, endothelial function, myocardial function, haemostasis, and thrombosis, amongst many other roles. NO is synthesised through the nitric oxide synthase (NOS)-dependent L-arginine-NO pathway, as well as the nitrate-nitrite-NO pathway. The three isoforms of NOS, namely neuronal (NOS1), inducible (NOS2), and endothelial (NOS3), have different localisation and functions in the human body, and are consequently thought to have differing pathophysiological roles. Furthermore, as we continue to develop a deepened understanding of the different roles of NOS isoforms in disease, the possibility of therapeutically modulating NOS activity has emerged. Indeed, impaired (or dysfunctional), as well as overactive (or dysregulated) NOS activity are attractive therapeutic targets in cardiovascular disease. This review aims to describe recent advances in elucidating the physiological role of NOS isoforms within the cardiovascular system, as well as mechanisms of dysfunctional and dysregulated NOS in cardiovascular disease. We then discuss the modulation of NO and NOS activity as a target in the development of novel cardiovascular therapeutics.

Keywords: nitric oxide synthase; endothelial NOS; neuronal NOS; inducible NOS; NOS inhibitor; cardiovascular disease

Citation: Roy, R.; Wilcox, J.; Webb, A.J.; O'Gallagher, K. Dysfunctional and Dysregulated Nitric Oxide Synthases in Cardiovascular Disease: Mechanisms and Therapeutic Potential. *Int. J. Mol. Sci.* **2023**, *24*, 15200. https://doi.org/10.3390/ijms242015200

Academic Editor: Anastasios Lymperopoulos

Received: 13 September 2023
Revised: 11 October 2023
Accepted: 13 October 2023
Published: 15 October 2023

1. Introduction

Nitric oxide (NO) plays an important signalling role in multiple organ systems, notably within the cardiovascular system. Whilst originally described as a vasoactive molecule that causes the relaxation of smooth muscles and thereby vasodilatation, NO has also been shown to exert a myriad of other effects, including contributing to endothelial function, myocardial function [1], and neuronal signalling [2].

The biosynthesis of NO in humans is via two pathways: the L-arginine-NO pathway, and the nitrate-nitrite-NO pathway. The L-arginine-NO pathway relies on nitric oxide synthases (NOSs) catalysing the conversion of L-arginine to L-citrulline, yielding NO in the process. Three isoforms of NOS have been described in humans, namely neuronal (nNOS, NOS1), inducible (iNOS, NOS2), and endothelial (eNOS, NOS3). The isoforms of NOS have distinct physiological roles within the human body. Broadly, within the cardiovascular system, eNOS and nNOS regulate arterial blood flow (with each isoform having varying effects dependent on the arterial size) and exert direct myocardial effects. nNOS also has a role in the central nervous system, regulating cerebral blood flow, synaptic plasticity and functional connectivity [3]. iNOS is primarily considered to be physiologically important in the host response to infection and inflammation [4].

An increasing body of evidence has described the physiological role of NOS isoforms and thus the ramifications of impaired or dysfunctional NOS activity, as well as the potential therapeutic benefits of NO-based therapies. Recent evidence has also highlighted the contribution of overactive and dysregulated NOS to a range of cardiovascular diseases, thus making this pathway a promising target for therapeutic interventions. This review will summarise the physiological role of NOS in the cardiovascular system, as well as the pathophysiological effects of dysfunctional and dysregulated NOS and how targeting these pathways may hold therapeutic potential (Figure 1).

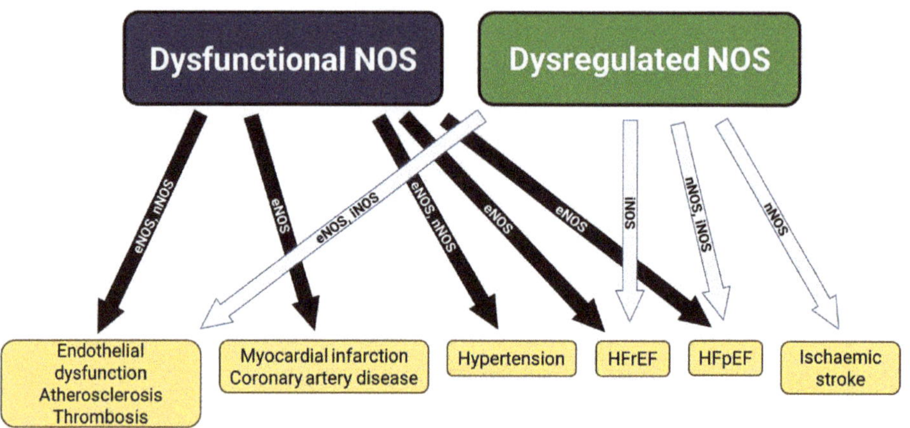

Figure 1. Overview of conditions with evidence for dysfunctional and dysregulated NOS activity. HFpEF: heart failure with preserved ejection fraction; HFrEF: heart failure with reduced ejection fraction.

2. Nitric Oxide Signalling

NO exerts widespread signalling activity, with well-described roles in vascular function, host defence against pathogens, the mediation of inflammation, redox homeostasis, and the regulation of metabolism. These varied roles make NO an important physiological molecule in systems throughout the human body, specifically the cardiovascular, cerebrovascular, respiratory, reproductive system, and many more.

NO primarily acts in a paracrine manner by diffusing into surrounding cells and binding to the haem moiety of soluble guanylate cyclase (sGC) [5]. This leads to an increase in cellular concentration of second messenger cyclic guanylate monophosphate (cGMP), which in turn activates protein kinase G (PKG). PKG subsequently phosphorylates L-type calcium channels, leading to a reduction in intracellular calcium concentration, the relaxation of smooth muscle cells, and vasodilatation. PKG also exerts its actions via the activation of a variety of transcription factors [6].

In addition, NO can exert cGMP-independent effects via the post-translational modification of proteins. S-nitrosylation, the process by which NO covalently binds to cysteine residues of proteins, is the best described mechanism of NO protein modification, and has been linked to various systems and implicated in pathophysiology of many diseases [7–9]. Given the large number of proteins that can undergo S-nitrosylation, this provides a mechanism by which NO can exert its many physiological effects. Other, less well described forms of NO-mediated post-translational modification include metal nitrosylation and nitration [7].

3. Structure and Function of Nitric Oxide Synthases

As described above, all NOS isoforms convert L-arginine to L-citrulline and NO. This process is haem-dependent and occurs in the presence of co-factors tetrahydrobiopterin (BH$_4$), nicotinamide adenine dinucleotide phosphate (NADPH), and O$_2$, involving dimerism of the NOS enzyme [10]. All three NOS isoforms depend on calmodulin binding to the enzyme in the transfer of electrons. For eNOS and nNOS, this process occurs in the context of high intracellular calcium concentrations (full activity at 500 nmol/L [11]), whereas iNOS binds to calmodulin very tightly and is therefore fully 'on' at resting calcium concentrations. iNOS is therefore considered calcium-independent, enabling it to produce large amounts of cytotoxic NO, which is important in its physiological function in the immune system [12]. NO synthesis by NOS is oxygen-dependent, with rates of synthesis proportional to the oxygen concentration. For nNOS, this occurs over a broad range, with a K$_m$O$_2$ value (the O$_2$ concentration at which the enzyme is at 50% Vmax) of ~400 µM, saturation at ~800 µM, and a Vmax value for NO synthesis of ~2.6 nmol/min, with a similar K$_m$O$_2$ for NADPH oxidation (~350 µM), with a corresponding Vmax value of ~6.0 nmol/min [13]. For purified iNOS, the K$_m$O$_2$ is about three-fold lower, ~120 µM, but with a three-fold greater, Vmax > 15 nmol/min, for NADPH consumption [14]. By contrast, eNOS has the weakest activity, with one-sixth and one-tenth the activity of nNOS and iNOS, respectively (due, at least in part, to a shorter hinge element with a unique composition that connects to the FMN module in the reductase domain) but with a much lower K$_m$O$_2$ of ~4 µM [15]. eNOS and nNOS are constitutively expressed and regulated by transcriptional, post-transcriptional, and post-translational modifications [16], whilst iNOS is induced primarily by gene transcription [17].

4. Physiological Role of Nitric Oxide Synthases in the Cardiovascular System

Despite their names, the isoforms of NOS are all expressed in various cell types throughout the body: nNOS is mostly found in neurons, but also in smooth, skeletal, and cardiac muscle; eNOS is typically expressed in endothelial cells, but is also found in smooth muscle, myocardium, and platelets; and iNOS is found in macrophages, smooth muscle, and the liver. Each NOS isoform plays a distinct physiological role within the cardiovascular system, which will be reviewed in detail below.

4.1. Endothelial NOS

As its name suggests, eNOS is constitutively expressed in the vascular endothelium, and thus has a crucial role in the regulation of vascular tone and endothelial function. eNOS-derived NO production is regulated by intracellular calcium concentrations, as described above, as well as by shear stress and agonist stimulation by acetylcholine and substance P [11,18–20].

Within the vasculature, the primary function of eNOS-derived NO is to vasodilate blood vessels via cGMP-dependent mechanisms as described above. eNOS-derived NO is therefore a significant determinant of tone in multiple vascular beds, contributing to control of cardiovascular haemodynamics.

eNOS-derived NO also plays an important physiological role in vascular protection against thrombosis and atherosclerosis [21]. NO inhibits platelet aggregation by a cGMP-mediated decrease in intracellular calcium flux and resulting in the negative regulation of glycoprotein IIb-IIIa (GP IIb-IIIa), protecting against thrombosis [22]. NO also inhibits leukocyte adhesion to the vascular wall by the inhibition of nuclear factor κB (NFκB), reducing atherosclerosis [23].

NO within the vasculature plays an important but complex role in redox homeostasis. At low levels, eNOS-derived NO acts as an anti-oxidant, reducing reactive oxygen species (ROS) production [22,24,25]. However, at higher concentrations, NO combining with superoxide to form peroxynitrite increases oxidative stress within cells, which has important pathological implications, as discussed below.

In addition to eNOS-derived NO from endothelial cells, eNOS (and to a lesser extent iNOS) contributes to platelet-derived NO [26]. Platelet-derived NO inhibits platelet activation, via cGMP-mediated pathways (decreasing intracellular calcium, inhibiting thromboxane A2 receptor function, and downregulating GP IIb-IIIa receptors [26,27]), but also via cGMP-independent mechanisms (the S-nitrosylation of N-ethylmaleimide-sensitive factor reducing platelet granule exocytosis [28]). Furthermore, through the inhibition of platelet adhesion to endothelial cells and leukocytes as well as promoting the disaggregation of platelets, platelet-derived NO plays an important role in modulating intravascular thrombus formation [26].

eNOS has an important role in the development of and protection against endothelial dysfunction, a state of elevated oxidative stress and inflammation within the endothelium that can lead to the abnormal vasodilatation and vasoconstriction of the vasculature. Endothelial dysfunction has been implicated in a multitude of cardiovascular diseases and diseases of other organ systems [21,29–31]. The role of eNOS and NO in endothelial dysfunction is complex; both a decrease and increase in eNOS function have been implicated in endothelial dysfunction. A large number of pre-clinical studies have illustrated the cardioprotective effects of eNOS and thereby its physiological role within the cardiovascular system. eNOS deficiency has been shown to cause a predisposition to atherosclerosis and endothelial dysfunction [32,33], coronary artery disease [32], aortic aneurysm and dissection [32], and hypertension [30]. In addition to its expression in the endothelium, eNOS has more recently been localised to adipocytes and the endothelial cells of perivascular adipose tissue, where it is thought to play a vasculoprotective role [34].

Finally, eNOS is also expressed constitutively in cardiomyocytes, where it primarily localises to the caveolae and plays an important role in mechanoregulation [35]. The effects of NO signalling on cardiac function are complex, depending on the balance in eNOS and nNOS activity within cardiomyocytes and endothelial cells, and also on the influence of β-adrenergic and cholinergic agonism [36]. Within the myocardium, eNOS and nNOS are expressed in different subcellular locations and are understood to have distinct physiological roles [1,37,38]. eNOS activity within the cardiomyocyte is regulated by β-adrenergic and cholinergic drive, and is thought to contribute to net negative chronotropic and inotropic, and positive lusitropic effects [24,35,38]. It is also implicated in the Anrep effect, which describes the positive inotropic effect of myocardial stretch caused by an increased afterload [35]. Beyond its acute effects on cardiac contractility and function, eNOS is also implicated in longer-term cardiac remodelling, acting under physiological circumstances to inhibit pro-hypertrophic stimuli [35].

4.2. Neuronal NOS

As implied by its name, nNOS is primarily located in the neurons of the central nervous system (CNS), but is also found in peripheral nitrergic nerves, endothelium, cardiac and skeletal muscle and other cell types. nNOS exerts important haemodynamic effects within the cardiovascular system. The isoform has been localised in the endothelium and/or smooth muscle of multiple vascular beds, including the coronary arteries [39], carotid arteries [40], aorta [40,41], kidneys (macula densa, collecting tubules, and neurons) [42] and microvasculature [19], and has been shown to regulate basal blood flow and vascular tone through these beds [18,19,43]. This regulation of vascular tone occurs through direct vasodilatory effects and via parasympathetic nitrergic nerves. Both nNOS and eNOS contribute to the regulation of arterial tone, with nNOS-derived NO regulating basal arterial tone and thereby systemic vascular resistance and blood pressure [19,44], whilst eNOS contributes to responses to changes in flow stimulated by shear stress or agonist stimulation, as described above [19].

nNOS localised to the cardiomyocyte has been shown to affect both inotropy and dromotropy via effects on intracellular calcium handling [45]. Within the heart, nNOS is found in the cardiomyocytes (sarcoplasmic reticulum, mitochondria, and plasma membrane), intrinsic neurons, and coronary arteries [46,47], contrary to the localisation of eNOS primarily within the caveolae of cardiomyocytes. nNOS regulates cardiac inotropy by increasing the sarcoplasmic reticulum Ca^{2+} ATPase (SERCA) reuptake of calcium and by modulating L-type calcium channel activity, overall reducing contractility and promoting lusitropy [38,47,48]. Beyond calcium handling, nNOS has a physiological role in the heart via its effects on ROS within the cardiomyocyte as well as on mitochondrial proteins (via the inhibition of the mitochondrial respiratory chain) [47].

In addition to its cardiovascular effects, nNOS also exerts important effects within the cerebrovascular system. nNOS-derived NO has been implicated in neurovascular coupling (NVC) [49] and dynamic cerebral autoregulation (dCA), two mechanisms contributing to regulation of cerebral blood flow. NVC describes the relationship between neuronal activity and local CNS blood flow, and is a mechanism by which neural activation leads to an increase in cerebral blood flow (otherwise described as functional hyperaemia). dCA refers to the ability to maintain stable cerebral blood flow despite variations in cerebral perfusion pressures. Cerebral blood flow regulation is complex and consists of a multitude of mechanisms, which act to regulate resistance via large conduit arteries (such as internal carotid arteries, vertebral arteries) but also at the neurovascular unit [50].

One such mechanism involves glutamate release at the synapse-activating N-methyl-D-aspartate (NMDA) receptors, leading to an influx of calcium and thereby nNOS activation. nNOS-derived NO then acts as a vasodilator via the sGC-cGMP-PKG pathway [51], increasing cerebral blood flow. It should be noted that additional nNOS-independent mechanisms of NVC have been described [51]. nNOS-derived NO has been shown to play a role in cerebral blood flow regulation in a variety of physiological and pathological settings, including functional neural activation [3], hypercapnia [52], and hypoxia [53].

Finally, at a neuronal level, nNOS contributes to synaptic plasticity and is thereby implicated in the development of memory and learning, but also in the regulation of neurogenesis [54]. nNOS also has an important role in the cardiovascular response to mental stress; healthy patients exposed to mental stress show nNOS-mediated coronary vasodilatation and increased blood flow [55], whilst patients with arterial hypertension show a blunted nNOS response to mental stress [56].

4.3. Inducible NOS

Initially purified from animal macrophages, iNOS has since been found in diverse cell types, including hepatocytes, smooth muscle cells, chondrocytes, CNS cells, and cardiac myocytes [57,58]. As opposed to the constitutively expressed eNOS and nNOS, iNOS has its expression upregulated by stimuli, usually proinflammatory cytokines and/or bacterial lipopolysaccharide [4]. Additionally, and in further contrast to the other two isoforms, iNOS exhibits activity at lower intracellular concentrations of the calcium/calmodulin (CaM) complex (due to tighter binding of CaM to the hinge region of its dimer) [57,59], thus locking iNOS in an "always on" position whereby calcium regulation is no longer relevant. This allows for the high-volume localised production of NO, which is physiologically important in an infection-initiated inflammatory response against multiple microbial pathogens (including viral, bacterial, protozoal and fungal infections) [60,61].

However, iNOS upregulation has been touted as both beneficial and detrimental, with the overproduction of NO via iNOS implicated in disease states such as sepsis [4,12], neurodegenerative disease and stroke [62,63], diabetes and obesity-induced insulin resistance [64], pain syndromes [4], and cancer [4]. Within the cardiovascular system, iNOS induced by inflammatory cytokines provides a mechanism by which inflammation contributes to pathophysiological states, such as atherosclerosis (discussed in more detail below) [65,66].

The inhibition of dysregulated iNOS has been attempted, but studies of in vivo treatment have been met with little success, including in septic shock [12,67] and pain [4]. However, early results in neurodegenerative disease, cancer, obesity-induced insulin resistance/diabetes, and lung disease show promise [4,68].

5. Dysfunctional NOS in Cardiovascular Disease

Dysfunctional NOS has been implicated in a variety of cardiovascular diseases. eNOS dysfunction in particular, has been demonstrated in various pathophysiological states, including endothelial dysfunction, atherosclerosis [66], arterial hypertension [30], cardiac hypertrophy [38], and heart failure [38]. The mechanism by which dysfunctional NOS contributes to pathophysiology is generally twofold: a reduction in NOS-derived NO, and NOS "uncoupling" leading to an increase in ROS and oxidative stress. Uncoupling is a state in which NOS generates ROS, including superoxide ($O_2{}^-$) and hydrogen peroxide (H_2O_2) [17]. It occurs under conditions of low L-arginine and BH_4 concentrations, and has previously been linked to the dissociation of the eNOS dimer to monomers [69,70], although this has more recently been disputed [71]. All three isoforms of NOS are prone to uncoupling, although it is most described in eNOS.

5.1. Hypertension

An abundance of evidence has linked hypertension to disruptions in NOS and NO signalling, with eNOS dysfunction and uncoupling particularly implicated [30,72,73]. Reduced vascular eNOS activity and subsequently reduced NO bioavailability leads to impaired vasodilatation and increased peripheral vascular resistance. In addition to changes seen in the regulation of vascular tone, both eNOS and nNOS localised to the myocardium are affected in hypertension; whilst eNOS activity is reduced in hypertensive hearts, nNOS-derived NO is increased and thought to play a protective role [74].

5.2. Endothelial Dysfunction

Abnormal eNOS function contributes to endothelial dysfunction, an early hallmark of atherosclerosis, which has been implicated in a wide variety of cardiovascular diseases, including hypertension [30], heart failure [31], and coronary artery disease [29], amongst many others. Endothelial dysfunction is characterised by reduced eNOS-derived NO within blood vessels leading to an impairment in endothelium-dependent relaxation [21]. The mechanisms impairing eNOS activity in disease are multiple, including inhibition by oxidised low-density lipoprotein [33,75] and lysophosphatidylcholine [76], as well as endogenous NOS inhibitors, such as asymmetric dimethylarginine (ADMA) [21,77]. In addition to the inhibition of eNOS-derived NO, the uncoupling of eNOS from its cofactor or substrate leads to production of ROS, further contributing to endothelial dysfunction [78]. nNOS has been shown to be protective against endothelial dysfunction and atherosclerosis, with studies demonstrating the development of accelerated atherosclerosis in nNOS knockout mice [79,80].

Elevated concentrations of arginase, which competes with eNOS for the substrate L-arginine, have been associated with endothelial dysfunction in a number of pathological states including hypertension, atherosclerosis and ischaemia/reperfusion injury [81]. Additionally, the S-glutathionylation of eNOS, promoted by a pro-oxidative environment within the endothelial cell, contributes to eNOS uncoupling [82]. However, eNOS is linked to endothelial dysfunction beyond the uncoupling phenomenon; proline-rich tyrosine kinase 2 (PYK2), activated by not only oxidative stress but also by angiotensin II and insulin, phosphorylates eNOS, inhibiting its effect and reducing NO bioavailability [83].

5.3. Myocardial Infarction and Ischaemia/Reperfusion Injury

Evidence has supported a cardioprotective role of both eNOS and nNOS after myocardial infarction (MI) and in ischaemia/reperfusion injury (IRI). Early studies showed that eNOS-knockout mice subjected to coronary artery ligation exhibited higher rates of cardiac remodelling, left ventricular dysfunction and death compared to wild-type mice [84]. Similarly, nNOS-knockout mice subjected to MI showed more cardiac remodelling and higher mortality rates than wild-type mice [85,86]. Furthermore, evidence has pointed towards a role for NOS-derived NO in ischaemic preconditioning (IPC); whilst eNOS may contribute to the early phase of IPC [87], iNOS upregulation is thought to contribute to late IPC [88,89].

A significant evidence base has investigated the use of NO donors in the IRI of the myocardium, implicating reduced NO bioavailability in its pathophysiology [90,91]. eNOS has been observed to be protective after an IRI, with eNOS-deficient mice showing exacerbated IRI compared to wild-type mice [92]. Furthermore, mice overexpressing eNOS have shown attenuated IRI, with beneficial effects of eNOS overexpression being reversed by the non-selective NOS inhibitor N^G-nitro-L-arginine methyl ester (L-NAME) [93].

5.4. Heart Failure

NOS dysfunction has been implicated in myocardial remodelling and heart failure [17,94], with early animal and human studies demonstrating that myocardial eNOS expression is decreased in hypertrophic cardiomyopathy [95], ischaemic cardiomyopathy [96], dilated cardiomyopathy [96,97], and diabetic cardiomyopathy [98]. Conversely, iNOS activity is increased in the failing heart and associated with decreased responsiveness to β-adrenergic stimulation [94,96]. Similarly, Nanos-derived NO is increased in patients with dilated cardiomyopathy [97], although the transgenic overexpression of nNOS in mice has been shown to be protective in IRIs [99]. It remains unclear whether this increase in nNOS activity is adaptive or maladaptive, although the translocation of nNOS to the plasma membrane seen under ischaemic conditions and in heart failure [97] has been postulated to limit myocardial remodelling caused by chronic β-adrenergic stimulation, implying an adaptive role [17]. Animal models have underlined the deleterious effects of eNOS dysfunction whilst highlighting the therapeutic potential of eNOS upregulation in HF. The upregulation of eNOS expression and/or activity in mice has been shown to be protective against post-MI left ventricular dysfunction and remodelling [24,100–102], oxidative stress [102] and survival [101], as well as infarct size following IRI [99]. In diabetic cardiomyopathy, studies have shown not only downregulated eNOS but also evidence of uncoupled eNOS activity [98,103]. Upregulated iNOS activity and subsequent increased NO and ROS production has also been associated with diabetic cardiomyopathy [98,103].

More recently, dysfunctional or uncoupled eNOS has been implicated in the pathophysiology of heart failure with preserved ejection (HFpEF). The currently accepted pathophysiological model of HFpEF involves co-morbidities driving systemic and arterial inflammation. At the level of the heart, this inflammation drives eNOS uncoupling in the coronary vasculature, resulting in impaired paracrine NO and cGMP signalling [104], with effects on the giant cytoskeletal protein titin, resulting in increased myocardial stiffness and cardiac diastolic dysfunction [105,106]. However, as will be discussed below, the differing effects of various NOS isoforms must be considered, with evidence of a pathological increase in myocardial NOS activity also described in HFpEF.

5.5. Therapeutic Potential

Attempts to target the NO signalling pathways in cardiovascular therapeutics are abundant. Efforts have been made to increase the bioavailability of NO using NO donors, as well as targeting downstream molecules in the NO signalling cascade (downstream targets have been reviewed elsewhere [17]). Additionally, the modulation of dysfunctional NOS has been a target of therapies of cardiovascular disease. Indeed, although not their primary mechanism of action, established cardiovascular drugs, such as those targeting the renin–angiotensin system [107], statins [108], calcium channel blockers [109], and β-adrenergic antagonists [110], have been shown to exert some of their cardio- and vasculoprotective effects via eNOS-dependent mechanisms. Attempts at targeting NO signalling are discussed below, with a focus on targeting dysfunctional NOS.

5.6. L-Arginine

As discussed above, the endogenous production of NO by NOS enzymes is dependent on the availability of L-arginine. Consequently, supplementation with L-arginine has been proposed as a means to increase eNOS-derived NO and reduce eNOS uncoupling, which could lead to increased NO bioavailability and reduced ROS production. However, clinical trials investigating L-arginine supplementation in a wide variety of cardiovascular diseases have failed to consistently demonstrate benefits on markers of endothelial function and on clinical outcomes, and have even been associated with possible harm [111–113].

One explanation for the lack of efficacy of L-arginine may be its relationship with the endogenous NOS inhibitor ADMA, which competes with L-arginine to bind with NOS, thereby reducing the availability of NOS-derived NO. Studies have shown increased levels of ADMA in patients with heart failure, but also conditions causing predisposition to heart failure, including hypertension, coronary artery disease, valvular disease, endothelial dysfunction, and others [114,115]. Furthermore, L-arginine supplementation may augment iNOS activity and associated deleterious effects via the overproduction of NO and ROS. Finally, human studies have shown impaired L-arginine transport in the failing myocardium and in hypertension, suggesting that the supplementation of L-arginine alone in the context of impaired transport may not be sufficient to improve the bioavailability of NOS-derived NO [116].

5.7. BH_4

Under physiological conditions, BH_4 acts as a cofactor, allowing the NOS enzyme to remain dimerized and thereby facilitating NOS-derived NO production and preventing uncoupling [24]. BH_4 supplementation has shown promise in preclinical models of atherosclerosis, inhibiting atherogenesis in apolipoprotein E knockout (ApoE-KO) mice fed a high-cholesterol diet [117], and reversing severe atherosclerosis in ApoE-KO mice with partial carotid ligation fed a high-fat diet [118]. Similarly, the oral supplementation of BH_4 in mice with transverse aortic constriction-induced heart failure led to recoupled eNOS activity and reversed hypertrophy and fibrosis [119]. Interestingly, a 2013 study showed that the co-administration of L-arginine and BH_4 was able to protect rats and pigs from IRI [120]. In ex vivo human coronary arterioles from patients undergoing coronary artery bypass grafting (CABG), treatment with sepiapterin (a substrate for BH_4 synthesis) was able to improve endothelium-dependent vasodilatation [121].

Despite promising preclinical results, clinical evidence of benefit from BH_4 supplementation is limited. One randomised controlled trial (RCT) in patients undergoing CABG showed that oral BH_4 had no effect on vascular redox state or endothelial function, potentially due to the oxidation of BH_4 to BH_2, which lacks eNOS cofactor activity [122]. This limitation of oral supplementation of BH_4 presents a major obstacle to its therapeutic use. Similarly, a RCT in patients with CADASIL (cerebral autosomal dominant arteriopathy with subcortical infarcts and leukoencephalopathy) showed no improvement in markers of endothelium-dependent vasodilatation with sapropterin treatment (a synthetic BH_4 analogue) [123].

5.8. NO Donors

NO donors, compounds that release NO and NO-related molecules, have been extensively studied for their potential therapeutic benefits in cardiovascular disease. Glyceryl trinitrate, isosorbide nitrates, and nicorandil are three common examples of NO donors used in cardiovascular disease, primarily for angina and acute MI. Similarly, a wealth of evidence has investigated the therapeutic use of dietary nitrate/nitrite in conditions including hypertension, heart failure (with reduced and preserved ejection fraction; HFrEF and HFpEF), as well as ischaemic heart disease [124,125]. These approaches aim to increase the bioavailability of NO in settings where NOS-derived NO may be deficient [126], as described above. Numerous ongoing studies are investigating the role of dietary nitrate and NO donors in various cardiovascular diseases, including in HFrEF, HFpEF, systemic hypertension, pulmonary hypertension, angina, and stroke [127].

Although the delivery of NO donors is primarily via the oral route in clinical practice, they can also be delivered sublingually, intranasally, transdermally, or intravenously. NO itself can be delivered to the lungs through inhalation, such as for the treatment of pulmonary artery hypertension. Sublingual and immediate-release orally absorbed NO donors tend to have a shorter duration of action compared to transdermal and slow-release oral NO donors. Importantly, barriers to therapeutic success for organic nitrates have included side effects, such as headaches and tolerance, requiring drug-free intervals, as well as endothelial dysfunction, which may develop with chronic use [128]. Slow-release oral nitrates are less likely to cause tolerance than fast release preparations. Furthermore, the NEAT-HFpEF trial found a dose-dependent decrease in daily activity levels in patients taking isosorbide mononitrate versus placebo, highlighting the limitations of organic nitrates [129]. Nitrite, on the other hand, carries the possible adverse effects of methaemoglobinaemia and conversion to carcinogenic nitrosamines [128], limiting its clinical use.

5.9. NOS Transcriptional Regulators/Enhancers

In an effort to enhance NOS function in disease states characterised by dysfunctional NOS, transcriptional regulators of NOS activity have been developed. Preclinical studies have demonstrated the potential of transcriptional enhancers for eNOS in improving left ventricular remodelling and contractile dysfunction in rats with experimental MI [130], as well as in reducing hypertrophy/fibrosis and improving diastolic function in a rat model of diastolic heart failure [131]. In addition, they have shown efficacy in reducing cardiac remodelling in mice subjected to aortic banding [132], and reducing platelet activation in rats with post-MI heart failure [133], amongst others [134,135]. Despite these promising preclinical results, the translation of these approaches into clinical research and improved outcomes has yet to be achieved.

5.10. Gene Therapy

NOS gene therapy has provided a platform for the improvement of NO bioavailability in cardiovascular diseases. For instance, the delivery of the eNOS gene via an adenovirus vector before MI in rats has been shown to be cardioprotective, reducing infarct size and improving contractility and left ventricular diastolic function [136]. Similarly, the adenovirus vector delivery of eNOS after MI protected against myocardial fibrosis and remodelling as well as apoptosis [102].

Furthermore, adenoviral vectors for the expression of dimethylarginine dimethylaminohydrolase (DDAH), an enzyme that degrades ADMA and thereby increases NO, have shown promise in improving vascular function in mouse carotid arteries [137]. This approach may offer a potential strategy to target endothelial dysfunction in a range of disorders. However, further research is needed to evaluate the efficacy and safety of NOS gene therapy in humans.

6. Dysregulated NOS in Cardiovascular Disease

6.1. Endothelial Dysfunction, Inflammation, and Oxidative Stress

Alongside evidence for dysfunctional eNOS function, preclinical studies have demonstrated that dysregulated or elevated iNOS activity may also contribute to endothelial dysfunction. Investigations showed that an iNOS-specific inhibitor reversed the impaired pressor responsiveness and endothelial function observed in rats with streptozotocin-induced diabetes, indicating that increased iNOS expression may play a role in diabetes-associated endothelial dysfunction [98]. Similarly, in mice with lipopolysaccharide-induced endothelial dysfunction, iNOS protein expression and plasma NO_x levels were increased, whereas iNOS knockout mice exhibited no changes in NO_x levels [138], supporting a link between iNOS activity and endothelial dysfunction. Furthermore, early work showed elevated macrophage iNOS expression in a rabbit model of post-MI inflammation [139].

Mechanisms linking inflammation and disease states characterised by inflammation to dysregulated iNOS activity are numerous. Inflammatory cytokines, including IL-1β and IFN-γ, induce iNOS activity [140], leading to the generation of ROS and reactive nitrogen species (RNS) [141]. Furthermore, iNOS-derived NO leads to the upregulation of cyclo-oxygenases (COX), a group of enzymes involved in the generation of prostanoids and thereby contributing to a pro-inflammatory state [65]. Finally, beyond iNOS, the activation of the pro-inflammatory cascade will cause the uncoupling of eNOS, further contributing to reduced NO bioavailability and driving oxidative stress within the cell. These effects lead to an unfavourable oxidant versus antioxidant balance, driving apoptosis, contractile dysfunction, and mitochondrial dysfunction amongst other effects [142].

6.2. NO and Mitochondria

In addition to endothelial function, NO and NOS have important effects on mitochondrial health. NO, either from NO donors or produced within the mitochondria itself, reversibly inhibits the action of cytochrome C oxidase, the last enzyme in the electron transport chain, thereby modulating mitochondrial oxygen consumption [143,144]. At higher concentrations, however, NO has harmful effects on the mitochondria. NO causes the inhibition of the mitochondrial respiratory chain, the generation of peroxynitrite, the nitration of mitochondrial proteins, and the release of cytochrome C into the cytosol [143,145]. Peroxynitrite will oxidise a wide variety of substrates (including proteins, tyrosine residues, and DNA, amongst others), inhibiting enzymes at multiple sites of the mitochondrial respiratory chain, and leading to cytochrome C release into the cytosol with subsequent signalling, leading to apoptosis [144,146].

Dysregulated NOS and subsequent excess NO production, therefore, will have detrimental effects on mitochondrial and cell function. iNOS, for example, has been associated with metabolic remodelling and cytokine production, contributing to the pro-inflammatory cascades within macrophages [145].

6.3. Heart Failure with Preserved Ejection Fraction (HFpEF)

As described above, inflammation and subsequent eNOS uncoupling within the coronary vasculature has been shown to contribute to the pathophysiology of HFpEF [104,105]. This focus on eNOS function, however, ignores the anatomical compartmentalisation and differing cardiac effects of the various NOS isoforms. Indeed, recent evidence has identified that pathological increases in myocardial NOS activity may be implicated in HFpEF pathophysiology. In a study reporting data from both animal models and human myocardial tissue in patients with cardiac diastolic dysfunction, upregulated/dysregulated nNOS function led to the S-nitrosylation of histone deacetylase, an enzyme known to be involved in the regulation of cardiac hypertrophy [147]. Similarly, animal and human data suggest that upregulated iNOS activity leads to the S-nitrosylation of proteins involved in the cardiomyocyte stress response [94]. Furthermore, it remains unclear how nNOS function in the coronary microvasculature affects myocardial function in a paracrine manner, suggesting possible mechanisms via cellular crosstalk.

The complexity of the pathophysiological role of the different NOS isoforms in HFpEF, as well as their anatomical compartmentalisation may explain why treating patients with NO donors is of limited clinical benefit. Targeting therapy to the dysfunction of specific isoforms with isoform specific NOS inhibitors may provide greater success.

Interestingly, sodium-glucose cotransporter 2 (SGLT2) inhibitors, recently shown to reduce cardiovascular death in patients with HFpEF [148], have been shown to improve NOS coupling [149]—further work will elucidate the extent to which clinical benefits of SGLT2 inhibitors depend upon the modulation of NO signalling pathways.

6.4. Coronary Microvascular Disease (CMD)

CMD is increasingly recognised as an important cause of angina in patients that were found to have unobstructed coronary arteries via coronary angiography. Recent work has classified CMD into two key endotypes: functional and structural [150]. In the functional endotype, there is elevated resting basal coronary blood flow. Given that nNOS (rather than eNOS) is responsible for the regulation of basal arterial tone, it has been postulated that the raised basal coronary flow seen in the functional endotype may be due to dysregulated nNOS activity, with studies ongoing to investigate this possibility [151].

6.5. Ischaemia-Reperfusion Injury in the Heart

NOS inhibition has been proposed as a potential therapy for IRI following acute coronary syndromes. As the myocardium is reperfused following injury, a complex set of mechanisms, including the generation of free radicals, pro-apoptotic and pro-inflammatory signalling cascades, and endothelial dysfunction, contribute to myocardial damage/cell death, arrhythmias, and myocardial stunning [152,153]. The role of NOS isoforms in IRI is complex: nNOS is upregulated in the myocardium following ischaemia [154] and is thought to mediate the protective effects of ischaemic post-conditioning against IRI. iNOS has been implicated in IRI through worse contractile function and increased oxidative stress via the production of peroxynitrite, but has also been suggested to have a protective role via iNOS-derived NO increasing TNF-α- and COX-2-dependent prostanoids, protecting the myocardium [155]. Recently, iNOS expression was seen to be increased in myocardial tissue from autopsies of patients with acute MI [156], although it is unclear if this is an adaptive or maladaptive change.

Early work investigated non-selective NOS inhibitors, such as N^G-nitro-L-arginine methyl ester (L-NAME) and N^G-monomethyl-L-arginine (L-NMMA), in rabbit and rat models of IRI, finding improved recovery of mechanical heart function during the reperfusion period with NOS inhibition [157,158] and reduced infarct size [159]. A subsequent work using a selective inhibitor of iNOS dimerization found improved contractile performance and reduced cell death in isolated perfused rat hearts after IRI [160]. A similar study demonstrated that selective iNOS inhibition and iNOS-knockout mice showed reduced apoptosis and infarct size after IRI in chronic β-adrenergic stimulation-induced cardiac damage in mice [161].

To date there are few clinical studies investigating the use of NOS inhibitors in patients with IRI. A group has investigated NOS inhibition in cardiogenic shock complicating acute MI, with initial studies suggesting that L-NMMA and L-NAME led to haemodynamic improvements [162,163]. However, a subsequent placebo-controlled study in 79 patients showed that L-NMMA resulted only in a modest, short-lived increase in mean arterial pressure at 15 min which was not sustained at two hours [164], and a larger RCT was terminated early for futility, finding no difference in mortality in patients with refractory cardiogenic shock after MI [165].

6.6. Post-Stroke Reperfusion Injury

In ischaemic stroke, a large proportion of the neurological damage comes not just from the initial ischaemic insult but also from excitotoxicity, i.e., dysregulated glutaminergic and NO signalling that occurs for several hours/days after the initial ischaemia [166,167]. Current established therapies in ischaemic stroke (e.g., tissue plasminogen activator and mechanical thrombectomy) are aimed at achieving the successful reperfusion of the ischaemic territory but play no direct role in diminishing the inevitable excitotoxicity that follows. Therefore, there remains a significant clinical need for effective treatments to reduce the levels of excitotoxicity and minimise infarct volumes and neurological injury following stroke [168].

Animal data suggest that nNOS contributes significantly to excitotoxicity via cellular crosstalk [169,170]. nNOS is bound to the glutaminergic NMDA receptor by postsynaptic density protein 95 (PSD95), creating a death-inducing signalling complex (Figure 2). NMDA receptor activation in the ischaemic penumbra therefore leads to neurotoxic levels of NO production [168]. Inhibiting nNOS protects against glutaminergic excitotoxicity in animal models of non-ischaemic neurological disease, while nNOS knockout mice are resistant to cerebral ischaemia [171–174].

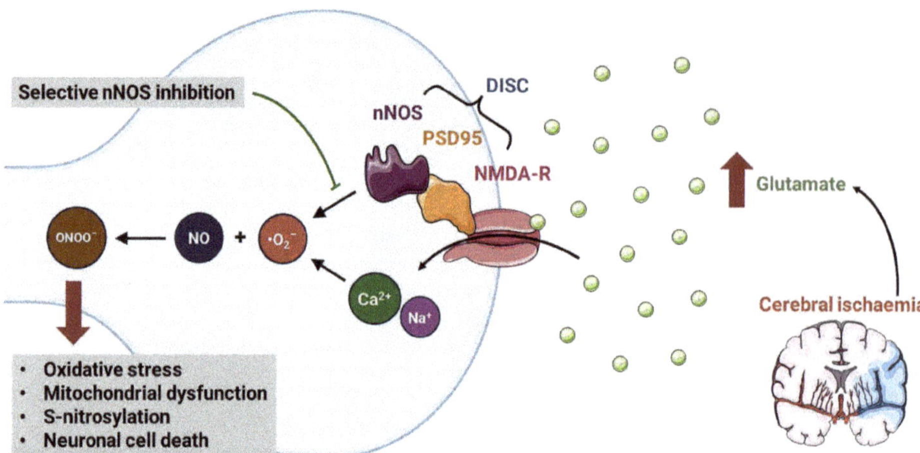

Figure 2. Mechanism of nNOS inhibition in post-stroke glutamate excitotoxicity: Cerebral ischaemia leads to an increase in release of the excitatory neurotransmitter glutamate. Glutamate activates the N-methyl-D-aspartate receptor (NMDA-R), with subsequent influx of calcium (Ca^{2+}) (and sodium, Na^+). Excess Ca^{2+} within the neuron contributes to the generation of superoxide ($\bullet O_2^-$), which reacts with nitric oxide (NO) to generate peroxynitrite ($ONOO^-$). Furthermore, neuronal nitric oxide synthase (nNOS) is bound to the NMDA-R by the postsynaptic density protein 95 (PSD95), forming a death-inducing signalling complex (DISC). This cascade leads to neuronal cell death via a number of mechanisms, including increased oxidative stress, mitochondrial dysfunction, and S-nitrosylation. Selective nNOS inhibition may provide a mechanism to interrupt this detrimental signalling cascade. Figure was partly generated using Servier Medical Art, provided by Servier, licensed under a Creative Commons Attribution 3.0 unported license.

Attempts have been made to indirectly interrupt nNOS signalling following acute stroke using the drug NA-1, which inhibits PSD95. In a phase II study, NA-1 was shown to reduce the number of ischaemic infarcts versus placebo in patients undergoing endovascular treatment for intracranial aneurysm [175]. However, in a phase III RCT in patients with ischaemic stroke, NA-1 did not improve functional outcome compared to placebo [176].

Beyond acute (within minutes) detrimental effects of increased nNOS activity, iNOS activity has also been seen to be upregulated in the later stages following stroke (within hours), where its effects are also suggested to be neurotoxic [177]. Conversely, eNOS-derived NO, which is acutely upregulated following stroke, is thought to be neuroprotective [177], underlining the importance of a targeted therapeutic approach.

6.7. Therapeutic Potential of NOS Inhibitors

To date, no assessment has been made of the effect of directly inhibiting nNOS in ischaemic stroke. This is largely because direct nNOS inhibition has previously been considered an unfavourable approach, based on behavioural effects seen in murine nNOS knock-out models [178]. However, direct nNOS inhibition with the synthetic L-arginine analogue S-methyl-L-thiocitrulline (SMTC) has been employed in a number of human mechanistic studies without adverse effect [3,18,44,49,55,56], suggesting that the acute systemic and local intra-arterial dosing of SMTC is safe. SMTC is 17 times more selective for nNOS in brain tissue than eNOS in vascular endothelium [179] and is also considered to cross the blood–brain barrier, as demonstrated through use of ^{11}C-labelled SMTC in rat and primate models [180]. In a human study assessing the effect of nNOS inhibition on cerebral blood flow responses, changes in functional connectivity between different brain regions were identified, strongly suggesting that SMTC crosses the blood–brain barrier to affect neuronal function rather than simply altering perfusion [3].

Although it may seem counter-intuitive to treat a condition that is primarily one of insufficient blood flow (i.e., ischaemic stroke) with a drug that acts to decrease brain perfusion, it is important to consider that (a) nNOS inhibition is being considered as a potential therapy for post-reperfusion excitotoxicity and would only be given once adequate reperfusion had been achieved, and (b) nNOS-derived NO is one of a number of molecules implicated in the regulation of cerebral blood flow and the effect of nNOS inhibition on cerebral blood flow is modest. In addition, as the interventional management of stroke becomes more widespread, the catheterisation of the cerebral arteries may afford the opportunity to deliver more anatomically targeted treatment, thereby improving the delivery of nNOS inhibitors to the ischaemic brain tissue most at risk of excitotoxicity, and limiting any systemic side effects.

Previous work has investigated the use of agmatine, a compound with pleiotropic effects, which include the modulation of NOS function, in ischaemic stroke. Agmatine, an amine synthesised by the decarboxylation of arginine, acts to inhibit nNOS activity [181,182] whilst enhancing eNOS activity [182,183]. Preclinical studies of agmatine use following stroke have shown promise as a neuroprotective strategy [181,184,185].

6.8. Safety Concerns with NOS Inhibitors

The safety profile of NOS inhibitors is critical as their use in conditions with evidence of dysregulated NOS is suggested. In addition to the aforementioned behavioural concerns in murine models of nNOS inhibitors, there have been concerns about multi-system effects. Dose-dependent acute kidney injury was seen in the NOSTRA trial, a study investigating the use of ronopterin in traumatic brain injury [186]; however, a subsequent RCT described this effect as only a mild, transient reduction in renal perfusion and glomerular function [187].

7. Conclusions, Limitations, and Future Directions

The vasoactive properties, and direct cardiac and cerebral effects of NO have long made it an attractive therapeutic target within these systems. As our understanding of the roles of NOS and its isoforms within the cardiovascular system has grown, the distinct roles played by eNOS, nNOS, and iNOS in the pathophysiology of major diseases has become clearer (Figure 1). Strategies to improve NOS activity and thereby NO signalling continue to be developed for use in conditions with a pathological lack of NO activity. Furthermore, disorders causing significant mortality and morbidity, including HFpEF and post-stroke excitotoxicity, now have evidence supporting distinct roles of the NOS isoforms in their pathophysiology, implying the possibility of selective NOS modulation. Future, studies should continue to elucidate the potential of modulating NOS activity, with a particular focus on deepening our understanding of anatomical and functional differences in NOS isoforms and harnessing this in the therapeutic approach to cardiovascular disease.

The primary limitation of this review is that we did not perform a systematic review of the topic. Nevertheless, by basing a non-systematic review of the literature on expertise of the authors, we were able to identify key articles in this field and synthesise them.

Author Contributions: Conceptualization: R.R., A.J.W. and K.O.; writing—original draft preparation: R.R., J.W., A.J.W. and K.O.; writing—review and editing: R.R., A.J.W. and K.O.; supervision: A.J.W. and K.O. All authors have read and agreed to the published version of the manuscript.

Funding: Dr O'Gallagher is funded by the British Heart Foundation Centre of Research Excellence, King's College London (RE19542).

Institutional Review Board Statement: Not applicable.

Informed Consent Statement: Not applicable.

Data Availability Statement: No new data were created or analysed in this study. Data sharing is not applicable to this article.

Conflicts of Interest: The authors declare no conflict of interest.

References

1. Barouch, L.A.; Harrison, R.W.; Skaf, M.W.; Rosas, G.O.; Cappola, T.P.; Kobeissi, Z.A.; Hobai, I.A.; Lemmon, C.A.; Burnett, A.L.; O'Rourke, B.; et al. Nitric Oxide Regulates the Heart by Spatial Confinement of Nitric Oxide Synthase Isoforms. *Nature* **2002**, *416*, 337–339. [CrossRef]
2. Lundberg, J.O.; Weitzberg, E. Nitric Oxide Signaling in Health and Disease. *Cell* **2022**, *185*, 2853–2878. [CrossRef]
3. O'Gallagher, K.; Puledda, F.; O'Daly, O.; Ryan, M.; Dancy, L.; Chowienczyk, P.J.; Zelaya, F.; Goadsby, P.J.; Shah, A.M. Neuronal Nitric Oxide Synthase Regulates Regional Brain Perfusion in Healthy Humans. *Cardiovasc. Res.* **2022**, *118*, 1321–1329. [CrossRef]
4. Cinelli, M.A.; Do, H.T.; Miley, G.P.; Silverman, R.B. Inducible Nitric Oxide Synthase: Regulation, Structure, and Inhibition. *Med. Res. Rev.* **2020**, *40*, 158–189. [CrossRef]
5. Ahmad, A.; Dempsey, S.; Daneva, Z.; Azam, M.; Li, N.; Li, P.-L.; Ritter, J. Role of Nitric Oxide in the Cardiovascular and Renal Systems. *Int. J. Mol. Sci.* **2018**, *19*, 2605. [CrossRef]
6. Lincoln, T.M.; Dey, N.; Sellak, H. Invited Review: CGMP-Dependent Protein Kinase Signaling Mechanisms in Smooth Muscle: From the Regulation of Tone to Gene Expression. *J. Appl. Physiol.* **2001**, *91*, 1421–1430. [CrossRef]
7. Fernando, V.; Zheng, X.; Walia, Y.; Sharma, V.; Letson, J.; Furuta, S. S-Nitrosylation: An Emerging Paradigm of Redox Signaling. *Antioxidants* **2019**, *8*, 404. [CrossRef]
8. Nakamura, T.; Oh, C.; Zhang, X.; Lipton, S.A. Protein S-Nitrosylation and Oxidation Contribute to Protein Misfolding in Neurodegeneration. *Free Radic. Biol. Med.* **2021**, *172*, 562–577. [CrossRef]
9. Zhang, Y.; Deng, Y.; Yang, X.; Xue, H.; Lang, Y. The Relationship Between Protein S-Nitrosylation and Human Diseases: A Review. *Neurochem. Res.* **2020**, *45*, 2815–2827. [CrossRef]
10. Gantner, B.N.; LaFond, K.M.; Bonini, M.G. Nitric Oxide in Cellular Adaptation and Disease. *Redox Biol.* **2020**, *34*, 101550. [CrossRef]
11. Förstermann, U.; Closs, E.I.; Pollock, J.S.; Nakane, M.; Schwarz, P.; Gath, I.; Kleinert, H. Nitric Oxide Synthase Isozymes. Characterization, Purification, Molecular Cloning, and Functions. *Hypertension* **1994**, *23*, 1121–1131. [CrossRef]
12. Spiller, F.; Oliveira Formiga, R.; Fernandes da Silva Coimbra, J.; Alves-Filho, J.C.; Cunha, T.M.; Cunha, F.Q. Targeting Nitric Oxide as a Key Modulator of Sepsis, Arthritis and Pain. *Nitric Oxide* **2019**, *89*, 32–40. [CrossRef]
13. Abu-Soud, H.M.; Rousseau, D.L.; Stuehr, D.J. Nitric Oxide Binding to the Heme of Neuronal Nitric-Oxide Synthase Links Its Activity to Changes in Oxygen Tension. *J. Biol. Chem.* **1996**, *271*, 32515–32518. [CrossRef] [PubMed]
14. Dweik, R.A.; Laskowski, D.; Abu-Soud, H.M.; Kaneko, F.T.; Hutte, R.; Stuehr, D.J.; Erzurum, S.C. Nitric Oxide Synthesis in the Lung. Regulation by Oxygen through a Kinetic Mechanism. *J. Clin. Investig.* **1998**, *101*, 660–666. [CrossRef]
15. Haque, M.M.; Panda, K.; Tejero, J.; Aulak, K.S.; Fadlalla, M.A.; Mustovich, A.T.; Stuehr, D.J. A Connecting Hinge Represses the Activity of Endothelial Nitric Oxide Synthase. *Proc. Natl. Acad. Sci. USA* **2007**, *104*, 9254–9259. [CrossRef] [PubMed]
16. Searles, C.D. Transcriptional and Posttranscriptional Regulation of Endothelial Nitric Oxide Synthase Expression. *Am. J. Physiol.-Cell Physiol.* **2006**, *291*, C803–C816. [CrossRef] [PubMed]
17. Farah, C.; Michel, L.Y.M.; Balligand, J.-L. Nitric Oxide Signalling in Cardiovascular Health and Disease. *Nat. Rev. Cardiol.* **2018**, *15*, 292–316. [CrossRef]
18. Seddon, M.; Melikian, N.; Dworakowski, R.; Shabeeh, H.; Jiang, B.; Byrne, J.; Casadei, B.; Chowienczyk, P.; Shah, A.M. Effects of Neuronal Nitric Oxide Synthase on Human Coronary Artery Diameter and Blood Flow In Vivo. *Circulation* **2009**, *119*, 2656–2662. [CrossRef]
19. Seddon, M.D.; Chowienczyk, P.J.; Brett, S.E.; Casadei, B.; Shah, A.M. Neuronal Nitric Oxide Synthase Regulates Basal Microvascular Tone in Humans In Vivo. *Circulation* **2008**, *117*, 1991–1996. [CrossRef]
20. Park, S.-K.; La Salle, D.T.; Cerbie, J.; Cho, J.M.; Bledsoe, A.; Nelson, A.; Morgan, D.E.; Richardson, R.S.; Shiu, Y.-T.; Boudina, S.; et al. Elevated Arterial Shear Rate Increases Indexes of Endothelial Cell Autophagy and Nitric Oxide Synthase Activation in Humans. *Am. J. Physiol.-Heart Circ. Physiol.* **2019**, *316*, H106–H112. [CrossRef]
21. Cyr, A.R.; Huckaby, L.V.; Shiva, S.S.; Zuckerbraun, B.S. Nitric Oxide and Endothelial Dysfunction. *Crit. Care Clin.* **2020**, *36*, 307–321. [CrossRef]
22. Costa, D.; Benincasa, G.; Lucchese, R.; Infante, T.; Nicoletti, G.F.; Napoli, C. Effect of Nitric Oxide Reduction on Arterial Thrombosis. *Scand. Cardiovasc. J.* **2019**, *53*, 1–8. [CrossRef]
23. Theofilis, P.; Sagris, M.; Oikonomou, E.; Antonopoulos, A.S.; Siasos, G.; Tsioufis, C.; Tousoulis, D. Inflammatory Mechanisms Contributing to Endothelial Dysfunction. *Biomedicines* **2021**, *9*, 781. [CrossRef]
24. Tran, N.; Garcia, T.; Aniqa, M.; Ali, S.; Ally, A.; Nauli, S.M. Endothelial Nitric Oxide Synthase (ENOS) and the Cardiovascular System: In Physiology and in Disease States. *Am. J. Biomed. Sci. Res.* **2022**, *15*, 153–177. [PubMed]
25. Tejero, J.; Shiva, S.; Gladwin, M.T. Sources of Vascular Nitric Oxide and Reactive Oxygen Species and Their Regulation. *Physiol. Rev.* **2019**, *99*, 311–379. [CrossRef]
26. Gkaliagkousi, E.; Ritter, J.; Ferro, A. Platelet-Derived Nitric Oxide Signaling and Regulation. *Circ. Res.* **2007**, *101*, 654–662. [CrossRef]
27. Gresele, P.; Momi, S.; Guglielmini, G. Nitric Oxide-Enhancing or -Releasing Agents as Antithrombotic Drugs. *Biochem. Pharmacol.* **2019**, *166*, 300–312. [CrossRef]

28. Morrell, C.N.; Matsushita, K.; Chiles, K.; Scharpf, R.B.; Yamakuchi, M.; Mason, R.J.A.; Bergmeier, W.; Mankowski, J.L.; Baldwin, W.M.; Faraday, N.; et al. Regulation of Platelet Granule Exocytosis by S-Nitrosylation. *Proc. Natl. Acad. Sci. USA* **2005**, *102*, 3782–3787. [CrossRef] [PubMed]

29. Corban, M.T.; Lerman, L.O.; Lerman, A. Endothelial Dysfunction. *Arterioscler. Thromb. Vasc. Biol.* **2019**, *39*, 1272–1274. [CrossRef]

30. Gallo, G.; Volpe, M.; Savoia, C. Endothelial Dysfunction in Hypertension: Current Concepts and Clinical Implications. *Front. Med.* **2022**, *8*, 798958. [CrossRef] [PubMed]

31. Zuchi, C.; Tritto, I.; Carluccio, E.; Mattei, C.; Cattadori, G.; Ambrosio, G. Role of Endothelial Dysfunction in Heart Failure. *Heart Fail. Rev.* **2020**, *25*, 21–30. [CrossRef]

32. Kuhlencordt, P.J.; Gyurko, R.; Han, F.; Scherrer-Crosbie, M.; Aretz, T.H.; Hajjar, R.; Picard, M.H.; Huang, P.L. Accelerated Atherosclerosis, Aortic Aneurysm Formation, and Ischemic Heart Disease in Apolipoprotein E/Endothelial Nitric Oxide Synthase Double-Knockout Mice. *Circulation* **2001**, *104*, 448–454. [CrossRef]

33. Hong, F.; Liang, X.; Liu, W.; Lv, S.; He, S.; Kuang, H.; Yang, S. Roles of ENOS in Atherosclerosis Treatment. *Inflamm. Res.* **2019**, *68*, 429–441. [CrossRef]

34. Man, A.W.C.; Zhou, Y.; Xia, N.; Li, H. Endothelial Nitric Oxide Synthase in the Perivascular Adipose Tissue. *Biomedicines* **2022**, *10*, 1754. [CrossRef] [PubMed]

35. Balligand, J.-L.; Feron, O.; Dessy, C. ENOS Activation by Physical Forces: From Short-Term Regulation of Contraction to Chronic Remodeling of Cardiovascular Tissues. *Physiol. Rev.* **2009**, *89*, 481–534. [CrossRef] [PubMed]

36. Massion, P.B.; Feron, O.; Dessy, C.; Balligand, J.-L. Nitric Oxide and Cardiac Function. *Circ. Res.* **2003**, *93*, 388–398. [CrossRef] [PubMed]

37. Seddon, M.; Shah, A.; Casadei, B. Cardiomyocytes as Effectors of Nitric Oxide Signalling. *Cardiovasc. Res.* **2007**, *75*, 315–326. [CrossRef] [PubMed]

38. Umar, S.; van der Laarse, A. Nitric Oxide and Nitric Oxide Synthase Isoforms in the Normal, Hypertrophic, and Failing Heart. *Mol. Cell Biochem.* **2010**, *333*, 191–201. [CrossRef] [PubMed]

39. Huang, A.; Sun, D.; Shesely, E.G.; Levee, E.M.; Koller, A.; Kaley, G. Neuronal NOS-Dependent Dilation to Flow in Coronary Arteries of Male ENOS-KO Mice. *Am. J. Physiol.-Heart Circ. Physiol.* **2002**, *282*, H429–H436. [CrossRef] [PubMed]

40. Buchwalow, I.B.; Podzuweit, T.; Böcker, W.; Samoilova, V.E.; Thomas, S.; Wellner, M.; Baba, H.A.; Robenek, H.; Schnekenburger, J.; Lerch, M.M. Vascular Smooth Muscle and Nitric Oxide Synthase. *FASEB J.* **2002**, *16*, 500–508. [CrossRef]

41. Cotter, M.A.; Cameron, N.E.; Nangle, M.R. An in Vitro Investigation of Aorta and Corpus Cavernosum from ENOS and NNOS Gene-Deficient Mice. *Pflug. Arch. Eur. J. Physiol.* **2004**, *448*, 139–145. [CrossRef]

42. Toda, N.; Okamura, T. Modulation of Renal Blood Flow and Vascular Tone by Neuronal Nitric Oxide Synthase-Derived Nitric Oxide. *J. Vasc. Res.* **2011**, *48*, 1–10. [CrossRef] [PubMed]

43. Melikian, N.; Seddon, M.D.; Casadei, B.; Chowienczyk, P.J.; Shah, A.M. Neuronal Nitric Oxide Synthase and Human Vascular Regulation. *Trends Cardiovasc. Med.* **2009**, *19*, 256–262. [CrossRef]

44. Shabeeh, H.; Khan, S.; Jiang, B.; Brett, S.; Melikian, N.; Casadei, B.; Chowienczyk, P.J.; Shah, A.M. Blood Pressure in Healthy Humans Is Regulated by Neuronal NO Synthase. *Hypertension* **2017**, *69*, 970–976. [CrossRef] [PubMed]

45. Ally, A.; Powell, I.; Ally, M.M.; Chaitoff, K.; Nauli, S.M. Role of Neuronal Nitric Oxide Synthase on Cardiovascular Functions in Physiological and Pathophysiological States. *Nitric Oxide* **2020**, *102*, 52–73. [CrossRef] [PubMed]

46. Danson, E.; Choate, J.; Paterson, D. Cardiac Nitric Oxide: Emerging Role for NNOS in Regulating Physiological Function. *Pharmacol. Ther.* **2005**, *106*, 57–74. [CrossRef] [PubMed]

47. Zhang, Y.H.; Jin, C.Z.; Jang, J.H.; Wang, Y. Molecular Mechanisms of Neuronal Nitric Oxide Synthase in Cardiac Function and Pathophysiology. *J. Physiol.* **2014**, *592*, 3189–3200. [CrossRef] [PubMed]

48. Seddon, M.; Looi, Y.H.; Shah, A.M. Oxidative Stress and Redox Signalling in Cardiac Hypertrophy and Heart Failure. *Heart* **2007**, *93*, 903–907. [CrossRef]

49. O'Gallagher, K.; Rosentreter, R.E.; Elaine Soriano, J.; Roomi, A.; Saleem, S.; Lam, T.; Roy, R.; Gordon, G.R.; Raj, S.R.; Chowienczyk, P.J.; et al. The Effect of a Neuronal Nitric Oxide Synthase Inhibitor on Neurovascular Regulation in Humans. *Circ. Res.* **2022**, *131*, 952–961. [CrossRef]

50. Claassen, J.A.H.R.; Thijssen, D.H.J.; Panerai, R.B.; Faraci, F.M. Regulation of Cerebral Blood Flow in Humans: Physiology and Clinical Implications of Autoregulation. *Physiol. Rev.* **2021**, *101*, 1487–1559. [CrossRef]

51. Phillips, A.A.; Chan, F.H.; Zheng, M.M.Z.; Krassioukov, A.V.; Ainslie, P.N. Neurovascular Coupling in Humans: Physiology, Methodological Advances and Clinical Implications. *J. Cereb. Blood Flow. Metab.* **2016**, *36*, 647–664. [CrossRef]

52. Okamoto, H.; Hudetz, A.G.; Roman, R.J.; Bosnjak, Z.J.; Kampine, J.P. Neuronal NOS-Derived NO Plays Permissive Role in Cerebral Blood Flow Response to Hypercapnia. *Am. J. Physiol.-Heart Circ. Physiol.* **1997**, *272*, H559–H566. [CrossRef]

53. Hudetz, A.G.; Shen, H.; Kampine, J.P. Nitric Oxide from Neuronal NOS Plays Critical Role in Cerebral Capillary Flow Response to Hypoxia. *Am. J. Physiol.-Heart Circ. Physiol.* **1998**, *274*, H982–H989. [CrossRef]

54. Kourosh-Arami, M.; Hosseini, N.; Mohsenzadegan, M.; Komaki, A.; Joghataei, M.T. Neurophysiologic Implications of Neuronal Nitric Oxide Synthase. *Rev. Neurosci.* **2020**, *31*, 617–636. [CrossRef]

55. Khan, S.G.; Melikian, N.; Shabeeh, H.; Cabaco, A.R.; Martin, K.; Khan, F.; O'Gallagher, K.; Chowienczyk, P.J.; Shah, A.M. The Human Coronary Vasodilatory Response to Acute Mental Stress Is Mediated by Neuronal Nitric Oxide Synthase. *Am. J. Physiol.-Heart Circ. Physiol.* **2017**, *313*, H578–H583. [CrossRef]

56. Khan, S.G.; Geer, A.; Fok, H.W.; Shabeeh, H.; Brett, S.E.; Shah, A.M.; Chowienczyk, P.J. Impaired Neuronal Nitric Oxide Synthase–Mediated Vasodilator Responses to Mental Stress in Essential Hypertension. *Hypertension* **2015**, *65*, 903–909. [CrossRef]

57. Hemmens, B.; Mayer, B. Enzymology of Nitric Oxide Synthases. In *Nitric Oxide Protocols*; Humana Press: Totowa, NJ, USA, 1998; pp. 1–32.

58. Hevel, J.M.; White, K.A.; Marletta, M.A. Purification of the Inducible Murine Macrophage Nitric Oxide Synthase. Identification as a Flavoprotein. *J. Biol. Chem.* **1991**, *266*, 22789–22791. [CrossRef]

59. Venema, R.C.; Sayegh, H.S.; Kent, J.D.; Harrison, D.G. Identification, Characterization, and Comparison of the Calmodulin-Binding Domains of the Endothelial and Inducible Nitric Oxide Synthases. *J. Biol. Chem.* **1996**, *271*, 6435–6440. [CrossRef]

60. Bath, P.M.; Coleman, C.M.; Gordon, A.L.; Lim, W.S.; Webb, A.J. Nitric Oxide for the Prevention and Treatment of Viral, Bacterial, Protozoal and Fungal Infections. *F1000Res* **2021**, *10*, 536. [CrossRef]

61. Chakravortty, D.; Hensel, M. Inducible Nitric Oxide Synthase and Control of Intracellular Bacterial Pathogens. *Microbes Infect.* **2003**, *5*, 621–627. [CrossRef]

62. Jayaraj, R.L.; Azimullah, S.; Beiram, R.; Jalal, F.Y.; Rosenberg, G.A. Neuroinflammation: Friend and Foe for Ischemic Stroke. *J. Neuroinflammation* **2019**, *16*, 142. [CrossRef]

63. Justo, A.F.O.; Suemoto, C.K. The Modulation of Neuroinflammation by Inducible Nitric Oxide Synthase. *J. Cell Commun. Signal* **2022**, *16*, 155–158. [CrossRef]

64. Król, M.; Kepinska, M. Human Nitric Oxide Synthase—Its Functions, Polymorphisms, and Inhibitors in the Context of Inflammation, Diabetes and Cardiovascular Diseases. *Int. J. Mol. Sci.* **2020**, *22*, 56. [CrossRef]

65. Lind, M.; Hayes, A.; Caprnda, M.; Petrovic, D.; Rodrigo, L.; Kruzliak, P.; Zulli, A. Inducible Nitric Oxide Synthase: Good or Bad? *Biomed. Pharmacother.* **2017**, *93*, 370–375. [CrossRef]

66. Chen, J.; Ye, Z.; Wang, X.; Chang, J.; Yang, M.; Zhong, H.; Hong, F.; Yang, S. Nitric Oxide Bioavailability Dysfunction Involves in Atherosclerosis. *Biomed. Pharmacother.* **2018**, *97*, 423–428. [CrossRef]

67. López, A.; Lorente, J.A.; Steingrub, J.; Bakker, J.; McLuckie, A.; Willatts, S.; Brockway, M.; Anzueto, A.; Holzapfel, L.; Breen, D.; et al. Multiple-Center, Randomized, Placebo-Controlled, Double-Blind Study of the Nitric Oxide Synthase Inhibitor 546C88: Effect on Survival in Patients with Septic Shock*. *Crit. Care Med.* **2004**, *32*, 21–30. [CrossRef]

68. Chung, A.W.; Anand, K.; Anselme, A.C.; Chan, A.A.; Gupta, N.; Venta, L.A.; Schwartz, M.R.; Qian, W.; Xu, Y.; Zhang, L.; et al. A Phase 1/2 Clinical Trial of the Nitric Oxide Synthase Inhibitor L-NMMA and Taxane for Treating Chemoresistant Triple-Negative Breast Cancer. *Sci. Transl. Med.* **2021**, *13*, eabj5070. [CrossRef]

69. Zou, M.-H.; Cohen, R.A.; Ullrich, V. Peroxynitrite and Vascular Endothelial Dysfunction in Diabetes Mellitus. *Endothelium* **2004**, *11*, 89–97. [CrossRef]

70. Yang, Y.-M.; Huang, A.; Kaley, G.; Sun, D. ENOS Uncoupling and Endothelial Dysfunction in Aged Vessels. *Am. J. Physiol. Heart Circ. Physiol.* **2009**, *297*, H1829–H1836. [CrossRef]

71. Gebhart, V.; Reiß, K.; Kollau, A.; Mayer, B.; Gorren, A.C.F. Site and Mechanism of Uncoupling of Nitric-Oxide Synthase: Uncoupling by Monomerization and Other Misconceptions. *Nitric Oxide* **2019**, *89*, 14–21. [CrossRef]

72. Griendling, K.K.; Camargo, L.L.; Rios, F.J.; Alves-Lopes, R.; Montezano, A.C.; Touyz, R.M. Oxidative Stress and Hypertension. *Circ. Res.* **2021**, *128*, 993–1020. [CrossRef]

73. Wu, Y.; Ding, Y.; Ramprasath, T.; Zou, M.-H. Oxidative Stress, GTPCH1, and Endothelial Nitric Oxide Synthase Uncoupling in Hypertension. *Antioxid. Redox Signal* **2021**, *34*, 750–764. [CrossRef]

74. Zhang, Y.H. Neuronal Nitric Oxide Synthase in Hypertension—An Update. *Clin. Hypertens.* **2016**, *22*, 20. [CrossRef]

75. Gliozzi, M.; Scicchitano, M.; Bosco, F.; Musolino, V.; Carresi, C.; Scarano, F.; Maiuolo, J.; Nucera, S.; Maretta, A.; Paone, S.; et al. Modulation of Nitric Oxide Synthases by Oxidized LDLs: Role in Vascular Inflammation and Atherosclerosis Development. *Int. J. Mol. Sci.* **2019**, *20*, 3294. [CrossRef] [PubMed]

76. Liu, P.; Zhu, W.; Chen, C.; Yan, B.; Zhu, L.; Chen, X.; Peng, C. The Mechanisms of Lysophosphatidylcholine in the Development of Diseases. *Life Sci.* **2020**, *247*, 117443. [CrossRef] [PubMed]

77. Mitchell, J.A.; Kirkby, N.S. Eicosanoids, Prostacyclin and Cyclooxygenase in the Cardiovascular System. *Br. J. Pharmacol.* **2019**, *176*, 1038–1050. [CrossRef]

78. Pong, T.; Huang, P.L. Effects of Nitric Oxide on Atherosclerosis. In *Atherosclerosis*; John Wiley & Sons, Inc.: Hoboken, NJ, USA, 2015; pp. 353–364.

79. Kuhlencordt, P.J.; Hötten, S.; Schödel, J.; Rützel, S.; Hu, K.; Widder, J.; Marx, A.; Huang, P.L.; Ertl, G. Atheroprotective Effects of Neuronal Nitric Oxide Synthase in Apolipoprotein E Knockout Mice. *Arterioscler. Thromb. Vasc. Biol.* **2006**, *26*, 1539–1544. [CrossRef]

80. Costa, E.D.; Rezende, B.A.; Cortes, S.F.; Lemos, V.S. Neuronal Nitric Oxide Synthase in Vascular Physiology and Diseases. *Front. Physiol.* **2016**, *7*, 206. [CrossRef] [PubMed]

81. Mahdi, A.; Kövamees, O.; Pernow, J. Improvement in Endothelial Function in Cardiovascular Disease—Is Arginase the Target? *Int. J. Cardiol.* **2020**, *301*, 207–214. [CrossRef]

82. Janaszak-Jasiecka, A.; Płoska, A.; Wierońska, J.M.; Dobrucki, L.W.; Kalinowski, L. Endothelial Dysfunction Due to ENOS Uncoupling: Molecular Mechanisms as Potential Therapeutic Targets. *Cell Mol. Biol. Lett.* **2023**, *28*, 21. [CrossRef]

83. Siragusa, M.; Fleming, I. The ENOS Signalosome and Its Link to Endothelial Dysfunction. *Pflug. Arch.* **2016**, *468*, 1125–1137. [CrossRef] [PubMed]

84. Scherrer-Crosbie, M.; Ullrich, R.; Bloch, K.D.; Nakajima, H.; Nasseri, B.; Aretz, H.T.; Lindsey, M.L.; Vançon, A.-C.; Huang, P.L.; Lee, R.T.; et al. Endothelial Nitric Oxide Synthase Limits Left Ventricular Remodeling After Myocardial Infarction in Mice. *Circulation* **2001**, *104*, 1286–1291. [CrossRef] [PubMed]
85. Saraiva, R.M.; Minhas, K.M.; Raju, S.V.Y.; Barouch, L.A.; Pitz, E.; Schuleri, K.H.; Vandegaer, K.; Li, D.; Hare, J.M. Deficiency of Neuronal Nitric Oxide Synthase Increases Mortality and Cardiac Remodeling After Myocardial Infarction. *Circulation* **2005**, *112*, 3415–3422. [CrossRef]
86. Dawson, D.; Lygate, C.A.; Zhang, M.-H.; Hulbert, K.; Neubauer, S.; Casadei, B. NNOS Gene Deletion Exacerbates Pathological Left Ventricular Remodeling and Functional Deterioration After Myocardial Infarction. *Circulation* **2005**, *112*, 3729–3737. [CrossRef]
87. Bell, R.M.; Yellon, D.M. The Contribution of Endothelial Nitric Oxide Synthase to Early Ischaemic Preconditioning: The Lowering of the Preconditioning Threshold. An Investigation in ENOS Knockout Mice. *Cardiovasc. Res.* **2001**, *52*, 274–280. [CrossRef] [PubMed]
88. Takano, H.; Manchikalapudi, S.; Tang, X.L.; Qiu, Y.; Rizvi, A.; Jadoon, A.K.; Zhang, Q.; Bolli, R. Nitric Oxide Synthase Is the Mediator of Late Preconditioning against Myocardial Infarction in Conscious Rabbits. *Circulation* **1998**, *98*, 441–449. [CrossRef]
89. Bolli, R. Cardioprotective Function of Inducible Nitric Oxide Synthase and Role of Nitric Oxide in Myocardial Ischemia and Preconditioning: An Overview of a Decade of Research. *J. Mol. Cell Cardiol.* **2001**, *33*, 1897–1918. [CrossRef]
90. Lee, H.-M.; Choi, J.W.; Choi, M.S. Role of Nitric Oxide and Protein S-Nitrosylation in Ischemia-Reperfusion Injury. *Antioxidants* **2021**, *11*, 57. [CrossRef]
91. Griffiths, K.; Lee, J.J.; Frenneaux, M.P.; Feelisch, M.; Madhani, M. Nitrite and Myocardial Ischaemia Reperfusion Injury. Where Are We Now? *Pharmacol. Ther.* **2021**, *223*, 107819. [CrossRef]
92. Jones, S.P.; Girod, W.G.; Palazzo, A.J.; Granger, D.N.; Grisham, M.B.; Jourd'Heuil, D.; Huang, P.L.; Lefer, D.J. Myocardial Ischemia-Reperfusion Injury Is Exacerbated in Absence of Endothelial Cell Nitric Oxide Synthase. *Am. J. Physiol. Heart Circ. Physiol.* **1999**, *276*, H1567–H1573. [CrossRef] [PubMed]
93. Jones, S.P.; Greer, J.J.M.; Kakkar, A.K.; Ware, P.D.; Turnage, R.H.; Hicks, M.; van Haperen, R.; de Crom, R.; Kawashima, S.; Yokoyama, M.; et al. Endothelial Nitric Oxide Synthase Overexpression Attenuates Myocardial Reperfusion Injury. *Am. J. Physiol. Heart Circ. Physiol.* **2004**, *286*, H276–H282. [CrossRef]
94. Schiattarella, G.G.; Altamirano, F.; Tong, D.; French, K.M.; Villalobos, E.; Kim, S.Y.; Luo, X.; Jiang, N.; May, H.I.; Wang, Z.V.; et al. Nitrosative Stress Drives Heart Failure with Preserved Ejection Fraction. *Nature* **2019**, *568*, 351–356. [CrossRef]
95. Piech, A.; Massart, P.E.; Dessy, C.; Feron, O.; Havaux, X.; Morel, N.; Vanoverschelde, J.-L.; Donckier, J.; Balligand, J.-L. Decreased Expression of Myocardial ENOS and Caveolin in Dogs with Hypertrophic Cardiomyopathy. *Am. J. Physiol. Heart Circ. Physiol.* **2002**, *282*, H219–H231. [CrossRef] [PubMed]
96. Drexler, H.; Kästner, S.; Strobel, A.; Studer, R.; Brodde, O.E.; Hasenfuß, G. Expression, Activity and Functional Significance of Inducible Nitric Oxide Synthase in the Failing Human Heart. *J. Am. Coll. Cardiol.* **1998**, *32*, 955–963. [CrossRef]
97. Damy, T.; Ratajczak, P.; Shah, A.M.; Camors, E.; Marty, I.; Hasenfuss, G.; Marotte, F.; Samuel, J.-L.; Heymes, C. Increased Neuronal Nitric Oxide Synthase-Derived NO Production in the Failing Human Heart. *Lancet* **2004**, *363*, 1365–1367. [CrossRef]
98. Nagareddy, P.R.; Xia, Z.; McNeill, J.H.; MacLeod, K.M. Increased Expression of INOS Is Associated with Endothelial Dysfunction and Impaired Pressor Responsiveness in Streptozotocin-Induced Diabetes. *Am. J. Physiol. Heart Circ. Physiol.* **2005**, *289*, H2144–H2152. [CrossRef]
99. Szelid, Z.; Pokreisz, P.; Liu, X.; Vermeersch, P.; Marsboom, G.; Gillijns, H.; Pellens, M.; Verbeken, E.; Werf, F.; Collen, D.; et al. Cardioselective Nitric Oxide Synthase 3 Gene Transfer Protects against Myocardial Reperfusion Injury. *Basic. Res. Cardiol.* **2010**, *105*, 169–179. [CrossRef]
100. Janssens, S.; Pokreisz, P.; Schoonjans, L.; Pellens, M.; Vermeersch, P.; Tjwa, M.; Jans, P.; Scherrer-Crosbie, M.; Picard, M.H.; Szelid, Z.; et al. Cardiomyocyte-Specific Overexpression of Nitric Oxide Synthase 3 Improves Left Ventricular Performance and Reduces Compensatory Hypertrophy After Myocardial Infarction. *Circ. Res.* **2004**, *94*, 1256–1262. [CrossRef] [PubMed]
101. Jones, S.P.; Greer, J.J.M.; van Haperen, R.; Duncker, D.J.; de Crom, R.; Lefer, D.J. Endothelial Nitric Oxide Synthase Overexpression Attenuates Congestive Heart Failure in Mice. *Proc. Natl. Acad. Sci. USA* **2003**, *100*, 4891–4896. [CrossRef] [PubMed]
102. Smith, R.S.; Agata, J.; Xia, C.-F.; Chao, L.; Chao, J. Human Endothelial Nitric Oxide Synthase Gene Delivery Protects against Cardiac Remodeling and Reduces Oxidative Stress after Myocardial Infarction. *Life Sci.* **2005**, *76*, 2457–2471. [CrossRef]
103. Khanna, S.; Singh, G.B.; Khullar, M. Nitric Oxide Synthases and Diabetic Cardiomyopathy. *Nitric Oxide* **2014**, *43*, 29–34. [CrossRef] [PubMed]
104. Sorop, O.; Heinonen, I.; van Kranenburg, M.; van de Wouw, J.; de Beer, V.J.; Nguyen, I.T.N.; Octavia, Y.; van Duin, R.W.B.; Stam, K.; van Geuns, R.-J.; et al. Multiple Common Comorbidities Produce Left Ventricular Diastolic Dysfunction Associated with Coronary Microvascular Dysfunction, Oxidative Stress, and Myocardial Stiffening. *Cardiovasc. Res.* **2018**, *114*, 954–964. [CrossRef]
105. O'Gallagher, K.; Shah, A.M. Modelling the Complexity of Heart Failure with Preserved Ejection Fraction. *Cardiovasc. Res.* **2018**, *114*, 919–921. [CrossRef] [PubMed]
106. Paulus, W.J.; Tschöpe, C. A Novel Paradigm for Heart Failure With Preserved Ejection Fraction. *J. Am. Coll. Cardiol.* **2013**, *62*, 263–271. [CrossRef] [PubMed]
107. Suvorava, T.; Metry, S.; Pick, S.; Kojda, G. Alterations in Endothelial Nitric Oxide Synthase Activity and Their Relevance to Blood Pressure. *Biochem. Pharmacol.* **2022**, *205*, 115256. [CrossRef] [PubMed]

108. Gorabi, A.M.; Kiaie, N.; Hajighasemi, S.; Banach, M.; Penson, P.E.; Jamialahmadi, T.; Sahebkar, A. Statin-Induced Nitric Oxide Signaling: Mechanisms and Therapeutic Implications. *J. Clin. Med.* **2019**, *8*, 2051. [CrossRef]

109. Ding, Y.; Vaziri, N.D. Nifedipine and Diltiazem but Not Verapamil Up-Regulate Endothelial Nitric-Oxide Synthase Expression. *J. Pharmacol. Exp. Ther.* **2000**, *292*, 606–609.

110. Daiber, A.; Xia, N.; Steven, S.; Oelze, M.; Hanf, A.; Kröller-Schön, S.; Münzel, T.; Li, H. New Therapeutic Implications of Endothelial Nitric Oxide Synthase (ENOS) Function/Dysfunction in Cardiovascular Disease. *Int. J. Mol. Sci.* **2019**, *20*, 187. [CrossRef] [PubMed]

111. Rodrigues-Krause, J.; Krause, M.; Rocha, I.; Umpierre, D.; Fayh, A. Association of L-Arginine Supplementation with Markers of Endothelial Function in Patients with Cardiovascular or Metabolic Disorders: A Systematic Review and Meta-Analysis. *Nutrients* **2018**, *11*, 15. [CrossRef] [PubMed]

112. Sun, T.; Zhou, W.; Luo, X.; Tang, Y.; Shi, H. Oral L-Arginine Supplementation in Acute Myocardial Infarction Therapy: A Meta-Analysis of Randomized Controlled Trials. *Clin. Cardiol.* **2009**, *32*, 649–652. [CrossRef]

113. Schulman, S.P.; Becker, L.C.; Kass, D.A.; Champion, H.C.; Terrin, M.L.; Forman, S.; Ernst, K.V.; Kelemen, M.D.; Townsend, S.N.; Capriotti, A.; et al. L-Arginine Therapy in Acute Myocardial Infarction. *JAMA* **2006**, *295*, 58. [CrossRef]

114. Liu, X.; Hou, L.; Xu, D.; Chen, A.; Yang, L.; Zhuang, Y.; Xu, Y.; Fassett, J.T.; Chen, Y. Effect of Asymmetric Dimethylarginine (ADMA) on Heart Failure Development. *Nitric Oxide* **2016**, *54*, 73–81. [CrossRef]

115. Gawrys, J.; Gajecki, D.; Szahidewicz-Krupska, E.; Doroszko, A. Intraplatelet L-Arginine-Nitric Oxide Metabolic Pathway: From Discovery to Clinical Implications in Prevention and Treatment of Cardiovascular Disorders. *Oxid. Med. Cell Longev.* **2020**, *2020*, 1015908. [CrossRef]

116. Chin-Dusting, J.P.F.; Willems, L.; Kaye, D.M. L-Arginine Transporters in Cardiovascular Disease: A Novel Therapeutic Target. *Pharmacol. Ther.* **2007**, *116*, 428–436. [CrossRef]

117. Hattori, Y.; Hattori, S.; Wang, X.; Satoh, H.; Nakanishi, N.; Kasai, K. Oral Administration of Tetrahydrobiopterin Slows the Progression of Atherosclerosis in Apolipoprotein E-Knockout Mice. *Arterioscler. Thromb. Vasc. Biol.* **2007**, *27*, 865–870. [CrossRef]

118. Li, L.; Chen, W.; Rezvan, A.; Jo, H.; Harrison, D.G. Tetrahydrobiopterin Deficiency and Nitric Oxide Synthase Uncoupling Contribute to Atherosclerosis Induced by Disturbed Flow. *Arterioscler. Thromb. Vasc. Biol.* **2011**, *31*, 1547–1554. [CrossRef]

119. Moens, A.L.; Takimoto, E.; Tocchetti, C.G.; Chakir, K.; Bedja, D.; Cormaci, G.; Ketner, E.A.; Majmudar, M.; Gabrielson, K.; Halushka, M.K.; et al. Reversal of Cardiac Hypertrophy and Fibrosis From Pressure Overload by Tetrahydrobiopterin. *Circulation* **2008**, *117*, 2626–2636. [CrossRef]

120. Tratsiakovich, Y.; Gonon, A.T.; Kiss, A.; Yang, J.; Böhm, F.; Tornvall, P.; Settergren, M.; Channon, K.M.; Sjöquist, P.-O.; Pernow, J. Myocardial Protection by Co-Administration of l-Arginine and Tetrahydrobiopterin during Ischemia and Reperfusion. *Int. J. Cardiol.* **2013**, *169*, 83–88. [CrossRef]

121. Tiefenbacher, C.P.; Bleeke, T.; Vahl, C.; Amann, K.; Vogt, A.; Kübler, W. Endothelial Dysfunction of Coronary Resistance Arteries Is Improved by Tetrahydrobiopterin in Atherosclerosis. *Circulation* **2000**, *102*, 2172–2179. [CrossRef]

122. Cunnington, C.; Van Assche, T.; Shirodaria, C.; Kylintireas, I.; Lindsay, A.C.; Lee, J.M.; Antoniades, C.; Margaritis, M.; Lee, R.; Cerrato, R.; et al. Systemic and Vascular Oxidation Limits the Efficacy of Oral Tetrahydrobiopterin Treatment in Patients With Coronary Artery Disease. *Circulation* **2012**, *125*, 1356–1366. [CrossRef]

123. De Maria, R.; Campolo, J.; Frontali, M.; Taroni, F.; Federico, A.; Inzitari, D.; Tavani, A.; Romano, S.; Puca, E.; Orzi, F.; et al. Effects of Sapropterin on Endothelium-Dependent Vasodilation in Patients With CADASIL. *Stroke* **2014**, *45*, 2959–2966. [CrossRef]

124. Webb, A.J.; Patel, N.; Loukogeorgakis, S.; Okorie, M.; Aboud, Z.; Misra, S.; Rashid, R.; Miall, P.; Deanfield, J.; Benjamin, N.; et al. Acute Blood Pressure Lowering, Vasoprotective, and Antiplatelet Properties of Dietary Nitrate via Bioconversion to Nitrite. *Hypertension* **2008**, *51*, 784–790. [CrossRef]

125. Banez, M.J.; Geluz, M.I.; Chandra, A.; Hamdan, T.; Biswas, O.S.; Bryan, N.S.; Von Schwarz, E.R. A Systemic Review on the Antioxidant and Anti-Inflammatory Effects of Resveratrol, Curcumin, and Dietary Nitric Oxide Supplementation on Human Cardiovascular Health. *Nutr. Res.* **2020**, *78*, 11–26. [CrossRef]

126. Liu, Y.; Croft, K.D.; Hodgson, J.M.; Mori, T.; Ward, N.C. Mechanisms of the Protective Effects of Nitrate and Nitrite in Cardiovascular and Metabolic Diseases. *Nitric Oxide* **2020**, *96*, 35–43. [CrossRef]

127. Petraina, A.; Nogales, C.; Krahn, T.; Mucke, H.; Lüscher, T.F.; Fischmeister, R.; Kass, D.A.; Burnett, J.C.; Hobbs, A.J.; Schmidt, H.H.H.W. Cyclic GMP Modulating Drugs in Cardiovascular Diseases: Mechanism-Based Network Pharmacology. *Cardiovasc. Res.* **2021**, *118*, 2085–2102. [CrossRef] [PubMed]

128. Münzel, T.; Daiber, A. Inorganic Nitrite and Nitrate in Cardiovascular Therapy: A Better Alternative to Organic Nitrates as Nitric Oxide Donors? *Vasc. Pharmacol.* **2018**, *102*, 1–10. [CrossRef] [PubMed]

129. Redfield, M.M.; Anstrom, K.J.; Levine, J.A.; Koepp, G.A.; Borlaug, B.A.; Chen, H.H.; LeWinter, M.M.; Joseph, S.M.; Shah, S.J.; Semigran, M.J.; et al. Isosorbide Mononitrate in Heart Failure with Preserved Ejection Fraction. *N. Engl. J. Med.* **2015**, *373*, 2314–2324. [CrossRef]

130. Fraccarollo, D.; Widder, J.D.; Galuppo, P.; Thum, T.; Tsikas, D.; Hoffmann, M.; Ruetten, H.; Ertl, G.; Bauersachs, J. Improvement in Left Ventricular Remodeling by the Endothelial Nitric Oxide Synthase Enhancer AVE9488 After Experimental Myocardial Infarction. *Circulation* **2008**, *118*, 818–827. [CrossRef]

131. Westermann, D.; Riad, A.; Richter, U.; Jäger, S.; Savvatis, K.; Schuchardt, M.; Bergmann, N.; Tölle, M.; Nagorsen, D.; Gotthardt, M.; et al. Enhancement of the Endothelial NO Synthase Attenuates Experimental Diastolic Heart Failure. *Basic. Res. Cardiol.* **2009**, *104*, 499–509. [CrossRef] [PubMed]

132. Chen, Y.; Chen, C.; Feng, C.; Tang, A.; Ma, Y.; He, X.; Li, Y.; He, J.; Dong, Y. AVE 3085, a Novel Endothelial Nitric Oxide Synthase Enhancer, Attenuates Cardiac Remodeling in Mice through the Smad Signaling Pathway. *Arch. Biochem. Biophys.* **2015**, *570*, 8–13. [CrossRef]

133. Schäfer, A.; Fraccarollo, D.; Widder, J.; Eigenthaler, M.; Ertl, G.; Bauersachs, J. Inhibition of Platelet Activation in Rats with Severe Congestive Heart Failure by a Novel Endothelial Nitric Oxide Synthase Transcription Enhancer. *Eur. J. Heart Fail.* **2009**, *11*, 336–341. [CrossRef]

134. Yang, Q.; Xue, H.-M.; Wong, W.-T.; Tian, X.-Y.; Huang, Y.; Tsui, S.K.; Ng, P.K.; Wohlfart, P.; Li, H.; Xia, N.; et al. AVE3085, an Enhancer of Endothelial Nitric Oxide Synthase, Restores Endothelial Function and Reduces Blood Pressure in Spontaneously Hypertensive Rats. *Br. J. Pharmacol.* **2011**, *163*, 1078–1085. [CrossRef] [PubMed]

135. Cheang, W.S.; Wong, W.T.; Tian, X.Y.; Yang, Q.; Lee, H.K.; He, G.-W.; Yao, X.; Huang, Y. Endothelial Nitric Oxide Synthase Enhancer Reduces Oxidative Stress and Restores Endothelial Function in Db/Db Mice. *Cardiovasc. Res.* **2011**, *92*, 267–275. [CrossRef] [PubMed]

136. Chen, L.-L.; Yin, H.; Huang, J. Inhibition of TGF-B1 Signaling by ENOS Gene Transfer Improves Ventricular Remodeling after Myocardial Infarction through Angiogenesis and Reduction of Apoptosis. *Cardiovasc. Pathol.* **2007**, *16*, 221–230. [CrossRef]

137. Torondel, B.; Nandi, M.; Kelly, P.; Wojciak-Stothard, B.; Fleming, I.; Leiper, J. Adenoviral-Mediated Overexpression of DDAH Improves Vascular Tone Regulation. *Vasc. Med.* **2010**, *15*, 205–213. [CrossRef] [PubMed]

138. Chauhan, S.D.; Seggara, G.; Vo, P.A.; Macallister, R.J.; Hobbs, A.J.; Ahluwalia, A. Protection against Lipopolysaccharide-induced Endothelial Dysfunction in Resistance and Conduit Vasculature of INOS Knockout Mice. *FASEB J.* **2003**, *17*, 773–775. [CrossRef]

139. Wildhirt, S.M.; Dudek, R.R.; Suzuki, H.; Bing, R.J. Involvement of Inducible Nitric Oxide Synthase in the Inflammatory Process of Myocardial Infarction. *Int. J. Cardiol.* **1995**, *50*, 253–261. [CrossRef]

140. Arstall, M.A.; Sawyer, D.B.; Fukazawa, R.; Kelly, R.A. Cytokine-Mediated Apoptosis in Cardiac Myocytes. *Circ. Res.* **1999**, *85*, 829–840. [CrossRef]

141. D'Oria, R.; Schipani, R.; Leonardini, A.; Natalicchio, A.; Perrini, S.; Cignarelli, A.; Laviola, L.; Giorgino, F. The Role of Oxidative Stress in Cardiac Disease: From Physiological Response to Injury Factor. *Oxid. Med. Cell Longev.* **2020**, *2020*, 5732956. [CrossRef] [PubMed]

142. Dubois-deruy, E.; Peugnet, V.; Turkieh, A.; Pinet, F. Oxidative Stress in Cardiovascular Diseases. *Antioxidants* **2020**, *9*, 864. [CrossRef]

143. Haynes, V.; Elfering, S.L.; Squires, R.J.; Traaseth, N.; Solien, J.; Ettl, A.; Giulivi, C. Mitochondrial Nitric-Oxide Synthase: Role in Pathophysiology. *IUBMB Life* **2003**, *55*, 599–603. [CrossRef] [PubMed]

144. Poderoso, J.J.; Helfenberger, K.; Poderoso, C. The Effect of Nitric Oxide on Mitochondrial Respiration. *Nitric Oxide* **2019**, *88*, 61–72. [CrossRef] [PubMed]

145. Bailey, J.D.; Diotallevi, M.; Nicol, T.; McNeill, E.; Shaw, A.; Chuaiphichai, S.; Hale, A.; Starr, A.; Nandi, M.; Stylianou, E.; et al. Nitric Oxide Modulates Metabolic Remodeling in Inflammatory Macrophages through TCA Cycle Regulation and Itaconate Accumulation. *Cell Rep.* **2019**, *28*, 218–230.e7. [CrossRef]

146. Ow, Y.L.P.; Green, D.R.; Hao, Z.; Mak, T.W. Cytochrome c: Functions beyond Respiration. *Nat. Rev. Mol. Cell Biol.* **2008**, *9*, 532–542. [CrossRef]

147. Yoon, S.; Kim, M.; Lee, H.; Kang, G.; Bedi, K.; Margulies, K.B.; Jain, R.; Nam, K.-I.; Kook, H.; Eom, G.H. S-Nitrosylation of Histone Deacetylase 2 by Neuronal Nitric Oxide Synthase as a Mechanism of Diastolic Dysfunction. *Circulation* **2021**, *143*, 1912–1925. [CrossRef]

148. Solomon, S.D.; McMurray, J.J.V.; Claggett, B.; de Boer, R.A.; DeMets, D.; Hernandez, A.F.; Inzucchi, S.E.; Kosiborod, M.N.; Lam, C.S.P.; Martinez, F.; et al. Dapagliflozin in Heart Failure with Mildly Reduced or Preserved Ejection Fraction. *N. Engl. J. Med.* **2022**, *387*, 1089–1098. [CrossRef] [PubMed]

149. Kondo, H.; Akoumianakis, I.; Badi, I.; Akawi, N.; Kotanidis, C.P.; Polkinghorne, M.; Stadiotti, I.; Sommariva, E.; Antonopoulos, A.S.; Carena, M.C.; et al. Effects of Canagliflozin on Human Myocardial Redox Signalling: Clinical Implications. *Eur. Heart J.* **2021**, *42*, 4947–4960. [CrossRef]

150. Rahman, H.; Demir, O.M.; Khan, F.; Ryan, M.; Ellis, H.; Mills, M.T.; Chiribiri, A.; Webb, A.; Perera, D. Physiological Stratification of Patients With Angina Due to Coronary Microvascular Dysfunction. *J. Am. Coll. Cardiol.* **2020**, *75*, 2538–2549. [CrossRef] [PubMed]

151. Sinha, A.; Rahman, H.; Webb, A.; Shah, A.M.; Perera, D. Untangling the Pathophysiologic Link between Coronary Microvascular Dysfunction and Heart Failure with Preserved Ejection Fraction. *Eur. Heart J.* **2021**, *42*, 4431–4441. [CrossRef]

152. He, J.; Liu, D.; Zhao, L.; Zhou, D.; Rong, J.; Zhang, L.; Xia, Z. Myocardial Ischemia/Reperfusion Injury: Mechanisms of Injury and Implications for Management (Review). *Exp. Ther. Med.* **2022**, *23*, 430. [CrossRef]

153. Heusch, G. Myocardial Ischaemia–Reperfusion Injury and Cardioprotection in Perspective. *Nat. Rev. Cardiol.* **2020**, *17*, 773–789. [CrossRef]

154. Damy, T.; Ratajczak, P.; Robidel, E.; Bendall, J.K.; Oliviéro, P.; Boczkowski, J.; Ebrahimian, T.; Marotte, F.; Samuel, J.-L.; Heymes, C. Up-regulation of Cardiac Nitric Oxide Synthase 1-derived Nitric Oxide after Myocardial Infarction in Senescent Rats. *FASEB J.* **2003**, *17*, 1–22. [CrossRef]
155. Yu, X.; Ge, L.; Niu, L.; Lian, X.; Ma, H.; Pang, L. The Dual Role of Inducible Nitric Oxide Synthase in Myocardial Ischemia/Reperfusion Injury: Friend or Foe? *Oxid. Med. Cell Longev.* **2018**, *2018*, 8364848. [CrossRef] [PubMed]
156. Wilmes, V.; Scheiper, S.; Roehr, W.; Niess, C.; Kippenberger, S.; Steinhorst, K.; Verhoff, M.A.; Kauferstein, S. Increased Inducible Nitric Oxide Synthase (INOS) Expression in Human Myocardial Infarction. *Int. J. Legal Med.* **2020**, *134*, 575–581. [CrossRef] [PubMed]
157. Schulz, R.; Wambolt, R. Inhibition of Nitric Oxide Synthesis Protects the Isolated Working Rabbit Heart from Ischaemia-Reperfusion Injury. *Cardiovasc. Res.* **1995**, *30*, 432–439. [CrossRef] [PubMed]
158. Yasmin, W. Generation of Peroxynitrite Contributes to Ischemia-Reperfusion Injury in Isolated Rat Hearts. *Cardiovasc. Res.* **1997**, *33*, 422–432. [CrossRef] [PubMed]
159. Patel, V.C.; Yellon, D.M.; Singh, K.J.; Neild, G.H.; Woolfson, R.G. Inhibition of Nitric Oxide Limits Infarct Size in the in Situ Rabbit Heart. *Biochem. Biophys. Res. Commun.* **1993**, *194*, 234–238. [CrossRef]
160. Ramasamy, R.; Hwang, Y.C.; Liu, Y.; Son, N.H.; Ma, N.; Parkinson, J.; Sciacca, R.; Albala, A.; Edwards, N.; Szabolcs, M.J.; et al. Metabolic and Functional Protection by Selective Inhibition of Nitric Oxide Synthase 2 During Ischemia-Reperfusion in Isolated Perfused Hearts. *Circulation* **2004**, *109*, 1668–1673. [CrossRef]
161. Hu, A.; Jiao, X.; Gao, E.; Koch, W.J.; Sharifi-Azad, S.; Grunwald, Z.; Ma, X.L.; Sun, J.-Z. Chronic β-Adrenergic Receptor Stimulation Induces Cardiac Apoptosis and Aggravates Myocardial Ischemia/Reperfusion Injury by Provoking Inducible Nitric-Oxide Synthase-Mediated Nitrative Stress. *J. Pharmacol. Exp. Ther.* **2006**, *318*, 469–475. [CrossRef]
162. Cotter, G.; Kaluski, E.; Blatt, A.; Milovanov, O.; Moshkovitz, Y.; Zaidenstein, R.; Salah, A.; Alon, D.; Michovitz, Y.; Metzger, M.; et al. L-NMMA (a Nitric Oxide Synthase Inhibitor) Is Effective in the Treatment of Cardiogenic Shock. *Circulation* **2000**, *101*, 1358–1361. [CrossRef]
163. Cotter, G.; Kaluski, E.; Milo, O.; Blatt, A.; Salah, A.; Hendler, A.; Krakover, R.; Golick, A.; Vered, Z. LINCS: L-NAME (a NO Synthase Inhibitor) In the Treatment of Refractory Cardiogenic Shock A Prospective Randomized Study. *Eur. Heart J.* **2003**, *24*, 1287–1295. [CrossRef] [PubMed]
164. Dzavik, V.; Cotter, G.; Reynolds, H.R.; Alexander, J.H.; Ramanathan, K.; Stebbins, A.L.; Hathaway, D.; Farkouh, M.E.; Ohman, E.M.; Baran, D.A.; et al. Effect of Nitric Oxide Synthase Inhibition on Haemodynamics and Outcome of Patients with Persistent Cardiogenic Shock Complicating Acute Myocardial Infarction: A Phase II Dose-Ranging Study. *Eur. Heart J.* **2007**, *28*, 1109–1116. [CrossRef]
165. The TRIUMPH Investigators. Effect of Tilarginine Acetate in Patients With Acute Myocardial Infarction and Cardiogenic Shock. *JAMA* **2007**, *297*, 1657. [CrossRef]
166. Shen, Z.; Xiang, M.; Chen, C.; Ding, F.; Wang, Y.; Shang, C.; Xin, L.; Zhang, Y.; Cui, X. Glutamate Excitotoxicity: Potential Therapeutic Target for Ischemic Stroke. *Biomed. Pharmacother.* **2022**, *151*, 113125. [CrossRef]
167. Yang, Q.; Huang, Q.; Hu, Z.; Tang, X. Potential Neuroprotective Treatment of Stroke: Targeting Excitotoxicity, Oxidative Stress, and Inflammation. *Front. Neurosci.* **2019**, *13*, 1036. [CrossRef]
168. Choi, D.W. Excitotoxicity: Still Hammering the Ischemic Brain in 2020. *Front. Neurosci.* **2020**, *14*, 579953. [CrossRef]
169. Zhu, L.-J.; Li, F.; Zhu, D.-Y. NNOS and Neurological, Neuropsychiatric Disorders: A 20-Year Story. *Neurosci. Bull.* **2023**, *39*, 1439–1453. [CrossRef]
170. Lai, T.W.; Zhang, S.; Wang, Y.T. Excitotoxicity and Stroke: Identifying Novel Targets for Neuroprotection. *Prog. Neurobiol.* **2014**, *115*, 157–188. [CrossRef] [PubMed]
171. Dawson, V.; Dawson, T.; Bartley, D.; Uhl, G.; Snyder, S. Mechanisms of Nitric Oxide-Mediated Neurotoxicity in Primary Brain Cultures. *J. Neurosci.* **1993**, *13*, 2651–2661. [CrossRef] [PubMed]
172. Dawson, V.L.; Dawson, T.M.; London, E.D.; Bredt, D.S.; Snyder, S.H. Nitric Oxide Mediates Glutamate Neurotoxicity in Primary Cortical Cultures. *Proc. Natl. Acad. Sci. USA* **1991**, *88*, 6368–6371. [CrossRef]
173. Dawson, V.; Kizushi, V.; Huang, P.; Snyder, S.; Dawson, T. Resistance to Neurotoxicity in Cortical Cultures from Neuronal Nitric Oxide Synthase-Deficient Mice. *J. Neurosci.* **1996**, *16*, 2479–2487. [CrossRef] [PubMed]
174. Huang, Z.; Huang, P.L.; Panahian, N.; Dalkara, T.; Fishman, M.C.; Moskowitz, M.A. Effects of Cerebral Ischemia in Mice Deficient in Neuronal Nitric Oxide Synthase. *Science* **1994**, *265*, 1883–1885. [CrossRef] [PubMed]
175. Hill, M.D.; Martin, R.H.; Mikulis, D.; Wong, J.H.; Silver, F.L.; terBrugge, K.G.; Milot, G.; Clark, W.M.; MacDonald, R.L.; Kelly, M.E.; et al. Safety and Efficacy of NA-1 in Patients with Iatrogenic Stroke after Endovascular Aneurysm Repair (ENACT): A Phase 2, Randomised, Double-Blind, Placebo-Controlled Trial. *Lancet Neurol.* **2012**, *11*, 942–950. [CrossRef] [PubMed]
176. Hill, M.D.; Goyal, M.; Menon, B.K.; Nogueira, R.G.; McTaggart, R.A.; Demchuk, A.M.; Poppe, A.Y.; Buck, B.H.; Field, T.S.; Dowlatshahi, D.; et al. Efficacy and Safety of Nerinetide for the Treatment of Acute Ischaemic Stroke (ESCAPE-NA1): A Multicentre, Double-Blind, Randomised Controlled Trial. *Lancet* **2020**, *395*, 878–887. [CrossRef]
177. NIWA, M.; INAO, S.; TAKAYASU, M.; KAWAI, T.; KAJITA, Y.; NIHASHI, T.; KABEYA, R.; SUGIMOTO, T.; YOSHIDA, J. Time Course of Expression of Three Nitric Oxide Synthase Isoforms After Transient Middle Cerebral Artery Occlusion in Rats. *Neurol. Med. Chir.* **2001**, *41*, 63–73. [CrossRef]

178. Nelson, R.J.; Demas, G.E.; Huang, P.L.; Fishman, M.C.; Dawson, V.L.; Dawson, T.M.; Snyder, S.H. Behavioural Abnormalities in Male Mice Lacking Neuronal Nitric Oxide Synthase. *Nature* **1995**, *378*, 383–386. [CrossRef]
179. Furfine, E.S.; Harmon, M.F.; Paith, J.E.; Knowles, R.G.; Salter, M.; Kiff, R.J.; Duffy, C.; Hazelwood, R.; Oplinger, J.A.; Garvey, E.P. Potent and Selective Inhibition of Human Nitric Oxide Synthases. Selective Inhibition of Neuronal Nitric Oxide Synthase by S-Methyl-L-Thiocitrulline and S-Ethyl-L-Thiocitrulline. *J. Biol. Chem.* **1994**, *269*, 26677–26683. [CrossRef]
180. Zhang, J.; Xu, M.; Dence, C.S.; Sherman, E.L.; McCarthy, T.J.; Welch, M.J. Synthesis, in Vivo Evaluation and PET Study of a Carbon-11-Labeled Neuronal Nitric Oxide Synthase (NNOS) Inhibitor S-Methyl-L-Thiocitrulline. *J. Nucl. Med.* **1997**, *38*, 1273–1278.
181. Kim, J.H.; Yenari, M.A.; Giffard, R.G.; Cho, S.W.; Park, K.A.; Lee, J.E. Agmatine Reduces Infarct Area in a Mouse Model of Transient Focal Cerebral Ischemia and Protects Cultured Neurons from Ischemia-like Injury. *Exp. Neurol.* **2004**, *189*, 122–130. [CrossRef]
182. Kotagale, N.R.; Taksande, B.G.; Inamdar, N.N. Neuroprotective Offerings by Agmatine. *Neurotoxicology* **2019**, *73*, 228–245. [CrossRef]
183. Yang, M.Z.; Mun, C.H.; Choi, Y.J.; Baik, J.H.; Park, K.A.; Lee, W.T.; Lee, J.E. Agmatine Inhibits Matrix Metalloproteinase-9 via Endothelial Nitric Oxide Synthase in Cerebral Endothelial Cells. *Neurol. Res.* **2007**, *29*, 749–754. [CrossRef]
184. Kim, D.J.; Kim, D.I.; Lee, S.K.; Suh, S.H.; Lee, Y.J.; Kim, J.; Chung, T.S.; Lee, J.E. Protective Effect of Agmatine on a Reperfusion Model after Transient Cerebral Ischemia: Temporal Evolution on Perfusion MR Imaging and Histopathologic Findings. *AJNR Am. J. Neuroradiol.* **2006**, *27*, 780–785. [PubMed]
185. Feng, Y.; Piletz, J.E.; Leblanc, M.H. Agmatine Suppresses Nitric Oxide Production and Attenuates Hypoxic-Ischemic Brain Injury in Neonatal Rats. *Pediatr. Res.* **2002**, *52*, 606–611. [CrossRef] [PubMed]
186. Tegtmeier, F.; Schinzel, R.; Beer, R.; Bulters, D.; LeFrant, J.-Y.; Sahuquillo, J.; Unterberg, A.; Andrews, P.; Belli, A.; Ibanez, J.; et al. Efficacy of Ronopterin (VAS203) in Patients with Moderate and Severe Traumatic Brain Injury (NOSTRA Phase III Trial): Study Protocol of a Confirmatory, Placebo-Controlled, Randomised, Double Blind, Multi-Centre Study. *Trials* **2020**, *21*, 80. [CrossRef] [PubMed]
187. Ott, C.; Bosch, A.; Winzer, N.; Friedrich, S.; Schinzel, R.; Tegtmeier, F.; Schmieder, R.E. Effects of the Nitric Oxide Synthase Inhibitor Ronopterin (VAS203) on Renal Function in Healthy Volunteers. *Br. J. Clin. Pharmacol.* **2019**, *85*, 900–907. [CrossRef] [PubMed]

International Journal of
Molecular Sciences

Review

Model Systems to Study the Mechanism of Vascular Aging

Janette van der Linden [1,2], Lianne Trap [3,4], Caroline V. Scherer [2], Anton J. M. Roks [1], A. H. Jan Danser [1], Ingrid van der Pluijm [2,5,*] and Caroline Cheng [6,7,*]

[1] Division of Vascular Medicine and Pharmacology, Department of Internal Medicine, Erasmus MC, 3015 GD Rotterdam, The Netherlands
[2] Department of Molecular Genetics, Cancer Genomics Center Netherlands, Erasmus MC, 3015 GD Rotterdam, The Netherlands
[3] Department of Pulmonary Medicine, Erasmus MC, 3015 GD Rotterdam, The Netherlands
[4] Department of Internal Medicine, Erasmus MC, 3015 GD Rotterdam, The Netherlands
[5] Department of Vascular Surgery, Cardiovascular Institute, Erasmus MC, 3015 GD Rotterdam, The Netherlands
[6] Division of Experimental Cardiology, Department of Cardiology, Erasmus MC, 3015 GD Rotterdam, The Netherlands
[7] Department of Nephrology and Hypertension, Division of Internal Medicine and Dermatology, University Medical Center Utrecht, 3584 CX Utrecht, The Netherlands
* Correspondence: i.vanderpluijm@erasmusmc.nl (I.v.d.P.); k.l.cheng-2@umcutrecht.nl (C.C.)

Abstract: Cardiovascular diseases are the leading cause of death globally. Within cardiovascular aging, arterial aging holds significant importance, as it involves structural and functional alterations in arteries that contribute substantially to the overall decline in cardiovascular health during the aging process. As arteries age, their ability to respond to stress and injury diminishes, while their luminal diameter increases. Moreover, they experience intimal and medial thickening, endothelial dysfunction, loss of vascular smooth muscle cells, cellular senescence, extracellular matrix remodeling, and deposition of collagen and calcium. This aging process also leads to overall arterial stiffening and cellular remodeling. The process of genomic instability plays a vital role in accelerating vascular aging. Progeria syndromes, rare genetic disorders causing premature aging, exemplify the impact of genomic instability. Throughout life, our DNA faces constant challenges from environmental radiation, chemicals, and endogenous metabolic products, leading to DNA damage and genome instability as we age. The accumulation of unrepaired damages over time manifests as an aging phenotype. To study vascular aging, various models are available, ranging from in vivo mouse studies to cell culture options, and there are also microfluidic in vitro model systems known as vessels-on-a-chip. Together, these models offer valuable insights into the aging process of blood vessels.

Keywords: vascular aging; genomic instability; mouse models; vessels-on-a-chip

Citation: van der Linden, J.; Trap, L.; Scherer, C.V.; Roks, A.J.M.; Danser, A.H.J.; van der Pluijm, I.; Cheng, C. Model Systems to Study the Mechanism of Vascular Aging. *Int. J. Mol. Sci.* **2023**, *24*, 15379. https://doi.org/10.3390/ijms242015379

Academic Editor: Silvia S. Barbieri

Received: 31 August 2023
Revised: 16 October 2023
Accepted: 17 October 2023
Published: 19 October 2023

1. Introduction

Cardiovascular diseases (CVDs) are the leading cause of death worldwide, especially in the elderly [1]. In 2019, CVD affected 523 million people and killed 18.6 million people globally. Modifiable risk factors for developing CVD include high systolic blood pressure, high fasting plasma glucose, high low-density lipoprotein cholesterol, and smoking [2]. Blood pressure and pulse wave velocity measurements, which assess arterial stiffness, are used to predict cardiovascular events and mortality, and increase in prevalence with age [3–5]. Arterial aging is a fundamental aspect of cardiovascular aging, as the structural and functional changes in arteries significantly contribute to the overall decline in cardiovascular health associated with the aging process. Unlike atherosclerosis, arteriosclerosis develops spontaneously during aging independently from cardiovascular risk factors other than age, although it is strongly accelerated by hypertension and hyperglycemia [6]. Arterial aging results in structural and functional changes in the arterial wall (Figure 1). Which involves chronic, low-grade inflammation, extracellular matrix (ECM) remodeling including elastin degradation, medial thickening, endothelial dysfunction, phenotypical switching of vascular smooth muscle cells (VSMCs), calcification, and fibrosis [7]. Various molecular and cellular mechanisms contribute to aging of the vasculature (Figure 1). Reactive oxygen species (ROS) and impaired bioavailability of nitric oxide (NO), together with reduced antioxidant defense systems, cause oxidative stress, leading to endothelial dysfunction and inflammation [8]. Vascular inflammation, mitochondrial dysfunction, cellular senescence, apoptosis, deregulated nutrient sensing, and genomic instability also play a major role in vascular aging [8]. Accordingly, most of these factors are considered to be hallmarks of aging [9,10].

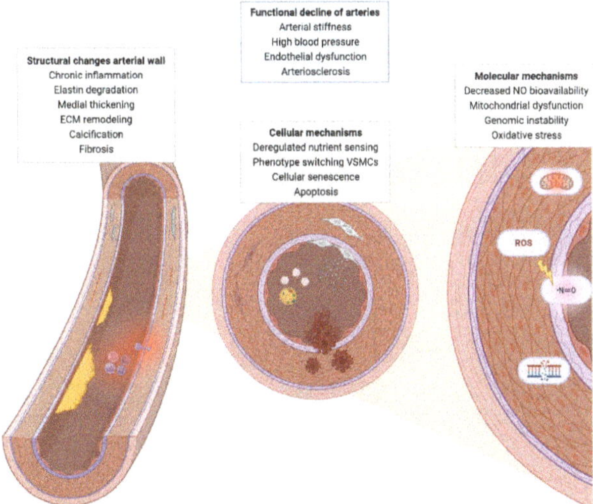

Figure 1. Structural and functional changes in vascular aging. Vascular aging is characterized by functional decline of the arteries. Structural changes in the arterial wall together with various cellular and molecular mechanisms contribute to arterial functional decline.

In this review, we describe different parameters of vascular aging and the way that different model systems are used to investigate the underlying molecular mechanism of vascular aging. We will focus on the different types of models to study vascular aging, ranging from in vivo mouse studies to cell culture options and microfluidic in vitro model systems, also known as vessels-on-a-chip.

2. Structural, Functional, and Pathological Changes in the Aged Vascular System

Aging of arteries is characterized by structural changes (Figure 1); these include increase in luminal diameter, intimal and medial thickening, loss of VSMCs, senescence, ECM remodeling, collagen and calcium deposition, and cellular remodeling [11]. These structural alterations result in functional decline of the vascular system. As a result of these structural changes, older adults display physiological changes, including endothelial dysfunction, arterial stiffening, and high blood pressure [12]. Pathological changes can also occur during aging; these include arteriosclerosis, aneurysms, and heart failure.

Cellular Changes in the Aged Vascular System

Mitochondria, being the primary source of ROS production, play a significant role in promoting oxidative stress and low-grade inflammation during vascular aging. With age, there is an increase in ROS production, decreased antioxidant capacity, and impaired cellular redox signaling. Excessive ROS may lead to endothelial dysfunction, inflammation, and structural damage to the vascular wall [8]. Chronic low-grade inflammation is a characteristic feature of aging. In vascular aging, upregulation of inflammatory markers, increased infiltration of immune cells into the vessel wall, and activation of pro-inflammatory signaling pathways are observed. Arterial stiffness is a hallmark of vascular aging and is partially attributed to changes in the elastic fibers and collagen content of the arterial walls. With aging, there is an accumulation of collagen and other extracellular matrix proteins in the arterial wall, leading to reduced arterial compliance and increased arterial stiffness. This contributes to impaired arterial distensibility and elevated systolic blood pressure. Arterial distensibility is also influenced by endothelial dysfunction, the inability of arteries to maintain proper vascular tone. The processes of vasodilation and its counterpart vasoconstriction are essential to regulate blood flow and the supply of oxygen and nutrients throughout the body. NO-cyclic guanosine monophosphate (cGMP) signaling is crucial for proper functioning of the vasculature. Endothelial cells produce NO, which then diffuses to VSMCs and activates soluble guanylyl cyclase (sGC)-mediated production of cGMP, finally leading to vasodilation, e.g., relaxation and widening of blood vessels (Figure 2). The eNOS protein, together with cofactor tetrahydrobiopterin (BH4), synthesizes NO through the oxidization of nicotinamide adenine dinucleotide phosphate (NADPH) and the conversion of L-Arginine. ROS, which are abundant in the case of oxidative stress, are capable of scavenging NO, thereby decreasing NO bioavailability. Endothelial dysfunction occurs when maintenance of vascular tone is impaired (Figure 2). This is caused by a shortage of endothelium-derived relaxing factors, like NO, or by a decrease in VSMC responsiveness to NO. A decrease in NO bioavailability likely results from uncoupling of endothelial nitric oxide synthase (eNOS), which results in decreased production of NO and increased generation of ROS. Furthermore, it increases NADPH oxidase (NOX) activity, which results in increased ROS production, or increased production of ROS by mitochondria. Phosphodiesterase (PDEs) catalyze hydrolysis of cGMP and cAMP, leading to decreased responsiveness to NO (Figure 2).

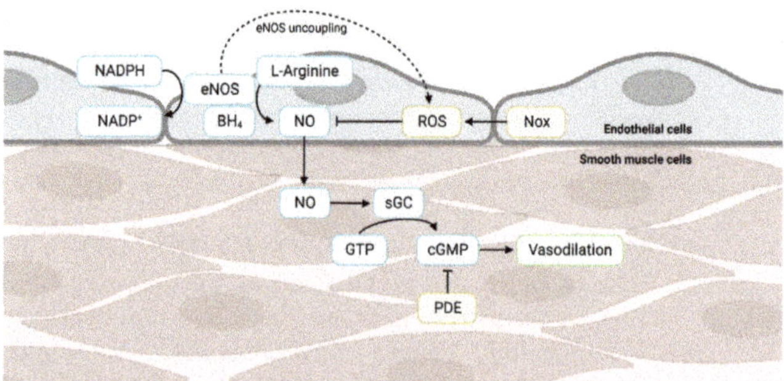

Figure 2. **Nitric oxide (NO)-cyclic guanosine monophosphate (cGMP) signaling and vascular dysfunction.** In blue, important mediators of NO-cGMP signaling; in yellow, factors capable of perturbing NO-cGMP signaling. GTP, guanosine triphosphate. In green, these processes are leading to vasodilation.

Other vasodilator signaling pathways of the endothelium are endothelium-derived hyperpolarization (EDHF) and prostaglandins (PG). The faith of EDHF in aging is unknown. For PG, it is known that in aging there is a switch from dilatory PG, such as prostacyclin, toward vasoconstrictive PG [13]. However, PG-mediated responses are often not observed in mouse models, and are, therefore, difficult to study in genetic aging models.

Furthermore, aging contributes to phenotypical switching of VSMCs. Several VSMC phenotypes have been described; contractile, synthetic, osteogenic, senescent, macrophage-like, mesenchymal stem cell-like, fibroblast-like and adipose-like (Figure 3) [14–17]. The contractile phenotype expresses contractile markers such as SMA, SM22, and MYH11 [18]. These cells can dedifferentiate towards a more proliferative and migrating phenotype, the synthetic VSMCs which secrete high levels of ECM, and are positive for vimentin, EREG, and MMP9 [18,19]. Osteogenic VSMCs are characterized by increased levels of RunX2, OPN, and ALP [20]. These cells contribute to vascular calcification. Macrophage-like VSMCs are positive for LGALS3, CD68, and VCAM1. These cells are capable of initiating an immune response and are associated with atherosclerosis [21]. The fibroblast-like VSMCs are associated with fibrosis of the vascular wall. They express LUM, BGN, and DCN [16]. Senescence is a hallmark of aging; senescent VSMCs express increased P16, P21, and senescence-associated secretory phenotype (SASP) markers [14]. Mesenchymal stem cell-like VSMC express SCA-, CD34, and CD44. The role of mesenchymal stem cell-like VSMC is not yet known [14]. The adipocyte-VSMCs are only described in single cell RNA-seq [16]. They express markers such as UCP1, DIO2, and PPARGC1. More research is needed to unravel the function of this phenotype.

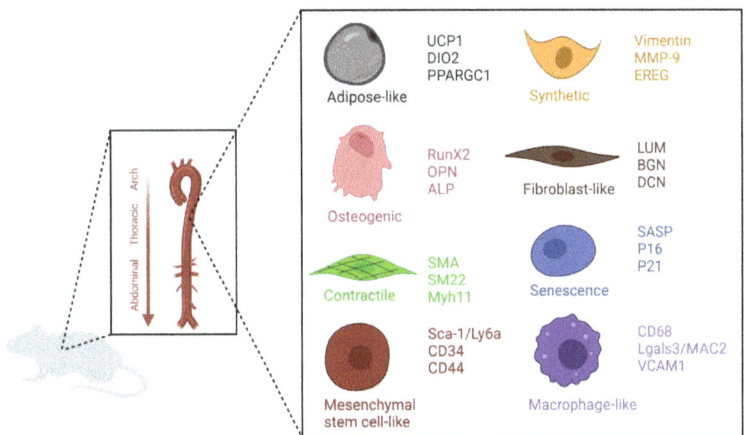

Figure 3. Vascular smooth muscle cell phenotypes. These phenotypes are adipose-like, osteogenic, contractile, mesenchymal, stem cell-like synthetic, fibroblast-like, senescent, and macrophage-like VSMCs.

The phenotypical switching of VSMCs occurs upon various stimuli. Endothelial denudation or atherogenic lesions have been well-studied in this respect. The impact of endothelial aging is not as clearly studied directly in aging [22,23]. However, decreased NO signaling, a hallmark of endothelial aging, has been studied. ECs synthesize and release NO, which induces relaxation in VSMCs while also preserving the quiescent state of VSMCs via inhibition of proliferation by cGMP, the main messenger of NO [24–26]. Furthermore, NO–cGMP signaling decreases VSMC-mediated fibrosis and prevents medial hypertrophy. Hence, improved NO signaling contributes to vasodilation and maintenance of vascular compliance, thus sustaining proper tonus and flow regulation via VSMC [27]. Loss of NO is likely to lead to switching to the pro-fibrotic, osteogenic, and synthetic VSMC phenotypes.

3. Vascular Aging in Human Progeria Syndromes

Genomic instability contributes to accelerated vascular aging in humans. Progeria syndromes are a group of rare genetic disorders that result in premature aging, with a range of health problems that are typically associated with advanced age. Cardiovascular disease is a major cause of morbidity and mortality in patients with progeria syndromes, and several types of progeria syndromes have been identified that manifest juvenile cardiovascular diseases.

Hutchinson–Gilford Progeria Syndrome (HGPS) is the most well-known progeria syndrome and is caused, in most patients, by a mutation in the LMNA gene, which encodes for lamin A/C. Lamins play a crucial role in providing structural support to the nucleus, and their expression levels are directly associated with nuclear stiffness, tissue rigidity, and plasticity. Furthermore, lamin A contributes to genomic instability. Lamin A participates in maintaining telomere homeostasis and plays a role in DNA repair processes by silencing DNA repair proteins like BRCA1 and RAD51 [28,29]. HGPS is characterized by premature aging, including accelerated cardiovascular disease, which is the leading cause of death in these patients. HGPS patients typically have loss of VSMCs, arterial stiffness, and calcification of the arterial walls. Interestingly, they do not develop fully formed plaques, but the initial phase of atherosclerosis. These pathological alterations can lead to myocardial infarction, stroke, and premature death, mainly caused by arteriosclerosis [30,31]. The average age of death for children with HGPS is 13 years [32].

Werner Syndrome (WS) is another progeria syndrome that is associated with increased risk of cardiovascular disease. WS is caused by a mutation in the WRN gene, which encodes for Werner syndrome ATP-dependent helicase. WRN is involved in DNA repair by unwinding $5'$ flaps and interacting with base-excision repair proteins [33]. WS patients often develop atherosclerosis, which increases risk of myocardial infarction and stroke. Furthermore, these patients show calcification of the aortic valve. WS patients also have a high incidence of hypertension and heart failure, which further increases their risk of cardiovascular disease [34].

4. Vascular Aging Mouse Models

The fundamental processes behind aging have also been studied by using invertebrate animal models systems such as roundworms (*Caenorhabditis elegans*) or the fruit fly (*Drosophila melanogaster*) [35]. While these models are inexpensive and available in great quantity, mammalian model organisms better approximate the complexity of human aging. However, while some long-lived species such as primates are closest to humans, their generational cycles are so long that it is difficult to use them in research [36]. Mice, specifically genetically modified fast-aging mice, are therefore currently most often used for aging research.

Genetically engineered mouse models have been established that mimic human progeria syndromes [35]. In these progeroid mice, genes involved in genome maintenance are genetically manipulated, which leads to accelerated aging and a shortened lifespan. These accelerated aging models can be divided into different groups: genomic instability, telomere attrition, epigenetic alterations, loss of proteostasis, mitochondrial dysfunction, stem cell exhaustion, and altered intercellular communication. The majority of the gene mutations in these accelerated aging mouse models result in genomic instability. Furthermore, genomic instability is linked to all hallmarks of aging, highlighting the importance of DNA damage repair in aging [9].

Hence, although different types of mouse models are used to study vascular abnormalities, in this review, we will specifically focus on mouse models with genome instability that show accelerated vascular aging (Table 1).

4.1. Nucleotide Excision Repair (NER) Deficient Mouse Models

NER is a DNA repair pathway which eliminates a wide range of DNA lesions including single strand lesions and helix distorting lesions induced by exogenous sources such as ultraviolet radiation or by endogenous sources such as reactive metabolites [37,38]. Currently, $Ercc1^{d/-}$ and Xpd^{TTD} mice are the only two NER deficient mouse models which have been described to show characteristics of vascular aging (Durik et al., 2012) [39]. DNA damage detection by NER occurs through two sub pathways: global genome NER (GG-NER), in which the whole genome is probed for DNA helix distortions, or transcription-coupled NER (TC-NER), which is activated by polymerase stalling due to lesions in the transcribed strand [39]. ERCC excision repair 1 (ERCC1) and Xeroderma Pigmentosum Group D (XPD) mediate the final part of the NER pathway, after which GG-NER and TC-NER converge. XPD is a helicase involved in verification of the DNA damage lesion and unwinding of the DNA around the lesion. The ERCC1 endonuclease forms a heterodimeric complex with Xeroderma Pigmentosum Group F endonuclease and excises the DNA strand $5'$ of the lesion, in combination with Xeroderma Pigmentosum Group G (XPG), which excises $3'$ of the lesion [38]. Ercc1 knockout mice lack expression of Ercc1 and live up to 38 days. On the other hand, $Ercc1^{d/-}$ mice–hemizygous for an Ercc1 knock-out allele and a delta -7 allele -resulting in ERCC1 protein with a seven amino acid truncation– live up to 6 months [40] and show accelerated age-related pathologies which chronologically follow changes in naturally aging wildtype mice [41]. These include neurodegeneration, muscular dystrophy, ataxia, brain atrophy, renal tubular degeneration, heart myocardial degeneration, peripheral nerve vacuolization, and several liver pathologies [42].

$Ercc1^{d/-}$ mice display increased aortic stiffness, blood pressure, aortic senescence, endothelial dysfunction, and decreased reactive hyperemia [39,41]—which is a measure for peripheral microvascular function used to predict cardiovascular morbidity and mortality.

Additionally, vascular aging has been studied in two tissue-specific Ercc1 knock-out mouse models. Endothelial (EC)-specific Ercc1 knockout mice are hemizygous for a knock-out Ercc1 allele and a floxed Ercc1 allele and express Cre-recombinase under the Tie2 promoter [43]. Tie2 is expressed in ECs and in monocytes with a pro-angiogenic function. These EC-specific Ercc1 knockout mice exhibit decreased endothelium-derived NO, altered vasoconstriction, arterial remodeling, and increased arterial stiffness. Vascular smooth muscle cell (VSMC)-specific Ercc1 knockout mice are hemizygous for a knock-out Ercc1 allele and a floxed Ercc1 allele and express Cre-recombinase under the SM22α promoter [44]. SM22α is involved in SMC contraction. These mice show diminished NO-mediated vasodilation, decreased reactive hyperemia, and increased arterial stiffness.

Only three patients have been described with an ERCC1 mutation and all three died before the age of 3 [45–47]. No vascular aging phenotypes were described. However, it is unknown if these patients were screened for vascular problems at this very young age. As the $Ercc1^{d/-}$ mouse model shows characteristics of vascular aging, this would suggest that likewise patients with Ercc1 mutations may experience vascular aging problems with age.

4.2. LMNA

Multiple mouse models exist with mutations in the LMNA gene. The most commonly used is the $Lmna^{G609G/G609G}$ with a c.1827C>T;p.Gly609Gly mutation, which mimics a human c.1824C>T;p.Gly608Gly mutation in exon 11 [48]. The c.1827C>T;p.Gly609Gly mutation causes abnormal splicing, leading to the production of a shortened form of prelamin A, which serves as the precursor for lamin A/C. This truncated protein is commonly referred to as progerin. Progerin accumulates in the nuclear lumina, and thereby disrupts the nuclear architecture [49,50].

One study showed that the stiffness and inward remodeling observed in $Lmna^{G609G/G609G}$ arteries are primarily due to damage to VSMCs caused by ubiquitous progerin [50]. This leads to an increase in medial collagen deposition and subsequent changes to elastin structure [50]. The $Ldlr^{-/-}Lmna^{G609G/G609G}$ mouse aortas showed similar results, including VSMC depletion, adventitial thickening, and changes to the elastin structure. They furthermore show atherosclerosis and structural abnormalities [51]. Coll-Bonfill et al. showed that progerin expression in VSMCs results in a phenotypical switch of the VSMCs (characterized by a decrease in contractile markers) towards a synthetic-like phenotype. This switch increases replication speed, resulting in replication stress, genomic instability and eventually VSMC depletion [52].

In the $Lmna^{HG}$ mouse model intron 10 and the last 150 nucleotide of intron 11 of the LMNA gene are deleted, resulting in the production of progerin [53]. Interestingly, both the $Lmna^{HG/+}$ (expressing progerin, lamin A/C) and the $Lmna^{HG/HG}$ mouse (expressing only progerin) do not show vascular aging [54]. This is in contrast with the c.1827C>T;p.Gly609Gly mutation, indicating a difference between the progerin protein resulting from the deletion of exon 10 and partially 11 and the c.1827C>T;p.Gly609Gly mutation, which mimics the most common human mutation, and gives a smaller deletion of only 50 amino acids within prelamin A, likely resulting in a differently folded protein.

Additionally, the process of vascular aging has been examined in mouse models that specifically target tissues of interest. To investigate the effect of VSMCs and ECs on vascular aging, del Campo et al. generated tissue specific mouse models, showing similar effects for lamin mutation in a VSMC specific mouse model ($Lmna^{LCS/LCS}SM22\alpha Cre^{+/tg}$) compared to the full body mutant, but not in the EC-specific mouse model ($Lmna^{LCS/LCS}Tie2Cre^{+/tg}$ mice) [50]. To further mimic the characteristics of HGPS, Hamczyk et al. crossed the $Apoe^{-/-}$ mouse with $Lmna^{G609G/G609G}$. These mice, when fed a high fat diet, showed medial VSMC loss, lipid retention, adventitial fibrosis, and accelerated atherosclerosis [55]. Additionally, they showed a similar phenotype in the VSMC-specific mutant ($Apoe^{-/-}Lmna^{LCS/LCS}SM22\alpha Cre$), but not in the macrophage-specific mutant ($Apoe^{-/-}Lmna^{LCS/LCS}LysMCre$) [55].

Taken together, this shows that the $Lmna^{G609G}$ mutation promotes vascular aging in mice, while the $Lmna^{HG}$ mutant mice do not exhibit signs of vascular aging. The tissue-specific mouse models highlight the importance of Lmna in VSMCs, but no phenotypes of vascular aging were observed in the EC and macrophage KO mice.

This could indicate that VSMC dysfunction precedes and might trigger endothelial and macrophage dysfunction. Overall, defects in the nuclear lamina may generate genomic instability, and thereby trigger vascular aging. The observed vascular pathology closely mimics the majority of cardiovascular disease symptoms seen in HGPS patients [55].

4.3. Bub1b

BUB1B is involved in mitotic checkpoint regulation and the DNA damage response. Mutations in this gene have been associated with premature aging syndromes, such as mosaic variegated aneuploidy (MVA) syndrome. MVA is a rare genetic disorder characterized by abnormal chromosome numbers (aneuploidy) and mosaic patterns of chromosomal variation within different cells of an individual's body. It is typically associated with severe developmental abnormalities, growth retardation, intellectual disability, and an increased risk of cancer. Studies in the BubR1 (Bub1b) mouse model have demonstrated that these animals display several features of accelerated aging, including in the vasculature.

Matsumoto et al. showed endothelial dysfunction, decreased levels of elastin, fibrosis, and thinning of both the arterial wall and the inner diameter, which was associated with a reduced number of VSMCs [56]. Another study showed increased senescence in the stromal vascular fraction, in particular, the fat progenitor cells [57]. Furthermore, *BubR1* mouse showed impaired angiogenesis [58]. Overall, these findings suggest a role of Bub1b, and thereby genomic instability, in vascular aging in mice, leading to endothelial dysfunction, ECM remodeling, and senescence.

4.4. Telomere Attrition

The Terc mouse is a commonly used model to study the effects of telomerase deficiency on aging and age-related diseases. Telomerase is an enzyme that maintains the length of telomeres, the protective caps at the end of chromosomes. Telomere shortening is a hallmark of cellular aging, and telomerase deficiency has been associated with accelerated aging and increased susceptibility to age-related diseases in humans. When telomeres become too short to function, it will trigger a DNA damage response in order to maintain genome stability.

It is important to mention the difference in telomere length between mice and humans, with mice having much longer telomeres. The generation and background of the Terc-deficient mice are crucial for vascular aging research, as phenotypes differ per generation [59]. The telomeres of the Terc-deficient mice shorten approximately 5 kb per generation. Most experiments are therefore performed in Terc-deficient mice after several generations, as older generations show more defects [60].

Several studies have investigated the effect of telomerase deficiency on vascular aging in Terc-deficient mice. One study found that *Terc*-deficient mice exhibit endothelial dysfunction and senescence, which is restored after antioxidant treatment, indicating a role for oxidative stress [61]. Another study showed that *Terc*-deficient mice experience hypertension as a result of increased endothelin-1 levels [62].

4.5. Epigenetic Alterations

SIRT6 is a protein that plays a crucial role in regulating cellular senescence and age-related cardiovascular diseases. SIRT6 is a histone deacetylase, and is involved in DNA repair by promoting the double strand break repair pathway non-homologous end joining (NHEJ), regulation of the expression of metabolic genes, and maintenance of genomic stability [63]. Studies have shown that SIRT6-deficient mice exhibit premature aging and accelerated vascular aging, characterized by endothelial dysfunction, inflammation, and increased oxidative stress. These mice also display increased susceptibility to age-related diseases, such as atherosclerosis and heart failure.

SIRT1, another deacetylase, protects against endothelial senescence and is involved in DNA damage recognition. SIRT1 decreases with aging, which is associated with endothelial dysfunction. Activators of SIRT1 such as Resveratrol are tested as intervention to prevent vascular aging [64].

SIRT6 is a key regulator of vasomotor function in conduit vessels, via tonic suppression of NAD(P)H oxidase expression and activation. It also contributes to endothelial dysfunction, growth arrest, and senescence. This phenotype can be rescued by overexpressing the transcription factor FOXM1, which is critical for cell cycle progression [65,66]. Another study showed that SIRT6 has a protective role in the vascular system, as downregulation results in vascular calcification [67]. Furthermore, $SIRT6^{+/-}ApoE^{-/-}$ mice form atherosclerotic lesions and express the proinflammatory cytokine VCAM-1 [68]. These findings suggest that Sirt6 plays a critical role in promoting vascular aging as well as in protecting against age-related cardiovascular diseases.

4.6. Limitations of Mouse Models for Vascular Aging

None of the mouse models perfectly resembles human vascular aging. However, Table 1 shows an overview of the vascular phenotypes that are observed for each mouse model and are consistent with human vascular aging. Depending on the research question, the most suitable mouse model can be chosen. The $ERCC1^{d/-}$ and *LMNA* mutant mice are the best characterized models and cell type specific mouse models are designed for both genes to further study the vascular aging.

Nevado et al. described premature vascular aging phenotypes in $Ldlr^{-/-}Lmna^{G609G/G609G}$ and $Apoe^{-/-}Lmna^{G609G/G609G}$ mouse models and observed a difference in atherosclerosis in normal aging and LMNA mice [51]. During normal aging, atheroma plaques start at one point and grow eccentrically, whereas in progeroid mice plaque formation occurs across almost the entire intimal surface, thereby making this mouse model less suitable to study atherosclerosis.

Matsumoto et al. describe the vascular phenotypes of *BubR1* mice, including endothelial dysfunction and VSMC loss (Table 1); however, only male mice were studied in detail [56].

A disadvantage of the *Terc*-deficient mice is that phenotypes differ per generation, making it more difficult to study vascular aging in this mouse model [59]. Furthermore, Terc-deficient mice display a shortened lifespan; however, they still live to over 1 year old, making them more time consuming compared to the other mouse models. On the other hand, *Sirt-6* deficient mice only live to 4 weeks of age, which makes intervention studies more difficult [69].

Table 1. Overview of vascular aging phenotypes in accelerated aging mouse models.

Mouse Model	Gene Targeted	Cell Type	Vascular Phenotypes	Reference
$ERCC1^{d/-}$	ERCC1	Full body	Increased aortic stiffness, blood pressure, aortic senescence, endothelial dysfunction, and decreased reactive hyperemia	[39–41]
ERCC1-Tie2Cre	ERCC1	EC	Decreased endothelium-derived NO, altered vasoconstriction, arterial remodeling, and arterial stiffness	[43]
ERCC1-SM22	ERCC1	SMCs	Diminished NO-mediated vasodilation, decreased reactive hyperemia, and arterial stiffness	[44]
$Lmna^{G609G/G609G}$	LMNA	Full body	Vascular stiffening, ECM remodeling	[48,50]
$Ldlr^{-/-}Lmna^{G609G/G609G}$	LMNA	Full body	VSMC depletion, adventitial thickening, and changes to the elastin structure, atherosclerosis	[51]
$Lmna^{LCS/LCS}SM22\alpha Cre^{+/tg}$	LMNA	SMCs	Vascular stiffening and ECM remodeling	[50]
$Lmna^{LCS/LCS}Tie2Cre^{+/tg}$	LMNA	EC	Unkown	[50]
$Apoe^{-/-}Lmna^{G609G/G609G}$	LMNA	Full body	VSMC depletion, adventitial thickening, and changes to the elastin structure	[55]
$Apoe^{-/-}Lmna^{LCS/LCS}SM22\alpha Cre$	LMNA	SMCs	VSMC depletion, adventitial thickening, and changes to the elastin structure	[55]
$Apoe^{-/-}Lmna^{LCS/LCS}LysMCre$	LMNA	Macrophage	Unkown	[55]
Bub1b	Bub1b	Full body	Endothelial dysfunction, decreased levels of elastin, fibrosis, thinning of both the arterial wall and the inner diameter, VSMC loss, and impaired angiogenesis	[56,58]
Terc	Terc	Full body	Endothelial dysfunction, senescence, and hypertension	[60–62]
Sirt6	Sirt6	Full body	Endothelial dysfunction and vascular calcification	[65–67]
$SIRT6^{+/-};ApoE^{-/-}$	Sirt6	Full body	Atherosclerosis and vascular inflammation	[68]

5. The Role of Genomic Instability in Molecular and Cellular Mechanisms Underlying Vascular Aging

Our DNA is constantly challenged by environmental radiation and chemicals, as well as endogenous metabolic products causing DNA damage resulting in genome instability as we age. When these damages remain unrepaired and accumulate over time [9,10,70], they cause an aging phenotype. This damage to the DNA is further exacerbated in combination with oxidative stress which can trigger apoptosis or senescence, leading to the loss of ECs and VSMCs, thereby resulting in endothelial dysfunction, arterial stiffness, and arteriosclerosis (Figure 4) [71]. Contributors to vascular stiffening such as phenotype switching are also triggered by genomic instability as this changes gene expression in the ECM and vascular cells directly [72].

Genomic instability has been further associated with the dedifferentiation of VSMCs where they lose their contractile phenotype and shift towards a synthetic and osteogenic phenotype. The connecting factor here is the loss of expression of smooth muscle-specific genes, such as alpha-smooth muscle actin (SMA), muscle myosin heavy chain (Myh11), and calponin (Cnn1) (Figure 4) [73]. Instead, dedifferentiated VSMCs acquire markers of synthetic cells, like proliferative factors and extracellular matrix remodeling enzymes, including collagen and fibronectin. An increase in the expression of RUNX2 has also been linked to vascular aging [74,75]. RunX2 promotes calcification and stiffening of the arterial wall by inducing the expression of genes that regulate osteogenic differentiation (mineralization) and extracellular matrix (ECM) remodeling [15]. The ECM is the non-cellular component of the vessel wall that provides structural support to vascular cells and regulates vascular function. Changes in gene expression of ECM remodeling enzymes, such as matrix metalloproteinases (MMPs) and tissue inhibitors of metalloproteinases (TIMPs) affect the overall composition of the ECM causing a decrease in elastin content and an increase in collagen deposition, further contributing to vascular stiffness.

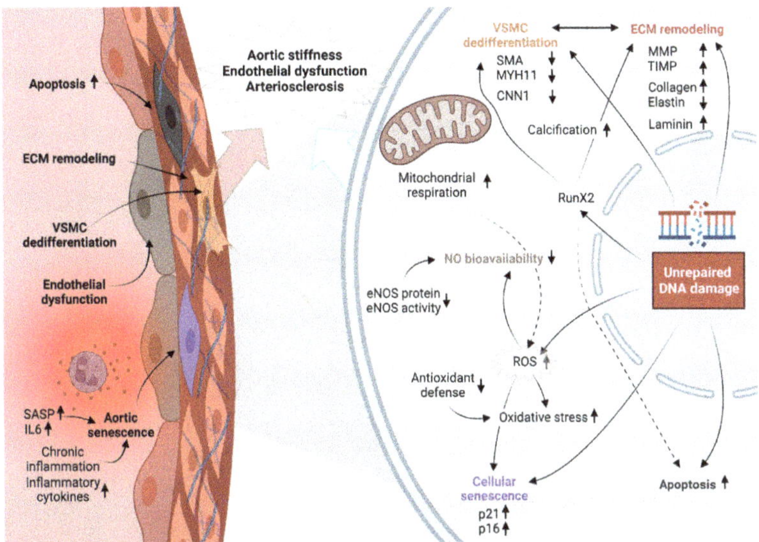

Figure 4. Parameters of vascular aging and the role of genomic instability. Accumulation of DNA damage results in structural and functional changes in the vascular cells and the arterial wall.

Dedifferentiation of VSMCs and ECM remodeling are interconnected processes. Dedifferentiated VSMCs secrete ECM remodeling enzymes, such as MMPs, that degrade the ECM and promote its remodeling. In turn, changes in ECM composition, such as collagen IV and laminin, affect VSMC phenotype and behavior by binding integrin receptors. Interaction between VSMCs, collagen IV, and laminin are important to keep the VSMCs in a quiescent state [76]. Furthermore, the interaction between VSMCs and ECM is affected by hemodynamic stress. Stiffening of the ECM leads to conformational changes in integrins, which plays an essential role in the response to pressure [77].

6. Models to Study Vascular Aging In Vitro

As mentioned earlier, mouse models exist with deficiencies in DNA repair mechanisms. Studying these models aids in comprehending the underlying mechanisms through which genomic instability leads to vascular aging. In order to gain more detailed insights at a cellular level and translate murine findings towards the human condition, it is necessary to utilize human in vitro models. In vitro models offer advantages such as easier scalability and the possibility to examine multiple parameters at the same time. Various approaches can be employed, such as utilizing human (patient derived) DNA repair-deficient vascular cells, as well as introducing endogenous DNA damage to these cells, in order to further investigate the mechanisms involved.

7. Cell Culture Models to Study Vascular Aging

Tissue-specific mouse models have significantly contributed to our understanding of vascular aging. These models are often designed to study specific cell types involved in the vasculature, such as ECs and VSMCs. ECs and VSMCs, in addition to pericytes and fibroblasts, are also commonly used in simple 2D cell culture assays, and, increasingly, also in complex 3D microfluidic tissue systems. Studying vascular aging in cell culture is an important approach to understand the molecular and cellular mechanisms underlying this complex process. Several parameters are commonly evaluated using (human) cell assays. Senescence is a hallmark of vascular aging and can be assessed through experiments such as the SA-β-gal activity assay or by detecting SASP markers [78]. Measuring the proliferation rate of cells can be useful to determine senescence and to investigate the phenotypical switching of VSMCs in vascular aging. Immunofluorescent staining and western blot are used to further evaluate the state of VSMCs, including the phenotype of the VSMCs by detecting, for example, SMA (contractile VSMC), RUNX2 (osteogenic VSMC), MAC2 (macrophage-like VSMC), and SCA1 (mesenchymal-like VSMC).

Structural changes in the ECM are also a hallmark of vascular aging and are studied in vitro by measuring the expression and activity of MMPs or other ECM components including collagen and fibronectin. Endothelial dysfunction, another common feature of vascular aging, is assessed in vitro by measuring the production of NO or endothelin-1 or by monitoring EC migration and proliferation [79]. Chronic low-grade inflammation, which is a characteristic feature of aging, is assessed in vitro by measuring the production of cytokines and chemokines by ECs and VSMCs [8]. In vitro models of vascular calcification, such as culturing VSMCs with calcifying media, are used to study the cellular and molecular mechanisms underlying this process. Vascular stiffening is evaluated by measuring the contraction force of VSMCs, displacement of the VSMCs relative to a fluorescent microbead, or by assessing migration defects in the cytoskeleton. Tube formation assays, where cells are grown on matrigel to form a tube structure, are used to evaluate EC culture as a model system of vascular aging [80].

Changes in cellular metabolism are another important hallmark of aging. Mitochondrial function is typically assessed using the Seahorse XF Analyzer (Agilent, Santa Clara, CA, USA), which measures cellular respiration and glycolysis in cell culture. Furthermore, gene expression analysis is often combined with the previously mentioned experiments to identify genes that are differentially expressed in aged versus young cells in 2D culture. These genes include markers for senescence, immune infiltration, phenotypical switching of VSMCs, and DNA damage.

Limitations Cell Culture Models to Study Vascular Aging

Overall, cell culture methods provide insight into cellular processes involved in aspects of vascular aging. However, in vitro culturing of cells can induce chemical and physical alterations, thereby not fully reflecting all processes that are involved in vascular aging [22]. Moreover, the vasculature comprises diverse cell types that engage in constant interactions and communication. Restricting studies to a single isolated cell type may oversimplify the complex nature of vascular aging. To better replicate the intricate dynamics of vascular aging, advanced microfluidic cell culture systems are employed. These systems offer a more comprehensive approach to investigate the complex interactions among different cell types within the vasculature. By integrating multiple cell types and simulating physiological conditions, microfluidic platforms provide a valuable tool for studying the multifaceted aspects of vascular aging.

8. Microfluidic Cell Culture Systems and Vascular Aging

The field of microfluidic technology is rapidly advancing, offering increasingly biomimetic in vitro models to complement animal models. Vascular microfluidic devices, which mimic blood flow, are highly adaptable in terms of shape, surface coating, and network complexity. These devices will be a great addition in studying vascular aging where cells with specific aging related mutations can be used to mimic vascular aging. While traditional in vitro culture models are comparatively simplistic, microfluidic models do require a certain level of complexity in their fabrication process. Co-culturing cells of different origins poses a challenge, especially when combined in a 3D culture setting, but is essential to capture the interactions between different cell types, especially in the context of the vascular system. The advantages of microfluidic in vitro systems include their introduction of hemodynamic stimuli (such as shear stress) and interaction with circulating cells, as well as their compatibility with multiple analytical techniques and high throughput data generation and processing.

Microfluidic systems aim to recapitulate the mechanical, electrical, chemical, and structural features of the in vivo microenvironment. Compared to current in vivo or 2D culture conditions, they allow for more precise control of factors such as blood flow, blood pressure, and chemical gradients while providing a reproducible system. Microfluidic cell culture systems can help generate valuable insights by measuring cellular changes with age such as proliferation, migration, or differentiation.

In this review, we will describe four different strategies of microfluidic modelling as depicted in Figure 5, namely, microfabricated channels (a), microfluidic vessel systems (b), self-assembly networks (c), and cell patterning microfluidics (d).

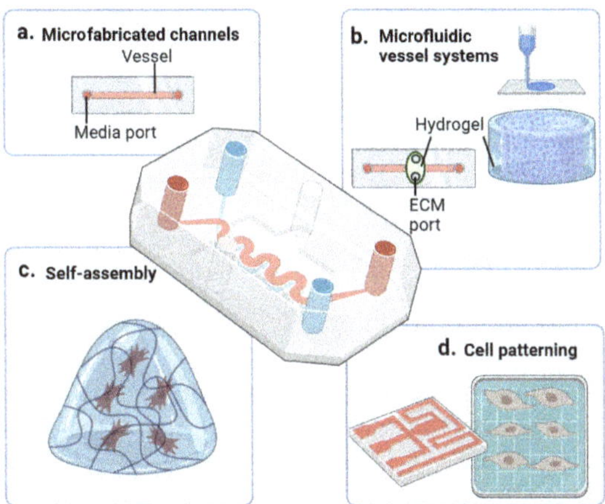

Figure 5. Schematic representation of microfluidic fabrication methods. (a) Microfabricated fluid-channels. (**b**) Microfabricated blood vessel systems, e.g., bioprinting and sacrificial molds. (c) Self-assembled network within hydrogel. (**d**) Patterned microfluidic networks.

8.1. Microfabricated Channels

Microfabricated channels are the most basic form of microfluidic modelling (Figure 5a), yet they still provide a more physiologically relevant environment for cells compared to traditional 2D cell culture models. To create artificial vessels that mimic the specific dimensions and geometries of blood vessels, small channels can be etched or molded into biocompatible substrates like glass or synthetic polymers. Glass, silicon, and polymers are substrates that are often used to make a channel surface to replace natural ECM for cell attachment. These systems are best suited to evaluate the endothelial response to variations in flow (shear stress) and can be coupled to the analysis of endothelium derived factors released in the circulating medium.

An example of such a system is the microvascular network of Borenstein et al., who created circular polystyrene channels with bifurcation in which they seeded ECs to form a monolayer, with the ECs remaining viable for at least 24 h [81].

Another example is the use of glass etching as shown by Rodriguez et al. In one of their approaches, they use a polydimethylsiloxane (PDMS) stamp to form the desired structures on a glass slide [82]. Once the scaffold is complete, ECs are cultured on a specified membrane, in a designed pattern. However, culture on the stiff surfaces of synthetic materials could be detrimental to natural vascular cell behavior and function.

8.2. Microfluidic Vessel Models

Recent advances in bio-fabrication techniques allow the engineering of complex vascular networks in vitro that aims to mimic in vivo vasculature as closely as possible [83]. Microfluidic vessel systems (Figure 5b) approximate the structure and function of blood vessels even closer than microfabricated channels. These systems are generated using materials such as (fully synthetic or semi synthetic) hydrogels and are designed to have specific dimensions and geometries that mimic the network structure of blood vessels (Figure 5b) [84]. These models allow for co-cultures of, e.g., ECs with VSMCs, pericytes, and fibroblasts in a 3D extracellular matrix environment. An example of such a model is van Dijk et al.'s PDMS device with a reservoir for a 3D fibrinogen gel with pericytes. This multi-cell type device allows the interaction of ECs with pericytes in full bio-matrix encased 3D vessel structures (neovessels) that can be subjected to continuous, unidirectional flow and perfusion with circulating immune cells [85]. Another example is presented by the research of Polacheck et al. in which they designed and fabricated microfluidic channels with dimensions mimicking small blood vessels. Similar to the van Dijk system, the channel surface provides the ECM environment for EC attachment and expansion to replicate the vascular endothelium. The fabricated microchannels are used to investigate the effects of fluid flow and shear stress on ECs, providing insights into vascular physiology and pathology [86]. Such systems are often produced as small units that can be flowed in serial or parallel configuration and can be scaled up to study many conditions in high throughput fashion.

8.3. Microfluidics Combined with Self-Assembly

The self-assembly (Figure 5c) technique differs from the other methods as it does not require a pre-designed structure to direct the growth of cells. For this approach VSMCs are seeded into hydrogel, where conditions are optimized to stimulate spontaneous sprouting and 3D vascular network formation, more closely mimicking the in vivo development of vessels.

An example of this system is the microfluidic device of Wang et al. [87], which uses two microfluidic channels which are coated with laminin and lined with a monolayer of ECs. These channels (representing artery and vein) are connected by a middle tissue chamber filled with an ECM-based hydrogel, in which a new capillary network will form via sprouting of the endothelium in the channels. This system can be perfused with only minimal leakage. Another self-assembly system is described by Vila Cuenca et al. in which multiple cell types are seeded in a fibrin hydrogel microenvironment. In this system, human iPSC-derived EC and VSMCs are used, with cells self-organizing in the fibrin matrix into lumenized and functional vessels. Notably, the created vessel responds to vasoactive stimulation, which can be measured [88].

8.4. Cell Patterning

Dependent on the parameters under investigation it can be advantageous to apply cell patterning rather than self-assembly. In the field of tissue engineering, the strategy of cell patterning (often) involves material driven solutions that allow predefining cell organization within an "aggregate" module, with multiple modules constructing the engineered tissue. On a higher design level, cell patterning can also involve the organization of either self-assembled cell units or "patterned" modules into a predesigned configuration to mimic organ structures or to allow more effective self-assembly into higher organization structures (Figure 5d). An example of this strategy is provided by Garvin et al., who used an ultrasound standing wave field to induce patterning of cells in an engineered tissue construct [89]. This technique can determine the speed of formation, density of the cells, and the morphology of the vascular network [90]. Another example of cell patterning is the use of a PDMS stamp to induce cell elongation. These stamps contain strips of fibronectin in a microfluidic channel. Cells are seeded and non-adhesive cells are washed off, resulting in alignment of the cells in one direction [91].

8.5. Scientific Outcomes Microfluidic Systems

Various applications have emerged for utilizing vasculature-on-a-chip in in vitro disease studies, particularly focusing on endothelial dysfunction. This vascular condition involves impaired vasomotion control, linked to increased inflammatory responses and vascular leakage. While chip systems mainly investigate inflammatory responses, thrombogenic reactions, flow disturbances, and endothelial permeability in endothelial dysfunction, there is a gap in exploring endothelial-dependent vasomotion dysregulation [89,92,93]. This gap is mainly due to the high stiffness of chip materials, limiting the contractile behavior of VMSCs. Adjusting substrate stiffness could pave the way for future chip designs addressing the defining aspect of this vascular disease. Another area of research in which vascular chips have proven useful is cancer research. Tumor angiogenesis signifies the orchestrated development of new blood vessels crucial for supplying nutrients and oxygen to cancer cells. Integral to tumor progression, it involves the formation of aberrant and intricate blood vessel networks within and surrounding the tumor tissue. These processes have been investigated with the use of various chip systems via tumor and vascular cell co-cultures [94–96]. Similarly, cancer metastasis, representing the spread of tumor cells from the primary tumor site to distant organs or tissues, has been studied in complex multi-organ-chip systems incorporating vascular cells to mimic trans-endothelial migration [97,98]. This multi-organ approach may provide a more representative system for drug screening, which has, thus far, used less complex microfluidic systems for visualizing biological and mechanical dynamic processes. These processes include adaptations in cell–cell junction stability, viability, and proliferation rate, as well as vascular cell response to altered shear stress when exploring the efficacy of pharmaceutical compounds [99–101].

Future chips should explore the integration of vasomotion regulation, complex perfused vessel building, and co-culture with tissue/tumor cells in multi-organ designs, in combination with an increase in scale and reproducibility. This will allow further implementation in pharmacological studies and application in wider research fields, including (vascular) aging.

8.6. Comparison Microfluidic Systems

Microfabricated channels are more straightforward compared to other systems, making them quicker and easier to use. They often use only one cell type, while the other systems allow for a co-culture, closely mimicking the in vivo situation. The other three systems allow the formation of multiple layers of cells in ECM. Microfluidic vessel systems focus on replicating the structure and function of blood vessels or vascular networks, primarily for vascular biology studies and tissue engineering applications. This system works very well for co-culturing larger structures and is, thereby, suitable to study arterioles. The self-assembly microfluidic system is more suitable for investigating the capillary system. Cell patterning microfluidic systems, on the other hand, are designed to precisely pattern and arrange cells, enabling the study of cell behavior, tissue development, and organ-on-a-chip platforms. A disadvantage of this system is that the majority of these only have a limited throughput [102].

9. Personalized Medicine

The use of cell culture and microfluidic devices also allows for precision medicine, in which we consider individual differences such as gender, genetics, and lifestyle [103,104]. With precision medicine, a large group of patients can be divided into clusters based on patient characteristics such as clinical phenotype, environment, and molecular analysis, and receive specialized treatment. Examples of how cell experiments and microfluidic devices can be used in future are by identifying new biomarkers, predict clinical outcome, and determine disease susceptibility.

Disease modelling and prevention based on human cells have the advantage that they do not require translation from a mammalian model organism to the human situation, and they can also potentially be personalized for each patient. This personalization of medicine can be facilitated using human induced pluripotent stem cells (hiPSC) derived from healthy individuals or patients. These hiPSCs can be utilized to study disease mechanisms and develop targeted therapies. By using CRISPR Cas9 technology, mutations causing DNA repair deficiencies can be introduced into control hiPSCs, allowing for the investigation of different mutations in an isogenic background. By differentiating these iPSCs into various cell types, such as human ECs, pericytes, and VSMCs, researchers can study the underlying mechanisms of genomic instability in vascular cells. The mutations of the previously mentioned mouse models are good candidate genes to target. Once the necessary cell types are generated, they can be transferred into microfluidic systems to compare specific drug responses within the controlled environment of a high throughput system [105,106]. The use of microfluidics systems can also be advantageous to model human age-related disorders that may manifest differently in humans compared to model organisms. Microfluidic systems are already used in preclinical drug testing. Luna et al. use a microfluidic device to assess blood thrombosis and study anti-thrombotic therapy [107]. They showed blood clotting under flow in tortious arteriolar vessels. Another advantage of using iPSCs in microfluidic devices is the opportunity to test the effect of a mutation in a specific cell type. This information can be used to determine which mouse models to develop. Studying vascular aging in a microfluidic system will be a new and interesting step forward in the use of microfluidic models. However, limitations exist for the use of microfluidic systems. A disadvantage of microfluidic systems compared to in vivo systems is the lack of whole-body integration, such as the lack of hormone circulation, an immune system, and the influence of the central nerve system. Furthermore, microfluidic systems are limited scale; the size of microfluidic vessel-on-a-chip systems is much smaller than actual blood vessels in the human body.

10. Conclusions

In summary, vascular aging is linked to genomic instability, as observed in patients with progeria. To better understand age-related vascular changes, researchers have developed mouse models with genome instability that exhibit accelerated vascular aging. These models offer an efficient experimental system to investigate the mechanisms, pathways, and interventions associated with vascular aging. They provide time efficiency, facilitate the discovery of molecular pathways, allow for intervention testing, and offer translational relevance and biomarker discovery, enhancing our understanding of vascular aging and potential therapeutic approaches. However, it is important to note that there are prominent differences between progeroid mice and naturally aged mice or humans. To address this, human vascular cells can be utilized in cell culture studies and microfluidic systems. High complexity microfluidic cell culture systems are powerful tools to study vascular aging as they resemble the structure and function of blood vessels and provide the possibility to work with human cells of the vessel wall. These systems provide precise control over the experimental environment and the cells under investigation. Combining research of microfluidic systems with accelerated aging mouse models offers the opportunity to uncover the underlying mechanisms of vascular aging, bridging the gap between mouse models and human.

Author Contributions: Conceptualization, J.v.d.L., I.v.d.P. and C.C.; writing—original draft preparation J.v.d.L., L.T., C.V.S., I.v.d.P. and C.C; writing—review and editing J.v.d.L., I.v.d.P., C.C., A.H.J.D. and A.J.M.R.; Visualization J.v.d.L., L.T. and C.V.S.; Supervision I.v.d.P., C.C., A.H.J.D. and A.J.M.R. All authors have read and agreed to the published version of the manuscript.

Funding: This research was funded by Erasmus Medical Center human disease model award 2018.

Acknowledgments: Figures were created using BioRender.com accessed on 29 August 2023.

Conflicts of Interest: The authors declare no conflict of interest.

References

1. Vos, T.; Lim, S.S.; Abbafati, C.; Abbas, K.M.; Abbasi, M.; Abbasifard, M.; Abbasi-Kangevari, M.; Abbastabar, H.; Abd-Allah, F.; Abdelalim, A.; et al. Global Burden of 369 Diseases and Injuries in 204 Countries and Territories, 1990–2019: A Systematic Analysis for the Global Burden of Disease Study 2019. *Lancet* **2020**, *396*, 1204–1222. [CrossRef]
2. Roth, G.A.; Mensah, G.A.; Johnson, C.O.; Addolorato, G.; Ammirati, E.; Baddour, L.M.; Barengo, N.C.; Beaton, A.Z.; Benjamin, E.J.; Benziger, C.P.; et al. Global Burden of Cardiovascular Diseases and Risk Factors, 1990-2019: Update From the GBD 2019 Study. *J. Am. Coll. Cardiol* **2020**, *76*, 2982–3021. [CrossRef] [PubMed]
3. Reference Values for Arterial Stiffness' Collaboration. Determinants of Pulse Wave Velocity in Healthy People and in the Presence of Cardiovascular Risk Factors: 'Establishing Normal and Reference Values'. *Eur. Heart. J.* **2010**, *31*, 2338–2350. [CrossRef] [PubMed]
4. Vlachopoulos, C.; Aznaouridis, K.; Stefanadis, C. Prediction of Cardiovascular Events and All-Cause Mortality with Arterial Stiffness: A Systematic Review and Meta-Analysis. *J. Am. Coll. Cardiol.* **2010**, *55*, 1318–1327. [CrossRef] [PubMed]
5. Fuchs, F.D.; Whelton, P.K. High Blood Pressure and Cardiovascular Disease. *Hypertension* **2020**, *75*, 285–292. [CrossRef]

6. Lakatta, E.G. Arterial and Cardiac Aging: Major Shareholders in Cardiovascular Disease Enterprises: Part III: Cellular and Molecular Clues to Heart and Arterial Aging. *Circulation* **2003**, *107*, 490–497. [CrossRef]
7. AlGhatrif, M.; Lakatta, E.G. The Conundrum of Arterial Stiffness, Elevated Blood Pressure, and Aging. *Curr. Hypertens. Rep.* **2015**, *17*, 12. [CrossRef] [PubMed]
8. Ungvari, Z.; Tarantini, S.; Donato, A.J.; Galvan, V.; Csiszar, A. Mechanisms of Vascular Aging. *Circ. Res.* **2018**, *123*, 849–867. [CrossRef] [PubMed]
9. López-Otín, C.; Blasco, M.A.; Partridge, L.; Serrano, M.; Kroemer, G. Hallmarks of Aging: An Expanding Universe. *Cell* **2023**, *186*, 243–278. [CrossRef]
10. Lopez-Otin, C.; Blasco, M.A.; Partridge, L.; Serrano, M.; Kroemer, G. The Hallmarks of Aging. *Cell* **2013**, *153*, 1194–1217. [CrossRef]
11. Thijssen, D.H.J.; Carter, S.E.; Green, D.J. Arterial Structure and Function in Vascular Ageing: Are You as Old as Your Arteries? *J. Physiol.* **2016**, *594*, 2275–2284. [CrossRef] [PubMed]
12. Dai, X.; Hummel, S.L.; Salazar, J.B.; Taffet, G.E.; Zieman, S.; Schwartz, J.B. Cardiovascular Physiology in the Older Adults. *J. Geriatr. Cardiol.* **2015**, *12*, 196–201. [CrossRef]
13. Lüscher, T.F.; Boulanger, C.M.; Dohi, Y.; Yang, Z.H. Endothelium-Derived Contracting Factors. *Hypertension* **1992**, *19*, 117–130. [CrossRef] [PubMed]
14. Sorokin, V.; Vickneson, K.; Kofidis, T.; Woo, C.C.; Lin, X.Y.; Foo, R.; Shanahan, C.M. Role of Vascular Smooth Muscle Cell Plasticity and Interactions in Vessel Wall Inflammation. *Front. Immunol.* **2020**, *11*, 599415. [CrossRef]
15. Cao, G.; Xuan, X.; Hu, J.; Zhang, R.; Jin, H.; Dong, H. How Vascular Smooth Muscle Cell Phenotype Switching Contributes to Vascular Disease. *Cell. Commun. And. Signal.* **2022**, *20*, 180. [CrossRef]
16. Yap, C.; Mieremet, A.; de Vries, C.J.M.; Micha, D.; de Waard, V. Six Shades of Vascular Smooth Muscle Cells Illuminated by KLF4 (Krüppel-Like Factor 4). *Arterioscler. Thromb. Vasc. Biol.* **2021**, *41*, 2693–2707. [CrossRef]
17. Tang, H.-Y.; Chen, A.-Q.; Zhang, H.; Gao, X.-F.; Kong, X.-Q.; Zhang, J.-J. Vascular Smooth Muscle Cells Phenotypic Switching in Cardiovascular Diseases. *Cells* **2022**, *11*, 4060. [CrossRef] [PubMed]
18. Rensen, S.S.M.; Doevendans, P.A.F.M.; van Eys, G.J.J.M. Regulation and Characteristics of Vascular Smooth Muscle Cell Phenotypic Diversity. *Neth. Heart J.* **2007**, *15*, 100–108. [CrossRef]
19. Zhao, D.; Li, J.; Xue, C.; Feng, K.; Liu, L.; Zeng, P.; Wang, X.; Chen, Y.; Li, L.; Zhang, Z.; et al. TL1A Inhibits Atherosclerosis in ApoE-Deficient Mice by Regulating the Phenotype of Vascular Smooth Muscle Cells. *J. Biol. Chem.* **2020**, *295*, 16314–16327. [CrossRef]
20. Durham, A.L.; Speer, M.Y.; Scatena, M.; Giachelli, C.M.; Shanahan, C.M. Role of Smooth Muscle Cells in Vascular Calcification: Implications in Atherosclerosis and Arterial Stiffness. *Cardiovasc. Res.* **2018**, *114*, 590–600. [CrossRef]
21. Depuydt, M.A.C.; Prange, K.H.M.; Slenders, L.; Örd, T.; Elbersen, D.; Boltjes, A.; de Jager, S.C.A.; Asselbergs, F.W.; de Borst, G.J.; Aavik, E.; et al. Microanatomy of the Human Atherosclerotic Plaque by Single-Cell Transcriptomics. *Circ. Res.* **2020**, *127*, 1437–1455. [CrossRef] [PubMed]
22. Rzucidlo, E.M.; Martin, K.A.; Powell, R.J. Regulation of Vascular Smooth Muscle Cell Differentiation. *J. Vasc. Surg.* **2007**, *45*, A25–A32. [CrossRef] [PubMed]
23. Chistiakov, D.A.; Orekhov, A.N.; Bobryshev, Y.V. Vascular Smooth Muscle Cell in Atherosclerosis. *Acta Physiol.* **2015**, *214*, 33–50. [CrossRef] [PubMed]
24. Gorog, P.; Kovacs, I.B. Inhibition of Vascular Smooth Muscle Cell Migration by Intact Endothelium Is Nitric Oxide-Mediated: Interference by Oxidised Low Density Lipoproteins. *J. Vasc. Res.* **1998**, *35*, 165–169. [CrossRef]
25. Sarkar, R.; Meinberg, E.G.; Stanley, J.C.; Gordon, D.; Clinton Webb, R. Nitric Oxide Reversibly Inhibits the Migration of Cultured Vascular Smooth Muscle Cells. *Circ. Res.* **1996**, *78*, 225–230. [CrossRef] [PubMed]
26. van den Oever, I.A.M.; Raterman, H.G.; Nurmohamed, M.T.; Simsek, S. Endothelial Dysfunction, Inflammation, and Apoptosis in Diabetes Mellitus. *Mediators. Inflamm.* **2010**, *2010*, 792393. [CrossRef]
27. Ataei Ataabadi, E.; Golshiri, K.; Jüttner, A.; Krenning, G.; Danser, A.H.J.; Roks, A.J.M. Nitric Oxide-CGMP Signaling in Hypertension. *Hypertension* **2020**, *76*, 1055–1068. [CrossRef]
28. Graziano, S.; Kreienkamp, R.; Coll-Bonfill, N.; Gonzalo, S. Causes and Consequences of Genomic Instability in Laminopathies: Replication Stress and Interferon Response. *Nucleus* **2018**, *9*, 258–275. [CrossRef]
29. Gonzalo, S. DNA Damage and Lamins. In *Cancer Biology and the Nuclear Envelope*; Springer: Berlin/Heidelberg, Germany, 2014; pp. 377–399. [CrossRef]
30. Olive, M.; Harten, I.; Mitchell, R.; Beers, J.K.; Djabali, K.; Cao, K.; Erdos, M.R.; Blair, C.; Funke, B.; Smoot, L.; et al. Cardiovascular Pathology in Hutchinson-Gilford Progeria: Correlation with the Vascular Pathology of Aging. *Arterioscler. Thromb. Vasc. Biol.* **2010**, *30*, 2301–2309. [CrossRef] [PubMed]
31. Ragnauth, C.D.; Warren, D.T.; Liu, Y.; McNair, R.; Tajsic, T.; Figg, N.; Shroff, R.; Skepper, J.; Shanahan, C.M. Prelamin A Acts to Accelerate Smooth Muscle Cell Senescence and Is a Novel Biomarker of Human Vascular Aging. *Circulation* **2010**, *121*, 2200–2210. [CrossRef]
32. Merideth, M.A.; Gordon, L.B.; Clauss, S.; Sachdev, V.; Smith, A.C.M.; Perry, M.B.; Brewer, C.C.; Zalewski, C.; Kim, H.J.; Solomon, B.; et al. Phenotype and Course of Hutchinson–Gilford Progeria Syndrome. *N. Engl. J. Med.* **2008**, *358*, 592–604. [CrossRef]
33. Brosh, R.M. DNA Helicases Involved in DNA Repair and Their Roles in Cancer. *Nat. Rev. Cancer* **2013**, *13*, 542–558. [CrossRef]

34. Valero, A.; Gellei, B. Retinitis Pigmentosa, Hypertension, and Uraemia in Werner's Syndrome. *BMJ* **1960**, *2*, 351–354. [CrossRef] [PubMed]
35. Folgueras, A.R.; Freitas-Rodríguez, S.; Velasco, G.; López-Otín, C. Mouse Models to Disentangle the Hallmarks of Human Aging. *Circ. Res.* **2018**, *123*, 905–924. [CrossRef] [PubMed]
36. Vanhooren, V.; Libert, C. The Mouse as a Model Organism in Aging Research: Usefulness, Pitfalls and Possibilities. *Ageing Res. Rev* **2013**, *12*, 8–21. [CrossRef]
37. Marteijn, J.A.; Lans, H.; Vermeulen, W.; Hoeijmakers, J.H.J. Understanding Nucleotide Excision Repair and Its Roles in Cancer and Ageing. *Nat. Rev. Mol. Cell. Biol* **2014**, *15*, 465–481. [CrossRef] [PubMed]
38. Gillet, L.C.; Scharer, O.D. Molecular Mechanisms of Mammalian Global Genome Nucleotide Excision Repair. *Chem. Rev.* **2006**, *106*, 253–276. [CrossRef] [PubMed]
39. Durik, M.; Kavousi, M.; van der Pluijm, I.; Isaacs, A.; Cheng, C.; Verdonk, K.; Loot, A.E.; Oeseburg, H.; Bhaggoe, U.M.; Leijten, F. Nucleotide Excision DNA Repair Is Associated with Age-Related Vascular Dysfunction. *Circulation* **2012**, *126*, 468–478. [CrossRef]
40. Weeda, G.; Donker, I.; de Wit, J.; Morreau, H.; Janssens, R.; Vissers, C.J.; Nigg, A.; van Steeg, H.; Bootsma, D.; Hoeijmakers, J.H.J. Disruption of Mouse ERCC1 Results in a Novel Repair Syndrome with Growth Failure, Nuclear Abnormalities and Senescence. *Curr. Biol.* **1997**, *7*, 427–439. [CrossRef]
41. Yousefzadeh, M.J.; Zhao, J.; Bukata, C.; Wade, E.A.; McGowan, S.J.; Angelini, L.A.; Bank, M.P.; Gurkar, A.U.; McGuckian, C.A.; Calubag, M.F.; et al. Tissue Specificity of Senescent Cell Accumulation during Physiologic and Accelerated Aging of Mice. *Aging Cell* **2020**, *19*, e13094. [CrossRef]
42. Dollé, M.E.T.; Kuiper, R.V.; Roodbergen, M.; Robinson, J.; de Vlugt, S.; Wijnhoven, S.W.P.; Beems, R.B.; de la Fonteyne, L.; de With, P.; van der Pluijm, I.; et al. Broad Segmental Progeroid Changes in Short-Lived Ercc1 −/Δ7 Mice. *Pathobiol. Aging Age-Relat. Dis.* **2011**, *1*, 7219. [CrossRef]
43. Bautista-Niño, P.K.; Portilla-Fernandez, E.; Rubio-Beltrán, E.; van der Linden, J.J.; de Vries, R.; van Veghel, R.; de Boer, M.; Durik, M.; Ridwan, Y.; Brandt, R.; et al. Local Endothelial DNA Repair Deficiency Causes Aging-Resembling Endothelial-Specific Dysfunction. *Clin. Sci.* **2020**, *134*, 727–746. [CrossRef] [PubMed]
44. Ataei Ataabadi, E.; Golshiri, K.; Van Der Linden, J.; De Boer, M.; Duncker, D.J.; Jüttner, A.; De Vries, R.; Van Veghel, R.; Van Der Pluijm, I.; Dutheil, S.; et al. Vascular Ageing Features Caused by Selective DNA Damage in Smooth Muscle Cell. *Oxid. Med. Cell. Longev.* **2021**, *2021*, 2308317. [CrossRef] [PubMed]
45. Kashiyama, K.; Nakazawa, Y.; Pilz, D.T.; Guo, C.; Shimada, M.; Sasaki, K.; Fawcett, H.; Wing, J.F.; Lewin, S.O.; Carr, L.; et al. Malfunction of Nuclease ERCC1-XPF Results in Diverse Clinical Manifestations and Causes Cockayne Syndrome, Xeroderma Pigmentosum, and Fanconi Anemia. *Am. J. Hum. Genet.* **2013**, *92*, 807–819. [CrossRef] [PubMed]
46. Jaspers, N.G.J.; Raams, A.; Silengo, M.C.; Wijgers, N.; Niedernhofer, L.J.; Robinson, A.R.; Giglia-Mari, G.; Hoogstraten, D.; Kleijer, W.J.; Hoeijmakers, J.H.J.; et al. First Reported Patient with Human ERCC1 Deficiency Has Cerebro-Oculo-Facio-Skeletal Syndrome with a Mild Defect in Nucleotide Excision Repair and Severe Developmental Failure. *Am. J. Hum. Genet.* **2007**, *80*, 457–466. [CrossRef]
47. Baer, S.; Obringer, C.; Julia, S.; Chelly, J.; Capri, Y.; Gras, D.; Baujat, G.; Felix, T.M.; Doray, B.; Sanchez del Pozo, J.; et al. Early-onset Nucleotide Excision Repair Disorders with Neurological Impairment: Clues for Early Diagnosis and Prognostic Counseling. *Clin. Genet.* **2020**, *98*, 251–260. [CrossRef]
48. Osorio, F.G.; Navarro, C.L.; Cadiñanos, J.; López-Mejía, I.C.; Quirós, P.M.; Bartoli, C.; Rivera, J.; Tazi, J.; Guzmán, G.; Varela, I.; et al. Splicing-Directed Therapy in a New Mouse Model of Human Accelerated Aging. *Sci. Transl. Med.* **2011**, *3*, 106–107. [CrossRef]
49. Cabral, W.A.; Tavarez, U.L.; Beeram, I.; Yeritsyan, D.; Boku, Y.D.; Eckhaus, M.A.; Nazarian, A.; Erdos, M.R.; Collins, F.S. Genetic Reduction of MTOR Extends Lifespan in a Mouse Model of Hutchinson-Gilford Progeria Syndrome. *Aging. Cell* **2021**, *20*, e13457. [CrossRef]
50. del Campo, L.; Sánchez-López, A.; Salaices, M.; von Kleeck, R.A.; Expósito, E.; González-Gómez, C.; Cussó, L.; Guzmán-Martínez, G.; Ruiz-Cabello, J.; Desco, M.; et al. Vascular Smooth Muscle Cell-specific Progerin Expression in a Mouse Model of Hutchinson–Gilford Progeria Syndrome Promotes Arterial Stiffness: Therapeutic Effect of Dietary Nitrite. *Aging. Cell* **2019**, *18*, e12936. [CrossRef]
51. Nevado, R.M.; Hamczyk, M.R.; Gonzalo, P.; Andrés-Manzano, M.J.; Andrés, V. Premature Vascular Aging with Features of Plaque Vulnerability in an Atheroprone Mouse Model of Hutchinson–Gilford Progeria Syndrome with Ldlr Deficiency. *Cells* **2020**, *9*, 2252. [CrossRef]
52. Coll-Bonfill, N.; Mahajan, U.; Shashkova, E.V.; Lin, C.-J.; Mecham, R.P.; Gonzalo, S. Progerin Induces a Phenotypic Switch in Vascular Smooth Muscle Cells and Triggers Replication Stress and an Aging-Associated Secretory Signature. *Geroscience* **2023**, *45*, 965–982. [CrossRef]
53. SH, Y.; MO, B.; JI, T.; Qiao, X.; Hu, Y.; Sandoval, S.; Meta, M.; Bendale, P.; MH, G.; SG, Y.; et al. Blocking protein farnesyltransferase improves nuclear blebbing in mouse fibroblasts with a targeted Hutchinson-Gilford progeria syndrome mutation. *Proc. Natl. Acad. Sci. USA* **2005**, *102*, 10291–10296. [CrossRef]
54. Benedicto, I.; Dorado, B.; Andrés, V. Molecular and Cellular Mechanisms Driving Cardiovascular Disease in Hutchinson-Gilford Progeria Syndrome: Lessons Learned from Animal Models. *Cells* **2021**, *10*, 1157. [CrossRef]

55. Hamczyk, M.R.; Villa-Bellosta, R.; Gonzalo, P.; Andrés-Manzano, M.J.; Nogales, P.; Bentzon, J.F.; López-Otín, C.; Andrés, V. Vascular Smooth Muscle–Specific Progerin Expression Accelerates Atherosclerosis and Death in a Mouse Model of Hutchinson-Gilford Progeria Syndrome. *Circulation* **2018**, *138*, 266–282. [CrossRef] [PubMed]

56. Matsumoto, T.; Baker, D.J.; d'Uscio, L.V.; Mozammel, G.; Katusic, Z.S.; van Deursen, J.M. Aging-Associated Vascular Phenotype in Mutant Mice with Low Levels of BubR1. *Stroke* **2007**, *38*, 1050–1056. [CrossRef]

57. Baker, D.J.; Weaver, R.L.; van Deursen, J.M. P21 Both Attenuates and Drives Senescence and Aging in BubR1 Progeroid Mice. *Cell. Rep* **2013**, *3*, 1164–1174. [CrossRef]

58. Okadome, J.; Matsumoto, T.; Yoshiya, K.; Matsuda, D.; Tamada, K.; Onimaru, M.; Nakano, K.; Egashira, K.; Yonemitsu, Y.; Maehara, Y. BubR1 Insufficiency Impairs Angiogenesis in Aging and in Experimental Critical Limb Ischemic Mice. *J. Vasc. Surg.* **2018**, *68*, 576–586.e1. [CrossRef]

59. Wong, L.S.M.; Oeseburg, H.; de Boer, R.A.; van Gilst, W.H.; van Veldhuisen, D.J.; van der Harst, P. Telomere Biology in Cardiovascular Disease: The TERC-/- Mouse as a Model for Heart Failure and Ageing. *Cardiovasc. Res.* **2008**, *81*, 244–252. [CrossRef]

60. Blasco, M.A.; Lee, H.-W.; Hande, M.P.; Samper, E.; Lansdorp, P.M.; DePinho, R.A.; Greider, C.W. Telomere Shortening and Tumor Formation by Mouse Cells Lacking Telomerase RNA. *Cell* **1997**, *91*, 25–34. [CrossRef] [PubMed]

61. Bhayadia, R.; Schmidt, B.M.W.; Melk, A.; Hömme, M. Senescence-Induced Oxidative Stress Causes Endothelial Dysfunction. *J. Gerontol. A Biol. Sci. Med. Sci.* **2016**, *71*, 161–169. [CrossRef] [PubMed]

62. Pérez-Rivero, G.; Ruiz-Torres, M.P.; Rivas-Elena, J.V.; Jerkic, M.; Díez-Marques, M.L.; Lopez-Novoa, J.M.; Blasco, M.A.; Rodríguez-Puyol, D. Mice Deficient in Telomerase Activity Develop Hypertension Because of an Excess of Endothelin Production. *Circulation* **2006**, *114*, 309–317. [CrossRef]

63. Klein, M.A.; Denu, J.M. Biological and Catalytic Functions of Sirtuin 6 as Targets for Small-Molecule Modulators. *J. Biol. Chem.* **2020**, *295*, 11021–11041. [CrossRef] [PubMed]

64. Labinskyy, N.; Csiszar, A.; Veress, G.; Stef, G.; Pacher, P.; Oroszi, G.; Wu, J.; Ungvari, Z. Vascular Dysfunction in Aging: Potential Effects of Resveratrol, an Anti- Inflammatory Phytoestrogen. *Curr. Med. Chem.* **2006**, *13*, 989–996. [CrossRef]

65. Greiten, L.E.; Zhang, B.; Roos, C.M.; Hagler, M.; Jahns, F.-P.; Miller, J.D. Sirtuin 6 Protects Against Oxidative Stress and Vascular Dysfunction in Mice. *Front. Physiol.* **2021**, *12*, 753501. [CrossRef]

66. Lee, O.-H.; Woo, Y.M.; Moon, S.; Lee, J.; Park, H.; Jang, H.; Park, Y.-Y.; Bae, S.-K.; Park, K.-H.; Heo, J.H.; et al. Sirtuin 6 Deficiency Induces Endothelial Cell Senescence via Downregulation of Forkhead Box M1 Expression. *Aging* **2020**, *12*, 20946–20967. [CrossRef] [PubMed]

67. Li, W.; Feng, W.; Su, X.; Luo, D.; Li, Z.; Zhou, Y.; Zhu, Y.; Zhang, M.; Chen, J.; Liu, B.; et al. SIRT6 Protects Vascular Smooth Muscle Cells from Osteogenic Transdifferentiation via Runx2 in Chronic Kidney Disease. *J. Clin. Investig.* **2022**, *132*, 1–15. [CrossRef] [PubMed]

68. Xu, S.; Yin, M.; Koroleva, M.; Mastrangelo, M.A.; Zhang, W.; Bai, P.; Little, P.J.; Jin, Z.G. SIRT6 Protects against Endothelial Dysfunction and Atherosclerosis in Mice. *Aging* **2016**, *8*, 1064–1082. [CrossRef] [PubMed]

69. Peshti, V.; Obolensky, A.; Nahum, L.; Kanfi, Y.; Rathaus, M.; Avraham, M.; Tinman, S.; Alt, F.W.; Banin, E.; Cohen, H.Y. Characterization of Physiological Defects in Adult SIRT6-/- Mice. *PLoS ONE* **2017**, *12*, e0176371. [CrossRef]

70. Moskalev, A.A.; Shaposhnikov, M.V.; Plyusnina, E.N.; Zhavoronkov, A.; Budovsky, A.; Yanai, H.; Fraifeld, V.E. The Role of DNA Damage and Repair in Aging through the Prism of Koch-like Criteria. *Ageing Res. Rev.* **2013**, *12*, 661–684. [CrossRef] [PubMed]

71. Bautista-Niño, P.K.; Portilla-Fernandez, E.; Vaughan, D.E.; Danser, A.H.; Roks, A.J. DNA Damage: A Main Determinant of Vascular Aging. *Int. J. Mol. Sci.* **2016**, *17*, 748. [CrossRef]

72. Lin, Z.; Ding, Q.; Li, X.; Feng, Y.; He, H.; Huang, C.; Zhu, Y. Targeting Epigenetic Mechanisms in Vascular Aging. *Front. Cardiovasc. Med.* **2022**, *8*, 806988. [CrossRef]

73. Farina, F.M.; Hall, I.F.; Serio, S.; Zani, S.; Climent, M.; Salvarani, N.; Carullo, P.; Civilini, E.; Condorelli, G.; Elia, L. MiR-128-3p Is a Novel Regulator of Vascular Smooth Muscle Cell Phenotypic Switch and Vascular Diseases. *Circ. Res.* **2020**, *126*, e120–e135. [CrossRef] [PubMed]

74. Lin, M.E.; Chen, T.M.; Wallingford, M.C.; Nguyen, N.B.; Yamada, S.; Sawangmake, C.; Zhang, J.; Speer, M.Y.; Giachelli, C.M. Runx2 deletion in smooth muscle cells inhibits vascular osteochondrogenesis and calcification but not atherosclerotic lesion formation. *Cardiovasc. Res.* **2016**, *112*, 606–616. [CrossRef]

75. Cobb, A.M.; Yusoff, S.; Hayward, R.; Ahmad, S.; Sun, M.; Verhulst, A.; D'Haese, P.C.; Shanahan, C.M. Runx2 (Runt-Related Transcription Factor 2) Links the DNA Damage Response to Osteogenic Reprogramming and Apoptosis of Vascular Smooth Muscle Cells. *Arterioscler. Thromb. Vasc. Biol.* **2021**, *41*, 1339–1357. [CrossRef] [PubMed]

76. Zhang, F.; Guo, X.; Xia, Y.; Mao, L. An Update on the Phenotypic Switching of Vascular Smooth Muscle Cells in the Pathogenesis of Atherosclerosis. *Cell. Mol. Life Sci.* **2022**, *79*, 6. [CrossRef]

77. Heerkens, E.H.J.; Izzard, A.S.; Heagerty, A.M. Integrins, Vascular Remodeling, and Hypertension. *Hypertension* **2007**, *49*, 1–4. [CrossRef]

78. Hartmann, C.; Herling, L.; Hartmann, A.; Köckritz, V.; Fuellen, G.; Walter, M.; Hermann, A. Systematic Estimation of Biological Age of in Vitro Cell Culture Systems by an Age-Associated Marker Panel. *Front. Aging* **2023**, *4*, 1129107. [CrossRef] [PubMed]

79. Jimenez Trinidad, F.R.; Arrieta Ruiz, M.; Solanes Batlló, N.; Vea Badenes, À.; Bobi Gibert, J.; Valera Cañellas, A.; Roqué Moreno, M.; Freixa Rofastes, X.; Sabaté Tenas, M.; Dantas, A.P.; et al. Linking In Vitro Models of Endothelial Dysfunction with Cell Senescence. *Life* **2021**, *11*, 1323. [CrossRef]

80. DeCicco-Skinner, K.L.; Henry, G.H.; Cataisson, C.; Tabib, T.; Gwilliam, J.C.; Watson, N.J.; Bullwinkle, E.M.; Falkenburg, L.; O'Neill, R.C.; Morin, A.; et al. Endothelial Cell Tube Formation Assay for the In Vitro Study of Angiogenesis. *J. Vis. Exp.* **2014**, *91*, e51312. [CrossRef]

81. Borenstein, J.T.; Tupper, M.M.; MacK, P.J.; Weinberg, E.J.; Khalil, A.S.; Hsiao, J.; García-Cardeña, G. Functional Endothelialized Microvascular Networks with Circular Cross-Sections in a Tissue Culture Substrate. *Biomed. Microdevices* **2010**, *12*, 71–79. [CrossRef] [PubMed]

82. Rodriguez, I.; Spicar-Mihalic, P.; Kuyper, C.L.; Fiorini, G.S.; Chiu, D.T. Rapid Prototyping of Glass Microchannels. *Anal. Chim. Acta* **2003**, *496*, 205–215. [CrossRef]

83. Pollet, A.M.A.O.; Den Toonder, J.M.J. Recapitulating the Vasculature Using Organ-on-Chip Technology. *Bioengineering* **2020**, *7*, 17. [CrossRef] [PubMed]

84. Mandrycky, C.J.; Howard, C.C.; Rayner, S.G.; Shin, Y.J.; Zheng, Y. Organ-on-a-Chip Systems for Vascular Biology. *J. Mol. Cell. Cardiol.* **2021**, *159*, 1–13. [CrossRef]

85. van Dijk, C.G.M.; Brandt, M.M.; Poulis, N.; Anten, J.; van der Moolen, M.; Kramer, L.; Homburg, E.; Louzao-Martinez, L.; Pei, J.; Krebber, M.M.; et al. A New Microfluidic Model That Allows Monitoring of Complex Vascular Structures and Cell Interactions in a 3D Biological Matrix. *Lab. Chip.* **2020**, *20*, 1827–1844. [CrossRef]

86. Polacheck, W.J.; Kutys, M.L.; Tefft, J.B.; Chen, C.S. Microfabricated Blood Vessels for Modeling the Vascular Transport Barrier. *Nat. Protoc* **2019**, *14*, 1425–1454. [CrossRef] [PubMed]

87. Wang, X.; Phan, D.T.T.; Sobrino, A.; George, S.C.; Hughes, C.C.W.; Lee, A.P. Engineering Anastomosis between Living Capillary Networks and Endothelial Cell-Lined Microfluidic Channels. *Lab. Chip.* **2016**, *16*, 282–290. [CrossRef]

88. Vila Cuenca, M.; Cochrane, A.; van den Hil, F.E.; de Vries, A.A.F.; Lesnik Oberstein, S.A.J.; Mummery, C.L.; Orlova, V.V. Engineered 3D Vessel-on-Chip Using HiPSC-Derived Endothelial- and Vascular Smooth Muscle Cells. *Stem. Cell Rep.* **2021**, *16*, 2159–2168. [CrossRef] [PubMed]

89. Tsai, M.; Kita, A.; Leach, J.; Rounsevell, R.; Huang, J.N.; Moake, J.; Ware, R.E.; Fletcher, D.A.; Lam, W.A. In Vitro Modeling of the Microvascular Occlusion and Thrombosis That Occur in Hematologic Diseases Using Microfluidic Technology. *J. Clin. Investig.* **2012**, *122*, 408–418. [CrossRef]

90. Garvin, K.A.; Dalecki, D.; Yousefhussien, M.; Helguera, M.; Hocking, D.C. Spatial Patterning of Endothelial Cells and Vascular Network Formation Using Ultrasound Standing Wave Fields. *J. Acoust. Soc. Am* **2013**, *134*, 1483–1490. [CrossRef]

91. Wu, C.-C.; Li, Y.-S.; Haga, J.H.; Kaunas, R.; Chiu, J.-J.; Su, F.-C.; Usami, S.; Chien, S. Directional Shear Flow and Rho Activation Prevent the Endothelial Cell Apoptosis Induced by Micropatterned Anisotropic Geometry. *Proc. Natl. Acad. Sci. USA* **2007**, *104*, 1254–1259. [CrossRef]

92. Thomas, A.; Daniel Ou-Yang, H.; Lowe-Krentz, L.; Muzykantov, V.R.; Liu, Y. Biomimetic Channel Modeling Local Vascular Dynamics of Pro-Inflammatory Endothelial Changes. *Biomicrofluidics* **2016**, *10*, 014101. [CrossRef] [PubMed]

93. Kim, Y.; Lobatto, M.E.; Kawahara, T.; Lee Chung, B.; Mieszawska, A.J.; Sanchez-Gaytan, B.L.; Fay, F.; Senders, M.L.; Calcagno, C.; Becraft, J.; et al. Probing Nanoparticle Translocation across the Permeable Endothelium in Experimental Atherosclerosis. *Proc. Natl. Acad. Sci. USA* **2014**, *111*, 1078–1083. [CrossRef]

94. Chen, M.B.; Whisler, J.A.; Jeon, J.S.; Kamm, R.D. Mechanisms of Tumor Cell Extravasation in an in Vitro Microvascular Network Platform. *Integr. Biol.* **2013**, *5*, 1262–1271. [CrossRef]

95. Bray, L.J.; Binner, M.; Holzheu, A.; Friedrichs, J.; Freudenberg, U.; Hutmacher, D.W.; Werner, C. Multi-Parametric Hydrogels Support 3D in Vitro Bioengineered Microenvironment Models of Tumour Angiogenesis. *Biomaterials* **2015**, *53*, 609–620. [CrossRef]

96. Stroock, A.D.; Fischbach, C. Microfluidic Culture Models of Tumor Angiogenesis. *Tissue Eng. Part A* **2010**, *16*, 2143–2146. [CrossRef]

97. Bersini, S.; Jeon, J.S.; Dubini, G.; Arrigoni, C.; Chung, S.; Charest, J.L.; Moretti, M.; Kamm, R.D. A Microfluidic 3D in Vitro Model for Specificity of Breast Cancer Metastasis to Bone. *Biomaterials* **2014**, *35*, 2454–2461. [CrossRef] [PubMed]

98. Jeon, J.S.; Bersini, S.; Gilardi, M.; Dubini, G.; Charest, J.L.; Moretti, M.; Kamm, R.D. Human 3D Vascularized Organotypic Microfluidic Assays to Study Breast Cancer Cell Extravasation. *Proc. Natl. Acad. Sci. USA* **2015**, *112*, 214–219. [CrossRef] [PubMed]

99. Huh, D.; Leslie, D.C.; Matthews, B.D.; Fraser, J.P.; Jurek, S.; Hamilton, G.A.; Thorneloe, K.S.; McAlexander, M.A.; Ingber, D.E. A Human Disease Model of Drug Toxicity-Induced Pulmonary Edema in a Lung-on-a-Chip Microdevice. *Sci. Transl. Med.* **2012**, *4*, 159ra147. [CrossRef] [PubMed]

100. Namdee, K.; Thompson, A.J.; Charoenphol, P.; Eniola-Adefeso, O. Margination Propensity of Vascular-Targeted Spheres from Blood Flow in a Microfluidic Model of Human Microvessels. *Langmuir* **2013**, *29*, 2530–2535. [CrossRef]

101. Kastrup, C.J.; Nahrendorf, M.; Figueiredo, J.L.; Lee, H.; Kambhampati, S.; Lee, T.; Cho, S.-W.; Gorbatov, R.; Iwamoto, Y.; Dang, T.T.; et al. Painting Blood Vessels and Atherosclerotic Plaques with an Adhesive Drug Depot. *Proc. Natl. Acad. Sci. USA* **2012**, *109*, 21444–21449. [CrossRef] [PubMed]

102. Wang, Z.; Lang, B.; Qu, Y.; Li, L.; Song, Z.; Wang, Z. Single-Cell Patterning Technology for Biological Applications. *Biomicrofluidics* **2019**, *13*, 61502. [CrossRef] [PubMed]

103. Leopold, J.A.; Loscalzo, J. Emerging Role of Precision Medicine in Cardiovascular Disease. *Circ. Res.* **2018**, *122*, 1302–1315. [CrossRef] [PubMed]
104. Strianese, O.; Rizzo, F.; Ciccarelli, M.; Galasso, G.; D'Agostino, Y.; Salvati, A.; Del Giudice, C.; Tesorio, P.; Rusciano, M.R. Precision and Personalized Medicine: How Genomic Approach Improves the Management of Cardiovascular and Neurodegenerative Disease. *Genes* **2020**, *11*, 747. [CrossRef] [PubMed]
105. Gray, K.M.; Stroka, K.M. Vascular Endothelial Cell Mechanosensing: New Insights Gained from Biomimetic Microfluidic Models. *Semin. Cell. Dev. Biol.* **2017**, *71*, 106–117. [CrossRef]
106. Hesh, C.A.; Qiu, Y.; Lam, W.A. Vascularized Microfluidics and the Blood-Endothelium Interface. *Micromachines* **2019**, *11*, 18. [CrossRef] [PubMed]
107. Luna, D.J.; Pandian, N.K.R.; Mathur, T.; Bui, J.; Gadangi, P.; Kostousov, V.V.; Hui, S.-K.R.; Teruya, J.; Jain, A. Tortuosity-Powered Microfluidic Device for Assessment of Thrombosis and Antithrombotic Therapy in Whole Blood. *Sci. Rep.* **2020**, *10*, 5742. [CrossRef]

International Journal of
Molecular Sciences

Article

Sex-Specific Features of the Correlation between GWAS-Noticeable Polymorphisms and Hypertension in Europeans of Russia

Tatiana Ivanova [1], Maria Churnosova [1], Maria Abramova [1], Denis Plotnikov [2], Irina Ponomarenko [1], Evgeny Reshetnikov [1], Inna Aristova [1], Inna Sorokina [1] and Mikhail Churnosov [1,*]

[1] Department of Medical Biological Disciplines, Belgorod State National Research University, 308015 Belgorod, Russia; 4602@bsu.edu.ru (T.I.); churnosovamary@gmail.com (M.C.); abramova_myu@bsu.edu.ru (M.A.); ponomarenko_i@bsu.edu.ru (I.P.); reshetnikov@bsu.edu.ru (E.R.); aristova@bsu.edu.ru (I.A.); sorokina@bsu.edu.ru (I.S.)
[2] Genetic Epidemiology Lab, Kazan State Medical University, 420012 Kazan, Russia; denis.plotnikov@kazangmu.ru
* Correspondence: churnosov@bsu.edu.ru

Citation: Ivanova, T.; Churnosova, M.; Abramova, M.; Plotnikov, D.; Ponomarenko, I.; Reshetnikov, E.; Aristova, I.; Sorokina, I.; Churnosov, M. Sex-Specific Features of the Correlation between GWAS-Noticeable Polymorphisms and Hypertension in Europeans of Russia. *Int. J. Mol. Sci.* **2023**, *24*, 7799. https://doi.org/10.3390/ijms24097799

Academic Editors: Yutang Wang and Dianna Magliano

Received: 25 March 2023
Revised: 13 April 2023
Accepted: 20 April 2023
Published: 25 April 2023

Abstract: The aim of the study was directed at studying the sex-specific features of the correlation between genome-wide association studies (GWAS)-noticeable polymorphisms and hypertension (HTN). In two groups of European subjects of Russia (n = 1405 in total), such as men (n = 821 in total: n = 564 HTN, n = 257 control) and women (n = 584 in total: n = 375 HTN, n = 209 control), the distribution of ten specially selected polymorphisms (they have confirmed associations of GWAS level with blood pressure (BP) parameters and/or HTN in Europeans) has been considered. The list of studied loci was as follows: (*PLCE1*) rs932764 A > G, (*AC026703.1*) rs1173771 G > A, (*CERS5*) rs7302981 G > A, (*HFE*) rs1799945 C > G, (*OBFC1*) rs4387287 C > A, (*BAG6*) rs805303 G > A, (*RGL3*) rs167479 T > G, (*ARHGAP42*) rs633185 C > G, (*TBX2*) rs8068318 T > C, and (*ATP2B1*) rs2681472 A > G. The contribution of individual loci and their inter-locus interactions to the HTN susceptibility with bioinformatic interpretation of associative links was evaluated separately in men's and women's cohorts. The men–women differences in involvement in the disease of the BP/HTN-associated GWAS SNPs were detected. Among women, the HTN risk has been associated with *HFE* rs1799945 C > G (genotype GG was risky; OR_{GG} = 11.15 p_{permGG} = 0.014) and inter-locus interactions of all 10 examined SNPs as part of 26 intergenic interactions models. In men, the polymorphism *BAG6* rs805303 G > A (genotype AA was protective; OR_{AA} = 0.30 p_{permAA} = 0.0008) and inter-SNPs interactions of eight loci in only seven models have been founded as HTN-correlated. HTN-linked loci and strongly linked SNPs were characterized by pronounced polyvector functionality in both men and women, but at the same time, signaling pathways of HTN-linked genes/SNPs in women and men were similar and were represented mainly by immune mechanisms. As a result, the present study has demonstrated a more pronounced contribution of BP/HTN-associated GWAS SNPs to the HTN susceptibility (due to weightier intergenic interactions) in European women than in men.

Keywords: sex; blood pressure; hypertension; GWAS; SNP; association

1. Introduction

HTN belongs to the group of the most common human diseases, a characteristic feature of which is high BP [1]. According to statistical reports in the world over the past decades (from 1990 to 2015), the number of persons with a systolic BP (SBP) of 140 mm Hg and above rose by 18.59% (from 17,307 to 20,526 per 100 thousand population), and the mortality and disability-adjusted life-years (DALYs) rates associated with an SBP of 140 mm Hg and higher increased by 8.58% (from 97.9 to 106.3 per 100,000 people) and 49.11% (from 95.9 million to 143.0 million), respectively [2]. A rise in SBP by 10 mm Hg leads to an increased risk of coronary heart disease by 45% and ischemic/hemorrhagic stroke by 63–66% in individuals aged 55–64 years [3]. The problem of HTN is no less important for the Russian population [4,5], among which the projected number of deaths and DALYs correlated with SBP $\geq$ 140 mm Hg increased by 25.6% (from 452.3 to 568.0 thousand) and 49.11% (from 8267.8 to 9672.7 thousand) over the period from 1990 to 2015, respectively [2]. The disease is more often ($\approx$20%) registered in men than in women (age-standardized prevalence rates of HTN in men and women are 24 and 20%, respectively) [6] and this pattern is also characteristic of the population of Russia [5]. In men, the indicator DALYs linked with SBP $\geq$ 140 mmHg (82.915 million) significantly exceeds (1.38 times) the same parameter in women (60.122 million) [2]. Even more pronounced differences (by 1.44 times) between men (125.124 million) and women (86.692 million) are observed in the value for DALYs correlated with SBP $\geq$ 110–115 mm Hg [2]. Significant differences in the prevalence of HTN and its impact on the state of health in men and women may be associated with both hormonal mechanisms [7] and environmental risk factors [8], which determine the features of gene–environmental interactions [9,10].

The influence of hereditary factors on the HTN occurrence has been studied for a long time and, at present, the genetic contribution to the disease formation has been confirmed in numerous works performed on the basis of twin/family [11–17], GWAS ([9,10,18–34], etc.) and other associative (based on the candidate genes' analysis) ([35–48], etc.) studies. The prevailing value of genetic factors in HTN development is confirmed by the fact that the presence of a family history burden increases the disease risk by 3.5–3.8 times compared to the general population [49,50]. Estimates of the heredity contribution to BP vary on average from 30% to 55% [11,12,14–17], reaching the level of 63–68% in some works [12,13]. At the same time, the BP heritability indicators, estimated on the basis of currently known GWAS data (SNP-wide heritability), amount to 19.4–21.3% [51] and only partially "reveal" genetic factors of HTN predisposition (from 1/3 to 2/3). In general, the currently known loci, detected by GWAS, explain only 2.87–5.66% of BP variance [51]. The above data indicate the presence of a problem of "hidden" (unidentified) heredity for BP (and, accordingly, HTN) [16] and dictate the need for further research aimed at solving it.

So, the available literature materials clearly indicate, on the one hand, a significant heredity contribution to HTN susceptibility [11,12,14–17]; but, on the other hand, demonstrate pronounced men–women differences in HTN incidence [4,6]. These facts stipulate the appropriateness of genetic and epidemiological studies "revealing" the genetic mechanisms underlying the differences in the genetic determination of HTN in men and women (as one of the factors determining sex-specific differences in the disorder incidence), including using genes/polymorphisms with a previously proven strong role (GWAS data) in HTN predisposition. In some previously performed studies (including the European population of Russia investigated in this paper), men–women differences in the involvement in the formation of HTN polymorphisms of a number of candidate genes (matrix metalloproteinases, tumor necrosis factors, etc.) are shown, including the taking into account of the action of sex-specific risk factors of the disease such as obesity, smoking, etc. [40,45–48] (etc.). In addition, significant men–women differences in sex hormone genetic traits [52–54], BMI genetic traits (obesity, waist/hip circumference, etc.) [55–59] and insulin/type 2 diabetes genetic traits [59,60] have been demonstrated at the GWAS level, which may be remarkable genetic predictors of sex-specific differences for cardiovascular diseases [61–65].

Our study was directed at studying the sex-specific features of the correlation between GWAS-noticeable polymorphisms and HTN.

2. Results

The characteristics of generalized participants (men and women) have been given in the Table 1. In both men (n = 821) and women (n = 584) cohorts, HTN subjects differed from the HTN-free subjects by higher values of body mass index (BMI), total cholesterol (TC), triglycerides (TG), low-density lipoprotein cholesterol (LDL-C), blood glucose, smokers and a lower high-density lipoprotein cholesterol (HDL-C) level (p < 0.05–0.001). Based on these results, the aforementioned features were introduced into genetic calculations as covariates in both men and women (Model 1). There were also differences between patients and controls (both in men and women) in such parameters as low physical activity and high fatty food consumption (in patients, these indicators were significantly higher, $p < 0.001$). In order to assess the effect of these indicators of lifestyle (physical activity) and diet (fatty food consumption) on genetic associations, we additionally included them in the analysis as confounders in Model 2.

The SNPs allele and genotype frequencies in the HTN and HTN-free groups of men (Table S1) and women (Table S2) have Hardy–Weinberg (H-W) equilibrium state, $p_{Bonferroni} \geq 0.025$ (Bonferroni's correction related to the amount of comparison pairs studied was accounted for, $n = 2$ [men and women]).

The men–women differences in involvement in the disease of the BP/HTN-associated GWAS SNPs were detected (both in Model 1 and Model 2). According to Model 1, among men, the polymorphism *BAG6* rs805303 G > A was founded as HTN-correlated (genotype AA was protective; $OR_{AA} = 0.30$; $p_{AA} = 0.0008$; $p_{permAA} = 0.0008$ [recessive model] statistical power = 98.75%) (Table 2). In women, the HTN risk was associated with *HFE* rs1799945 C > G (genotype GG was risky; $OR_{GG} = 11.15$; $p_{GG} = 0.011$; $p_{permGG} = 0.014$ [recessive model] statistical power = 99.99%) (Table 2). Almost similar results were obtained in Model 2: the SNP *BAG6* rs805303 G > A was HTN-associated in men ($OR_{AA} = 0.31$; $p_{AA} = 0.001$; $p_{permAA} = 0.001$; statistical power = 98.43% [recessive model]) (Table 2) and locus *HFE* rs1799945 C > G was HTN-involved in women ($OR_{GG} = 10.96$; $p_{GG} = 0.012$; $p_{permGG} = 0.014$; statistical power = 99.99% [recessive model]) (Table 2). The great similarity in the results of our evaluation of genetic associations in Model 1 and Model 2 may be due to the fact that, apparently, the effects of additional confounders included in Model 2 (low physical activity and high fatty food consumption) have already been "taken into account" in the HTN effects of confounders of Model 1. For example, the high consumption of fatty foods (confounder Model 2) obviously directly determines the lipid profile of the body such as TC, TG, LDL-C and HDL-C levels (confounders Model 1), and low physical activity (confounder Model 2) will largely correlate with BMI (confounder Model 1). Accordingly, at the next stage of the work, when analyzing the associations of inter-locus interactions with HTN, we used a list of confounders of Model 1.

Pronounced sex-specific differences in engagement in the HTN of GWAS-noticeable polymorphisms were revealed during the evaluation of inter-locus communications. In the men's group the inter-SNP interactions of eight loci in only seven models have been registered as HTN-involved ($p_{perm} \leq 0.022$, polymorphic loci of two genes—*ARHGAP42* rs633185 C > G, *ATP2B1* rs2681472 A > G, not included in these models) (Table 3). Polymorphisms of four genes such as *CERS5* rs7302981 G > A, *AC026703.1* rs1173771 G > A, *HFE* rs1799945 C > G and *PLCE1* rs932764 A > G were represented in more than 50% of all models. The best possible genetic model is a four-gene model (*PLCE1* rs932764 A > G × *CERS5* rs7302981 G > A × *HFE* rs1799945 C > G × *TBX2* rs8068318 T > C) with Wald stat. = 32.12 ($p_{perm} = 0.001$). Twenty-three HTN-associated combinations of genotypes were established in men, practically all of which were disorder-protective (22 out of 23, 95.65%). Only one genotype combination (three SNPs) such as rs805303-GG (*BAG6*) × rs1173771-AA (*AC026703.1*) × rs4387287-AA (*OBFC1*) increases the HTN risk in men (beta = 2.29; $p = 0.029$). Among the HTN-protective combinations of genotypes, the brightly expressed effect (*beta* values were maximum) has been demonstrated by rs932764-GG (*PLCE1*) × rs8068318-TC (*TBX2*) × rs805303-GG (*BAG6*) × rs1173771-GA (*AC026703.1*) (*beta* = −2.63; $p = 0.017$), rs7302981-GG (*CERS5*) × rs1799945-CC (*HFE*) × rs805303-GG (*BAG6*) × rs167479-GG (*RGL3*) (*beta* = −2.24; $p = 0.042$), rs932764-GG (*PLCE1*) × rs7302981-GA (*CERS5*) × rs1799945-CG (*HFE*) × rs8068318-CC (*TBX2*) (*beta* = −2.13; $p = 0.002$), rs932764-AA (*PLCE1*) × rs7302981-GG (*CERS5*) × rs1799945-CC (*HFE*) × rs1173771-GA (*AC026703.1*) (*beta* = −2.01; $p = 0.019$) (Table S3).

Table 1. Phenotypic characteristics of the study participants.

Parameters	Men (n = 821)			Women (n = 584)		
	HTN, Mean ± SD, % (n)	Controls, Mean ± SD, % (n)	p	HTN, Mean ± SD, % (n)	Controls, Mean ± SD, % (n)	p
N	564	257	-	375	209	-
Age (years)	57.60 ± 8.36	57.54 ± 9.73	0.86	58.80 ± 9.64	58.17 ± 9.30	0.43
BMI (kg/m^2)	30.76 ± 4.52	25.04 ± 2.86	<0.001	30.81 ± 5.84	24.83 ± 3.41	<0.001
SBP (mmHg)	180.45 ± 27.40	123.89 ± 11.92	<0.001	185.53 ± 29.26	120.98 ± 10.76	<0.001
DBP (mmHg)	104.76 ± 12.75	78.19 ± 6.76	<0.001	107.47 ± 14.35	76.98 ± 7.09	<0.001
TC (mM)	5.65 ± 1.32	5.24 ± 0.98	<0.001	5.79 ± 1.23	5.28 ± 1.09	<0.001
HDL-C (mM)	1.35 ± 0.44	1.48 ± 0.41	<0.001	1.33 ± 0.37	1.56 ± 0.42	<0.001
LDL-C (mM)	3.78 ± 0.73	3.18 ± 0.66	<0.001	3.77 ± 1.15	3.25 ± 0.79	<0.001
TG (mM)	2.00 ± 1.01	1.28 ± 0.81	<0.001	1.80 ± 1.04	1.18 ± 0.62	<0.001
Blood glucose (mM)	5.82 ± 1.20	4.93 ± 1.12	<0.001	6.08 ± 1.91	4.81 ± 0.67	<0.001
Smoking	65.58 (318)	34.96 (74)	<0.001	9.75 (35)	4.78 (10)	0.04
Alcohol abuse	9.30 (52)	2.29 (13)	0.17	0.28 (1)	0 (0)	1.00
Antihypertensive medication use	82.80 (467)	-	-	80.27 (301)	-	-
HTN Grade: Grade 1	18.44 (104)	-	-	12.80 (48)	-	-
Grade 2	48.23 (272)	-	-	45.60 (171)	-	-
Grade 3	33.33 (188)	-	-	41.60 (156)	-	-
HTN Stage: Stage 1	11.88 (67)	-	-	7.20 (27)	-	-
Stage 2	31.21 (176)	-	-	30.40 (114)	-	-
Stage 3	56.91 (321)	-	-	62.40 (234)	-	-
Stroke incident	35.64 (201)	-	-	27.20 (102)	-	-
Coronary artery disease	29.08 (164)	-	-	32.00 (120)	-	-
Type 2 diabetes mellitus	33.51 (189)	-	-	30.13 (113)	-	-
Low physical activity	57.62 (325)	24.12 (62)	<0.001	60.27 (226)	31.58 (66)	<0.001
Low fruit/vegetable consumption	11.52 (65)	7.39 (19)	0.09	11.20 (42)	6.70 (14)	0.10
High fatty food consumption	24.47 (138)	10.12 (26)	<0.001	25.06 (94)	10.53 (22)	<0.001
High sodium consumption	16.49 (93)	14.01 (36)	0.42	17.07 (64)	12.44 (26)	0.17

Note: Clinical characteristics of age, BMI, SBP, DBP, HDL-C, LDL-C, TG and TC are given as means ± SD and other values as number of individuals; BMI—body mass index; SBP—systolic blood pressure; DBP—diastolic blood pressure; TC—total cholesterol; HDL-C—high-density lipoprotein cholesterol; TG—triglycerides; LDL-C—low-density lipoprotein cholesterol; HTN Grade and Stage are presented according to [1]: Grade 1 SBP = 140–159 mmHg and/or DBP = 90–99 mmHg; Grade 2 SBP = 160–179 mmHg and/or DBP = 100–109 mmHg; Grade 3 SBP > 180 mmHg and/or DBP > 110 mmHg).

Table 2. Associations of the studied gene polymorphisms with HTN in men and women.

Gene (SNP, Major/Minor Alleles)	n	Allelic Model				Additive Model				Dominant Model				Recessive Model			
		OR	95%CI L95	95%CI U95	p	OR	95%CI L95	95%CI U95	p	OR	95%CI L95	95%CI U95	p	OR	95%CI L95	95%CI U95	p
Model 1																	
Men																	
AC026703.1 (rs1173771, G/A)	774	0.91	0.73	1.13	0.383	0.88	0.61	1.29	0.524	0.67	0.38	1.17	0.161	1.25	0.61	2.55	0.542
HFE (rs1799945, C/G)	800	1.16	0.89	1.52	0.281	1.30	0.85	1.99	0.220	1.38	0.80	2.37	0.248	1.54	0.54	4.38	0.421
BAG6 (rs805303, G/A)	789	0.95	0.76	1.19	0.665	0.66	0.46	0.96	0.028	0.82	0.49	1.39	0.466	**0.30**	**0.15**	**0.61**	**0.0008**
PLCE1 (rs932764, A/G)	775	1.23	0.99	1.53	0.056	1.24	0.85	1.80	0.267	1.30	0.71	2.36	0.392	1.36	0.72	2.55	0.339
OBFC1 (rs4387287, C/A)	751	0.88	0.66	1.17	0.380	0.84	0.50	1.41	0.497	0.80	0.44	1.46	0.469	0.88	0.17	4.72	0.884
ARHGAP42 (rs633185, C/G)	800	1.01	0.79	1.28	0.964	1.18	0.80	1.74	0.418	1.14	0.68	1.90	0.631	1.58	0.65	3.86	0.316
CERS5 (rs7302981, G/A)	767	0.93	0.74	1.16	0.499	0.67	0.45	0.99	0.044	0.70	0.40	1.22	0.211	0.44	0.21	0.92	0.028
ATP2B1 (rs2681472, A/G)	778	1.12	0.82	1.54	0.479	1.64	0.91	2.97	0.103	1.61	0.83	3.09	0.157	5.61	0.42	74.38	0.191
TBX2 (rs8068318, T/C)	762	1.14	0.89	1.46	0.299	1.33	0.87	2.05	0.194	1.45	0.85	2.49	0.173	1.33	0.46	3.83	0.596
RGL3 (rs167479, T/G)	783	0.98	0.79	1.21	0.833	0.82	0.57	1.18	0.283	0.90	0.49	1.64	0.722	0.66	0.37	1.18	0.161
Women																	
AC026703.1 (rs1173771, G/A)	543	0.91	0.71	1.17	0.458	0.92	0.64	1.31	0.631	0.85	0.50	1.45	0.554	0.95	0.52	1.76	0.877
HFE (rs1799945, C/G)	573	0.71	0.53	0.96	0.026	0.75	0.48	1.17	0.210	0.60	0.37	0.98	0.040	**11.15**	**1.15**	**97.68**	**0.011**
BAG6 (rs805303, G/A)	560	0.99	0.77	1.28	0.938	1.08	0.76	1.51	0.675	1.06	0.66	1.72	0.801	1.18	0.60	2.35	0.630
PLCE1 (rs932764, A/G)	544	0.95	0.74	1.21	0.663	0.83	0.59	1.17	0.286	0.59	0.34	1.03	0.061	1.07	0.59	1.91	0.832
OBFC1 (rs4387287, C/A)	509	0.97	0.72	1.30	0.891	0.88	0.57	1.35	0.551	0.89	0.52	1.50	0.650	0.70	0.21	2.25	0.545
ARHGAP42 (rs633185, C/G)	577	1.04	0.79	1.37	0.759	1.03	0.71	1.49	0.868	1.05	0.65	1.68	0.850	1.02	0.42	2.46	0.965
CERS5 (rs7302981, G/A)	535	1.20	0.93	1.55	0.152	1.19	0.84	1.68	0.326	1.03	0.79	2.15	0.306	1.20	0.62	2.30	0.591
ATP2B1 (rs2681472, A/G)	551	0.96	0.68	1.35	0.809	0.89	0.55	1.45	0.648	0.94	0.55	1.59	0.808	0.45	0.08	2.62	0.372
TBX2 (rs8068318, T/C)	530	1.06	0.80	1.40	0.704	1.04	0.70	1.55	0.841	1.03	0.63	1.68	0.918	1.16	0.42	3.18	0.772
RGL3 (rs167479, T/G)	550	1.15	0.90	1.47	0.266	1.29	0.92	1.82	0.138	1.44	0.83	2.50	0.200	1.38	0.79	2.41	0.252

Table 2. *Cont.*

Gene (SNP, Major/Minor Alleles)	n	Allelic Model				Additive Model				Dominant Model				Recessive Model			
		OR	95%CI		p	OR	95%CI		p	OR	95%CI		p	OR	95%CI		p
			L95	U95			L95	U95			L95	U95			L95	U95	
Model 2*																	
Men																	
AC026703.1 (rs1173771, G/A)						0.90	0.62	1.30	0.560	0.68	0.39	1.19	0.176	1.25	0.62	2.51	0.529
HFE (rs1799945, C/G)						1.33	0.87	2.01	0.187	1.42	0.83	2.43	0.204	1.55	0.56	4.29	0.403
BAG6 (rs805303, G/A)						0.66	0.46	0.95	0.026	0.80	0.47	1.34	0.395	**0.31**	**0.16**	**0.64**	**0.001**
PLCE1 (rs932764, A/G)						1.21	0.84	1.75	0.311	1.27	0.71	2.29	0.425	1.31	0.71	2.44	0.391
OBFC1 (rs4387287, C/A)						0.90	0.54	1.50	0.681	0.86	0.48	1.56	0.629	1.03	0.19	5.64	0.974
ARHGAP42 (rs633185, C/G)						1.14	0.77	1.67	0.510	1.10	0.66	1.83	0.727	1.49	0.62	3.76	0.376
CERS5 (rs7302981, G/A)						0.67	0.46	1.00	0.047	0.70	0.40	1.22	0.213	0.46	0.22	0.94	0.033
ATP2B1 (rs2681472, A/G)						1.59	0.89	2.83	0.118	1.55	0.82	2.95	0.179	5.11	0.42	61.98	0.200
TBX2 (rs8068318, T/C)						1.29	0.85	1.97	0.233	1.40	0.82	2.38	0.220	1.33	0.48	3.75	0.584
RGL3 (rs167479, T/G)						0.80	0.56	1.15	0.237	0.90	0.49	1.65	0.740	0.63	0.35	1.11	0.110
Women																	
AC026703.1 (rs1173771, G/A)						0.92	0.65	1.31	0.649	0.87	0.51	1.47	0.595	0.95	0.51	1.74	0.858
HFE (rs1799945, C/G)						0.77	0.50	1.19	0.241	0.61	0.37	0.99	0.043	**10.96**	**1.12**	**89.68**	**0.012**
BAG6 (rs805303, G/A)						1.05	0.75	1.47	0.781	1.05	0.65	1.69	0.848	1.10	0.56	2.19	0.776
PLCE1 (rs932764, A/G)						0.80	0.57	1.13	0.211	0.58	0.33	1.00	0.050	1.00	0.56	1.77	0.997
OBFC1 (rs4387287, C/A)						0.89	0.58	1.37	0.593	0.91	0.54	1.54	0.726	0.67	0.21	2.16	0.502
ARHGAP42 (rs633185, C/G)						1.02	0.71	1.48	0.898	1.03	0.65	1.65	0.895	1.03	0.43	2.47	0.951
CERS5 (rs7302981, G/A)						1.20	0.85	1.68	0.305	1.34	0.81	2.21	0.250	1.16	0.61	2.21	0.649
ATP2B1 (rs2681472, A/G)						0.98	0.61	1.56	0.920	1.00	0.60	1.68	0.989	0.70	0.13	3.78	0.678
TBX2 (rs8068318, T/C)						1.07	0.72	1.58	0.754	1.03	0.63	1.68	0.899	1.31	0.48	3.54	0.602
RGL3 (rs167479, T/G)						1.20	0.86	1.68	0.274	1.32	0.77	2.27	0.312	1.25	0.72	2.16	0.435

Note: * For Model 2, calculations of the allelic model were not performed because their results are identical to those of Model 1 (covariates are not used when calculating the allelic model); all results were obtained after adjustment for covariates; list covariates for Model 1: BMI, TC, TG, LDL-C, HDL-C, blood glucose, smokers; list covariates for Model 2: BMI, TC, TG, LDL-C, HDL-C, blood glucose, smokers, low physical activity, high fatty food consumption; OR—odds ratio; 95% CI-95% confidence interval; p values $\leq$ 0.025 are shown in bold.

Table 3. SNP × SNP interactions significantly associated with HTN in men and women.

N	SNP × SNP Interaction Models	NH	betaH	WH	NL	betaL	WL	P$_{perm}$
	Men							
	Three-order interaction models							
1	rs805303 BAG6 × rs1173771 AC026703.1 × rs4387287 OBFC1	1	2.287	4.73	4	−0.815	16.76	0.008
2	rs932764 PLCE1 × rs7302981 CERS5 × rs1799945 HFE	-	-	-	3	−1.118	18.67	0.010
3	rs7302981 CERS5 × rs1173771 AC026703.1 × rs167479 RGL3	-	-	-	4	−0.933	18.13	0.016
	Four-order interaction models							
1	rs932764 PLCE1 × rs7302981 CERS5 × rs1799945 HFE × rs8068318 TBX2	1	2.096	3.75	5	−1.325	32.12	0.001
2	rs932764 PLCE1 × rs7302981 CERS5 × rs1799945 HFE × rs1173771 AC026703.1	0	-	-	5	−1.637	28.51	0.008
3	rs932764 PLCE1 × rs8068318 TBX2 × rs805303 BAG6 × rs1173771 AC026703.1	0	-	-	6	−1.265	27.17	0.016
4	rs7302981 CERS5 × rs1799945 HFE × rs805303 BAG6 × rs167479 RGL3	1	0.531	3.07	5	−1.261	24.40	0.022
	Women							
	Two-order interaction models							
1	rs8068318 TBX2 × rs167479 RGL3	3	0.623	10.18	3	−0.880	21.92	<0.001
2	rs932764 PLCE1 × rs1799945 HFE	2	-	-	2	−0.801	10.83	0.013
	Three-order interaction models							
1	rs1799945 HFE × rs8068318 TBX2 × rs167479 RGL3	2	1.140	21.30	4	−0.898	20.87	<0.001
2	rs2681472 ATP2B1 × rs8068318 TBX2 × rs167479 RGL3	2	0.804	13.59	2	−0.870	18.15	<0.001
3	rs932764 PLCE1 × rs7302981 CERS5 × rs805303 BAG6	2	1.011	8.23	2	−1.311	20.96	0.001
4	rs932764 PLCE1 × rs805303 BAG6 × rs4387287 OBFC1	4	1.394	20.28	1	−1.419	3.43	0.001
5	rs8068318 TBX2 × rs4387287 OBFC1 × rs167479 RGL3	3	0.778	15.03	4	−0.817	18.36	0.001
6	rs2681472 ATP2B1 × rs805303 BAG6 × rs1173771 AC026703.1	1	1.404	7.81	3	−1.573	16.48	0.001
7	rs932764 PLCE1 × rs8068318 TBX2 × rs167479 RGL3	2	1.323	10.47	2	−0.966	18.01	0.001
8	rs932764 PLCE1 × rs1799945 HFE × rs805303 BAG6	0	-	-	3	−1.218	17.93	0.001
9	rs932764 PLCE1 × rs1799945 HFE × rs8068318 TBX2	1	0.679	8.01	3	−0.849	15.49	0.010
10	rs932764 PLCE1 × rs1799945 HFE × rs167479 RGL3	0	-	-	3	−1.102	15.68	0.016

Table 3. *Cont.*

N	SNP × SNP Interaction Models	NH	*beta*H	WH	NL	*beta*L	WL	P$_{perm}$
	Four-order interaction models							
1	rs932764 *PLCE1* × rs8068318 *TBX2* × rs1173771 *AC026703.1* × rs167479 *RGL3*	1	1.446	3.33	7	−1.275	33.53	<0.001
2	rs1799945 *HFE* × rs8068318 *TBX2* × rs805303 *BAG6* × rs167479 *RGL3*	4	1.045	18.51	7	−1.229	32.64	<0.001
3	rs932764 *PLCE1* × rs1799945 *HFE* × rs1173771 *AC026703.1* × rs167479 *RGL3*	3	1.182	9.42	6	−1.840	31.40	<0.001
4	rs2681472 *ATP2B1* × rs1799945 *HFE* × rs8068318 *TBX2* × rs167479 *RGL3*	4	1.235	27.98	6	−1.276	31.20	<0.001
5	rs1799945 *HFE* × rs8068318 *TBX2* × rs1173771 *AC026703.1* × rs167479 *RGL3*	3	0.995	14.27	7	−1.356	30.93	<0.001
6	rs932764 *PLCE1* × rs1799945 *HFE* × rs8068318 *TBX2* × rs167479 *RGL3*	3	1.256	17.70	5	−1.287	29.85	<0.001
7	rs1799945 *HFE* × rs8068318 *TBX2* × rs4387287 *OBFC1* × rs167479 *RGL3*	3	1.410	26.21	2	−0.707	7.43	<0.001
8	rs2681472 *ATP2B1* × rs633185 *ARHGAP42* × rs7302981 *CERS5* × rs805303 *BAG6*	0	-	-	5	−1.199	26.05	<0.001
9	rs2681472 *ATP2B1* × rs633185 *ARHGAP42* × rs8068318 *TBX2* × rs167479 *RGL3*	2	0.813	9.37	4	−1.192	24.91	<0.001
10	rs2681472 *ATP2B1* × rs8068318 *TBX2* × rs1173771 *AC026703.1* × rs167479 *RGL3*	1	1.157	4.12	4	−1.207	24.60	<0.001
11	rs633185 *ARHGAP42* × rs932764 *PLCE1* × rs7302981 *CERS5* × rs805303 *BAG6*	2	1.503	7.55	2	−1.238	25.86	0.001
12	rs932764 *PLCE1* × rs7302981 *CERS5* × rs1799945 *HFE* × rs167479 *RGL3*	0	-	-	5	−1.477	25.01	0.001
13	rs7302981 *CERS5* × rs1799945 *HFE* × rs8068318 *TBX2* × rs167479 *RGL3*	3	0.995	14.27	5	−1.242	24.28	0.001
14	rs633185 *ARHGAP42* × rs7302981 *CERS5* × rs8068318 *TBX2* × rs167479 *RGL3*	1	0.943	3.36	4	−1.376	24.37	0.010

Note: The results were obtained using the MB-MDR method with adjustment for covariates (Model 1); NH—number of significant high-risk genotypes in the interaction; beta H—regression coefficient for high-risk exposition in the step 2 analysis; WH—Wald statistic for high-risk category; NL—number of significant low-risk genotypes in the interaction; beta L—regression coefficient for low-risk exposition in the step 2 analysis; WL—Wald statistic for low-risk category; P$_{perm}$—permutation *p*-value for the interaction model (1.000 permutations).

In the women's cohort, the HTN genetic risk was associated with inter-locus interactions of all 10 examined SNPs as part of 26 intergenic interaction models ($p_{perm} \leq 0.016$) (Table 3). Three genes/polymorphisms such as *RGL3* rs167479 T > G (it is included in 18 models), *TBX2* rs8068318 T > C (16 models), *HFE* rs1799945 C > G (13 models) make the prevailing contribution to HTN susceptibility (they were part of 50% of models and more). The four-gene/locus model, including *PLCE1* rs932764 A > G × *TBX2* rs8068318 T > C × *AC026703.1* rs1173771 G > A × *RGL3* rs167479 T > G, is the top (Wald stat. = 33.53; p_{perm} < 0.001). Ninety-four different HTN-associated genotypes combinations were modeled (Table S4). Among them, the large majority of genetic combinations ($n = 67, 71.28\%$) have a protective orientation (*beta* < 0) and only less than 1/3 combinations ($n = 27, 28.72\%$) are risky (*beta* > 0). Genotype combinations with the maximum phenotypic effects in relation to the HTN risk (characterized by the highest *beta* indicators) were established both up-predisposition (risky) [rs932764-AA (*PLCE1*) × rs8068318-TC (*TBX2*) × rs167479-GG (*RGL3*) (*beta* = 2.45; *p* = 0.024)] and down-predisposition (protective) [rs633185-GG (*ARHGAP42*) × rs932764-GG (*PLCE1*) × rs7302981-GA (*CERS5*) × rs805303-GA (*BAG6*) (*beta* = −3.11; *p* = 0.003); rs932764-AG (*PLCE1*) × rs1799945-GG (*HFE*) × rs8068318-CC (*TBX2*) × rs167479-TG (*RGL3*) (*beta* = −2.88; *p* = 0.012); rs932764-AG (*PLCE1*) × rs1799945-CG (*HFE*) × rs8068318-CC (*TBX2*) (*beta* = −2.87; *p* = 0.012), rs932764-AG (*PLCE1*) × rs1799945-GG (*HFE*) × rs1173771-GA (*AC026703.1*) × rs167479-TG (*RGL3*) (*beta* = −2.79; *p* = 0.016)] orientation.

The results of the MDR analysis showed that the greatest contribution to the HTN susceptibility in men was made by two-gene/locus synergetic interactions *HFE* (rs1799945)–*TBX2* (rs8068318) (0.76% of HTN entropy), *CERS5* (rs7302981)–*RGL3* (rs167479) (0.64%), *CERS5* (rs7302981)–*BAG6* (rs805303) (0.49%), *BAG6* (rs805303)–*AC026703.1* (rs1173771) (0.46%), *HFE* (rs1799945)–*AC026703.1* (rs1173771) (0.46%) (Figure 1). In women, the HTN predisposition was determined by the strongly pronounced influence of three two-gene/locus synergetic interactions *TBX2* (rs8068318)–*RGL3* (rs167479) (2.26%), *PLCE1* (rs932764)–*CERS5* (rs7302981) (1.31%) and *CERS5* (rs7302981)–*BAG6* (rs805303) (1.06%) (Figure 2). Interestingly, the percentage of disorder entropy explained by the HTN-important pair interactions between genes/loci in women (1.06–2.26%) is significantly higher (2–3 times) than in men (0.46–0.76%), which indicates a more significant role of intergenic/inter-locus interactions in the HTN formation in women compared to men.

2.1. Intended Functionality of HTN-Associated SNPs in Men and Women Cohorts

In this section of our work, we evaluated the alleged functionality of all 10 GWAS loci examined and 125 strongly linked SNPs associated with HTN in women (in total, information about 135 loci was considered), and 8 GWAS loci and 96 LD SNPs correlated with HTN in men (data on 104 loci were studied).

2.1.1. Prediction of the Possible SNPs Link with Amino Acid Substitution and Resulting with Human Protein Structure/Function

Among the HTN-involved loci in both men and women, nucleotide changes [T > G (rs167479); C > G (rs1799945); G > A (rs7302981); G > A (rs1046089); and C > G (rs1057987)] in the exons of the five genes (RGL3; HFE; CERS5; PRRC2A; TBX2, respectively) were non-synonymous and led to certain substitutions of amino acids (P162H; H63D; C9R; R1740H; S609R, respectively) in five corresponding proteins with potentially different predictor capability such as "deleterious/probably damaging" (SIFT/PolyPhen categories) [P162H; R1740H/P162H; H63D; R1740H] and "tolerated/benign" (SIFT/PolyPhen categories) [H63D; C9R; S609R/C9R; S609R] (Table S5).

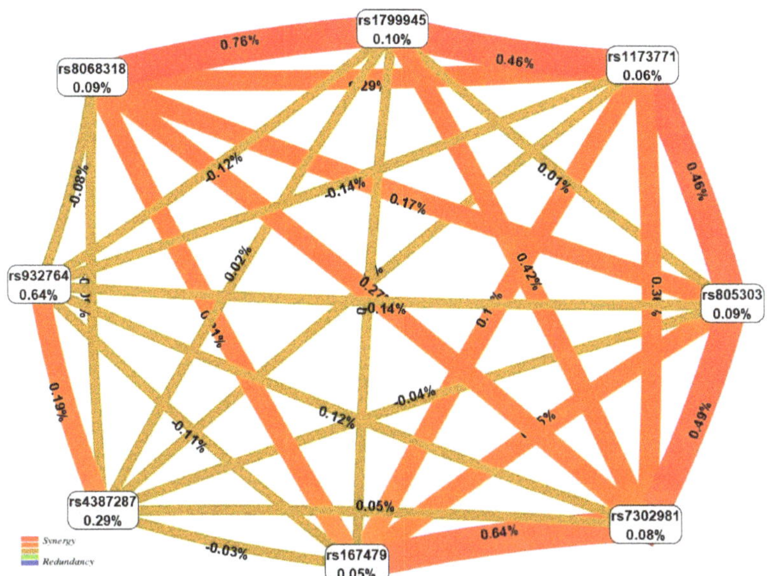

Figure 1. The entropy graph of the SNP×SNP interactions with HTN in men based on the MDR analysis. Positive values of entropy indicate synergistic interactions while the negative values indicate redundancy. The red and orange colors denote strong and moderate synergism, respectively, the brown color denotes the independent effect, the green and blue colors denote moderate and strong antagonism.

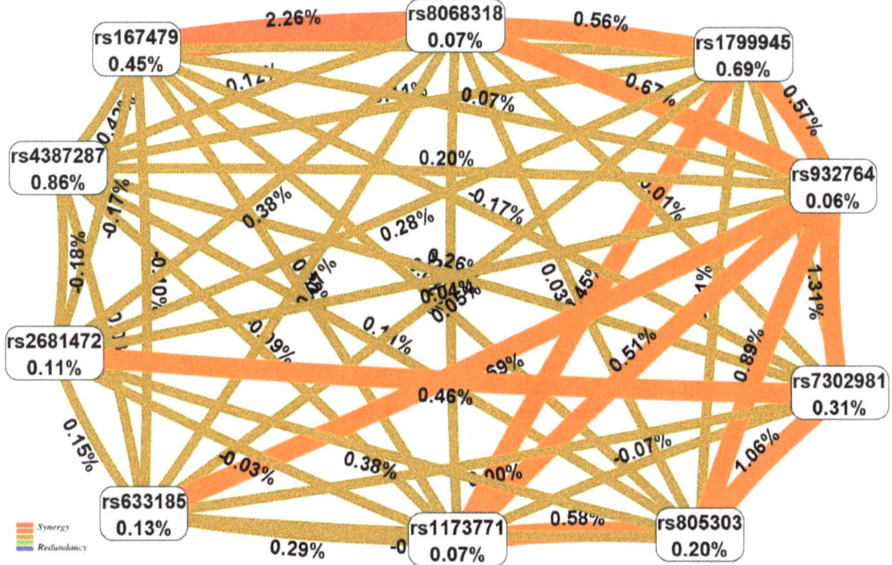

Figure 2. The entropy graph of the SNP×SNP interactions with HTN in women based on the MDR analysis. Positive values of entropy indicate synergistic interactions while the negative values indicate redundancy. The red and orange colors denote strong and moderate synergism, respectively, the brown color denotes the independent effect, the green and blue colors denote moderate and strong antagonism.

2.1.2. Epigenetic Changes of DNA Determined by HTN-Related Loci

Haploreg epigenomic annotations were used to identify regulatory variants among HTN-related loci in men (n = 104) and women (n = 135). The proposed functionality of the enormous proportion of analyzed loci [100/104, 96.15% (men); 130/135, 96.29% (women)] was found (Table S6), and these loci were located in such functionally active DNA sequences as enhancers [51/104, 49.04% (men); 69/135, 51.11% (women)], promoters [30/104, 28.85% (men); 30/135, 22.22% (women)], DNAase hyper-sensitive chromatin [47/104, 45.19% (men); 64/135, 47.41% (women)], evolutionary conservation nucleotide sequences [14/104, 13.46% (men); 19/135, 14.07% (women)], sites of DNA binding to regulatory proteins [23/104, 22.12% (men); 29/135, 21.48% (women)] and transcription factors [87/104, 83.65% (men); 113/135, 83.70% (women)]. More than half of all estimated loci were located in introns [62/104, 59.62% (men); 75/135, 55.55% (women)], about 5% in coding regions of the genome [6/104, 5.77% (men); 6/135, 4.44% (women)], about 4% in 3′-UTR gene regions [5/104, 4.81% (men); 5/135, 3.70% (women)] and less than 1% in 5′-UTR gene regions of the genome [1/104, 0.96% (men); 1/135, 0.74% (women)] (Table S6).

In aggregate, HTN-significant loci were functional with respect to 25 genes in men (AC026703.1; BAG6; CERS5; BCAS3; C12orf62; COX14; GPD1; CSNK2B; HIST1H2AC; GPD1; HFE; HIST1H1T; HIST1H2BC; LIMA1; HIST1H4C; PRRC2A; LY6G5B; RP11-411N4.1 LASS5; NPR3; PLCE1; RGL3; OBFC1; TBX2; C17orf82) and 30 genes in women (BCAS3; HIST1H1T; ATP2B1; AC026703.1; BAG6; ARHGAP42; CERS5; C12orf62; CSNK2B; COX14; GPD1; NPR3; HIST1H2AC; GPD1; LASS5; HFE; HIST1H2BC; LIMA1; HIST1H4C; LOC338758; PLCE1; RP11-981P6.1; LY6G5B; RGL3; OBFC1; POC1B-GALNT4; PRRC2A; RP11-411N4.1; TBX2; C17orf82) (Table S6).

The connection of several HTN-causal genes/loci such as *AC026703.1* rs1173771 G > A, *CERS5* rs7302981 G > A, *OBFC1* rs4387287 C > A, *BAG6* rs805303 G > A with the DNA epigenetic modifications characteristic of promoters (H3K4me3/H3K9ac) and enhancers (H3K4me1/H3K27ac) in disorder target organs—heart (fetal [rs4387287; rs7302981] and adult [rs1173771; rs805303; rs4387287; rs7302981]), aorta [rs4387287], is very interesting and important for understanding the involvement of the investigated polymorphisms in the HTN pathophysiology.

2.1.3. Expression Quantitative Traits (eQTL) Associated with HTN-Significant SNPs

The materials of the Blood eQTL browser specify the connection of a number of HTN-significant loci (four causal SNPs [*CERS5* rs7302981 G > A, *HFE* rs1799945 C > G, *OBFC1* rs4387287 C > A, *BAG6* rs805303 G > A] and 23 LD SNPs in men and five causal SNPs [*CERS5* rs7302981 G > A, *HFE* rs1799945 C > G, *OBFC1* rs4387287 C > A, *BAG6* rs805303 G > A, *ATP2B1* rs2681472 A > G] and 33 proxies in women) with the mRNA production of several genes (nine genes in men [*LY6G5C, SLK, LIMA1, HCP5, HSPA1B, LASS5, AIF1, TRIM38, ALAS2,*] and ten genes in women [*WDR51B, LIMA1, LASS5, LY6G5C, SLK, AIF1, TRIM38, ALAS2, HCP5, HSPA1B*]) in blood (Tables S7 and S8).

The data from a comprehensive public GTEx resource show a tissue-specific link between HTN-involved SNPs and gene expression in non-diseased tissue. All 8 HTN-causal SNPs and 88 of 96 coupled loci (91.67%) in men, and all 10 HTN-causal SNPs and 114 out 125 linked loci (91.20%) in women have allele-specific eQTL effects (Tables S9 and S10). The HTN-impact eQTL-dependent gene list in men and women has considerable similarities. Eighty HTN-associated genes due to the eQTL effects of HTN-related loci are the same in men and women. These are genes such as *NPR3, NEU1, NCR3, MPIG6B, MIR6891, MICB, LY6G6F, LY6G6E, LY6G6D, CYP21A2, CYP21A1P, CTC-510F12.3, CSNK2B, LY6G5C, LY6G5B, LIMA1, LARP4, ZNF322, ZBTB12, XXbac-BPG248L24.12, WASF5P, VWA7, UQCRHP1, U91328.19, TRIM38, TBX2-AS1, TBX2, STN1, STK19B, STK19, SMARCD1, SLK, SLC17A3, RNF5, SLC17A1, SH3PXD2A-AS1, SH3PXD2A, RP4-605O3.4, RP3-405J10.3, RP11-541N10.3, RP11-457M11.5, RGL3, PRRC2A, POU5F1, HSPA1A, HLA-S, HLA-DRB6, HLA-DRB5, HLA-DQB1, HLA-B, HIST1H3E, HFE, HDAC1P1, HCP5, HCG22, GUSBP2, GSTO1, GPD1, GPANK1, DXO, DIP2B, DDAH2, COX14, CLIC1, CERS5, ATF6B, ATF1,*

ASIC1, AQP5, APOM, AIF1, BAG6, C4B, C4A, CCHCR1, C6orf48, BTN2A3P, ATP6V1G2, LY6G6C, ABHD16A. Nine genes such as *TMEM133, RP11-981P6.1, ARHGAP42, POC1B-AS1, ATP2B1, GALNT4, ATP2B1-AS1, POC1B, RP11-567C2.1* are added to the HTN-impact genes list in women due to the eQTL effects of two additional HTN-related polymorphisms (*ARHGAP42* rs633185 C > G, *ATP2B1* rs2681472 A > G).

We found interesting SNP-eQTL correlations in organs (heart/aorta/coronary arteries, etc.) that are targets for HTN. For example, among HTN-causal SNPs with gene transcription in the heart have been associated rs1799945 (*HIST1H3E*), rs805303 (*LY6G5B, HLA-DRB5, DDAH2, CSNK2B, C4A, CYP21A1P*), rs633185 (*ARHGAP42, TMEM133*), rs7302981 (*RP4-605O3.4, CERS5*), rs2681472 (*RP11-981P6.1*), rs8068318 (*TBX2-AS1, TBX2*) (Table S9). In another example, SNP-eQTL connections in the aorta demonstrated loci such as rs1799945 (*HFE*), rs805303 (*LY6G5B, HLA-DRB5, BAG6, LY6G5C, ABHD16A, VWA7, C4A, CYP21A1P*), rs633185 (*ARHGAP42, TMEM133*), rs7302981 (*ATF1, RP4-605O3.4, COX14, SMARCD1*), rs2681472 (*RP11-981P6.1, ATP2B1*), rs8068318 (*TBX2-AS1*) (Table S9). Another clear example of SNP-eQTL correlations significant for the HTN pathophysiology is the connection of some disease-causal loci with mRNA levels in coronary arteries: rs805303 (*LY6G5B, LY6G5C, BAG6, HLA-DRB5*), rs633185 (*ARHGAP42, TMEM133*), rs7302981 (*RP4-605O3.4*) (Table S9).

Strongly coupled loci also have significant eQTL potential in the above-mentioned HTN target organs (Table S10). Linked SNPs have a special eQTL significance in relation to 18 genes in the heart (*TMEM133; TBX2-AS1; TBX2; STN1; STK19B; RP4-605O3.4; RP11-981P6.1; NCR3; LY6G5B; HLA-DRB5; HIST1H3E; DDAH2; CYP21A1P; CERS5; CSNK2B; C4A; ATF6B; ARHGAP42*), 21 genes in the aorta (*VWA7; TMEM133; TBX2; TBX2-AS1; SMARCD1; SLK; RP4-605O3.4; RP11-981P6.1; LY6G5C; LY6G5B; HLA-DRB6; HLA-DRB5; HFE; CYP21A1P; COX14; C4A; BAG6; ATP2B1; ATF1; ARHGAP42; ABHD16A*), 7 genes in coronary arteries (*TMEM133; RP4-605O3.4; LY6G5B; HLA-DRB5; BAG6; ARHGAP42; LY6G5C*), 24 genes in other arteries (*ZNF322; TMEM133; TBX2-AS1; TBX2; STK19; SLK; RP4-605O3.4; RP11-981P6.1; POC1B; LY6G5C; LY6G5B; LIMA1; HLA-DRB5; HFE; CYP21A2; C4A; BAG6; COX14; CYP21A1P; ATP2B1-AS1; ATP2B1; ATF1; ARHGAP42; ABHD16A*) (Table S10).

2.1.4. Splicing Quantitative Traits (sQTL) Associated with HTN-Significant SNPs

The presumable splicing regulation by investigated heritable DNA variations were detected. The tissue-specific SNP-splicing associations were recognized for five HTN-causal loci [*CERS5* rs7302981 G > A, *HFE* rs1799945 C > G, *OBFC1* rs4387287 C > A, *BAG6* rs805303 G > A, *TBX2* rs8068318 T > C] and 65 of 96 proxies SNPs (67.71%) in men and six HTN-causal polymorphisms [*CERS5* rs7302981 G > A, *HFE* rs1799945 C > G, *OBFC1* rs4387287 C > A, *BAG6* rs805303 G > A, *TBX2* rs8068318 T > C, *ATP2B1* rs2681472 A > G] and 67 out 125 linked loci (53.60%) in women (Tables S11 and S12). The HTN-significant sQTL-dependent gene list in men (78 genes) and women (79 genes) has an impressive likeness due to the similarity of the sQTL-correlated polymorphism list (five of six loci were the same). A list of the same sQTL-dependent genes for men and women such as ($n = 78$) (*VARS; TBX2-AS1; TBX2; STK19B; STK19; SMARCD1; SH3PXD2A-AS1; RP4-605O3.4; RP11-332H18.9; RP11-332H18.8; RP11-332H18.7; RP11-332H18.6; RP11-332H18.5; RP11-332H18.47; GPANK1; RP11-332H18.46; RP11-332H18.45; RP11-332H18.44; RP11-332H18.43; RP11-332H18.42; RP11-332H18.41; RP11-332H18.40; RP11-332H18.39; RP11-332H18.38; RP11-332H18.37; RP11-332H18.36; RP11-332H18.35; RP11-332H18.34; RP11-332H18.33; RP11-332H18.32; RP11-332H18.31; RP11-332H18.30; RP11-332H18.29; RP11-332H18.28; RP11-332H18.27; RP11-332H18.26; RP11-332H18.25; RP11-332H18.24; RP11-332H18.23; RP11-332H18.22; RP11-332H18.21; RP11-332H18.20; RP11-332H18.19; RP11-332H18.18; RP11-332H18.17; RP11-332H18.16; RP11-332H18.15; RP11-332H18.14; RP11-332H18.13; RP11-332H18.12; RP11-332H18.11; RP11-332H18.10; PRRC2A; MICA; LY6G6C; LY6G5C; LY6G5B; LST1; LSM2; HLA-DRB6; HLA-DRB5; HLA-DRB1; HLA-DQA1; HFE; FLOT1; FAM186A; DDX39B; COX14; CYP21A2; CYP21A1P; CERS5; CCHCR1; C6orf48; BCAS3;*

BAG6; ATF6B; ATF1; AIF1) (Tables S11 and S12). In women, only one additional sQTL dependent gene (*POC1B-AS1*) was added to this list.

Multiple SNP-sQTL connections, which are quite essential for HTN pathophysiology, were identified by us as they manifest themselves in organs that are targeted for HTN, such as the heart (rs805303 [*GPANK1; HLA-DRB1; STK19B; HLA-DRB5; LY6G5C; HLA-DRB6; LY6G5C*], rs7302981 [*CERS5*], rs8068318 [*RP11-332H18.13; RP11-332H18.14; RP11-332H18.19; RP11-332H18.20; RP11-332H18.44*]), aorta (rs805303 [*LSM2; BAG6; HLA-DRB1; STK19B; HLA-DRB5; HLA-DRB6*], rs7302981 [*CERS5; COX14; RP4-605O3.4*], rs8068318 [*RP11-332H18.5; RP11-332H18.6*]), coronary artery (rs805303 [*BAG6; HLA-DRB1; HLA-DRB5; HLA-DRB6*], rs7302981 [*CERS5; COX14; RP4-605O3.4*], rs8068318 [*RP11-332H18.11; RP11-332H18.12*]), other arterial vessels (rs805303 [*ATF6B; BAG6, GPANK1; HLA-DRB1; HLA-DRB5; HLA-DRB6*], rs7302981 [*CERS5*]), rs8068318 [*RP11-332H18.5; TBX2-AS1*]) (Table S11). More than 60 linked loci also exhibit their sQTL effects in target organs: the heart (ten genes such as *TBX2-AS1; TBX2; STK19B; RP11-332H18.5; LY6G5C; HLA-DRB6; HLA-DRB5; HLA-DRB1; GPANK1; CERS5*), aorta (twelve genes such as *TBX2-AS1; TBX2; STK19B; RP4-605O3.4; RP11-332H18.5; LSM2; HLA-DRB6; HLA-DRB5; BAG6; HLA-DRB1; COX14; CERS5;*), coronary artery (nine genes such as *TBX2-AS1; RP4-605O3.4; RP11-332H18.5; HLA-DRB6; HLA-DRB5; HLA-DRB1; COX14; CERS5; BAG6*), and other arterial vessels (thirteen genes such as *TBX2-AS1; TBX2; RP4-605O3.4; RP11-332H18.5; HLA-DRB6; HLA-DRB5; HLA-DRB1; GPANK1; COX14; CERS5; CCHCR1; BAG6; ATF6B*) (Table S12).

2.1.5. HTN-Associated Gene Pathways

As a result of the evaluation of the functionality at HTN-correlated loci (135 SNPs in women [10 causal loci and 125 strongly linked SNPs] and 104 SNPs in men [8 causal loci and 96 LD SNPs]), 159 genes were found to be involved in the HTN susceptibility in women and 147 genes in men. The substantial similarity in the HTN-involved genes list in men and women determines the almost identical biological pathways in them [in total, using Gene Ontology enrichment analysis tools, more than 140 different pathways were identified in both men (n = 145, Table S13) and women (n = 141, Table S14)] and were largely represented by pathways associated with the involvement of many different immune reactions/processes. Among both men and women, two biological pathways had the highest rates of statistical significance: MHC (major histocompatibility complex) protein (PANTHER Protein Class ID 00149) ($p_{(fdr)}$ equal 5.30×10^{-11} in men and 8.33×10^{-11} in women) and antigen processing & presentation (PANTHER Slim Biological Process ID 0019882) ($p_{(fdr)}$ equal 6.00×10^{-10} in men and 9.44×10^{-10} in women) (Tables S13 and S14).

The estimates of the mechanisms of intergenic interactions of HTN-significant genes in men and women, derived with help from the Genemania bioinformatic resource, also turned out to be almost the same. In both men (Figure 3) and women (Figure 4), the main biological mechanisms of the influence of HTN-significant genes on the disease development were co-expression [49.29% (men); 48.28% (women)], physical interactions [29.93% (men); 31.09% (women)], common protein domains [7.68% (men); 7.69% (women)], co-localization [7.33% (men); 7.10% (women)], predicted interactions [5.77% (men); 5.84% (women)]. Cooperation between genes such as *HLA-DQB1—HLA-DQA1, CSNK2A1—CSNK2B, LSM3—LSM2*, demonstrated the very strong contribution (their weight coefficient was maximum and equal to 1) to the HTN susceptibility in both men (Table S15) and women (Table S16).

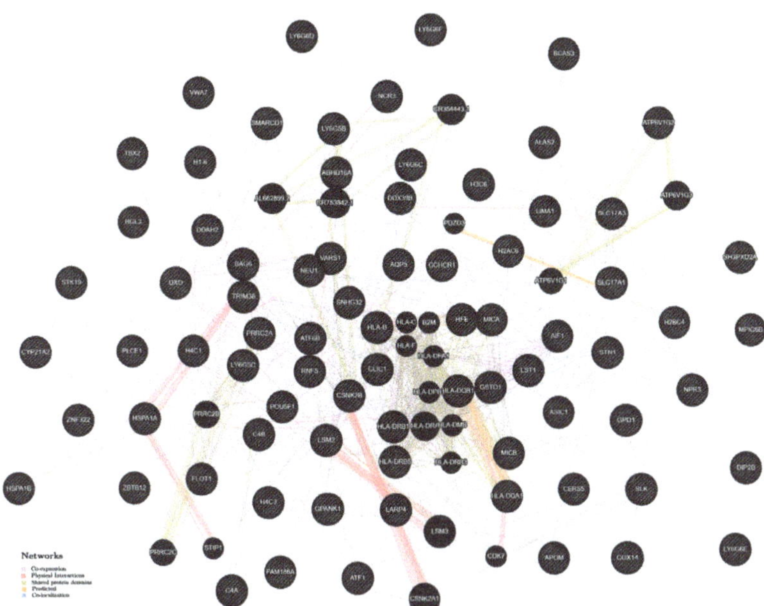

Figure 3. The interaction networks of the candidate genes for HTN in men in various tissues/organs inferred using GeneMANIA (http://genemania.org (accessed on 19 December 2022)). The candidate genes are cross-shaded.

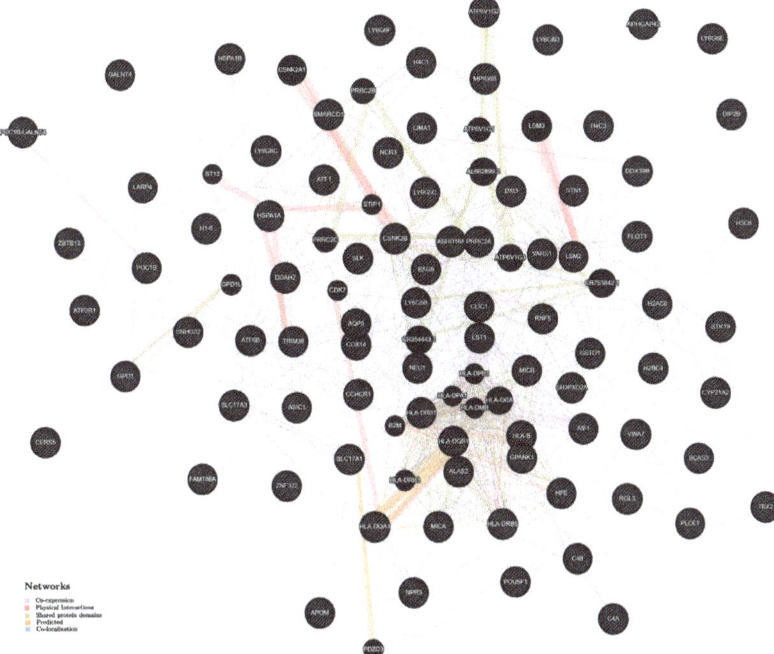

Figure 4. The interaction networks of the candidate genes for HTN in women in various tissues/organs inferred using GeneMANIA (http://genemania.org (accessed on 19 December 2022)). The candidate genes are cross-shaded.

3. Discussion

The present study identified the men–women differences in Europe in the HTN involvement of the BP/HTN-associated GWAS SNPs. Among women, the HTN risk was determined by *HFE* rs1799945 C > G and inter-locus interactions of all 10 examined SNPs as part of 26 intergenic interactions models, whereas in men, the locus *BAG6* rs805303 G > A and inter-SNPs interactions of eight loci in only seven models were correlated with HTN. The strongly pronounced functionality of HTN-correlated loci (135 SNPs in women [10 causal loci and 125 strongly related SNPs] and 104 SNPs in men [8 causal loci and 96 LD SNPs]) determines the involvement of 159 genes in women and 147 genes in men in the disease susceptibility architecture. A significant similarity in the list of genes involved in HTN in men and women determines almost identical signaling pathways (mainly due to immune mechanisms) in them.

The results of our study showed that the presence of the GG genotype (rs1799945 *HFE*) in a woman substantially increases the chance of HTN developing by more than 10 times (OR-11.15). The HTN-dangerous (high BP) effect of allele G and HTN-safe value (low BP) of allele C polymorphism *HFE* rs1799945 C > G have been detected in earlier studies (GWAS and etc.) [9,18,66–70], which is undividedly compatible with our data in the women's cohorts.

The polymorphism HFE rs1799945 C > G and the seven proxy loci exhibit pronounced functionality in relation to fifteen genes (U91328.19; RP11-457M11.5; HIST1H4C; ALAS2; SLC17A3; HIST1H1T; BTN2A3P; SLC17A1; HIST1H2AC; GUSBP2; ZNF322; HFE; HIST1H3E; TRIM38; HIST1H2BC) (our in silico data) and are of paramount importance in the regulation of iron metabolism (serum concentration of such iron status biomarkers as iron/transferrin/ferritin/transferrin saturation, total iron binding capacity) (literary GWAS data [71,72]) and related metabolic pathways which are HTN-important (hemoglobin concentration, red cells parameters, glucose homeostasis, glycated hemoglobin levels, etc.) [72–79]. SNP HFE rs1799945 C > G has been correlated with medication agents acting on the renin–angiotensin system [80]. Importantly, Gill et al. used a Mendelian randomization analysis of GWAS (48, 972 European/Genetics of Iron Status Consortium) and PheWAS (424, 439 European/UK Biobank) summary data and founded a causal link between genetically determined levels of serum iron and a hazard of hypercholesterolemia and anemia [81]. The risk value of hypercholesterolemia for HTN (and in common for cardiovascular diseases) is well-known [6,82] and it is also revealed in the sample studied in this work. So, data from the literature and this study material displayed visible HTN-impact pleiotropic effects of HFE rs1799945 C > G.

In the studied men, the genotype AA of *BAG6* rs805303 G > A dramatically reduced the danger of HTN (OR = 0.30). In two early GWAS allele G rs805303 was HTN/high BP-risky and allele A was linked with low BP [18,76]. The higher systolic/diastolic BP in Ugandan adolescents having allele G rs805303 was detected by Lule et al. [83]. So, it can be noted that the same orientation of allelic variants *BAG6* rs805303 G > A is associated with HTN/BP in our study and previously performed works. According to the in silico study results, the locus (with ten proxy SNPs) is the major regulator (the so—called "master regulator") of epigenetic/expression/splicing traits at fifty-five genes, including immune system genes (e.g., *HLA, LY6, HSP* gene clusters) which strongly correlated with HTN [84–88] (detailed information about the connection of the immunity-importance above-mentioned genes with HTN is given below).

Large-scale epidemiological studies indicate the presence of visible men–women differences in BP [2,6,8,82,89,90]. At that, these differences were more noticeable in high-income countries and those in Central/Eastern Europe than in the countries of other regions [89]. It is believed that the incidence of HTN in men is 1.2 times higher in comparison with women and the DALYs linked with SBP $\geq$ 140 mmHg (1.38 times) and SBP $\geq$ 110–115 mmHg (1.44 times) are greater by1.4 times [2,6]. One of the reasons determining the higher incidence of HTN in men (in comparison with women) may be the prevalence of such risk factors for cardiovascular diseases (including HTN) in them as a diet high in sodium and low in fruit, smoking, alcohol and drug use, and high fasting blood glucose levels [8]. Thus, a more pronounced exposure of environmental risk factors in men than in women, on the one hand, may have an independent HTN risk value, while on the other hand have a significant modifying effect on the realization of hereditary predisposition to the disease. The significant role of gene–environment interactions (SNPs with smoking, alcohol intake) in the nature of candidate gene polymorphism associations with HTN, including those considered in this work *ARHGAP42* rs633185 C > G, *HFE* rs1799945 C > G, *AC026703.1* rs1173771 G > A was shown in previously conducted GWAS [9,10,65]. It should be noted that a number of the above-mentioned HTN risk factors involved in SNP-environment interactions significant for the disease (in accordance with the GWAS data) were also registered in the studied sample of patients, both men and women (higher values of blood glucose, smokers, hypercholesterolemia, etc.).

Together with this, an important value in the HTN susceptibility is played by sex hormones, which, firstly, are directly involved in the process regulation of vasoconstriction/vasodilation [90–93]; secondly, they have a pronounced influence on a number of cardiovascular/HTN risk factors (distribution of adipose tissue in the body, the development of obesity and metabolic syndrome, the formation of obesity-dependent/independent insulin resistance, etc.) [7]; thirdly, they can be significant "modifiers" in the phenotypic manifestation of potential genetic determinants of HTN by modulating various nuclear and extra-nuclear pathways that control the expression of multiple genes, post-translational modifications of protein molecules, various HTN-impact signaling pathways, etc. [94,95]. Obviously, estrogens cause a BP decrease in premenopausal women and, accordingly, have a protective effect on HTN development [90–93]. Estrogens realize their HTN-protective phenotypic effects through vasoconstriction/vasodilation mechanisms due to the regulation of the renin–angiotensin–aldosterone system, the production of catecholamines, endothelins and angiotensin II [91–93]. After the menopause, the HTN-protective effects of estrogen in women decrease and the risk of raised BP increases [90]. It is assumed that testosterone, by increasing the activity of the renin–angiotensin–aldosterone system, promotes the development of oxidative stress, leading to the increased production of vasoconstrictors and a decrease in the effects of vasodilators (nitric oxide), which predetermines a higher blood pressure level and, accordingly, a higher HTN risk in men [92]. With age, testosterone levels in men decrease; however, the HTN risk does not decrease as expected, but rather increases due to a decrease in the regulatory effect of testosterone on adipose tissue (such as the suppression of adipocyte proliferation, decreased stromal vascular growth, androgen receptor deficiency, etc.) [7,96]), which, in turn, determines an increased risk of abdominal (visceral) obesity in men [7,97,98]. Thus, an age-dependent decrease in testosterone levels in men increases the risk of visceral obesity and thereby increases the HTN risk [7]. The data on the effect of testosterone on the development of obesity in women are ambiguous; there is evidence of a connection between high testosterone levels and both low visceral fat [99] and high fat content [100].

There is convincing evidence of correlations of sex-specific differences in the functioning of the autonomic nervous system and related features of the immune status/reactions of the body with the HTN risk [90,101]. It has been recorded that, in women compared with men, with age and also with obesity, the activity of the sympathetic nervous system is increased [90]. A change in the activity of the sympathetic nervous system has a direct regulatory effect on T cells, which in turn activates various signaling pathways of innate and adaptive immunity (production of various cytokines of pro-inflammatory action, pro-/anti- inflammatory cytokines signaling, interferon-γ-mediated reactions, activation of natural killer cells and monocytes, vascular inflammation, etc.) [85,87,101,102]. Along with the sympathetic nervous system, sex hormones are also involved in the regulation of the innate/adaptive immune system molecular mechanisms (estrogens have an immunostimulating effect while androgens have an immunosuppressive effect) [95]. Interestingly, the materials derived by us from data bioinformatic analysis indicate the paramount importance of immune mechanisms/processes (such as MHC proteins, antigen processing and presentation, and more than 100 other different immune pathways) in the HTN susceptibility in both men and women.

Our in silico data indicate a connection of HTN with a number of genes that control the organism's immune responses. These are genes such as *HLA* system (*HLA-DRB6; HLA-DRB5; HLA-DRB1; HLA-DQA1; HLA-S; HLA-B*), *HSP* (*heat shock protein:HSPA1A; HCP5; HSPA1B*), *LY6* (*lymphocyte antigen 6:LY6G5C; LY6G5B; LY6G6C; LY6G6E; LY6G6D; LY6G6F*), etc., the expression/splicing of which is regulated by HTN-associated polymorphisms in various organs, including those that are targets for disease (heart/aorta/coronary and other arteries). It is believed that the HLA background shapes the T cell receptor (TCR) repertoire by neglecting/favoring specific T cell subgroups represented by T cell receptor variable beta chain (TCRBV) usage [84]. Different HLAs present auto-/foreign antigens to TCRBV-specific subpopulations of T cells in different ways (better/worse), which determines the individual specific features of interactions in the HLA–TCR system and is important in the formation of susceptibility to various immuno-significant diseases, including hypertension [84,85]. It seems important to point out the presence of clearly expressed differences between men and women regarding the effect of HLA genes on TCRBV transcripts (CD8-T cells in men affected by autoimmune disorders had the ability to multiply in the absence of TCR expression with similarities in key HLA-binding regions), which are supposed to be based on hormone-mediated mechanisms [84]. HSPs act as regulators of the organism's immune responses, and they are produced when the body is exposed to various damaging/stressful factors (mechanical/oxidative stress, cytokine influences, etc.) and they are the "protective" response of the cells (including cells of the arterial wall) to these effects [86,87]. T cells, interacting with HSPs (primarily HSP60, HSP70), form a regulatory T-cell response of anti-inflammatory directivity [88]. So, momentary effects of HSPs in HTN are protective due to suppression of the NF-κB pathway and improve the BP reply to angiotensin II [88]. Nonetheless chronic hyperexpression of HSP70 probably has prohypertensive value due to its capacity to evoke autoimmune processes [88]. Data from the literature on the increased formation of some HSPs (HSP70, HSP72) in HTN patients (including arteries (adventitial areas) and kidney) are presented [87]. In the work by Li et al., the relationship of a number of polymorphisms of three *HSP70* genes (*HSPA1A; HSPA1B; HSPA1L*) both independently and as part of individual haplotypes with HTN, was shown in Uygur [86]. Of importance, the involvement of two of these genes (such as *HSPA1A* and *HSPA1B*) in the disease was also established by our in silico analysis. LY6 protein family members (including those considered in our work; gene products such as *LY6G5C; LY6G5B; LY6G6C; LY6G6E; LY6G6D; LY6G6F*) interacting with various endogenic regulatory factors (interferon-γ, type I and II interferons, retinoic acid, β2-integrins, matrix metalloproteinases, lymphotoxin alpha, etc.) affect immunity-significant cellular functions (T cell activation/proliferation, CD8+ T cell migration, cell adhesion, B cell specification, neutrophil recruitment, etc.) which are essential for the wide range of HTN-involved processes such as inflammation progression, activity of complement, neuronal activity, angiogenesis, etc. [103]. Despite the presence of certain sex-specific features of the immune pathways involved in HTN biology, as indicated in the literature (women, in comparison with men, are more likely to have inflammatory/autoimmune disorders that increase the HTN risk; have greater numbers of circulating IgM and more CD4 T cells; higher infiltration of the kidneys by T cells with an increase in the number of Th17 cells; an increase in the content of regulatory T cells in adipose with weight gain in women, and a pattern of reverse orientation in men; differences in HLA-mediated T-cell selection/expansion, predestining the beta chain of the determining the features of the beta chain of the T cell receptor, differences in the pro-/anti- inflammatory cytokines signaling, etc.) [84,85,102], according to our bioinformatic data, almost all of the detected HTN-significant immune pathways were the same in men and women (due to the almost identical HTN susceptibility gene list in men [147 genes] and women [159 genes], determined by us on the basis of in silico data).

Importantly, all of the above-described HTN-important environmental risk factors, biological processes/mechanisms (state of the autonomic nervous system, immune status, hormonal background, etc.) are closely correlated with each other and represent a sophisticated multi-level, multi-stage and multi-directional sex-specific system of regulation of BP-related traits (phenotypes) in HTN. There is no doubt that all these factors discussed above can be powerful epigenetic modifiers of the phenotypic manifestation of HTN predisposition genes and determine men–women differences in the involvement of genetic determinants in susceptibility to the disease. This study shows differences in the nature of the HTN genetic determination in men and women both within the framework of the main effects of BP/HTN-associated GWAS SNPs and their intergenic interactions: in women, susceptibility to the disease is determined by the polymorphism *HFE* rs1799945 C > G and strongly pronounced inter-locus interactions of all 10 examined SNPs (26 intergenic interactions were identified models), whereas in men, predisposition to the disease was associated with *BAG6* rs805303 G > A polymorphism and significantly less pronounced interactions between only 8 considered loci (just 7 models were founded).

It is quite interesting that the data of this work are largely consistent with our previously published materials in which, during an associative study of a women sample with pre-eclampsia/without pre-eclampsia from the identical population of Europeans in Central Russia (the list of studied loci was analogous), it was found that the *HFE* rs1799945 C > G increased the pre-eclampsia risk (for allele G OR = 2.24) and *BAG6* rs805303 G > A decreased the pre-eclampsia risk (for allele A OR = 0.55–0.78) [104] including this pregnancy complication risk in women with BMI ≥ 25 (for allele A OR = 0.36–0.66) [105]. So, the polymorphic variant G rs1799945 *HFE* increases both the HTN risk in women (data from this work, OR = 11.15 for genotype GG) and the pre-eclampsia risk in women (a specific symptom of this complication of pregnancy is increased BP (OR= 2.24 [104])), which may indicate the "universality" of this genetic marker as a risk factor for the development of hypertensive conditions in European women in Central Russia, and allows us to recommend its use in practical medicine in order to distinguish women with a high risk of developing elevated BP. Concurrently, there are some discrepancies in the results of these studies: SNP *BAG6* rs805303 G > A has a protective value for HTN in men (not in women!) (OR = 0.30), but in parallel, this polymorphism marks a low risk of developing pre-eclampsia in pregnant women as a whole (OR = 0.55–0.78) [104] and among women with a BMI ≥ 25 (OR = 0.36–0.66) [105]. One of the possible reasons for these sex-specific differences in the value of the *BAG6* rs805303 G > A locus in the development of pathology with elevated BP may be the modifying effect on the phenotypic manifestation of this polymorphism of an unequal confounder factor list taken into account in these studies (BMI, TC, TG, HDL-C, LDL-C, blood glucose, smokers—present study; age, family history of PE, pre-pregnancy BMI, obesity, number of gravidities, spontaneous/induced abortions, stillbirths, smokers—previous study [104]). At the same time, there is an obvious need to continue studies on the association of SNP *BAG6* rs805303 G > A with diseases associated with elevated BP in the studied population, in order to finally establish its predictive potential (including their sex-specific features).

This study has a number of limitations. Firstly, experimental confirmations of the functional effects of HTN-significant GWAS loci identified in silico by us (influence on expression, splicing, epigenetic modifications of genes) are needed. Secondly, it is necessary to confirm (in silico and experimentally) men–women differences in the functional effects of GWAS-significant loci. Thirdly, an increase in the number of samples of men and women under consideration would allow the identification of phenotypic effects (association with the disease) and others which were "weaker" for this population of GWAS loci. Fourthly, the expansion of the panel of studied GWAS loci would allow to expand the data on men–women features of the genetic determination of HTN. It should be noted that conducting similar studies in other ethno-territorial groups of the population can apparently "show" other patterns (different from our results) since these studies (as well as our work) will be replicative and their results will be largely determined by various features (the structure of the gene pool and the associated features of the "main" effects of genes and the nature of intergenic interactions, the spectrum and severity of environmental risk factors and the associated features of gene–environmental interactions, etc.) of those ethno-territorial groups of the population that will be studied.

4. Materials and Methods

4.1. Study Subjects

Two groups of European subjects of Russia ($n = 1405$ in total, all participants were born in Central Russia and have Russian origin (self-reported) [106,107], such as men ($n = 821$ in total: $n = 564$ HTN, $n = 257$ control) and women ($n = 584$ in total: $n = 375$ HTN, $n = 209$ control) were included in this "case-control" association study. The clinical examination of participants was performed during the 2013–2016 period in the "cardiology department" at the "St. Joasaph Belgorod Regional Clinical Hospital". Diagnosis of HTN (or absence of HTN) was carried out by qualified cardiologists according to the standards set out in the WHO methodological guidelines [1] (information about this has been presented previously [42]). BP indicators were confirmed by Korotkov (auscultative method using a sphygmomanometer) [108]. BP was measured at least twice within a few days. Thirty minutes before the procedure, the subjects did not consume caffeine/smoke/exercise. The measurement was carried out in the patient's sitting position after a five-minute rest. BP was measured on both hands; at least two measurements were taken with an interval of one to two minutes between measurements. As an indicator of an individual's BP, the average value for two measurements taken at least twice was taken. The HTN group was formed from the clinic's (cardiology department) patients. All HTN patients had a clinical history of disorder for one year or more and 81.79% (82.80% men and 80.27% women) received antihypertensive drugs. The absence of HTN (parameters BP were lower than 140mmHg for SBP and lower than 90mmHg for DBP), coronary artery disease and type 2 diabetes mellitus were the basis for inclusion in the control group. Persons who do not suffer from HTN (control group) were recruited during regular (annual) medical examinations at the aforementioned clinical hospital (these examinations were carried out by doctors of various specialties, including qualified cardiologists). All of the participants (HTN and HTN-free) did not have severe chronic allergic/autoimmune/hematological/oncological pathology [109]. Blood specimens of the subjects for defining TC, TG, LDL-C, HDL-C and blood glucose were obtained in the morning (7–9 h) after an eight-hour fast. The implementation of this study was supervised by the Ethics Committee (Human Investigation Committee) at "Belgorod State University" and was accompanied by the written consent of all the subjects.

Information about lifestyle and diet was collected for each subject (patient/control). The consumption of vegetables and fruits in an amount of less than 400 g daily (excluding salted/pickled vegetables, and starchy vegetables (potatoes)) was considered "low fruit/vegetable consumption" [110]. The average weekly physical activity at work and at home related to transport and recreation (including walking, running, fitness club classes, etc.) less than 150 min of moderate-intensity physical activity (for example, brisk walking) for 30 min or longer, at least five times a week) was considered "low physical activity". [111,112]. The average daily intake of fatty foods of more than 10% of the total food consumed (daily energy consumption due to fatty foods in the total amount of daily energy) was considered "high fatty food consumption" [110]. Daily salt intake (sodium chloride) of 5 g (teaspoon) or more per day was considered "high sodium consumption" [110].

4.2. Laboratory DNA Testing

We gathered peripheral blood (leukocytes) to extract genomic DNA [113] (the phenol/chloroform DNA extraction methodology was presented earlier [114]). The ten specially selected polymorphisms (confirmed associations of GWAS level with blood pressure (BP) parameters and/or HTN in European (Table S17) and assumed functional ability [104,105,115] (HaploReg information was regarded [116]) (Table S18)) were considered. The list of studied loci was as follows: (*PLCE1*) rs932764 A > G, (*AC026703.1*) rs1173771 G > A, (*CERS5*) rs7302981 G > A, (*HFE*) rs1799945 C > G, (*OBFC1*) rs4387287

C > A, (*BAG6*) rs805303 G > A, (*RGL3*) rs167479 T > G, (*ARHGAP42*) rs633185 C > G, (*TBX2*) rs8068318 T > C and (*ATP2B1*) rs2681472 A > G. All ten loci were correlated with BP in Europeans and all ten SNPs were associated with HTN: eight SNPs were HTN-linked in Europeans and two loci (rs4387287 *OBFC1* and rs2681472 *ATP2B1*) were disorder-associated in the sample with a predominance (>85%) of Europeans (Table S17). Nine loci out of ten (excluding rs4387287 *OBFC1*) were associated with HTN/BP in two or more GWAS (Table S17). All selected ten SNPs have significant functionality (Table S18). One of the generally accepted methods of genotyping (the allelic discrimination method) and the CFX96 RT System device (Bio-Rad Laboratories, Hercules, CA, USA) were used for laboratory genetic studies [117]. The participants case/control status was masked throughout the laboratory genetic analysis. Genotyping of a random duplicated sample (near 4–6% from all sample) was utilized as an independent internal control to provide the individual genotyping data quality assurance [118,119]. No genotyping errors were registered.

4.3. Association Statistical Analysis

The genotypes of the examined 10 loci were tested for H–W equilibrium [120,121]. The sex-specific disorder-impact genetic association of individual polymorphisms [additive/recessive/dominant/allelic common model was calculated [122]] and interaction between SNPs [123] and HTN was appreciated based on the results (OR with 95%CI was evaluated [124,125]) obtained in the gPLINK [126], MDR [127,128], MB-MDR [129,130] genetic programs. In the logistic regression calculations, we took into account covariates (BMI, TC, TG, LDL-C, HDL-C, blood glucose, smokers (Model 1) and BMI, TC, TG, LDL-C, HDL-C, blood glucose, smokers, low physical activity, high fatty food consumption (Model 2) in both men and women (Table 1)), performed permutation procedures (in order to minimize the probability of false positive results [131,132]) and the results' analysis was carried out on the basis of a stronger level of statistical significance, $p_{\text{perm-Bonferroni}} \leq 0.025$ (Bonferroni's correction related to the amount of comparison pairs studied was accounted, $n = 2$ [men and women]). For individual SNPs statistical power was estimated by Quanto (v.1.2.4) [133].

4.4. Definition of the Alleged Functional Ability of HTN-Related Polymorphisms and Genes

For the purpose of biological interpretation of the identified associations (establishing the mechanisms underlying these associations), we used in silico information on the assumed functional ability of HTN-related polymorphisms (taking into account strongly coupled loci with a coupling strength of at least 0.80 [134,135]) and genes. We used such seven bioinformatics databases as (1) HaploReg [116] (determination of the regulatory potential of polymorphisms: location in putative promoters/enhancers, association with transcription factors/regulatory proteins, localization in regions of open chromatin and evolutionarily conservative DNA sites), (2) SIFT [136] and (3) PolyPhen-2 [137] (identification and evaluation of predictive potential of non-synonymous SNPs), (4) GTEx [138] (correlation of loci with gene expression and alternative splicing in 54 different organs/tissues), (5) Blood eQTL browser [139] (the relationship of SNPs with gene expression in peripheral blood), (6) Gene Ontology [140] (identification of HTN-associated genes pathways), (7) GeneMANIA [141] (estimation and visualization of the mechanisms of intergenic interactions of HTN-significant genes).

5. Conclusions

This study showed a more strongly pronounced contribution of BP/HTN-associated GWAS SNPs to HTN susceptibility (due to weightier intergenic interactions) in European women than in men.

Supplementary Materials: The supporting information can be downloaded at: https://www.mdpi.com/article/10.3390/ijms24097799/s1.

Author Contributions: Conceptualization, T.I. and I.P.; Data curation, M.A. and I.S.; Formal analysis, E.R., I.A., D.P. and I.P.; Project administration, M.C. (Mikhail Churnosov); Writing—original draft, T.I., M.A. and M.C. (Maria Churnosova); Writing—review and editing, M.C. (Mikhail Churnosov), E.R., D.P. and I.A. All authors provided final approval of the version to be published. All authors are accountable for all aspects of the work in ensuring that questions related to the accuracy or integrity of any part of the work are appropriately investigated and resolved. All authors have read and agreed to the published version of the manuscript.

Funding: This research received no external funding.

Institutional Review Board Statement: This study was conducted according to the guidelines of the Declaration of Helsinki, and approved by the Local Ethical Committee of the Belgorod State University (21 January 2013, No. 1).

Informed Consent Statement: Informed consent was obtained from all subjects involved in this study.

Data Availability Statement: The data generated in the present study are available from the corresponding author upon reasonable request.

Conflicts of Interest: The authors declare no conflict of interest.

References

1. Williams, B.; Mancia, G.; Spiering, W.; Agabiti Rosei, E.; Azizi, M.; Burnier, M.; Clement, D.L.; Coca, A.; de Simone, G.; Dominiczak, A.; et al. 2018 ESC/ESH Guidelines for the management of arterial hypertension. *Eur. Heart J.* **2018**, *39*, 3021–3104. [CrossRef] [PubMed]
2. Forouzanfar, M.H.; Liu, P.; Roth, G.A.; Ng, M.; Biryukov, S.; Marczak, L.; Alexander, L.; Estep, K.; Hassen Abate, K.; Akinyemiju, T.F.; et al. Global Burden of Hypertension and Systolic Blood Pressure of at Least 110 to 115 mm Hg, 1990–2015. *JAMA* **2017**, *317*, 165–182. [CrossRef] [PubMed]
3. Singh, G.M.; Danaei, G.; Farzadfar, F.; Stevens, G.A.; Woodward, M.; Wormser, D.; Kaptoge, S.; Whitlock, G.; Qiao, Q.; Lewington, S.; et al. The age-specific quantitative effects of metabolic risk factors on cardiovascular diseases and diabetes: A pooled analysis. *PLoS ONE* **2013**, *8*, e65174. [CrossRef] [PubMed]
4. Muromtseva, G.A.; Kontsevaya, A.V.; Konstantinov, V.V.; Artamonova, G.V.; Gatagonova, T.M.; Duplyakov, D.V.; Efanov, A.Y.; Zhernakova, Y.V.; Il'in, V.A.; Konradi, A.O.; et al. The prevalence of non-infectious diseases risk factors in Russian population in 2012–2013 years. The results of ECVDRF. *Cardiovasc. Ther. Prev.* **2014**, *13*, 4–11. (In Russian) [CrossRef]
5. Boytsov, S.A.; Balanova, Y.A.; Shalnova, S.A.; Deev, A.D.; Artamonova, G.V.; Gatagonova, T.M.; Duplyakov, D.V.; Efanov, A.Y.; Zhernakova, Y.V.; Konradi, A.O.; et al. Arterial hypertension among individuals of 25–64 years old: Preva-lence, awareness, treatment and control. By the data from ECCD. *Cardiovasc. Ther. Prev.* **2014**, *13*, 4–14. (In Russian) [CrossRef]
6. NCD Risk Factor Collaboration. Worldwide trends in blood pressure from 1975 to 2015: A pooled analysis of 1479 population-based measurement studies with 19.1 million participants. *Lancet* **2017**, *389*, 37–55. [CrossRef]
7. Faulkner, J.L.; Belin de Chantemèle, E.J. Sex hormones, aging and cardiometabolic syndrome. *Biol. Sex Differ.* **2019**, *10*, 30. [CrossRef]
8. GBD 2015 Risk Factors Collaborators. Global, regional, and national comparative risk assessment of 79 behavioural, environmental and occupational, and metabolic risks or clusters of risks, 1990–2015: A systematic analysis for the Global Burden of Disease Study 2015. *Lancet* **2016**, *388*, 1659–1724. [CrossRef]
9. Sung, Y.J.; Winkler, T.W.; de Las Fuentes, L.; Bentley, A.R.; Brown, M.R.; Kraja, A.T.; Schwander, K.; Ntalla, I.; Guo, X.; Franceschini, N.; et al. A Large-Scale Multi-ancestry Genome-wide Study Accounting for Smoking Behavior Identifies Multiple Significant Loci for Blood Pressure. *Am. J. Hum. Genet.* **2018**, *102*, 375–400. [CrossRef]
10. Feitosa, M.F.; Kraja, A.T.; Chasman, D.I.; Sung, Y.J.; Winkler, T.W.; Ntalla, I.; Guo, X.; Franceschini, N.; Heng, C.K.; Sim, X.; et al. Novel genetic associations for blood pressure identified via gene-alcohol interaction in up to 570K individuals across multi-pleancestries. *PLoS ONE* **2018**, *13*, e0198166. [CrossRef]
11. Rotimi, C.N.; Cooper, R.S.; Cao, G.; Ogunbiyi, O.; Ladipo, M.; Owoaje, E.; Ward, R. Maximum-likelihood generalized heritability estimate for blood pressure in Nigerian families. *Hypertension* **1999**, *33*, 874–878. [CrossRef]
12. Kupper, N.; Willemsen, G.; Riese, H.; Posthuma, D.; Boomsma, D.I.; de Geus, E.J. Heritability of daytime ambulatory blood pressure in an extended twin design. *Hypertension* **2005**, *45*, 80–85. [CrossRef] [PubMed]
13. Tobin, M.D.; Raleigh, S.M.; Newhouse, S.; Braund, P.; Bodycote, C.; Ogleby, J.; Cross, D.; Gracey, J.; Hayes, S.; Smith, T.; et al. Association of WNK1 gene polymorphisms and haplotypes with ambulatory blood pressure in the general population. *Circulation* **2005**, *112*, 3423–3429. [CrossRef] [PubMed]
14. Gu, D.; Rice, T.; Wang, S.; Yang, W.; Gu, C.; Chen, C.S.; Hixson, J.E.; Jaquish, C.E.; Yao, Z.J.; Liu, D.P.; et al. Heritability of blood pressure responses to dietary sodium and potassium intake in a Chinese population. *Hypertension* **2007**, *50*, 116–122. [CrossRef]

15. Ehret, G.B. Genome-wide association studies: Contribution of genomics to understanding blood pressure and essential hypertension. *Curr. Hypertens. Rep.* **2010**, *12*, 17–25. [CrossRef] [PubMed]
16. Ehret, G.B.; Caulfield, M.J. Genes for blood pressure: An opportunity to understand hypertension. *Eur. Heart J.* **2013**, *34*, 951–961. [CrossRef]
17. Chen, J.; Wang, W.; Li, Z.; Xu, C.; Tian, X.; Zhang, D. Heritability and genome-wide association study of blood pressure in Chinese adult twins. *Mol. Genet. Genomic. Med.* **2021**, *9*, e1828. [CrossRef]
18. International Consortium for Blood Pressure Genome-Wide Association Studies; Ehret, G.B.; Munroe, P.B.; Rice, K.M.; Bochud, M.; Johnson, A.D.; Chasman, D.I.; Smith, A.V.; Tobin, M.D.; Verwoert, G.C.; et al. Genetic variants in novel pathways influence blood pressure and cardiovascular disease risk. *Nature* **2011**, *478*, 103–109. [CrossRef]
19. Giri, A.; Hellwege, J.N.; Keaton, J.M.; Park, J.; Qiu, C.X.; Warren, H.R.; Torstenson, E.S.; Kovesdy, C.P.; Sun, Y.V.; Wilson, O.D.; et al. Trans-ethnic association study of blood pressure determinants in over 750,000 individuals. *Nat. Genet.* **2019**, *51*, 51–62. [CrossRef]
20. German, C.A.; Sinsheimer, J.S.; Klimentidis, Y.C.; Zhou, H.; Zhou, J.J. Ordered multinomial regression for genetic association analysis of ordinal phenotypes at Biobank scale. *Genet. Epidemiol.* **2019**, *44*, 248–260. [CrossRef]
21. Jeong, H.; Jin, H.-S.; Kim, S.-S.; Shin, D. Identifying Interactions between Dietary Sodium, Potassium, Sodium–Potassium Ratios, and *FGF5* rs16998073 Variants and Their Associated Risk for Hypertension in Korean Adults. *Nutrients* **2020**, *12*, 2121. [CrossRef] [PubMed]
22. Lee, S.B.; Park, B.; Hong, K.W.; Jung, D.H. Genome-Wide Association of New-Onset Hypertension According to Renin Concentration: The Korean Genome and Epidemiology Cohort Study. *J. Cardiovasc. Dev. Dis.* **2022**, *9*, 104. [CrossRef]
23. Hartiala, J.A.; Han, Y.; Jia, Q.; Hilser, J.R.; Huang, P.; Gukasyan, J.; Schwartzman, W.S.; Cai, Z.; Biswas, S.; Trégouët, D.A.; et al. Genome-wide analysis identifies novel susceptibility loci for myocardial infarction. *Eur. Heart J.* **2021**, *42*, 919–933. [CrossRef] [PubMed]
24. Kato, N.; Loh, M.; Takeuchi, F.; Verweij, N.; Wang, X.; Zhang, W.; Kelly, T.N.; Saleheen, D.; Lehne, B.; Leach, I.M.; et al. Trans-ancestry genome-wide association study identifies 12 genetic loci influencing blood pressure and implicates a role for DNA methylation. *Nat. Genet.* **2015**, *47*, 1282–1293. [CrossRef]
25. Koyama, S.; Ito, K.; Terao, C.; Akiyama, M.; Horikoshi, M.; Momozawa, Y.; Matsunaga, H.; Ieki, H.; Ozaki, K.; Onouchi, Y.; et al. Population-specific and trans-ancestry genome-wide analyses identify distinct and shared genetic risk loci for coronary artery disease. *Nat. Genet.* **2020**, *52*, 1169–1177. [CrossRef] [PubMed]
26. Levy, D.; Ehret, G.B.; Rice, K.; Verwoert, G.C.; Launer, L.J.; Dehghan, A.; Glazer, N.L.; Morrison, A.C.; Johnson, A.D.; Aspelund, T.; et al. Genome-wide association study of blood pressure and hypertension. *Nat. Genet.* **2009**, *41*, 677–687. [CrossRef]
27. Nelson, C.P.; Goel, A.; Butterworth, A.S.; Kanoni, S.; Webb, T.R.; Marouli, E.; Zeng, L.; Ntalla, I.; Lai, F.Y.; Hopewell, J.C.; et al. Association analyses based on false discovery rate implicate new loci for coronary artery disease. *Nat. Genet.* **2017**, *49*, 1385–1391. [CrossRef]
28. Nikpay, M.; Goel, A.; Won, H.H.; Hall, L.M.; Willenborg, C.; Kanoni, S.; Saleheen, D.; Kyriakou, T.; Nelson, C.P.; Hopewell, J.C.; et al. A comprehensive 1,000 Genomes-based genome-wide association meta-analysis of coronary artery disease. *Nat. Genet.* **2015**, *47*, 1121–1130. [CrossRef]
29. Shungin, D.; Winkler, T.W.; Croteau-Chonka, D.C.; Ferreira, T.; Locke, A.E.; Mägi, R.; Strawbridge, R.J.; Pers, T.H.; Fischer, K.; Justice, A.E.; et al. New genetic loci link adipose and insulin biology to body fat distribution. *Nature* **2015**, *518*, 187–196. [CrossRef]
30. Tachmazidou, I.; Süveges, D.; Min, J.L.; Ritchie, G.R.S.; Steinberg, J.; Walter, K.; Iotchkova, V.; Schwartzentruber, J.; Huang, J.; Memari, Y.; et al. Whole-Genome Sequencing Coupled to Imputation Discovers Genetic Signals for Anthropometric Traits. *Am. J. Hum. Genet.* **2017**, *100*, 865–884. [CrossRef]
31. Takeuchi, F.; Akiyama, M.; Matoba, N.; Katsuya, T.; Nakatochi, M.; Tabara, Y.; Narita, A.; Saw, W.Y.; Moon, S.; Spracklen, C.N.; et al. Interethnic analyses of blood pressure loci in populations of East Asian and European descent. *Nat. Commun.* **2018**, *9*, 5052. [CrossRef] [PubMed]
32. Wain, L.V.; Verwoert, G.C.; O'Reilly, P.F.; Shi, G.; Johnson, T.; Johnson, A.D.; Bochud, M.; Rice, K.M.; Henneman, P.; Smith, A.V.; et al. Genome-wide association study identifies six new loci influencing pulse pressure and mean arterial pressure. *Nat. Genet.* **2011**, *43*, 1005–1011. [CrossRef]
33. Zhou, W.; Nielsen, J.B.; Fritsche, L.G.; Dey, R.; Gabrielsen, M.E.; Wolford, B.N.; LeFaive, J.; VandeHaar, P.; Gagliano, S.A.; Gifford, A.; et al. Efficiently controlling for case-control imbalance and sample relatedness in large-scale genetic association studies. *Nat. Genet.* **2018**, *50*, 1335–1341. [CrossRef] [PubMed]
34. Zhu, Z.; Wang, X.; Li, X.; Lin, Y.; Shen, S.; Liu, C.L.; Hobbs, B.D.; Hasegawa, K.; Liang, L.; International COPD Genetics Consortium; et al. Genetic overlap of chronic obstructive pulmonary disease and cardiovascular disease-related traits: A large-scale genome-wide cross-trait analysis. *Respir. Res.* **2019**, *20*, 64. [CrossRef]
35. Bushueva, O.; Solodilova, M.; Churnosov, M.; Ivanov, V.; Polonikov, A. The Flavin-containing monooxygenase 3 gene and essential hypertension: The joint effect of polymorphism E158K and cigarette smoking on disease susceptibility. *Int. J. Hypertens.* **2014**, *2014*, 712169. [CrossRef] [PubMed]
36. Polonikov, A.V.; Ushachev, D.V.; Ivanov, V.P.; Churnosov, M.I.; Freidin, M.B.; Ataman, A.V.; Harbuzova, V.Y.; Bykanova, M.A.; Bushueva, O.Y.; Solodilova, M.A. Altered erythrocyte membrane protein composition mirrors pleiotropic effects of hypertension susceptibility genes and disease pathogenesis. *J. Hypertens.* **2015**, *33*, 2265–2277. [CrossRef]

37. Polonikov, A.V.; Bushueva, O.Y.; Bulgakova, I.V.; Freidin, M.B.; Churnosov, M.I.; Solodilova, M.A.; Shvetsov, Y.D.; Ivanov, V.P. A comprehensive contribution of genes for aryl hydrocarbon receptor signaling pathway to hypertension susceptibility. *Pharm. Genom.* **2017**, *27*, 57–69. [CrossRef]

38. Polonikov, A.; Bykanova, M.; Ponomarenko, I.; Sirotina, S.; Bocharova, A.; Vagaytseva, K.; Stepanov, V.; Churnosov, M.; Bushueva, O.; Solodilova, M.; et al. The contribution of CYP2C gene subfamily involved in epoxygenase pathway of arachidonic acids metabolism to hypertension susceptibility in Russian population. *Clin. Exp. Hypertens.* **2017**, *39*, 306–311. [CrossRef]

39. Fan, W.; Qu, X.; Li, J.; Wang, X.; Bai, Y.; Cao, Q.; Ma, L.; Zhou, X.; Zhu, W.; Liu, W.; et al. Associations between polymorphisms of the ADIPOQ gene and hypertension risk: A systematic and meta-analysis. *Sci. Rep.* **2017**, *9*, 41683. [CrossRef]

40. Moskalenko, M.I.; Milanova, S.N.; Ponomarenko, I.V.; Polonikov, A.V.; Churnosov, M.I. Study of associations of polymorphism of matrix metalloproteinases genes with the development of arterial hypertension in men. *Kardiologiia* **2019**, *59*, 31–39. [CrossRef]

41. Zhang, H.; Pushkarev, B.; Zhou, J.; Mu, Y.; Bolshakova, O.; Shrestha, S.; Wang, N.; Jian, B.; Jin, M.; Zhang, K.; et al. CACNA1C rs1006737 SNP increases the risk of essential hypertension in both Chinese Han and ethnic Russian people of Northeast Asia. *Medicine* **2021**, *100*, e24825. [CrossRef] [PubMed]

42. Moskalenko, M.; Ponomarenko, I.; Reshetnikov, E.; Dvornyk, V.; Churnosov, M. Polymorphisms of the matrix metalloproteinase genes are associated with essential hypertension in a Caucasian population of Central Russia. *Sci. Rep.* **2021**, *11*, 5224. [CrossRef]

43. Takahashi, Y.; Yamazaki, K.; Kamatani, Y.; Kubo, M.; Matsuda, K.; Asai, S. A genome-wide association study identifies a novel candidate locus at the DLGAP1 gene with susceptibility to resistant hypertension in the Japanese population. *Sci. Rep.* **2021**, *11*, 19497. [CrossRef] [PubMed]

44. Alsamman, A.M.; Almabrazi, H.; Zayed, H. Whole-Genome Sequencing of 100 Genomes Identifies a Distinctive Genetic Susceptibility Profile of Qatari Patients with Hypertension. *J. Pers. Med.* **2022**, *12*, 722. [CrossRef]

45. Moskalenko, M.I.; Polonikov, A.V.; Sorokina, I.N.; Yakunchenko, T.I.; Krikun, Y.N.; Ponomarenko, I.V. Gene-environment interactions between polymorphic loci of MMPs and obesity in essential hypertension in women. *Probl. Endocrinol.* **2019**, *65*, 425–435. [CrossRef]

46. Moskalenko, M.I.; Ponomarenko, I.V.; Polonikov, A.V.; Sorokina, I.N.; Batlutskaya, I.V.; Churnosov, M.I. The role of obesity in the implementation of genetic predisposition to the development of essential hypertension in men. *Obes. Metab.* **2019**, *16*, 66–72. [CrossRef]

47. Moskalenko, M.I.; Ponomarenko, I.V.; Verzilina, I.N.; Efremova, O.A.; Polonikov, A.V. The role of gene-gene and geneenvironment interactions of polymorphic locuses of MMPs in the formation of hypertension in women. *Arter. Gipertenz. Arter. Hypertens.* **2020**, *26*, 518–525. [CrossRef]

48. Moskalenko, M.I.; Ponomarenko, I.V.; Milanova, S.N.; Verzilina, I.N.; Efremova, O.A.; Polonikov, A.V. Polymorphic locus rs1061624 of the TNFR2 gene is associated with the development of arterial hypertension in males. *Kardiologiia* **2020**, *60*, 78–83. [CrossRef]

49. Williams, R.R.; Hunt, S.C.; Hopkins, P.N.; Wu, L.L.; Hasstedt, S.J.; Berry, T.D.; Barlow, G.K.; Stults, B.M.; Schumacher, M.C.; Ludwig, E.H. Genetic basis of familial dyslipidemia and hypertension: 15-year results from Utah. *Am. J. Hypertens.* **1993**, *6*, 319S–327S. [CrossRef]

50. Brown, M.J. The causes of essential hypertension. *Br. J. Clin. Pharmacol.* **1996**, *42*, 21–27. [CrossRef]

51. Evangelou, E.; Warren, H.R.; Mosen-Ansorena, D.; Mifsud, B.; Pazoki, R.; Gao, H.; Ntritsos, G.; Dimou, N.; Cabrera, C.P.; Karaman, I.; et al. Genetic analysis of over 1 million people identifies 535 new loci associated with blood pressure traits. *Nat. Genet.* **2018**, *50*, 1412–1425. [CrossRef] [PubMed]

52. Ruth, K.S.; Day, F.R.; Tyrrell, J.; Thompson, D.J.; Wood, A.R.; Mahajan, A.; Beaumont, R.N.; Wittemans, L.; Martin, S.; Busch, A.S.; et al. Using human genetics to understand the disease impacts of testosterone in men and women. *Nat. Med.* **2020**, *26*, 252–258. [CrossRef] [PubMed]

53. Sinnott-Armstrong, N.; Naqvi, S.; Rivas, M.; Pritchard, J.K. GWAS of three molecular traits highlights core genes and pathways alongside a highly polygenic background. *Elife* **2021**, *10*, e58615. [CrossRef]

54. Schmitz, D.; Ek, W.E.; Berggren, E.; Höglund, J.; Karlsson, T.; Johansson, Å. Genome-wide Association Study of Estradiol Levels and the Causal Effect of Estradiol on Bone Mineral Density. *J. Clin. Endocrinol. Metab.* **2021**, *106*, e4471–e4486, Correction appears in *J. Clin. Endocrinol. Metab.* 26 October 2021. [CrossRef] [PubMed]

55. Heid, I.M.; Jackson, A.U.; Randall, J.C.; Winkler, T.W.; Qi, L.; Steinthorsdottir, V.; Thorleifsson, G.; Zillikens, M.C.; Speliotes, E.K.; Mägi, R.; et al. Meta-analysis identifies 13 new loci associated with waist-hip ratio and reveals sexual dimorphism in the genetic basis of fat distribution. *Nat. Genet.* **2010**, *42*, 949–960, correction appears in *Nat. Genet.* **2011**, *43*, 1164. [CrossRef]

56. Randall, J.C.; Winkler, T.W.; Kutalik, Z.; Berndt, S.I.; Jackson, A.U.; Monda, K.L.; Kilpeläinen, T.O.; Esko, T.; Mägi, R.; Li, S.; et al. Sex-stratified genome-wide association studies including 270,000 individuals show sexual dimorphism in genetic loci for anthropometric traits. *PLoS Genet.* **2013**, *9*, e1003500. [CrossRef]

57. Locke, A.E.; Kahali, B.; Berndt, S.I.; Justice, A.E.; Pers, T.H.; Day, F.R.; Powell, C.; Vedantam, S.; Buchkovich, M.L.; Yang, J.; et al. Genetic studies of body mass index yield new insights for obesity biology. *Nature* **2015**, *518*, 197–206. [CrossRef]

58. Pulit, S.L.; Karaderi, T.; Lindgren, C.M. Sexual dimorphisms in genetic loci linked to body fat distribution. *Biosci. Rep.* **2017**, *37*, BSR20160184. [CrossRef]

59. Flynn, E.; Tanigawa, Y.; Rodriguez, F.; Altman, R.B.; Sinnott-Armstrong, N.; Rivas, M.A. Sex-specific genetic effects across biomarkers. *Eur. J. Hum. Genet.* **2021**, *29*, 154–163. [CrossRef] [PubMed]

60. Schroeder, P.H.; Cole, J.B.; Leong, A.; Florez, J.C.; Mercader, J.M. 139-OR: Large-scale sex-stratified additive and recessive GWAS identifies novel large-effect variants and improves polygenic prediction for type 2 diabetes. *Diabetes* **2022**, *71*, 139-OR. [CrossRef]
61. Zhao, J.V.; Luo, S.; Schooling, C.M. Sex-specific Mendelian randomization study of genetically predicted insulin and cardiovascular events in the UK Biobank. *Commun. Biol.* **2019**, *2*, 332. [CrossRef] [PubMed]
62. Zhao, J.V.; Schooling, C.M. Genetically predicted sex hormone binding globulin and ischemic heart disease in men and women: A univariable and multivariable Mendelian randomization study. *Sci. Rep.* **2021**, *11*, 23172. [CrossRef] [PubMed]
63. Li, H.; Konja, D.; Wang, L.; Wang, Y. Sex Differences in Adiposity and Cardiovascular Diseases. *Int. J. Mol. Sci.* **2022**, *23*, 9338. [CrossRef]
64. Leinone, J.T.; Mars, N.; Lehtonen, L.E.; Ahola-Olli, A.; Ruotsalainen, S.; Lehtimäki, T.; Kähönen, M.; Raitakari, O.; FinnGen Consortium; Piltonen, T.; et al. Genetic analyses implicate complex links between adult testosterone levels and health and disease. *Commun. Med.* **2023**, *3*, 4. [CrossRef] [PubMed]
65. de Ruiter, S.C.; Schmidt, A.F.; Grobbee, D.E.; den Ruijter, H.M.; Peters, S.A.E. Sex-specific Mendelian randomisation to assess the causality of sex differences in the effects of risk factors and treatment: Spotlight on hypertension. *J. Hum. Hypertens.* **2023**, *2023*, 1–7. [CrossRef]
66. Surendran, P.; Drenos, F.; Young, R.; Warren, H.; Cook, J.P.; Manning, A.K.; Grarup, N.; Sim, X.; Barnes, D.R.; Witkowska, K.; et al. Trans-ancestry meta-analyses identify rare and common variants associated with blood pressure and hypertension. *Nat. Genet.* **2016**, *48*, 1151–1161. [CrossRef]
67. Liu, C.; Kraja, A.T.; Smith, J.A.; Brody, J.A.; Franceschini, N.; Bis, J.C.; Rice, K.; Morrison, A.C.; Lu, Y.; Weiss, S.; et al. Meta-analysis identifies common and rare variants influencing blood pressure and overlapping with metabolic trait loci. *Nat. Genet.* **2016**, *48*, 1162–1170. [CrossRef]
68. Ehret, G.B.; Ferreira, T.; Chasman, D.I.; Jackson, A.U.; Schmidt, E.M.; Johnson, T.; Thorleifsson, G.; Luan, J.; Donnelly, L.A.; Kanoni, S.; et al. The genetics of blood pressure regulation and its target organs from association studies in 342,415 individuals. *Nat. Genet.* **2016**, *48*, 1171–1184. [CrossRef]
69. Hoffmann, T.J.; Ehret, G.B.; Nandakumar, P.; Ranatunga, D.; Schaefer, C.; Kwok, P.Y.; Iribarren, C.; Chakravarti, A.; Risch, N. Genome-wide association analyses using electronic health records identify new loci influencing blood pressure variation. *Nat. Genet.* **2017**, *49*, 54–64. [CrossRef]
70. Wain, L.V.; Vaez, A.; Jansen, R.; Joehanes, R.; van der Most, P.J.; Erzurumluoglu, A.M.; O'Reilly, P.F.; Cabrera, C.P.; Warren, H.R.; Rose, L.M.; et al. Novel Blood Pressure Locus and Gene Discovery Using Genome-Wide Association Study and Expression Data Sets from Blood and the Kidney. *Hypertension* **2017**, *70*, e4–e19. [CrossRef]
71. Pichler, I.; Minelli, C.; Sanna, S.; Tanaka, T.; Schwienbacher, C.; Naitza, S.; Porcu, E.; Pattaro, C.; Busonero, F.; Zanon, A.; et al. Identification of a common variant in the TFR2 gene implicated in the physiological regulation of serum iron levels. *Hum. Mol. Genet.* **2011**, *20*, 1232–1240. [CrossRef] [PubMed]
72. Raffield, L.M.; Louie, T.; Sofer, T.; Jain, D.; Ipp, E.; Taylor, K.D.; Papanicolaou, G.J.; Avilés-Santa, L.; Lange, L.A.; Laurie, C.C.; et al. Genome-wide association study of iron traits and relation to diabetes in the Hispanic Community Health Study/Study of Latinos (HCHS/SOL): Potential genomic intersection of iron and glucose regulation? *Hum. Mol. Genet.* **2017**, *26*, 1966–1978. [CrossRef] [PubMed]
73. Astle, W.J.; Elding, H.; Jiang, T.; Allen, D.; Ruklisa, D.; Mann, A.L.; Mead, D.; Bouman, H.; Riveros-Mckay, F.; Kostadima, M.A.; et al. The Allelic Landscape of Human Blood Cell Trait Variation and Links to Common Complex Disease. *Cell* **2016**, *167*, 1415–1429.e19. [CrossRef]
74. Pilling, L.C.; Atkins, J.L.; Duff, M.O.; Beaumont, R.N.; Jones, S.E.; Tyrrell, J.; Kuo, C.L.; Ruth, K.S.; Tuke, M.A.; Yaghootkar, H.; et al. Red blood cell distribution width: Genetic evidence for aging pathways in 116,666 volunteers. *PLoS ONE* **2017**, *12*, e0185083. [CrossRef] [PubMed]
75. Kanai, M.; Akiyama, M.; Takahashi, A.; Matoba, N.; Momozawa, Y.; Ikeda, M.; Iwata, N.; Ikegawa, S.; Hirata, M.; Matsuda, K.; et al. Genetic analysis of quantitative traits in the Japanese population links cell types to complex human diseases. *Nat. Genet.* **2018**, *50*, 390–400. [CrossRef]
76. Oskarsson, G.R.; Oddsson, A.; Magnusson, M.K.; Kristjansson, R.P.; Halldorsson, G.H.; Ferkingstad, E.; Zink, F.; Helgadottir, A.; Ivarsdottir, E.V.; Arnadottir, G.A.; et al. Predicted loss and gain of function mutations in ACO1 are associated with erythropoiesis. *Commun. Biol.* **2020**, *3*, 189. [CrossRef]
77. Vuckovic, D.; Bao, E.L.; Akbari, P.; Lareau, C.A.; Mousas, A.; Jiang, T.; Chen, M.H.; Raffield, L.M.; Tardaguila, M.; Huffman, J.E.; et al. The Polygenic and Monogenic Basis of Blood Traits and Diseases. *Cell* **2020**, *182*, 1214–1231.e11. [CrossRef]
78. Sakaue, S.; Kanai, M.; Tanigawa, Y.; Karjalainen, J.; Kurki, M.; Koshiba, S.; Narita, A.; Konuma, T.; Yamamoto, K.; Akiyama, M.; et al. A cross-population atlas of genetic associations for 220 human phenotypes. *Nat. Genet.* **2021**, *53*, 1415–1424. [CrossRef] [PubMed]
79. Chen, J.; Spracklen, C.N.; Marenne, G.; Varshney, A.; Corbin, L.J.; Luan, J.; Willems, S.M.; Wu, Y.; Zhang, X.; Horikoshi, M.; et al. The trans-ancestral genomic architecture of glycemic traits. *Nat. Genet.* **2021**, *53*, 840–860. [CrossRef]
80. Wu, Y.; Byrne, E.M.; Zheng, Z.; Kemper, K.E.; Yengo, L.; Mallett, A.J.; Yang, J.; Visscher, P.M.; Wray, N.R. Genome-wide association study of medication-use and associated disease in the UK Biobank. *Nat. Commun.* **2019**, *10*, 1891. [CrossRef] [PubMed]

81. Gill, D.; Benyamin, B.; Moore, L.S.P.; Monori, G.; Zhou, A.; Koskeridis, F.; Evangelou, E.; Laffan, M.; Walker, A.P.; Tsilidis, K.K.; et al. Associations of genetically determined iron status across the phenome: A mendelian randomization study. *PLoS Med.* **2019**, *16*, e1002833. [CrossRef] [PubMed]

82. Lu, W.; Pikhart, H.; Tamosiunas, A.; Kubinova, R.; Capkova, N.; Malyutina, S.; Pająk, A.; Bobak, M. Prevalence, awareness, treatment and control of hypertension, diabetes and hypercholesterolemia, and associated risk factors in the Czech Republic, Russia, Poland and Lithuania: A cross-sectional study. *BMC Public Health* **2022**, *22*, 883. [CrossRef]

83. Lule, S.A.; Mentzer, A.J.; Namara, B.; Muwenzi, A.G.; Nassanga, B.; Kizito, D.; Akurut, H.; Lubyayi, L.; Tumusiime, J.; Zziwa, C.; et al. A genome-wide association and replication study of blood pressure in Ugandan early adolescents. *Mol. Genet. Genomic. Med.* **2019**, *7*, e00950. [CrossRef] [PubMed]

84. Schneider-Hohendorf, T.; Görlich, D.; Savola, P.; Kelkka, T.; Mustjoki, S.; Gross, C.C.; Owens, G.C.; Klotz, L.; Dornmair, K.; Wiendl, H.; et al. Sex bias in MHC I-associated shaping of the adaptive immune system. *Proc. Natl. Acad. Sci. USA* **2018**, *115*, 2168–2173. [CrossRef]

85. Mikolajczyk, T.P.; Guzik, T.J. Adaptive Immunity in Hypertension. *Curr. Hypertens. Rep.* **2019**, *21*, 68. [CrossRef]

86. Li, J.X.; Tang, B.P.; Sun, H.P.; Feng, M.; Cheng, Z.H.; Niu, W.Q. Interacting contribution of the five polymorphisms in three genes of Hsp70 family to essential hypertension in Uygur ethnicity. *Cell Stress Chaperones* **2009**, *14*, 355–362. [CrossRef] [PubMed]

87. Rodriguez-Iturbe, B.; Pons, H.; Johnson, R.J. Role of the Immune System in Hypertension. *Physiol. Rev.* **2017**, *97*, 1127–1164. [CrossRef]

88. Rodriguez-Iturbe, B.; Lanaspa, M.A.; Johnson, R.J. The role of autoimmune reactivity induced by heat shock protein 70 in the pathogenesis of essential hypertension. *Br. J. Pharmacol.* **2019**, *76*, 1829–1838. [CrossRef]

89. Zhou, B.; Perel, P.; Mensah, G.A.; Ezzati, M. Global epidemiology, health burden and effective interventions for elevated blood pressure and hypertension. *Nat. Rev. Cardiol.* **2021**, *18*, 785–802. [CrossRef]

90. Gerdts, E.; Sudano, I.; Brouwers, S.; Borghi, C.; Bruno, R.M.; Ceconi, C.; Cornelissen, V.; Diévart, F.; Ferrini, M.; Kahan, T.; et al. Sex differences in arterial hypertension. *Eur. Heart J.* **2022**, *43*, 4777–4788. [CrossRef]

91. Mendelsohn, M.E.; Karas, R.H. The protective effects of estrogen on the cardiovascular system. *N. Engl. J. Med.* **1999**, *340*, 1801–1811. [CrossRef]

92. Reckelhoff, J.F. Gender differences in the regulation of blood pressure. *Hypertension* **2001**, *37*, 1199–1208. [CrossRef]

93. Colafella, K.M.M.; Denton, K.M. Sex-specific differences in hypertension and associated cardiovascular disease. *Nat. Rev. Nephrol.* **2018**, *14*, 185–201. [CrossRef] [PubMed]

94. EUGenMed Cardiovascular Clinical Study Group. Gender in cardiovascular diseases: Impact on clinical manifestations, management, and outcomes. *Eur. Heart J.* **2016**, *37*, 24–34. [CrossRef]

95. Moulton, V.R. Sex Hormones in Acquired Immunity and Autoimmune Disease. *Front. Immunol.* **2018**, *9*, 2279. [CrossRef]

96. Fujioka, K.; Kajita, K.; Wu, Z.; Hanamoto, T.; Ikeda, T.; Mori, I.; Okada, H.; Yamauchi, M.; Uno, Y.; Morita, H.; et al. Dehydroepiandrosterone reduces preadipocyte proliferation via androgen receptor. *Am. J. Physiol. Endocrinol. Metab.* **2012**, *302*, E694–E704. [CrossRef] [PubMed]

97. Nardozza Junior, A.; Szelbracikowski Sdos, S.; Nardi, A.C.; Almeida, J.C. Age-related testosterone decline in a Brazilian cohort of healthy military men. *Int. Braz. J. Urol.* **2011**, *37*, 591–597. [CrossRef]

98. Rotter, I.; Ryl, A.; Grzesiak, K.; Szylińska, A.; Pawlukowska, W.; Lubkowska, A.; Sipak-Szmigiel, O.; Pabisiak, K.; Laszczyńska, M. Cross-Sectional Inverse Associations of Obesity and Fat Accumulation Indicators with Testosterone in Non-Diabetic Aging Men. *Int. J. Environ. Res. Public Health* **2018**, *15*, 1207. [CrossRef]

99. Ibanez, L.; Diaz, M.; Sebastiani, G.; Marcos, M.V.; Lopez-Bermejo, A.; de Zegher, F. Oral contraception vs insulin sensitization for 18 months in nonobese adolescents with androgen excess: Posttreatment differences in C-reactive protein, intima-media thickness, visceral adiposity, insulin sensitivity, and menstrual regularity. *J. Clin. Endocrinol. Metab.* **2013**, *98*, E902–E907. [CrossRef]

100. Díaz, M.; Chacón, M.R.; López-Bermejo, A.; Maymó-Masip, E.; Salvador, C.; Vendrell, J.; de Zegher, F.; Ibáñez, L. Ethinyl estradiol-cyproterone acetate versus low-dose pioglitazone-flutamide-metformin for adolescent girls with androgen excess: Divergent effects on CD163, TWEAK receptor, ANGPTL4, and LEPTIN expression in subcutaneous adipose tissue. *J. Clin. Endocrinol. Metab.* **2012**, *97*, 3630–3638. [CrossRef] [PubMed]

101. Carnagarin, R.; Matthews, V.; Zaldivia, M.T.K.; Peter, K.; Schlaich, M.P. The bidirectional interaction between the sympathetic nervous system and immune mechanisms in the pathogenesis of hypertension. *Br. J. Pharmacol.* **2019**, *176*, 1839–1852. [CrossRef]

102. Ji, H.; Zheng, W.; Li, X.; Liu, J.; Wu, X.; Zhang, M.A.; Umans, J.G.; Hay, M.; Speth, R.C.; Dunn, S.E.; et al. Sex-specific T-cell regulation of angiotensin II-dependent hypertension. *Hypertension* **2014**, *64*, 573–582. [CrossRef]

103. Loughner, C.L.; Bruford, E.A.; McAndrews, M.S.; Delp, E.E.; Swamynathan, S.; Swamynathan, S.K. Organization, evolution and functions of the human and mouse Ly6/uPAR family genes. *Hum. Genom.* **2016**, *10*, 10. [CrossRef] [PubMed]

104. Churnosov, M.; Abramova, M.; Reshetnikov, E.; Lyashenko, I.V.; Efremova, O.; Churnosova, M.; Ponomarenko, I. Polymorphisms of hypertension susceptibility genes as a risk factors of preeclampsia in the Caucasian population of central Russia. *Placenta* **2022**, *129*, 51–61. [CrossRef] [PubMed]

105. Abramova, M.; Churnosova, M.; Efremova, O.; Aristova, I.; Reshetnikov, E.; Polonikov, A.; Churnosov, M.; Ponomarenko, I. Effects of pre-pregnancy over-weight/obesity on the pattern of association of hypertension susceptibility genes with preeclampsia. *Life* **2022**, *12*, 2018. [CrossRef] [PubMed]

106. Litovkina, O.; Nekipelova, E.; Dvornyk, V.; Polonikov, A.; Efremova, O.; Zhernakova, N.; Reshetnikov, E.; Churnosov, M. Genes involved in the regulation of vascular homeostasis determine renal survival rate in patients with chronic glomerulo-nephritis. *Gene* **2014**, *546*, 112–116. [CrossRef]
107. Reshetnikov, E.; Zarudskaya, O.; Polonikov, A.; Bushueva, O.; Orlova, V.; Krikun, E.; Dvornyk, V.; Churnosov, M. Genetic markers for inherited thrombophilia are associated with fetal growth retardation in the population of Central Russia. *J. Obstet. Gynaecol. Res.* **2017**, *43*, 1139–1144. [CrossRef]
108. Hoeper, M.M.; Bogaard, H.J.; Condliffe, R.; Frantz, R.; Khanna, D.; Kurzyna, M.; Langleben, D.; Manes, A.; Satoh, T.; Torres, F.; et al. Defnitions and diagnosis of hypertension. *J. Am. Coll. Cardiol.* **2013**, *62*, 42–50. [CrossRef]
109. Ivanova, T.A. Polymorphic loci of AC026703.1 and HFE genes are associated with se vere hypertension. *Res. Results Biomed.* **2023**, *9*, 22–38. [CrossRef]
110. World Health Organization. *Diet, Nutrition, and the Prevention of Chronic Diseases: Report of a Joint WHO/FAO Expert 802 Consultation*; World Health Organization: Geneva, Switzerland, 2003; 148p.
111. Haskell, W.L.; Lee, I.M.; Pate, R.R.; Powell, K.E.; Blair, S.N.; Franklin, B.A.; Macera, C.A.; Heath, G.W.; Thompson, P.D.; Bauman, A. Physical activity and public health: Updated recommendation for adults from the American College of Sports Medicine and the American Heart Association. *Med. Sci. Sports. Exerc.* **2007**, *39*, 1423–1434. [CrossRef]
112. Lanier, J.B.; Bury, D.C.; Richardsonm, S.W. Diet and Physical Activity for Cardiovascular Disease Prevention. *Am. Fam. Physician* **2016**, *93*, 919–924. [PubMed]
113. Ponomarenko, I.; Reshetnikov, E.; Altuchova, O.; Polonikov, A.; Sorokina, I.; Yermachenko, A.; Dvornyk, V.; Golovchenko, O.; Churnosov, M. Association of genetic polymorphisms with age at menarche in Russian women. *Gene* **2019**, *686*, 228–236. [CrossRef] [PubMed]
114. Eliseeva, N.; Ponomarenko, I.; Reshetnikov, E.; Dvornyk, V.; Churnosov, M. LOXL1 gene polymorphism candidates for exfoliation glaucoma are also associated with a risk for primary open-angle glaucoma in a Caucasian population from central Russia. *Mol. Vis.* **2021**, *27*, 262–269. [PubMed]
115. Abramova, M. Genetic markers of severe preeclampsia. *Res. Results Biomed.* **2022**, *8*, 305–316. [CrossRef]
116. Ward, L.D.; Kellis, M. HaploReg v4: Systematic mining of putative causal variants, cell types, regulators and target genes for human complex traits and disease. *Nucleic Acids Res.* **2016**, *44*, D877–D881. [CrossRef]
117. Novakov, V.; Novakova, O.; Churnosova, M.; Sorokina, I.; Aristova, I.; Polonikov, A.; Reshetnikov, E.; Churnosov, M. Intergenic Interactions of SBNO1, NFAT5 and GLT8D1 Determine the Susceptibility to Knee Osteoarthritis among Europeans of Russia. *Life* **2023**, *13*, 405. [CrossRef]
118. Golovchenko, O.; Abramova, M.; Ponomarenko, I.; Reshetnikov, E.; Aristova, I.; Polonikov, A.; Dvornyk, V.; Churnosov, M. Functionally significant polymorphisms of ESR1and PGR and risk of intrauterine growth restriction in population of Central Russia. *Eur. J. Obstet. Gynecol. Reprod. Biol.* **2020**, *253*, 52–57. [CrossRef]
119. Starikova, D.; Ponomarenko, I.; Reshetnikov, E.; Dvornyk, V.; Churnosov, M. Novel Data about Association of the Functionally Significant Polymorphisms of the MMP9 Gene with Exfoliation Glaucoma in the Caucasian Population of Central Russia. *Ophthalmic Res.* **2021**, *64*, 458–464. [CrossRef]
120. Sirotina, S.; Ponomarenko, I.; Kharchenko, A.; Bykanova, M.; Bocharova, A.; Vagaytseva, K.; Stepanov, V.; Churnosov, M.; Solodilova, M.; Polonikov, A. A Novel Polymorphism in the Promoter of the CYP4A11 Gene Is Associated with Susceptibility to Coronary Artery Disease. *Dis. Markers* **2018**, *2018*, 5812802. [CrossRef]
121. Tikunova, E.; Ovtcharova, V.; Reshetnikov, E.; Dvornyk, V.; Polonikov, A.; Bushueva, O.; Churnosov, M. Genes of tumor necrosis factors and their receptors and the primary open angle glaucoma in the population of Central Russia. *Int. J. Ophthalmol.* **2017**, *10*, 1490–1494. [CrossRef]
122. Pavlova, N.; Demin, S.; Churnosov, M.; Reshetnikov, E.; Aristova, I.; Churnosova, M.; Ponomarenko, I. The Modifying Effect of Obesity on the Association of Matrix Metalloproteinase Gene Polymorphisms with Breast Cancer Risk. *Biomedicines* **2022**, *10*, 2617. [CrossRef] [PubMed]
123. Ponomarenko, I.; Reshetnikov, E.; Polonikov, A.; Verzilina, I.; Sorokina, I.; Elgaeva, E.E.; Tsepilov, Y.A.; Yermachenko, A.; Dvornyk, V.; Churnosov, M. Candidate genes for age at menarche are associated with endometriosis. *Reprod. Biomed. Online* **2020**, *41*, 943–956. [CrossRef] [PubMed]
124. Reshetnikov, E.; Ponomarenko, I.; Golovchenko, O.; Sorokina, I.; Batlutskaya, I.; Yakunchenko, T.; Dvornyk, V.; Polonikov, A.; Churnosov, M. The VNTR polymorphism of the endothelial nitric oxide synthase gene and blood pressure in women at the end of pregnancy. *Taiwan J. Obstet. Gynecol.* **2019**, *58*, 390–395. [CrossRef] [PubMed]
125. Minyaylo, O.; Ponomarenko, I.; Reshetnikov, E.; Dvornyk, V.; Churnosov, M. Functionally significant polymorphisms of the MMP-9 gene are associated with peptic ulcer disease in the Caucasian population of Central Russia. *Sci. Rep.* **2021**, *11*, 13515. [CrossRef] [PubMed]
126. Purcell, S.; Neale, B.; Todd-Brown, K.; Thomas, L.; Ferreira, M.A.R.; Bender, D.; Maller, J.; Sklar, P.; de Bakker, P.I.W.; Daly, M.J.; et al. PLINK: A Tool Set for Whole-Genome Association and Population-Based Linkage Analyses. *Am. J. Hum. Genet.* **2007**, *81*, 559–575. [CrossRef] [PubMed]
127. Sourceforge. Available online: http://sourceforge.net/projects/mdr (accessed on 22 November 2022).

128. Pavlova, N.; Demin, S.; Churnosov, M.; Reshetnikov, E.; Aristova, I.; Churnosova, M.; Ponomarenko, I. Matrix Metalloproteinase Gene Polymorphisms Are Associated with Breast Cancer in the Caucasian Women of Russia. *Int. J. Mol. Sci.* **2022**, *23*, 12638. [CrossRef]

129. Calle, M.L.; Urrea, V.; Malats, N.; Van Steen, K. Mbmdr: An R package for exploring gene-gene interactions associated with binary or quantitative traits. *Bioinformatics* **2010**, *26*, 2198–2199. [CrossRef]

130. Ponomarenko, I.V. Using the method of Multifactor Dimensionality Reduction (MDR) and its modifications for analysis of gene-gene and gene-environment interactions in genetic-epidemiological studies (review). *Res. Results Biomed.* **2019**, *5*, 4–21. [CrossRef]

131. Che, R.; Jack, J.R.; Motsinger-Reif, A.A.; Brown, C.C. An adaptive permutation approach for genome-wide association study: Evaluation and recommendations for use. *BioData Min.* **2014**, *7*, 9. [CrossRef]

132. Ponomarenko, I.; Reshetnikov, E.; Polonikov, A.; Sorokina, I.; Yermachenko, A.; Dvornyk, V.; Churnosov, M. Candidate genes for age at menarche are associated with endometrial hyperplasia. *Gene* **2020**, *757*, 4933. [CrossRef]

133. Gauderman, W.J.; Morrison, J.M. QUANTO 1.1: A Computer Program for Power and Sample Size Calculations for Genetic-Epidemiology Studies. Available online: https://hydra.usc.edu/gxe2006 (accessed on 2 December 2022).

134. Ponomarenko, I.; Reshetnikov, E.; Polonikov, A.; Verzilina, I.; Sorokina, I.; Yermachenko, A.; Dvornyk, V.; Churnosov, M. Candidate genes for age at menarche are associated with uterine leiomyoma. *Front. Genet.* **2021**, *11*, 512940. [CrossRef] [PubMed]

135. Golovchenko, I.; Aizikovich, B.; Golovchenko, O.; Reshetnikov, E.; Churnosova, M.; Aristova, I.; Ponomarenko, I.; Churnosov, M. Sex Hormone Candidate Gene Polymorphisms Are Associated with Endometriosis. *Int. J. Mol. Sci.* **2022**, *23*, 13691. [CrossRef] [PubMed]

136. Kumar, P.; Henikoff, S.; Ng, P.C. Predicting the effects of coding non-synonymous variants on protein function using the SIFT algorithm. *Nat. Protoc.* **2009**, *7*, 1073–1081. [CrossRef] [PubMed]

137. Adzhubei, I.; Jordan, D.M.; Sunyaev, S.R. Predicting functional effect of human missense mutations using PolyPhen-2. *Curr. Protoc. Hum. Genet.* **2013**, *76*, 7–20. [CrossRef] [PubMed]

138. GTEx Consortium. The GTEx Consortium atlas of genetic regulatory effects across human tissues. *Science* **2020**, *369*, 1318–1330. [CrossRef]

139. Westra, H.J.; Peters, M.J.; Esko, T.; Yaghootkar, H.; Schurmann, C.; Kettunen, J.; Christiansen, M.W.; Fairfax, B.P.; Schramm, K.; Powell, J.E.; et al. Systematic identification of trans eQTLs as putative drivers of known disease associations. *Nat. Genet.* **2013**, *45*, 1238–1243. [CrossRef]

140. Gene Ontology Consortium. The Gene Ontology resource: Enriching a GOld mine. *Nucleic Acids Res.* **2021**, *49*, D325–D334. [CrossRef]

141. Franz, M.; Rodriguez, H.; Lopes, C.; Zuberi, K.; Montojo, J.; Bader, G.D.; Morris, Q. GeneMANIA update 2018. *Nucleic Acids Res.* **2018**, *46*, W60–W64. [CrossRef]

International Journal of
Molecular Sciences

Article

SERPINE1 mRNA Binding Protein 1 Is Associated with Ischemic Stroke Risk: A Comprehensive Molecular–Genetic and Bioinformatics Analysis of *SERBP1* SNPs

Irina Shilenok [1,2], Ksenia Kobzeva [1], Tatiana Stetskaya [3], Maxim Freidin [4,5], Maria Soldatova [1], Alexey Deykin [6,7], Vladislav Soldatov [6,7], Mikhail Churnosov [8], Alexey Polonikov [3,9] and Olga Bushueva [1,9,*]

[1] Laboratory of Genomic Research, Research Institute for Genetic and Molecular Epidemiology, Kursk State Medical University, 305041 Kursk, Russia
[2] Division of Neurology, Kursk Emergency Hospital, 305035 Kursk, Russia
[3] Laboratory of Statistical Genetics and Bioinformatics, Research Institute for Genetic and Molecular Epidemiology, Kursk State Medical University, 305041 Kursk, Russia
[4] Department of Biology, School of Biological and Behavioural Sciences, Queen Mary University of London, London E1 4NS, UK
[5] Laboratory of Population Genetics, Research Institute of Medical Genetics, Tomsk National Research Medical Center, Russian Academy of Science, 634050 Tomsk, Russia
[6] Laboratory of Genome Editing for Biomedicine and Animal Health, Belgorod State National Research University, 308015 Belgorod, Russia
[7] Department of Pharmacology and Clinical Pharmacology, Belgorod State National Research University, 308015 Belgorod, Russia
[8] Department of Medical Biological Disciplines, Belgorod State University, 308015 Belgorod, Russia
[9] Department of Biology, Medical Genetics and Ecology, Kursk State Medical University, 305041 Kursk, Russia
* Correspondence: olga.bushueva@inbox.ru

Citation: Shilenok, I.; Kobzeva, K.; Stetskaya, T.; Freidin, M.; Soldatova, M.; Deykin, A.; Soldatov, V.; Churnosov, M.; Polonikov, A.; Bushueva, O. SERPINE1 mRNA Binding Protein 1 Is Associated with Ischemic Stroke Risk: A Comprehensive Molecular–Genetic and Bioinformatics Analysis of SERBP1 SNPs. *Int. J. Mol. Sci.* **2023**, *24*, 8716. https://doi.org/10.3390/ijms24108716

Academic Editors: Yutang Wang and Dianna Magliano

Received: 5 April 2023
Revised: 5 May 2023
Accepted: 10 May 2023
Published: 13 May 2023

Abstract: The *SERBP1* gene is a well-known regulator of SERPINE1 mRNA stability and progesterone signaling. However, the chaperone-like properties of *SERBP1* have recently been discovered. The present pilot study investigated whether *SERBP1* SNPs are associated with the risk and clinical manifestations of ischemic stroke (IS). DNA samples from 2060 unrelated Russian subjects (869 IS patients and 1191 healthy controls) were genotyped for 5 common SNPs—rs4655707, rs1058074, rs12561767, rs12566098, and rs6702742 *SERBP1*—using probe-based PCR. The association of SNP rs12566098 with an increased risk of IS (risk allele C; $p = 0.001$) was observed regardless of gender or physical activity level and was modified by smoking, fruit and vegetable intake, and body mass index. SNP rs1058074 (risk allele C) was associated with an increased risk of IS exclusively in women ($p = 0.02$), non-smokers ($p = 0.003$), patients with low physical activity ($p = 0.04$), patients with low fruit and vegetable consumption ($p = 0.04$), and BMI ≥ 25 ($p = 0.007$). SNPs rs1058074 ($p = 0.04$), rs12561767 ($p = 0.01$), rs12566098 ($p = 0.02$), rs6702742 ($p = 0.036$), and rs4655707 ($p = 0.04$) were associated with shortening of activated partial thromboplastin time. Thus, *SERBP1* SNPs represent novel genetic markers of IS. Further studies are required to confirm the relationship between *SERBP1* polymorphism and IS risk.

Keywords: ischemic stroke; hero; heat-resistant obscure; chaperones; SERPINE1 mRNA binding protein 1; SERBP1; SNP; activated partial thromboplastin time; body mass index; gene-environmental interaction

1. Introduction

Ischemic stroke (IS), which accounts for ~80% of the total cases of stroke [1], is the second leading cause of death and the third leading cause of disability worldwide [2]. IS is caused by an interruption in cerebral blood flow, induced by thrombosis or embolism [3].

In this regard, the main factors contributing to IS are atherogenesis, instability of atherosclerotic plaque, and increased platelet aggregation. Accordingly, genetic studies revealed numerous IS-associated polymorphisms among the genes related to lipids turnover, inflammatory response, oxidative stress, and platelets activity [4–7]. Nevertheless, further genetic studies may help to reveal other molecular contributors, shedding light on the pathogenesis of IS.

Brain damage during acute ischemia is a very multifaceted and complex process. However, among the major diversity of molecular events, the following scenario can be distinguished as key: unbalanced functioning of the mitochondrion and reduced ATP production cause the accumulation of H^+ and Ca^{2+} ions as well as free radicals. These toxic molecules directly damage cellular structure and provoke pathological responses displayed as excessive glutamate release and a switch to an inflammatory phenotype [8,9]. Finally, reaching the point of no return, the cell launches the "suicide programme" or worse, perishes via necrosis [10–12].

Under ischemia as well as under other pathological conditions, the cell utilizes all accessible resources to resist before it dies. One of the most powerful tools to decrease cellular damage is chaperone machinery. Chaperones refer to a functionally related group of proteins that provide proper folding of newly synthesized and unfolded proteins. The biological tasks of chaperones include providing correct protein folding, translocation of proteins across the membranes, helping in the assembly and disassembly of protein complexes, and participating in their degradation [13,14]. Obviously, the tremendously important role of chaperones determines their crucial significance for the course of brain ischemia. Moreover, being widely involved in the regulation of cellular proteostasis, the chaperones are important players in the development of IS [15].

SERPINE1 mRNA binding protein 1 (SERBP1) has first been identified as a protein bound to plasminogen activator inhibitor type 1 (PAI-1) mRNA [16] and which regulates the stability of the transcript. A significant role of PAI-1 in the increased risk of thrombosis and IS has been shown by a lot of studies [17–19]. Subsequent studies discovered that SERBP1 is involved in progesterone signaling via interactions with progesterone receptor membrane component 1 (PGRMC1) [20].

In 2020, Tsuboyama K et al. revealed SERBP1 among other small heat-resistant proteins in drosophila cell culture lysate that display chaperone-like activity even after being boiled [21]. As well as the other members, SERBP1 was named "heat-resistant obscure", or, acronymically, "Hero". The authors have also found mammalian orthologs of all the Hero-proteins and confirmed their chaperone-like properties. Our previous study has already shown an association of one of the "Hero" genes with the risk of IS [22].

Summarizing the biological roles of *SERBP1* may suggest its probable link with IS. Indeed, the involvement of *SERBP1* in the regulation of PAI-1, one of the main drivers of atherothrombosis, as well as the recently discovered "chaperone" properties of *SERBP1*, make this gene very attractive in determining the factors leading to the risk of IS. As far as such studies have not been previously carried out, we aimed to analyze the associations of functionally significant *SERBP1* genetic variants with the risk and clinical manifestations of IS.

2. Results

2.1. Associations of SERBP1 SNPs with the Risk of Ischemic Stroke

The genotype frequencies of rs4655707, rs1058074, rs12561767, rs12566098, and rs6702742 *SERBP1* in study groups are presented in Table S1. The distribution of genotype frequencies of all studied SNPs corresponded to the Hardy–Weinberg equilibrium in both the control and case groups ($p > 0.05$).

The analysis of the total sample revealed a significant association between SNP rs12566098 *SERBP1* and IS risk (OR = 1.28, 95% CI 1.10–1.49, $p = 0.001$) (Table 1).

Table 1. Results of the analysis of associations between *SERBP1* gene SNPs and ischemic stroke risk.

Genetic Variant	Effect Allele	Other Allele	N	OR (95% CI) [1]	*p* [2]
rs4655707	T	C	1883	1.07 [0.93; 1.23]	0.35
rs1058074	C	T	1915	1.15 [0.99; 1.35]	0.071
rs12561767 *	G	A	1857	1.11 [0.96; 1.27]	0.16
rs12566098 *	C	G	2060	**1.28 [1.10; 1.49]**	**0.001**
rs6702742	A	G	1898	1.04 [0.91; 1.20]	0.54

All calculations were performed relative to the minor alleles (Effect allele) with adjustment for sex, age, smoking; [1]—odds ratio and 95% confidence interval; [2]—*p*-value; tag SNPs are marked with an asterisk; statistically significant differences are marked in bold.

Owing to a potential sex difference in the associations of genetic variations with the risk of IS, the relationship between *SERBP1* polymorphisms and the emergence of IS was examined separately in males and females. The genetic variant rs12566098 was found to affect the risk of the disease in both males (OR = 1.25; 95% CI = 1.02–1.53; $p = 0.03$) and females (OR = 1.39; 95% CI = 1.14–1.71; $p = 0.002$) (Table 2). Meanwhile, rs1058074 was associated with the increased risk of IS exclusively in females (OR = 1.29, 95% CI 1.05–1.58, $p = 0.02$) (Table 2).

The analysis of relationships between *SERBP1* SNPs and the risk of IS depending on the smoking status revealed that rs1058074 (OR = 1.32, 95%CI 1.10–1.59, $p = 0.003$) and rs12566098 (OR = 1.43, 95%CI 1.20–1.72, $p = 1 \times 10^{-4}$) were associated with an increased risk of IS only in non-smokers (Table 2). Since revealed association between the genetic polymorphism and smoking-status may be biased by the disequilibrial proportion of smokers between males and females, we further performed separated analysis accounting both factors. The analysis has confirmed that smoking status has its own impact because we found that the association of rs1058074 and rs12566098 with the higher risk of IS in non-smokers still took place even after the exclusion of males (Table S2). Moreover, we found that two other *SERBP1* SNPs, rs12561767 and rs6702742, were associated with a higher risk of IS in non-smoking men (OR = 1.36, 95% CI 1.05–1.77, $p = 0.02$, and OR = 1.33, 95% CI 1.03–1.73, $p = 0.03$, respectively) (Table S3).

The influence of other environmental factors that may have a substantial impact on the development of IS, such as physical activity and dietary intake of fresh fruits and vegetables, was also taken into account for evaluating associations. Yet, independent of the effect of these factors, rs12566098 *SERBP1* was found to be related with the development of IS. (Table 2). Meanwhile, rs1058074 *SERBP1* showed associations with an increased risk of IS in patients who indicated a low level of physical activity before cerebrovascular events (OR = 1.26, 95% CI 1.04–1.54, $p = 0.02$, Pbonf = 0.04) (Table 2) and in patients who noted insufficient consumption of fresh vegetables and fruits (OR = 1.24, 95% CI 1.04–1.48, $p = 0.02$, Pbonf = 0.04) (Table 2), suggesting that these environmental risk factors, along with smoking, modify the association of rs1058074 with IS risk.

Moreover, we examined how the body mass index (BMI) of IS patients affected the link between the researched SNPs and the emergence of the condition. It was found that the carriage of the minor allele C rs1058074 was associated with a 1.5-fold increase in the risk of IS in a patient with a BMI $\geq$ 25 (OR = 1.47, 95% CI 1.14–1.89, $p = 0.0045$, Pbonf = 0.007). Similar data were obtained in the analysis of rs12566098: OR = 1.52, 95% CI 1.19–1.95, $p = 0.001$, Pbonf = 0.002 (Table 2).

Table 2. Subgroups analysis of associations between *SERBP1* SNPs and IS risk depending on sex, smoking status, body mass index, physical activity level, and fruit and vegetable intake.

Genetic Variant	Effect Allele	Other Allele	N	OR (95% CI) [1]	p [2] (P_{bonf})	N	OR (95% CI) [1]	p [2] (P_{bonf})
				Males			Females	
rs4655707	T	C	929	1.08 [0.90; 1.31]	0.39	954	1.07 [0.88; 1.29]	0.51
rs1058074	C	T	950	1.12 [0.91; 1.38]	0.29	965	**1.29 [1.05; 1.58]**	**0.02**
rs12561767 *	G	A	918	1.12 [0.93; 1.35]	0.23	939	1.09 [0.90; 1.32]	0.39
rs12566098 *	C	G	1026	**1.25 [1.02; 1.53]**	**0.03**	1124	**1.39 [1.14; 1.71]**	**0.002**
rs6702742	A	G	939	1.10 [0.92; 1.32]	0.29	959	1.00 [0.82; 1.20]	0.97
				Nonsmokers			Smokers	
rs4655707	T	C	1281	1.11 [0.94; 1.31]	0.22	602	0.93 [0.73; 1.20]	0.59
rs1058074	C	T	1317	**1.32 [1.10; 1.59]**	**0.003**	598	0.87 [0.67; 1.14]	0.32
rs12561767 *	G	A	1277	1.18 [1.00; 1.39]	0.054	580	0.94 [0.73; 1.21]	0.62
rs12566098 *	C	G	1400	**1.43 [1.20; 1.72]**	**1×10^{-4}**	660	1.02 [0.79; 1.32]	0.89
rs6702742	A	G	1305	1.09 [0.93; 1.29]	0.29	593	0.92 [0.72; 0.72]	0.49
				Low physical activity (f+)			Normal physical activity level (f-)	
rs4655707	T	C	1387	1.07 [0.89; 1.28]	0.46 (0.92)	1546	1.06 [0.90; 1.24]	0.48 (0.96)
rs1058074	C	T	1400	**1.26 [1.04; 1.54]**	**0.02 (0.04)**	1567	1.14 [0.96; 1.36]	0.14 (0.28)
rs12561767 *	G	A	1364	1.17 [0.98; 1.40]	0.086 (0.172)	1526	1.05 [0.90; 1.23]	0.54 (1.0)
rs12566098 *	C	G	1525	**1.39 [1.15; 1.68]**	**9×10^{-4} (0.002)**	1692	**1.26 [1.06; 1.49]**	**0.009 (0.02)**
rs6702742	A	G	1411	1.10 [0.93; 1.31]	0.3 (0.6)	1564	1.00 [0.85; 1.17]	0.99 (1.0)
				Low fruit/vegetable intake (f+)			Normal fruit/vegetable intake (f-)	
rs4655707	T	C	1492	1.08 [0.92; 1.28]	0.32 (0.64)	1441	1.04 [0.87; 1.23]	0.68 (1.0)
rs1058074	C	T	1514	**1.24 [1.04; 1.48]**	**0.02 (0.04)**	1453	1.13 [0.93; 1.37]	0.21 (0.42)
rs12561767 *	G	A	1468	1.08 [0.92; 1.28]	0.33 (0.66)	1422	1.11 [0.94; 1.32]	0.22 (0.44)
rs12566098*	C	G	1638	**1.37 [1.15; 1.63]**	**4×10^{-4} (8×10^{-4})**	1579	**1.24 [1.02; 1.49]**	**0.03 (0.06)**
rs6702742	A	G	1510	1.04 [0.89; 1.22]	0.62 (1.0)	1465	1.04 [0.87; 1.23]	0.67 (1.0)
				BMI ≥ 25			BMI < 25	
rs4655707	T	C	1234	1.20 [0.95; 1.53]	0.13 (0.26)	1443	1.01 [0.85; 1.20]	0.91 (1.0)
rs1058074	C	T	1252	**1.47 [1.14; 1.89]**	**0.0045 (0.007)**	1456	1.03 [0.85; 1.25]	0.74 (1.0)
rs12561767 *	G	A	1213	1.25 [0.98; 1.59]	0.071 (0.142)	1430	1.03 [0.87; 1.22]	0.71 (1.0)
rs12566098 *	C	G	1366	**1.52 [1.19; 1.95]**	**0.001 (0.002)**	1579	1.17 [0.97; 1.41]	0.096 (0.192)
rs6702742	A	G	1272	1.22 [0.97; 1.54]	0.091 (0.182)	1478	1.00 [0.84; 1.18]	0.99 (1.0)

[1]—odds ratio and 95% confidence interval; [2]—p-value (p-value with Bonferroni correction). All calculations were performed relative to the minor allele (Effect allele). Statistically significant differences are marked in bold; tag SNPs are marked with an asterisk.

2.2. Associations of SERBP1 with the Clinical Features of IS

All of the tested SNPs showed associations with the Activated Partial Thromboplastin Time (APTT). Interestingly, homozygotes for major alleles (protective against IS) had the highest APTT rates, while heterozygotes were characterized by intermediate APTT indicators. Homozygotes for minor alleles (risk alleles) had the lowest APTT values, suggesting that coagulation has been stimulated and that these patients had a tendency towards thrombosis. It is significant that consumption of fresh vegetables and fruits affects the relationship between rs1058074, rs12561767, rs12566098, and rs6702742 and the APTT (Table S3, Figure 1).

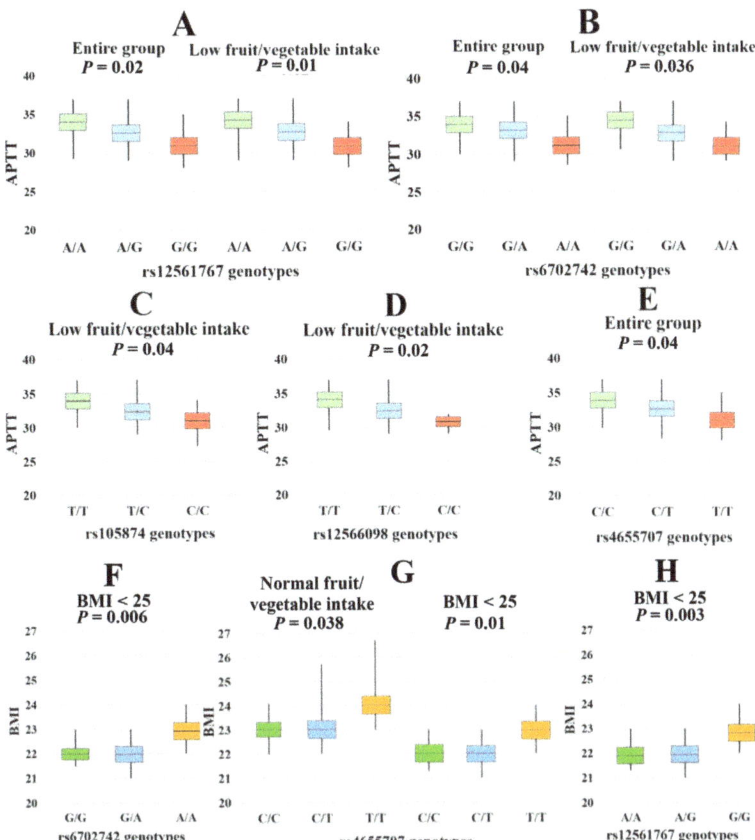

Figure 1. Associations of *SERBP1* genotypes with activated partial thromboplastin time and body mass index. (**A**)—APTT values for rs12561767 in the entire group ($p = 0.02$) and in the group of patients with low fruit/vegetable intake ($p = 0.01$); (**B**)—APTT values for rs6702742 in the entire group ($p = 0.04$) and in the group of patients with low fruit/vegetable intake ($p = 0.036$); (**C**)—APTT values for rs105874 in the group of patients with low fruit/vegetable intake ($p = 0.04$); (**D**)—APTT values for rs12566098 in the group of patients with low fruit/vegetable intake ($p = 0.02$); (**E**)—APTT values for rs4655707 in the entire group ($p = 0.04$); (**F**)—BMI values for rs6702742 in the group of patients with BMI < 25 ($p = 0.006$); (**G**)—BMI values for rs4655707 in the group of patients with normal fruit/vegetable intake ($p = 0.038$) and in the group of patients with BMI < 25 ($p = 0.01$); (**H**)—BMI values for rs12561767 in the group of patients with BMI < 25 ($p = 0.0023$).

An analysis of the influence of the studied *SERBP1* genotypes on body mass index established a relationship between rs4655707, rs12561767, and rs6702742 with this indicator only in patients with normal BMI, suggesting that *SERBP1* genetic variants lose their regulatory effects on BMI in patients with overweight. It is noteworthy that the association of rs4655707 with BMI was also observed in the group of patients with normal fruit and vegetable intake (Table S3, Figure 1).

2.3. Bioinformatics Analysis

GTEx Portal claims that the *SERBP1* gene is expressed in brain tissues, blood vessels, and whole blood. In brain tissues, *SERBP1* expression levels measured as median TPM (Transcripts Per Million) vary from 18.16 to 43.22; in blood vessels, from 70.52 to 246.5; and in whole blood, MeTPM = 79.88 (https://gtexportal.org/home/gene/SERBP1 (accessed on 21 January 2023)).

2.3.1. QTL-Effects

The results of the eQTL analysis for the studied *SERBP1* SNPs are shown in Table 3. According to the GTEx Portal browser, all studied SNPs are associated with expression of the *SERBP1* gene in the arteries (with cis-eQTL-mediated up-regulating effects of risk alleles) (Table 3). In the meantime, *SERBP1* SNPs exhibit bidirectional effects on the cis-eQTL-mediated *IL12RB2* level; they have a positive correlation with *IL12RB2* expression in the arteries and aorta and a negative correlation with *IL12RB2* expression in whole blood (Table 3).

Table 3. Relationship between *SERBP1* SNPs and cis-eQTL-mediated expression levels of genes in IS-related tissues (blood vessels, brain, whole blood) (according to browsers' GTEx Portal, eQTLGene).

SNP	GTEx Portal Data (https://gtexportal.org (Accessed on 21 January 2023))					eQTLGene Browser Data (https://www.eqtlgen.org/ (Accessed on 21 January 2023))			
	Effect Allele	Gene Expressed	*p*-Value	Effect (NES)	Tissue	Effect Allele	Gene Expressed	Z-Score	*p*-Value
rs4655707 *SERBP1* (T/C)	C	*IL12RB2*	6.2×10^{-10}	↑(0.28)	Artery—Tibial	T	*IL12RB2*	↑(25.93)	2.93×10^{-148}
		SERBP1	1.8×10^{-6}	↓(−0.091)	Artery—Tibial				
		IL12RB2	3.8×10^{-8}	↓(−0.13)	Whole Blood				
rs1058074 *SERBP1* (C/T)	T	*IL12RB2*	1.5×10^{-11}	↑(0.30)	Artery—Tibial	C	*IL12RB2*	↑(16.31)	9.01×10^{-60}
		SERBP1	8.7×10^{-10}	↓(−0.12)	Artery—Tibial				
		IL12RB2	5.6×10^{-7}	↓(−0.12)	Whole Blood				
		IL12RB2	3.6×10^{-5}	↑(0.25)	Artery—Aorta				
rs12561767 * *SERBP1* (G/A)	A	*IL12RB2*	2.8×10^{-10}	↑(0.28)	Artery—Tibial	G	*IL12RB2*	↑(26.20)	2.92×10^{-151}
		IL12RB2	9.1×10^{-9}	↓(−0.14)	Whole Blood				
		SERBP1	2.3×10^{-6}	↓(−0.09)	Artery—Tibial				
rs12566098 * *SERBP1* (C/G)	G	*SERBP1*	1.3×10^{-9}	↓(−0.12)	Artery—Tibial	C	*IL12RB2*	↑(17.40)	8.60×10^{-68}
		IL12RB2	3.7×10^{-7}	↓(−0.13)	Whole Blood				
		IL12RB2	2.3×10^{-11}	↑(0.30)	Artery—Tibial				
rs6702742 *SERBP1* (A/G)	G	*IL12RB2*	3.9×10^{-10}	↑(0.27)	Artery—Tibial	A	*IL12RB2*	↑(26.29)	2.42×10^{-152}
		IL12RB2	1.0×10^{-9}	↓(−0.14)	Whole Blood				
		SERBP1	9.5×10^{-7}	↓(−0.092)	Artery—Tibial				

tag SNPs are marked with an asterisk.

Similar data were obtained in the analysis of eQTL effects using the eQTLGene browser data resource, which presents the results of the analysis of cis-eQTL effects in the blood: risk alleles of *SERBP1* SNPs are associated with a cis-eQTL-mediated increase in *IL12RB2* expression in blood (Table 3). Thus, the minor alleles of the studied *SERBP1* SNPs (risk alleles) affect the increase in *SERBP1* expression in the artery-tibial, the decrease in the expression of *IL12RB2* in the artery-tibial, and the increase in the expression of *IL12RB2* in the blood.

Analysis of cis-mQTL effects established a cis-mQTL-mediated effect of the risk alleles C rs1058074 and C rs12566098 on the decrease in methylation of cg24364144 (chr1:68102479) in the brain prefrontal cortex. At the same time, protective alleles C rs4655707 and G rs6702742 influenced the increase in methylation of cg24364144 (chr1:68102479) in the brain prefrontal cortex (Table 4). Given the fact that a decrease in methylation leads to an increase in gene expression, it can be concluded that the risk alleles of all IS-linked *SERBP1* SNPs are associated with a decrease in cg24364144 (chr1:68102479) methylation and, accordingly, an increase in *SERBP1* expression in the brain prefrontal cortex.

Table 4. Established associations of the studied *SERBP1* SNPs with the cis-mQTL-mediated effect on the level of methylation of CpG-sites.

Trait	Effect Allele	Tissue	Effect Size (Beta)	FDR
		rs4655707 *SERBP1*		
cg24364144 (chr1:68102479)	C	Brain-Prefrontal Cortex	0.02	9.5×10^{-14}
		rs1058074 *SERBP1*		
cg24364144 (chr1:68102479)	C	Brain-Prefrontal Cortex	−0.01	1.6×10^{-10}
		rs12561767 *SERBP1* *		
cg24364144 (chr1:68102479)	A	Brain-Prefrontal Cortex	0.02	9.5×10^{-14}
		rs12566098 *SERBP1* *		
cg24364144 (chr1:68102479)	C	Brain-Prefrontal Cortex	−0.01	5.1×10^{-11}
		rs6702742 *SERBP1*		
cg24364144 (chr1:68102479)	G	Brain-Prefrontal Cortex	0.01	6.5×10^{-13}

Note: mQTL with an undetermined effect allele was excluded from the analysis. tag SNPs are marked with an asterisk.

2.3.2. Analysis of Transcription Factors

The analysis of transcription factors revealed that the protective allele C rs4655707 *SERBP1* creates DNA-binding sites for 60 TFs, co-controlling positive regulation of cysteine-type endopeptidase activity involved in the apoptotic signaling pathway (GO:0006919), response to hypoxia (GO:0001666), and cellular response to growth factor stimulus (GO:0071363) (Table S4).

Protective allele T rs1058074 *SERBP1* creates DNA-binding sites for 127 TFs that are jointly involved in 24 overrepresented GO controlling neurogenesis, cardio- and vasculogenesis, apoptosis, and cell signaling: cerebral cortex GABAergic interneuron fate commitment (GO:0021892); neurogenesis (GO:0022008); regulation of cell death (GO:0010941); cardiac muscle tissue regeneration (GO:0061026); forebrain dorsal/ventral pattern formation (GO:0021798); BMP signaling pathway involved in heart development (GO:0061312); cellular response to growth factor stimulus (GO:0071363); forebrain neuron fate commitment (GO:0021877); hypothalamus development (GO:0021854); neuroblast differentiation (GO:0014016); forebrain morphogenesis (GO:0048853); cardiac pacemaker cell development (GO:0060926); negative regulation of oligodendrocyte differentiation (GO:0048715); negative regulation of neurogenesis (GO:0050768); ventricular cardiac muscle cell development (GO:0055015); cardiac atrium morphogenesis (GO:0003209); neuron fate specification (GO:0048665); oligodendrocyte differentiation (GO:0048709); neural precursor cell proliferation (GO:0061351); heart valve development (GO:0003170); neuron migration (GO:0001764); hippocampus development (GO:0021766); regulation of neural precursor cell proliferation (GO:2000177); and blood vessel morphogenesis (GO:0048514) (Table S5). Meanwhile, the risk allele C rs1058074 *SERBP1* creates DNA-binding sites for 18 TFs jointly involved in the regulation of protein stability (GO:0031647), neurogenesis processes such as forebrain neuron development (GO:0021884), neuron fate commitment (GO:0048663), neuron migration (GO:0001764) (Table S5).

Protective allele A rs12561767 *SERBP1* creates DNA-binding sites for 38 TFs characterized by the following overrepresented GO involved in cell signaling, apoptosis, and neurogenesis: steroid hormone mediated signaling pathway (GO:0043401); apoptotic process (GO:0006915); programmed cell death (GO:0012501); cell death (GO:0008219); neuron migration (GO:0001764); nervous system development (GO:0007399); and the generation of neurons (GO:0048699) (Table S6). Risk allele G rs12561767 *SERBP1* creates DNA-binding sites for 15 TFs involved in the processes of programmed cell death: regulation of the apoptotic process (GO:0042981); and regulation of programmed cell death (GO:0043067) (Table S6).

Risk allele C rs12566098 *SERBP1* creates DNA-binding sites for 62 TFs, co-controlling cellular response to growth factor stimulus (GO:0071363) and positive regulation of cytokine production (GO:0001819). At once, no IS-specific GO was found for nine TFs, binding to the protective allele G rs12566098 *SERBP1* (Table S7).

Risk allele A rs6702742 *SERBP1* creates DNA-binding sites for 41 TFs that jointly involved in the positive regulation of neuron apoptotic process (GO:0043525), positive regulation of cytokine production (GO:0001819), and negative regulation of apoptotic process (GO:0043066). Meanwhile, protective allele G rs6702742 *SERBP1* creates DNA-binding sites for 23 TFs characterized by the following overrepresented GO: positive regulation of neuron death (GO:1901216); and regulation of neurogenesis (GO:0050767) (Table S8).

2.3.3. Bioinformatic Analysis of the Associations of SERBP1 SNPs with Cerebral Stroke and Intermediate Phenotypes

According to the bioinformatic resources, the Cerebrovascular Disease Knowledge Portal (CDKP) and the Cardiovascular Disease Knowledge Portal (CVDKP), which combine and analyze the results of genetic associations of the largest consortiums for the study of cardio- and cerebrovascular diseases, the IS-related *SERBP1* SNPs are associated with stroke and a number of stroke-related intermediate phenotypes (Table 5).

Table 5. Results of aggregated bioinformatic analyses of associations between SNPs in *SERBP1* gene cerebrovascular diseases and their intermediate phenotypes.

№	SNPs	Phenotype	*p*-Value	Beta (OR)	Sample Size
1.		[1] TOAST other undetermined	0.01	OR ▼ 0.7757	9487
2.		[1] TOAST cardio-aortic embolism	0.02	OR ▼ 0.7946	9470
3.	rs4655707 (T/**C**)	[1] Lacunar stroke	0.043	OR ▼ 0.9611	28,530
4.		[2] LDL cholesterol	0.015	Beta ▼ −0.0031	2,028,070
5.		[2] Leptin	0.038	Beta ▲ 0.0097	36,525
6.		[1] Lacunar stroke	0.020	OR ▼ 0.9546	28,530
7.		[1] TOAST other undetermined	0.021	OR ▼ 0.9546	9487
8.		[2] Arm fat ratio	0.007	Beta ▼ −0.0013	232,276
9.	rs1058074 (C/**T**)	[2] Dyslipidemia	0.020	OR ▲ 1.0322	56,375
10.		[2] LDL cholesterol	0.020	Beta ▼ −0.0029	2,010,200
11.		[2] Leptin	0.037	Beta ▲ 0.0108	32,169
12.		[2] Non-HDL cholesterol	0.042	Beta ▼ −0.0032	1,096,010
13.		[2] Total cholesterol	0.049	Beta ▼ −0.0026	1,921,240

Table 5. *Cont.*

№	SNPs	Phenotype	*p*-Value	Beta (OR)	Sample Size
14.		[1] TOAST other undetermined	0.010	OR ▼ 0.7775	9487
15.	rs12561767 * (G/**A**)	[2] LDL cholesterol	0.010	Beta ▼ −0.0032	2,014,850
16.		[2] Dyslipidemia	0.040	OR ▲ 1.0264	56,375
17.		[2] Leptin	0.036	Beta ▲ 0.0098	36,521
18.	rs12566098 * (**C**/G)	[1] TOAST small artery occlusion	0.047	OR ▲ 2.5943	254,558
19.		[2] LDL cholesterol	0.010	Beta ▼ −0.0032	2,021,740
20.		[2] Arm fat ratio	0.010	Beta ▼ −0.0112	232,276
21.		[2] Dyslipidemia	0.020	OR ▲ 1.0315	56,375
22.		[2] Leptin	0.020	Beta ▲ 0.0118	34,005
23.		[2] Non-HDL cholesterol	0.030	Beta ▼ −0.0035	1,108,320
24.	rs6702742 (A/G)	[1] TOAST other undetermined	0.009	OR ▼ 0.7726	9487
25.		[1] TOAST cardio-aortic embolism	0.025	OR ▼ 0.7952	9470
26.		[2] Leptin	0.002	Beta ▲ 0.0188	29,651
27.		[2] HDL3 cholesterol	0.021	Beta ▲ 0.0336	10,984
28.		[2] LDL cholesterol	0.024	Beta ▼ −0.0027	2,144,190
29.		[2] Leptin adj BMI	0.028	Beta ▲ 0.0170	22,924
30.		[2] Dyslipidemia	0.044	OR ▲ 1.0264	56,375

[1]—data obtained using the bioinformatic resource Cerebrovascular Disease Knowledge Portal (https://cd.hugeamp.org/ (accessed on 24 January 2023)); [2]—data obtained using the bioinformatic resource Cardiovascular Disease Knowledge Portal (https://cvd.hugeamp.org/ (accessed on 24 January 2023)). Effect alleles are marked in bold; tag SNPs are marked with an asterisk.

2.3.4. Protein–Protein Interactions

Using the STRING database, we identified 10 proteins characterized by the most pronounced interactions with SERBP1 (proteins of the first shell of interactors): PGRMC1, RPL21, RPL23, RPL5, RPL8, RPS19, RPS24, RPS25, RPS29, and RPS6 (PPI enrichment *p*-value = 1.54×10^{-10}) (Figure 2). These proteins are characterized predominantly by co-expression. Moreover, the interactions between these genes were established experimentally and using the textmining approaches. It is noteworthy that most of the main interaction partners of SERBP1 (RPS29, RPS24, RPL21, RPL5, RPS6, RPL23, RPS25, RPL8, RPS19) are directly involved in protein synthesis since they are structural components of ribosomes (Table S9). The role of the PGRMC1 protein in the most significant interactions with SERBP1 should also be noted, since it is involved in the metabolism of progesterone and may play an essential role in the manifestation of the sex-specific effects of *SERBP1* (Table S9).

Together, SERBP1 and its main interaction partners are involved in 14 biological processes and 6 molecular functions, reflecting not only the regulation of protein synthesis but also cotranslation, transport, ubiquitination, regulation of gene expression, the nitrogen compound biosynthetic process, and organic cyclic compound metabolism (Table S10).

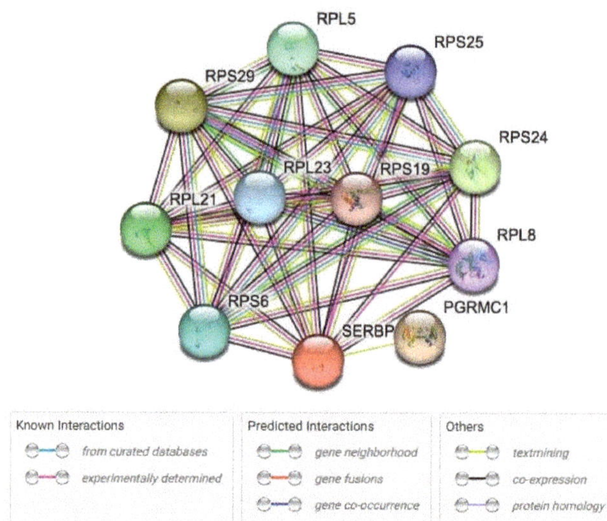

Figure 2. Predicted Interaction Partners of SERBP1 (https://string-db.org/ (accessed on 22 January 2023)).

3. Discussion

The present study shows for the first time that polymorphisms in the gene encoding SERPINE1 mRNA-binding protein 1 (*SERBP1*) are associated with the risk and clinical features of IS. In particular, tag SNP rs12566098 (risk allele C) was found to be associated with an elevated risk of IS in the total study population, in the group of males and in the group of females, regardless of the level of physical activity (both in patients with low physical activity and in patients with a normal level of physical activity). Nonetheless, we noticed that cigarette smoking and consumption of fresh vegetables and fruits act as modifiers of the associations of rs12566098 with IS: this genetic variant is associated with IS only in non-smokers and individuals with low fruit and vegetable consumption. In addition, the analysis based on the BMI revealed that rs12566098 correlates with IS among overweight individuals (BMI $\geq$ 25). Most likely, this genetic variant imparts more risk as compared to other SNPs linked to it; their risk effects were modified by all the studied environmental factors, sex and the BMI. In particular, rs1058074 (risk allele C) is associated with an increased risk of IS exclusively in women, non-smokers, individuals with low physical activity, individuals with low fruit and vegetable consumption, and individuals with the BMI $\geq$ 25. SNPs rs12561767 (risk allele G) and rs6702742 (risk allele A) proved to be associated with an elevated risk of IS only among non-smoking males.

3.1. SERBP1 SNPs and the IS Risk: Underlying Mechanisms

SERBP1 is expressed in vessels, brain tissues, and blood. *SERBP1* regulates mRNA degradation and formation of RNA–protein complexes, affects mRNA stability, and controls the mRNA levels of target genes [16]. According to GTex Portal data, risk alleles of all IS-associated SNPs (rs12561767, rs12566098, rs6702742, and rs1058074) are linked to cis-eQTL-mediated up-regulation of *SERBP1*. Further evidence of associations between risk alleles and *SERBP1* overexpression comes from the finding that risk alleles of all IS-related SNPs are associated with mQTLs, thus contributing to low methylation of CpG site cg24364144 (chr1:68102479) in the prefrontal cortex.

Because it binds to an AU-rich element in the 3′-untranslated region of PAI-1 mRNA and controls its stability, *SERBP1* causes PAI-1 protein levels to increase. Accordingly, it has been previously known as PAI-1 mRNA-binding protein 1 [16]. Plasminogen activator inhibitor 1 (PAI-1), which is a SERPIN inhibitor, is well known as a regulator of fibrinolysis. A high serum level of PAI-1 correlates with enhanced thrombosis by reducing fibrin degradation [23] and is seen in stroke patients [24]. Moreover, PAI-1 is involved in mechanisms of atherosclerotic plaque formation [25] and stabilization [26]. J.H. Heaton and co-authors have proposed that nuclear SERBP1 stabilizes PAI-1 mRNA and that cytoplasmic SERBP1 destabilizes it [27]. Most likely, the carriage of risk alleles associated with higher *SERBP1* expression contributes to the up-regulation of PAI-1. This notion is partially confirmed by our finding of the lowest APTT values in homozygotes for risk alleles of all the studied SNPs. It has been previously shown that hypofibrinolytic conditions are associated with shorter APTT and greater PAI-1 expression [28,29] and that PAI-1 overexpression correlates with a decrease in APTT [30]. Thus, our study indicates that risk alleles of *SERBP1* SNPs may be risk factors for excessive thrombosis. Nonetheless, their role in the PAI-1–mediated mechanisms of thrombosis in IS remains incompletely understood, and to assess the influence of *SERBP1* genotypes on the level of PAI-1 expression, functional studies are required: an analysis of the magnitude of PAI-1 expression depending on *SERBP1* genotypes.

Because, to date, there have been no studies evaluating an association of *SERBP1* SNPs with the IS risk, a bioinformatic analysis was carried out here to interpret the functional effects of genetic variants. First, it should be noted that protective alleles of SNPs that correlated with IS in our study also showed an association with a reduced risk of lacunar stroke (rs1058074), small artery occlusion (rs12566098), and cardio-aortic embolism (rs6702742), according to the Cerebrovascular Disease Knowledge Portal, which provides comprehensive genetic data on large numbers of patients with stroke from across the world (https://cd.hugeamp.org/ (accessed on 24 January 2023)). Second, data obtained using the bioinformatic resource, the Cardiovascular Disease Knowledge Portal (https://cvd.hugeamp.org/ (accessed on 24 January 2023)), showed that protective alleles of IS-associated SNPs reduce LDL cholesterol (rs1058074, rs12561767, rs12566098, and rs6702742), non-HDL cholesterol (rs12566098), and total cholesterol (rs1058074) levels and increase HDL3 cholesterol levels, thus pointing to the potentially significant participation of *SERBP1* polymorphic variants in the development of atherosclerosis and thrombosis. Third, all IS-associated SNPs seem to contribute to the expression of *IL12RB2* in the blood and arteries through cis-eQTL effects. IL12RB2 is involved in IL-35 control, which is important for atherosclerosis and inflammation [31] and has been shown to inhibit ischemia/hypoxia-induced angiogenesis, suggesting that this anti-inflammatory cytokine plays new roles at the recovery stage of angiogenesis [32].

Fourth, IS-associated SNPs affect transcriptional modulation by TFs and exert pronounced actions such as a loss or gain of function. For instance, carriage of the risk allele C rs1058074 leads to loss of DNA binding to the TFs that are predominantly involved in neurogenesis, cardio- and vasculogenesis, apoptosis, and cell signaling and creates binding sites for the TFs that jointly regulate "neuron migration", "neuron fate commitment" and "regulation of protein stability". The risk allele G rs12561767 causes a loss of binding to the TFs involved in the steroid hormone signaling pathway and neurogenesis (such as "neuron migration" and "formation of neurons") and creates binding sites for the TFs driving processes of programmed cell death. Risk allele C rs12566098 *SERBP1* creates binding sites for the TFs, which co-control the "cellular response to a growth factor stimulus" and "positive regulation of cytokine production". The rs6702742 *SERBP1* risk allele A generates binding sites for TFs that promote neuronal apoptosis and cytokine production. Allele C of SNP rs4655707 provides binding sites for TFs taking part in positive regulation of "cysteine-type endopeptidase activity involved in apoptotic signaling pathways", "response to hypoxia", and "cellular response to growth factor stimulus". The above-mentioned biological processes may be implicated both in the risk of IS and in processes of ischemia-reperfusion.

3.2. Sex-Specific Correlates of SERBP1

Our study revealed that rs1058074 is associated with IS only in females. Other research has already shown that genetic markers have well-pronounced sexual dimorphism in their correlations with stroke [33–36]. First, sex hormones progesterone and 17β-estradiol both regulate *SERBP1* mRNA levels [37]. On the other hand, according to the STRING database, one of the most important interaction partners of SERBP1 is PGRMC1 (progesterone receptor membrane component 1), which mediates the anti-apoptotic action of progesterone and whose mRNA is a target of SERBP1 [38]. Thus, *SERBP1* is both involved in the metabolism of female sex hormones and subject to regulation by these hormones, which may mediate sex-specific effects in the association of rs1058074 (in *SERBP1*) with the IS risk.

3.3. Smoking-Associated Correlates of SERBP1

Our study indicates that rs1058074 and rs12566098 (in *SERBP1*) are associated with an increased risk of IS only among non-smokers. On the one hand, smoking is a substantial environmental risk factor for the IS, and may contribute more significantly to the disease's onset than *SERBP1* genetic variants. As a result, the link between SNPs and the IS risk is clearly shown in non-smokers. On the other hand, smoking can have a major impact on *SERBP1* expression. By means of the bioinformatics resource, the Comparative Toxigenomic Database, it has been found that tobacco smoke pollution affects the expression of the SERBP1 protein in humans [39]. Nevertheless, the direction of this influence remains unclear. To further understand how SERBP1 interacts with various components of cigarette smoke, more research on *SERBP1* expression levels in response to smoking status is required.

3.4. Low-Physical-Activity–Associated Correlates of SERBP1

We observed that rs1058074 is associated with an increased risk of IS only among individuals with low physical activity. The latter is a known risk factor for IS and may affect inflammatory, thrombotic, and oxidative markers [40], whose levels are modulated by chaperones [41]. Evidence for this is an observation that aerobic exercise training decreases plasma SERPINE1 levels in humans [42]. This lifestyle intervention may help to reduce the pathological effects of *SERBP1* risk alleles potentially associated with elevated SERPINE1 (PAI-1) levels. Moreover, the level of SERPINE1, which is regulated by SERBP1, can be modulated by the magnitude of physical activity [43].

3.5. Low Fruit/Vegetable Intake–Related Correlates of SERBP1

The association of SNP rs1058074 with IS was found to be modified by a low intake of fresh vegetables and fruits. Nonetheless, the most important finding is that four SNPs showed an association with APTT in the presence of a risk factor such as low consumption of fresh vegetables and fruits, suggesting that the prothrombotic effects of *SERBP1* risk alleles are significantly modified by this risk factor. Low consumption of fresh vegetables and fruits primarily correlates with oxidative stress [44,45], the role of which in the risk of IS and IS-related phenotypes has been demonstrated in a substantial number of studies [6,46–50]. Plant-based diets usually contain phenolic acids, flavonoids, and carotenoids, which have strong antioxidant properties and therefore eliminate an excess of active oxygen from the body and protect cells from damage, thereby reducing the risk of cardiovascular diseases [51]. It should be pointed out that the vasculo-protective influence of polyphenols (the main contributors to the antioxidant capacity of fruits and vegetables) may be linked to their antithrombotic action [52]. Antithrombotic mechanisms may be related to inhibition of platelet aggregation [53] and lower plasma homocysteine levels [54]. We believe that the manifestation of the risk effects of *SERBP1* during low consumption of vegetables and fruits is primarily due to oxidative-stress–induced thrombosis, which is enhanced by the pro-thrombotic effects of risk alleles of *SERBP1*.

3.6. SERBP1 SNPs and the BMI

We noticed that rs1058074 and rs12566098 *SERBP1* were associated with higher IS risk exclusively in individuals with a BMI of 25 or more. By contrast, no such correlations were found among individuals with a normal BMI. Moreover, *SERBP1* SNPs were associated with a higher BMI among patients with a BMI < 25. This relation can be explained in terms of connections between obesity, leptin and the level of *SERBP1* expression.

Leptin is a relevant adipokine that is involved in the regulation of food intake [55] and manifests its physiological actions by inhibiting food intake via actions on certain receptors in the brain [56]. In particular, considerable evidence implicates leptin in the conveying of satiety signals to the feeding center [57,58]. A bioinformatic analysis showed that the risk alleles of all the IS-linked SNPs contribute to lower leptin levels and hence weaker inhibitory effects of leptin on appetite, thereby increasing body weight. Nonetheless, these regulatory effects may only be seen in patients of normal body weight. In overweight individuals, leptin resistance (due to mutations in the genes encoding leptin or its receptors [59]) represents an inability of leptin to exert its anorexigenic actions [60]. Naturally, *SERBP1*-mediated regulation of the BMI was observed in a patient with a normal BMI who does not have leptin resistance.

In addition, obesity promotes this propensity for thrombosis. PAI-1 levels, regulated by *SERBP1*, positively correlate with obesity [61] and are significantly reduced by weight loss in obese individuals [62].

4. Materials and Methods

The outline of the study design is shown in Figure 3.

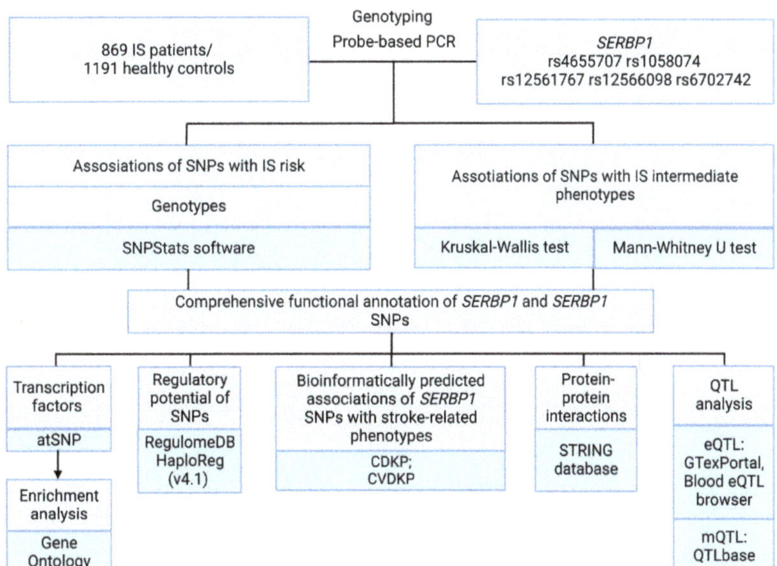

Figure 3. Study design.

The study was carried out on an ethnically homogeneous sample of unrelated residents of Central Russia (mainly natives of the Kursk region) of Russian nationality, with a total number of 2060 (869 patients with IS and 1191 healthy individuals). The Ethical Review Committee of Kursk State Medical University approved the study protocol. All the participants gave written informed consent before their enrollment in this study, subject to the following inclusion criteria: self-declared Russian descent, with a birthplace inside of Central Russia [63].

Baseline and clinical characteristics of the study population are listed in Table 6.

Table 6. Baseline and clinical characteristics of the studied groups.

Baseline and Clinical Characteristics		IS Patients (n = 869)	Controls (n = 1191)	p-Value
Age, Me [Q1; Q3]		62 [55; 70]	59 [53; 66]	**<0.001**
Gender	Males, N (%)	482 (55.5%)	544 (45.7%)	**<0.001**
	Females, N (%)	387 (44.5%)	647 (54.3%)	
Smoking	Yes, N (%)	419 (48.2%)	241 (20.2%)	**<0.001**
	No, N (%)	450 (51.8%)	950 (79.8%)	
Low physical activity	Yes, N (%)	334 (40.0%)	ND	
	No, N (%)	501 (60.0%)		
Low fruit/vegetable consumption	Yes, N (%)	447 (53.53%)	ND	
	No, N (%)	388 (46.47%)		
Coronary artery disease	Yes, N (%)	264 (26.72%)	-	
	No, N (%)	724 (73.28%)	-	
Type 2 diabetes mellitus	Yes, N (%)	101 (12.11%)	-	
	No, N (%)	733 (87.89%)	-	
Body mass index, Me [Q1; Q3]		23 [22; 26] (n = 563)	-	
Overweight	Normal weight (BMI = 18.5–24.9), N (%)	388 (68.92%)	ND	
	Overweight (BMI of 25–29.9), N (%)	118 (20.96%)		
	Obesity (BMI of 30 or greater), N (%)	57 (10.12%)		
Family history of cerebrovascular diseases	Yes, N (%)	311 (34.44%)	ND	
	No, N (%)	592 (65.56%)	ND	
Age at onset of stroke, Me [Q1; Q3]		61 [53; 69] (n = 851)	-	
Number of strokes including event in question	1, N (%)	751 (88.35%)	-	
	2, N (%)	86 (10.12%)	-	
	3, N (%)	13 (1.53%)	-	
Type of stroke	Atherothrombotic, N (%)	616 (70.89%)	-	
	Cardioembolic, N (%)	160 (18.41%)	-	
	Unspecified, N (%)	93 (10.70%)	-	
Stroke localization	Right/left middle cerebral artery basin, N (%)	705 (83.04%)	-	
	Vertebrobasilar basin, N (%)	144 (16.96%)	-	
Area of lesion in stroke, mm^2, Me [Q1; Q3]		99.50 [30; 461] (n = 776)	-	
Total cholesterol, mmol/L, Me [Q1; Q3]		5.2 [4.4; 5.8] (n = 577)	ND	
Triglycerides, mmol/L, Me [Q1; Q3]		1.3 [1.09; 1.80] (n = 571)	ND	
Glucose level, mmol/L, Me [Q1; Q3]		4.8 [4.3; 5.5] (n = 840)	ND	
Prothrombin time, seconds, Me [Q1; Q3]		10.79 [10.14; 11.70] (n = 827)	ND	
International normalized ratio, Me [Q1; Q3]		1 [0.93; 1.09] (n = 566)	ND	
Activated partial thromboplastin time, seconds, Me [Q1; Q3]		32.6 [29; 37] (n = 569)	ND	
Alanin aminotransferase, IU/L		21.9 [18; 31.2] (n = 646)	ND	
Aspartate aminotransferase, IU/L		28.2 [20.5; 37.4] (n = 646)	ND	

Statistically significant differences between groups are indicated in bold; ND—no data.

The patients were enrolled in the study in two periods: at the Regional Vascular Center of Kursk Regional Clinical Hospital between 2015 and 2017 and at the Neurology Clinics of Kursk Emergency Medicine Hospital between 2010 and 2012 [64]. All the patients were examined by qualified neurologists. The results of the brain computed tomography and/or magnetic resonance imaging were used to make the diagnosis of IS during the acute phase of stroke. The patients were recruited consecutively. The IS patients were enrolled under the following exclusion criteria: hepatic or renal failure; endocrine, autoimmune, oncological, or other diseases that can cause an acute cerebrovascular event; intracerebral hemorrhage; hemodynamic or dissection-related stroke; and traumatic brain injury. All the patients with IS had a history of hypertension and received antihypertensive therapy.

According to the WHO recommendations, low fruit and vegetable consumption was defined as an intake of less than 400 g per day (excluding potatoes and other starchy tubers) [65]. Insufficient physical activity was defined as less than 180 min per week of physical activities of moderate to vigorous intensity. We included both physical activities performed in leisure time (for example, walking and running) and fitness club activities (for example, running on a treadmill, aerobics, or resistance training). It is this level of physical activity that is considered protective against the development of cardiovascular diseases by most authors [66–68]. According to the WHO criteria, overweight is defined as patients with a body mass index of 25 or over [69].

Healthy volunteers who had normal blood pressure without receiving antihypertensive treatment and no clinical signs of cardiovascular, cerebrovascular, or other serious illnesses comprised the control group. A healthy person met the criteria for inclusion in the control group if their systolic blood pressure was less than 130 mm Hg and their diastolic blood pressure was less than 85 mm Hg on at least three separate measurements. Control subjects were enrolled from Kursk hospitals during periodic medical examinations at public institutions and industrial enterprises of Kursk region [70,71]. This group was recruited from the same population and during the same period.

The selection of SNPs was based on the following criteria: the SNP must have a minor allele frequency of at least 0.05 in the European population and be characterized by a high regulatory potential. According to the bioinformatic tools SNPinfo Web Server (https://snpinfo.niehs.nih.gov/ (accessed on 19 March 2022)) and LD TAG SNP Selection (TagSNP), which were used to select SNPs based on the reference haplotypic structure of the Caucasian population (CEU) of the project HapMap, the *SERBP1* (SERPINE1 MRNA Binding Protein 1, ID:26135) gene contains five SNPs (rs4655707, rs1058074, rs12561767, rs12566098, rs6702742). SNPs rs4655707 and rs1058074 are located in three prime UTR; rs12561767, rs12566098, and rs6702742 are located in introns.

The regulatory potential of these SNPs was evaluated using a variety of bioinformatic tools. RegulomeDB tool showed that rs4655707, rs1058074, and rs6702742 SERBP1 are characterized by a regulatory coefficient of 5 (TF binding or DNase peak); rs12566098, by a regulatory coefficient of 4 (TF binding + DNase peak); rs12561767, by a regulatory coefficient of 1f (eQTL + TF binding / DNase peak) (http://regulome.stanford.edu/ (accessed on 20 March 2022)) [72]. HaploReg (v4.1) database showed that this SNPs are characterized by enhancer histone marks in different tissues (rs4655707, rs12561767, rs12566098, rs6702742), regions of hypersensitivity for DNAse 1 (rs12561767, rs12566098, rs6702742), binding site for regulatory proteins (rs4655707, rs12561767), and DNA regulatory motifs (rs1058074, rs12561767, rs6702742) (http://archive.broadinstitute.org/mammals/haploreg/haploreg. php (accessed on March 20, 2022)) [73].

According to the data presented by the NCBI source (https://www.ncbi.nlm.nih. gov/snp/ (accessed on 20 January 2023)), these genetic variants are characterized by an average frequency of the minor alleles in European populations of 0.30 (rs12566098) to 0.46 (rs4655707) (https://www.ensembl.org/ (on 20 March 2022)). Thus, all five SNPs were selected for our research, which meets the necessary criteria for inclusion in the study.

4.1. Genetic Analysis

DNA analysis was carried out at the Laboratory of Genomic Research of Research Institute for Genetic and Molecular Epidemiology of Kursk State Medical University (Kursk, Russia). Approximately 5 mL of venous blood from the cubital vein of each participant was collected into EDTA-coated tubes and maintained at −20 oC until processed. Genomic DNA was extracted from thawed blood samples by the standard procedure of phenol/chloroform extraction and ethanol precipitation. A NanoDrop spectrophotometer (Thermo Fisher Scientific, Waltham, MA, USA) was used to evaluate the extracted DNA solution's purity, quality, and concentration.

Genotyping of the SNPs was conducted using allele-specific probe-based real-time polymerase chain reaction assays according to the protocols designed in the Laboratory of Genomic Research, Research Institute for Genetic and Molecular Epidemiology. Primers and probes were designed using the Primer3 program online (http://primer3.ut.ee/ (accessed on 30 March 2022)) [74], selected, and then synthesized by the Syntol company (Moscow, Russia).

The primers and probes used for genotyping the polymorphisms are presented in Table S. 11. A real-time PCR was conducted in a 25-mL reaction mixture containing 1.5 U of Hot Start Taq DNA polymerase (Biolabmix, Novosibirsk, Russia), about 1 μg of DNA, 0.25 μM each primer, 250 μM dNTPs, 3.0 mM $MgCl_2$ (for rs4655707), 2.5 mM $MgCl_2$ (for rs1058074, rs12566098, rs6702742), 1.5 mM $MgCl_2$ (for rs12561767), and 1xPCR buffer (67 mM Tris-HCl, pH 8.8, 16.6 mM $(NH_4)2SO_4$, 0.01% Tween-20). The amplification reaction consisted of an initial denaturation for 10 min at 95 °C, followed by 39 cycles of 92 °C for 30 s and 64 °C, 49 °C, 64 °C, 56 °C, 53 °C for 1 min (for rs4655707, rs1058074, rs12561767, rs12566098, and rs6702742, respectively). To ensure quality control, 10% of DNA samples were genotyped in duplicates, blinded to the case-control status. The concordance rate was >99%.

4.2. Statistical and Bioinformatics Analysis

The genetic association study power calculator, accessible online at http://csg.sph.umich.edu/abecasis/gas_power_calculator/ (accessed on 15 January 2023), was used to calculate the statistical power for the study. Association analysis between the *SERBP1* gene polymorphisms and IS risk could detect the genotype relative risk of 1.19–1.45 assuming 0.80 power and a 5% type-I error ($\alpha = 0.05$) on the sample size of 869 cases and 1191 controls.

All statistical analyses were performed using the STATISTICA software (v13.3, USA). The distribution of quantitative data was tested for normality using Shapiro–Wilk's test. Since the biochemical parameters and BMI showed a deviation from normal distribution, they were expressed as the median (Me) and first and third quartiles (Q1 and Q3). The Kruskal–Wallis test was used to compare quantitative variables among three independent groups. Following that, groups were contrasted pairwise using the Mann–Whitney test with FDR adjustment [75]. For categorical variables, the statistical significance of differences was evaluated by Pearson's chi-squared test with Yates's correction for continuity.

Compliance of the genotypes' distributions with Hardy–Weinberg equilibrium was assessed using Fisher's exact test. Genotype frequencies in the study groups and their associations with the disease risk were analyzed using SNPStats software (https://www.snpstats.net/start.htm (accessed on 15 January 2023)) [76]. For the analysis of associations among genotypes, additive models were considered. Associations in the entire group of IS patients/controls were adjusted for age, gender, and smoking status. In cases where there was no information about the environmental risk factor in the control group, associations were analyzed depending on the presence or absence of the risk factor in the group of patients, compared with the total control group. In this case, the Bonferroni correction was additionally introduced [77].

The following bioinformatics resources were used to analyze the functional effects of *SERBP1* SNPs:

- The bioinformatic tool GTExportal (http://www.gtexportal.org/ (accessed on 21 January 2023)) was used to analyze the expression levels of the studied genes in the brain, whole blood, and blood vessels, as well as to analyze expression quantitative trait loci (eQTLs) (The GTEx Consortium, 2020). Additionally, the eQTLGene browser (https://www.eqtlgen.org/ (accessed on 21 January 2023)) was used to analyze the cis-eQTL-mediated effects of *SERBP1* SNPs in blood [78];

- The methylation quantitative trait loci (mQTLs) in the brain, whole blood, and blood vessels were examined using QTLbase (http://www.mulinlab.org/qtlbase/index.html (accessed on 21 January 2023)) [79];

- Bioinformatic tools of the STRING database (https://string-db.org/ (accessed on 22 January 2023)) were used for the analysis of the main interaction partners of SERBP1 [80]. Analysis of biological processes and molecular functions reflecting interactions with main functionally related proteins also was carried out in STRING database;

- The atSNP Function Prediction online tool (http://atsnp.biostat.wisc.edu/search (accessed on 22 January 2023)) was used to evaluate the impact of *SERBP1* SNPs on the binding of transcription factors (TFs) to DNA depending on the carriage of the reference/alternative alleles [81]. TFs were included based on the degree of influence of SNPs on the interaction of TFs with DNA, calculated on the basis of a positional weight matrix;

- Using the Gene Ontology online tool (http://geneontology.org/ (accessed on 23 January 2023)), it was feasible to analyze the joint involvement of TFs linked to the reference/SNP alleles in overrepresented biological processes directly related to the pathogenesis of IS [82]. Biological functions controlled by transcription factors associated with *SERBP1* SNPs were used as functional groups;

- The Comparative Toxicogenomics Database (CTD) resource (http://ctdbase.org (accessed on 24 January 2023)) was used for the interpretation of environment-associated correlates of *SERBP1*. CTD provides the ability to analyze specific interactions between genes and chemicals in vertebrates and invertebrates based on data obtained from published scientific studies worldwide [83]. This tool was used to analyze binary interactions involving one chemical and one gene or protein;

- The Cerebrovascular Disease Knowledge Portal (CDKP) (https://cd.hugeamp.org/ (accessed on 24 January 2023)) and Cardiovascular Disease Knowledge Portal (https://cvd.hugeamp.org/ (accessed on 24 January 2023)) online tools were used for bioinformatic analyses of the associations of *SERBP1* SNPs with atherosclerosis-associated diseases, intermediate phenotypes, and risk factors for IS (such as total cholesterol, LDL, BMI, etc.).

Supplementary Materials: The following supporting information can be downloaded at: https://www.mdpi.com/article/10.3390/ijms24108716/s1, Table S1: Distribution of *SERBP1* genotypes in ischemic stroke patients/healthy controls and their correspondence to the Hardy—Weinberg equilibrium; Table S2: Results of the analysis of associations between *SERBP1* gene SNPs and ischemic stroke risk depending on sex and smoking status; Table S3: Established statistically significant associations of *SERBP1* genotypes with clinical and biological characteristics of IS patients; Table S4: Analysis of the effect of rs4655707 *SERBP1* on the binding of DNA to transcription factors; Table S5: Analysis of the effect of rs1058074 *SERBP1* on the binding of DNA to transcription factors; Table S6: Analysis of the effect of rs12561767 *SERBP1* on the binding of DNA to transcription factors; Table S7: Analysis of the effect of rs12566098 *SERBP1* on the binding of DNA to transcription factors. Table S8: Analysis of the effect of rs6702742 *SERBP1* on the binding of DNA to transcription factors; Table S9: Main functional characteristics of predicted functional partners of *SERBP1* (STRING database); Table S10: Functional enrichments of *SERBP1* network; Table S11: Primers and probes were designed for this study.

Author Contributions: I.S. wrote the initial draft. I.S., K.K., M.S., and T.S. collected the data and conducted experiments. O.B. conceived and developed the study, supervised the experiments, and reviewed and edited the manuscript. M.F. participated in data analysis and interpretation and data curation. A.P. and M.C. administrated the project and interpreted and validated data. K.K. and O.B. developed the methodology. V.S., A.D., and M.F. were involved in funding and original writing. I.S. and O.B. confirm the authenticity of all raw data. All authors have read and agreed to the published version of the manuscript.

Funding: This research was funded by Russian Science Foundation (№ 22-15-00288, https://rscf.ru/en/project/22-15-00288/ (accessed on 23 January 2023)).

Institutional Review Board Statement: The study was conducted according to the guidelines of the Declaration of Helsinki and was approved by the Ethical Review Committee of Kursk State Medical University, Russia (Protocol No. 12 from 7.12.2015). All the participants gave written informed consent before the enrollment in this study.

Informed Consent Statement: Informed consent was obtained from all subjects involved in the study.

Data Availability Statement: The data presented in this study are available upon request from corresponding author.

Conflicts of Interest: The authors declare no conflict of interest.

Study Limitations: Our study has several limitations. First, we did not study the level of *SERBP1* expression, which does not allow us to establish the effect of the studied SNPs on the level of SERBP1 mRNA. Secondly, no analysis of PAI-1 expression was performed; therefore, it is impossible to establish the influence of *SERBP1* SNPs on PAI-1 expression. Thirdly, we did not conduct a biochemical analysis to determine the individual components of the fibrinolytic system (plasmin, plasminogen), which would have allowed us to more definitely testify to the relationship of the studied genetic variants to fibrinolysis, which is directly regulated by PAI-1. Fourthly, the number of subjects on which the study is based is too small to draw an unambiguous conclusion about the use of *SERBP1* SNPs as a new markers of IS. Replicative studies are required in various populations around the world, involving large research samples.

References

1. Chugh, C. Acute Ischemic Stroke: Management Approach. *Indian J. Crit. Care Med. Peer-Rev. Off. Publ. Indian Soc. Crit. Care Med.* **2019**, *23* (Suppl. S2), S140–S146. [CrossRef]
2. Johnson, W.; Onuma, O.; Owolabi, M.; Sachdev, S. Stroke: A Global Response Is Needed. *Bull. World Health Organ.* **2016**, *94*, 634. [CrossRef]
3. Qin, C.; Yang, S.; Chu, Y.-H.; Zhang, H.; Pang, X.-W.; Chen, L.; Zhou, L.-Q.; Chen, M.; Tian, D.-S.; Wang, W. Signaling Pathways Involved in Ischemic Stroke: Molecular Mechanisms and Therapeutic Interventions. *Signal Transduct. Target. Ther.* **2022**, *7*, 215. [CrossRef]
4. Lin, J.; Wang, Y.; Wang, Y.; Pan, Y. Inflammatory Biomarkers and Risk of Ischemic Stroke and Subtypes: A 2-Sample Mendelian Randomization Study. *Neurol. Res.* **2020**, *42*, 118–125. [CrossRef]
5. Polonikov, A.; Rymarova, L.; Klyosova, E.; Volkova, A.; Azarova, I.; Bushueva, O.; Bykanova, M.; Bocharova, I.; Zhabin, S.; Churnosov, M.; et al. Matrix Metalloproteinases as Target Genes for Gene Regulatory Networks Driving Molecular and Cellular Pathways Related to a Multistep Pathogenesis of Cerebrovascular Disease. *J. Cell. Biochem.* **2019**, *120*, 16467–16482. [CrossRef] [PubMed]
6. Bushueva, O.; Barysheva, E.; Markov, A.; Belykh, A.; Koroleva, I.; Churkin, E.; Polonikov, A.; Ivanov, V.; Nazarenko, M. DNA Hypomethylation of the MPO Gene in Peripheral Blood Leukocytes Is Associated with Cerebral Stroke in the Acute Phase. *J. Mol. Neurosci.* **2021**, *71*, 1914–1932. [CrossRef]
7. Cui, J.; Li, H.; Chen, Z.; Dong, T.; He, X.; Wei, Y.; Li, Z.; Duan, J.; Cao, T.; Chen, Q.; et al. Thrombo-Inflammation and Immunological Response in Ischemic Stroke: Focusing on Platelet-Tregs Interaction. *Front. Cell. Neurosci.* **2022**, *16*, 955385. [CrossRef]
8. Janardhan, V.; Qureshi, A.I. Mechanisms of Ischemic Brain Injury. *Curr. Cardiol. Rep.* **2004**, *6*, 117–123. [CrossRef] [PubMed]
9. Singh, V.; Mishra, V.N.; Chaurasia, R.N.; Joshi, D.; Pandey, V. Modes of Calcium Regulation in Ischemic Neuron. *Indian J. Clin. Biochem.* **2019**, *34*, 246–253. [CrossRef]
10. Yamashita, T.; Abe, K. Pathophysiology of Neuronal Cell Death After Stroke. In *Stroke Revisited: Pathophysiology of Stroke: From Bench to Bedside*; Lee, S.-H., Ed.; Stroke Revisited; Springer: Singapore, 2020; pp. 235–241. [CrossRef]
11. Ludhiadch, A.; Sharma, R.; Muriki, A.; Munshi, A. Role of Calcium Homeostasis in Ischemic Stroke: A Review. *CNS Neurol. Disord.-Drug Targets-CNS Neurol. Disord.* **2022**, *21*, 52–61. [CrossRef]

12. Chan, P.H. Reactive Oxygen Radicals in Signaling and Damage in the Ischemic Brain. *J. Cereb. Blood Flow Metab.* **2001**, *21*, 2–14. [CrossRef]
13. Giffard, R.G.; Yenari, M.A. Many Mechanisms for Hsp70 Protection from Cerebral Ischemia. *J. Neurosurg. Anesthesiol.* **2004**, *16*, 53–61. [CrossRef] [PubMed]
14. Frydman, J. Folding of Newly Translated Proteins In Vivo: The Role of Molecular Chaperones. *Annu. Rev. Biochem.* **2001**, *70*, 603–647. [CrossRef] [PubMed]
15. Madrigal-Matute, J.; de Bruijn, J.; van Kuijk, K.; Riascos-Bernal, D.F.; Diaz, A.; Tasset, I.; Martín-Segura, A.; Gijbels, M.J.J.; Sander, B.; Kaushik, S.; et al. Protective Role of Chaperone-Mediated Autophagy against Atherosclerosis. *Proc. Natl. Acad. Sci. USA* **2022**, *119*, e2121133119. [CrossRef]
16. Heaton, J.H.; Dlakic, W.M.; Dlakic, M.; Gelehrter, T.D. Identification and CDNA Cloning of a Novel RNA-Binding Protein That Interacts with the Cyclic Nucleotide-Responsive Sequence in the Type-1 Plasminogen Activator Inhibitor MRNA. *J. Biol. Chem.* **2001**, *276*, 3341–3347. [CrossRef] [PubMed]
17. Griemert, E.-V.; Recarte Pelz, K.; Engelhard, K.; Schäfer, M.K.; Thal, S.C. PAI-1 but Not PAI-2 Gene Deficiency Attenuates Ischemic Brain Injury After Experimental Stroke. *Transl. Stroke Res.* **2019**, *10*, 372–380. [CrossRef]
18. Hu, X.; Zan, X.; Xie, Z.; Li, Y.; Lin, S.; Li, H.; You, C. Association Between Plasminogen Activator Inhibitor-1 Genetic Polymorphisms and Stroke Susceptibility. *Mol. Neurobiol.* **2017**, *54*, 328–341. [CrossRef]
19. Jood, K.; Ladenvall, P.; Tjärnlund-Wolf, A.; Ladenvall, C.; Andersson, M.; Nilsson, S.; Blomstrand, C.; Jern, C. Fibrinolytic Gene Polymorphism and Ischemic Stroke. *Stroke* **2005**, *36*, 2077–2081. [CrossRef] [PubMed]
20. Taraborrelli, S. Physiology, Production and Action of Progesterone. *Acta Obstet. Gynecol. Scand.* **2015**, *94*, 8–16. [CrossRef]
21. Tsuboyama, K.; Osaki, T.; Matsuura-Suzuki, E.; Kozuka-Hata, H.; Okada, Y.; Oyama, M.; Ikeuchi, Y.; Iwasaki, S.; Tomari, Y. A Widespread Family of Heat-Resistant Obscure (Hero) Proteins Protect against Protein Instability and Aggregation. *PLoS Biol.* **2020**, *18*, e3000632. [CrossRef]
22. Kobzeva, K.A.; Shilenok, I.V.; Belykh, A.E.; Gurtovoy, D.E.; Bobyleva, L.A.; Krapiva, A.B.; Stetskaya, T.A.; Bykanova, M.A.; Mezhenskaya, A.A.; Lysikova, E.A.; et al. C9orf16 (BBLN) Gene, Encoding a Member of Hero Proteins, Is a Novel Marker in Ischemic Stroke Risk. *Res. Results Biomed.* **2022**, *8*, 278–292. [CrossRef]
23. Schafer, K.; Müller, K.; Hecke, A.; Mounier, E.; Goebel, J.; Loskutoff, D.J.; Konstantinides, S. Enhanced Thrombosis in Atherosclerosis-Prone Mice Is Associated with Increased Arterial Expression of Plasminogen Activator Inhibitor-1. *Arterioscler. Thromb. Vasc. Biol.* **2003**, *23*, 2097–2103. [CrossRef] [PubMed]
24. de Paula Sabino, A.; Ribeiro, D.D.; Domingueti, C.P.; dos Santos, M.S.; Gadelha, T.; Dusse LM, S.; das Graças Carvalho, M.; Fernandes, A.P. Plasminogen Activator Inhibitor-1 4G/5G Promoter Polymorphism and PAI-1 Plasma Levels in Young Patients with Ischemic Stroke. *Mol. Biol. Rep.* **2011**, *38*, 5355–5360. [CrossRef]
25. Stefansson, S.; Lawrence, D.A.; Argraves, W.S. Plasminogen Activator Inhibitor-1 and Vitronectin Promote the Cellular Clearance of Thrombin by Low Density Lipoprotein Receptor-Related Proteins 1 and 2. *J. Biol. Chem.* **1996**, *271*, 8215–8220. [CrossRef] [PubMed]
26. Chomiki, N.; Henry, M.; Alessi, M.C.; Anfosso, F.; Juhan-Vague, I. Plasminogen Activator Inhibitor-1 Expression in Human Liver and Healthy or Atherosclerotic Vessel Walls. *Thromb. Haemost.* **1994**, *72*, 44–53. [CrossRef]
27. Heaton, J.H.; Dlakic, W.M.; Gelehrter, T.D. Posttranscriptional Regulation of PAI-1 Gene Expression. *Thromb. Haemost.* **2003**, *89*, 959–966.
28. Lijnen, H.R. Pleiotropic Functions of Plasminogen Activator Inhibitor-1. *J. Thromb. Haemost.* **2005**, *3*, 35–45. [CrossRef]
29. Alturfan, A.A.; Eralp, L.; Emekli, N. Investigation of Inflammatory and Hemostatic Parameters in Female Patients Undergoing Total Knee Arthroplasty Surgery. *Inflammation* **2008**, *31*, 414–421. [CrossRef] [PubMed]
30. Jeon, Y.J.; Kim, Y.R.; Lee, B.E.; Choi, Y.S.; Kim, J.H.; Shin, J.E.; Rah, H.; Cha, S.H.; Lee, W.S.; Kim, N.K. Genetic Association of Five Plasminogen Activator Inhibitor-1 (PAI-1) Polymorphisms and Idiopathic Recurrent Pregnancy Loss in Korean Women. *Thromb. Haemost.* **2013**, *110*, 742–750. [CrossRef] [PubMed]
31. Li, X.; Fang, P.; Yang, W.Y.; Wang, H.; Yang, X. IL-35, as a Newly Proposed Homeostasis-Associated Molecular Pattern, Plays Three Major Functions Including Anti-Inflammatory Initiator, Effector, and Blocker in Cardiovascular Diseases. *Cytokine* **2019**, *122*, 154076. [CrossRef]
32. Fu, H.; Yu, J.; Choi, E.T.; Wang, H.; Yang, X. IL-35 Inhibits Ischemia/Hypoxia-Induced Angiogenesis, Suggesting That This Anti-Inflammatory Cytokine Plays New Roles in the Recovery Stage of Angiogenesis. *Circulation* **2018**, *138* (Suppl. S1), 12666.
33. Bushueva, O.Y.; Stetskaya, T.A.; Polonikov, A.V.; Ivanov, V.P. The relationship between polymorphism 640A>G of the CYBA gene with the risk of ischemic stroke in the population of the Central Russia. *Zhurnal Nevrol. I Psikhiatrii Im. SS Korsakova* **2015**, *115 Pt 2*, 38–41. [CrossRef] [PubMed]
34. Fava, C.; Montagnana, M.; Danese, E.; Almgren, P.; Hedblad, B.; Engström, G.; Berglund, G.; Minuz, P.; Melander, O. Homozygosity for the EPHX2 K55R Polymorphism Increases the Long-Term Risk of Ischemic Stroke in Men: A Study in Swedes. *Pharmacogenet. Genomics* **2010**, *20*, 94–103. [CrossRef] [PubMed]
35. Vaura, F.; Palmu, J.; Aittokallio, J.; Kauko, A.; Niiranen, T. Genetic, Molecular, and Cellular Determinants of Sex-Specific Cardiovascular Traits. *Circ. Res.* **2022**, *130*, 611–631. [CrossRef]

36. Stetskaia, T.A.; Bushueva, O.I.; Bulgakova, I.V.; Vialykh, E.K.; Shuteeva, T.V.; Biriukov, A.E.; Ivanov, V.P.; Polonikov, A.V. Association of T174M polymorphism of the angiotensinogen gene with the higher risk of cerebral stroke in women. *Ter. Arkh.* **2014**, *86*, 66–71. [CrossRef] [PubMed]

37. Intlekofer, K.A.; Petersen, S.L. 17β-Estradiol and Progesterone Regulate Multiple Progestin Signaling Molecules in the Anteroventral Periventricular Nucleus, Ventromedial Nucleus and Sexually Dimorphic Nucleus of the Preoptic Area in Female Rats. *Neuroscience* **2011**, *176*, 86–92. [CrossRef] [PubMed]

38. Intlekofer, K.A.; Petersen, S.L. Distribution of mRNAs Encoding Classical Progestin Receptor, Progesterone Membrane Components 1 and 2, Serpine mRNA Binding Protein 1, and Progestin and ADIPOQ Receptor Family Members 7 and 8 in Rat Forebrain. *Neuroscience* **2011**, *172*, 55–65. [CrossRef] [PubMed]

39. Ishikawa, S.; Matsumura, K.; Kitamura, N.; Takanami, Y.; Ito, S. Multi-Omics Analysis: Repeated Exposure of a 3D Bronchial Tissue Culture to Whole-Cigarette Smoke. *Toxicol. Vitro Int. J. Publ. Assoc. BIBRA* **2019**, *54*, 251–262. [CrossRef]

40. Panagiotakos, D.B.; Kokkinos, P.; Manios, Y.; Pitsavos, C. Physical Activity and Markers of Inflammation and Thrombosis Related to Coronary Heart Disease. *Prev. Cardiol.* **2004**, *7*, 190–194. [CrossRef]

41. Diteepeng, T.; del Monte, F.; Luciani, M. The Long and Winding Road to Target Protein Misfolding in Cardiovascular Diseases. *Eur. J. Clin. Invest.* **2021**, *51*, e13504. [CrossRef]

42. Santos-Parker, J.R.; Santos-Parker, K.S.; McQueen, M.B.; Martens, C.R.; Seals, D.R. Habitual Aerobic Exercise and Circulating Proteomic Patterns in Healthy Adults: Relation to Indicators of Healthspan. *J. Appl. Physiol.* **2018**, *125*, 1646–1659. [CrossRef]

43. Needham, E.J.; Humphrey, S.J.; Cooke, K.C.; Fazakerley, D.J.; Duan, X.; Parker, B.L.; James, D.E. Phosphoproteomics of Acute Cell Stressors Targeting Exercise Signaling Networks Reveal Drug Interactions Regulating Protein Secretion. *Cell Rep.* **2019**, *29*, 1524–1538.e6. [CrossRef]

44. Yoshizaki, T.; Ishihara, J.; Kotemori, A.; Yamamoto, J.; Kokubo, Y.; Saito, I.; Yatsuya, H.; Yamagishi, K.; Sawada, N.; Iwasaki, M.; et al. Association of Vegetable, Fruit, and Okinawan Vegetable Consumption with Incident Stroke and Coronary Heart Disease. *J. Epidemiol.* **2020**, *30*, 37–45. [CrossRef] [PubMed]

45. Wang, J.; Li, J.; Liu, F.; Huang, K.; Yang, X.; Liu, X.; Cao, J.; Chen, S.; Shen, C.; Yu, L.; et al. Genetic Predisposition, Fruit Intake and Incident Stroke: A Prospective Chinese Cohort Study. *Nutrients* **2022**, *14*, 5056. [CrossRef] [PubMed]

46. Polonikov, A.; Vialykh, E.; Vasil'eva, O.; Bulgakova, I.; Bushueva, O.; Illig, T.; Solodilova, M. Genetic Variation in Glutathione S-Transferase Genes and Risk of Nonfatal Cerebral Stroke in Patients Suffering from Essential Hypertension. *J. Mol. Neurosci.* **2012**, *47*, 511–513. [CrossRef]

47. Bushueva, O.Y.U.; Bulgakova, I.V.; Ivanov, V.P.; Polonikov, A.V. Association of Flavin Monooxygenase Gene E158K Polymorphism with Chronic Heart Disease Risk. *Bull. Exp. Biol. Med.* **2015**, *159*, 776–778. [CrossRef] [PubMed]

48. Sorokin, A.V.; Kotani, K.; Bushueva, O.Y.; Polonikov, A.V. Antioxidant-Related Gene Polymorphisms Associated with the Cardio-Ankle Vascular Index in Young Russians. *Cardiol. Young* **2016**, *26*, 677–682. [CrossRef] [PubMed]

49. Polonikov, A.; Bocharova, I.; Azarova, I.; Klyosova, E.; Bykanova, M.; Bushueva, O.; Polonikova, A.; Churnosov, M.; Solodilova, M. The Impact of Genetic Polymorphisms in Glutamate-Cysteine Ligase, a Key Enzyme of Glutathione Biosynthesis, on Ischemic Stroke Risk and Brain Infarct Size. *Life* **2022**, *12*, 602. [CrossRef]

50. Sorokin, A.V.; Kotani, K.; Bushueva, O.Y. Association of Matrix Metalloproteinase 3 and γ-Glutamyltransferase 1 Gene Polymorphisms with the Cardio-Ankle Vascular Index in Young Russians. *Cardiol. Young* **2016**, *26*, 1238–1240. [CrossRef]

51. Guan, R.; Van Le, Q.; Yang, H.; Zhang, D.; Gu, H.; Yang, Y.; Sonne, C.; Lam, S.S.; Zhong, J.; Jianguang, Z.; et al. Review of Dietary Phytochemicals and Their Relation to Oxidative Stress and Human Diseases. *Chemosphere* **2021**, *271*, 129499. [CrossRef]

52. Pacifici, F.; Rovella, V.; Pastore, D.; Bellia, A.; Abete, P.; Donadel, G.; Santini, S.; Beck, H.; Ricordi, C.; Daniele, N.D.; et al. Polyphenols and Ischemic Stroke: Insight into One of the Best Strategies for Prevention and Treatment. *Nutrients* **2021**, *13*, 1967. [CrossRef]

53. Vita, J.A. Polyphenols and Cardiovascular Disease: Effects on Endothelial and Platelet Function. *Am. J. Clin. Nutr.* **2005**, *81*, 292S–297S. [CrossRef] [PubMed]

54. Bazzano, L.A.; He, J.; Ogden, L.G.; Loria, C.; Vupputuri, S.; Myers, L.; Whelton, P.K. Dietary Intake of Folate and Risk of Stroke in US Men and Women: NHANES I Epidemiologic Follow-up Study. National Health and Nutrition Examination Survey. *Stroke* **2002**, *33*, 1183–1188. [CrossRef]

55. Margetic, S.; Gazzola, C.; Pegg, G.G.; Hill, R.A. Leptin: A Review of Its Peripheral Actions and Interactions. *Int. J. Obes.* **2002**, *26*, 1407–1433. [CrossRef] [PubMed]

56. Funahashi, H.; Yada, T.; Suzuki, R.; Shioda, S. Distribution, Function, and Properties of Leptin Receptors in the Brain. *Int. Rev. Cytol.* **2003**, *224*, 1–27. [CrossRef] [PubMed]

57. Cao, W.; Liu, F.; Li, R.W.; Yang, R.; Wang, Y.; Xue, C.; Tang, Q. Triacylglycerol Rich in Docosahexaenoic Acid Regulated Appetite via the Mediation of Leptin and Intestinal Epithelial Functions in High-Fat, High-Sugar Diet-Fed Mice. *J. Nutr. Biochem.* **2022**, *99*, 108856. [CrossRef]

58. Bungau, S.; Behl, T.; Tit, D.M.; Banica, F.; Bratu, O.G.; Diaconu, C.C.; Nistor-Cseppento, C.D.; Bustea, C.; Aron, R.A.C.; Vesa, C.M. Interactions between Leptin and Insulin Resistance in Patients with Prediabetes, with and without NAFLD. *Exp. Ther. Med.* **2020**, *20*, 197. [CrossRef]

59. Gruzdeva, O.; Borodkina, D.; Uchasova, E.; Dyleva, Y.; Barbarash, O. Leptin Resistance: Underlying Mechanisms and Diagnosis. *Diabetes Metab. Syndr. Obes.* **2019**, *12*, 191–198. [CrossRef]

60. Izquierdo, A.G.; Crujeiras, A.B.; Casanueva, F.F.; Carreira, M.C. Leptin, Obesity, and Leptin Resistance: Where Are We 25 Years Later? *Nutrients* **2019**, *11*, 2704. [CrossRef]

61. Loskutoff, D.J.; Samad, F. The Adipocyte and Hemostatic Balance in Obesity: Studies of PAI-1. *Arterioscler. Thromb. Vasc. Biol.* **1998**, *18*, 1–6. [CrossRef]

62. Kershaw, E.E.; Flier, J.S. Adipose Tissue as an Endocrine Organ. *J. Clin. Endocrinol. Metab.* **2004**, *89*, 2548–2556. [CrossRef] [PubMed]

63. Sergeeva, C.N.; Sokorev, S.N.; Efremova, O.A.; Sorokina, I.N. Analysis of the level of population endogamia as the basis of population-genetic and medical-genetic studies. *Res. Results Biomed.* **2021**, *7*, 375–387. [CrossRef]

64. Vialykh, E.K.; Solidolova, M.A.; Bushueva, O.I.; Bulgakova, I.V.; Polonikov, A.V. [Catalase gene polymorphism is associated with increased risk of cerebral stroke in hypertensive patients]. *Zhurnal Nevrol. I Psikhiatrii Im. SS Korsakova* **2012**, *112 Pt 2*, 3–7.

65. World Health Organization. *Diet, Nutrition, and the Prevention of Chronic Diseases: Report of a Joint WHO/FAO Expert Consultation*; World Health Organization: Geneva, Switzerland, 2003; Volume 916.

66. Gutin, B.; Owens, S. The Influence of Physical Activity on Cardiometabolic Biomarkers in Youths: A Review. *Pediatr. Exerc. Sci.* **2011**, *23*, 169–185. [CrossRef]

67. Fernandes, R.A.; Zanesco, A. Early Physical Activity Promotes Lower Prevalence of Chronic Diseases in Adulthood. *Hypertens. Res. Off. J. Jpn. Soc. Hypertens.* **2010**, *33*, 926–931. [CrossRef]

68. Colbert, L.H.; Visser, M.; Simonsick, E.M.; Tracy, R.P.; Newman, A.B.; Kritchevsky, S.B.; Pahor, M.; Taaffe, D.R.; Brach, J.; Rubin, S.; et al. Physical Activity, Exercise, and Inflammatory Markers in Older Adults: Findings from The Health, Aging and Body Composition Study. *J. Am. Geriatr. Soc.* **2004**, *52*, 1098–1104. [CrossRef]

69. World Health Organization. *WHO European Regional Obesity Report 2022*; World Health Organization. Regional Office for Europe: Geneva, Switzerland, 2022.

70. Bushueva, O.; Solodilova, M.; Ivanov, V.; Polonikov, A. Gender-Specific Protective Effect of the −463G>A Polymorphism of Myeloperoxidase Gene against the Risk of Essential Hypertension in Russians. *J. Am. Soc. Hypertens.* **2015**, *9*, 902–906. [CrossRef] [PubMed]

71. Bushueva, O. Single Nucleotide Polymorphisms in Genes Encoding Xenobiotic Metabolizing Enzymes Are Associated with Predisposition to Arterial Hypertension. *Res. Results Biomed.* **2020**, *6*, 447–456. [CrossRef]

72. Dong, S.; Boyle, A.P. Predicting Functional Variants in Enhancer and Promoter Elements Using RegulomeDB. *Hum. Mutat.* **2019**, *40*, 1292–1298. [CrossRef]

73. Ward, L.D.; Kellis, M. HaploReg: A Resource for Exploring Chromatin States, Conservation, and Regulatory Motif Alterations within Sets of Genetically Linked Variants. *Nucleic Acids Res.* **2012**, *40*, D930–D934. [CrossRef]

74. Koressaar, T.; Remm, M. Enhancements and Modifications of Primer Design Program Primer3. *Bioinforma. Oxf. Engl.* **2007**, *23*, 1289–1291. [CrossRef]

75. Strimmer, K. A Unified Approach to False Discovery Rate Estimation. *BMC Bioinform.* **2008**, *9*, 303. [CrossRef]

76. Solé, X.; Guinó, E.; Valls, J.; Iniesta, R.; Moreno, V. SNPStats: A Web Tool for the Analysis of Association Studies. *Bioinforma. Oxf. Engl.* **2006**, *22*, 1928–1929. [CrossRef] [PubMed]

77. Andrade, C. Multiple Testing and Protection Against a Type 1 (False Positive) Error Using the Bonferroni and Hochberg Corrections. *Indian J. Psychol. Med.* **2019**, *41*, 99–100. [CrossRef] [PubMed]

78. Võsa, U.; Claringbould, A.; Westra, H.-J.; Bonder, M.J.; Deelen, P.; Zeng, B.; Kirsten, H.; Saha, A.; Kreuzhuber, R.; Kasela, S.; et al. Unraveling the Polygenic Architecture of Complex Traits Using Blood EQTL Metaanalysis. *bioRxiv* **2018**. [CrossRef]

79. Zheng, Z.; Huang, D.; Wang, J.; Zhao, K.; Zhou, Y.; Guo, Z.; Zhai, S.; Xu, H.; Cui, H.; Yao, H.; et al. QTLbase: An Integrative Resource for Quantitative Trait Loci across Multiple Human Molecular Phenotypes. *Nucleic Acids Res.* **2020**, *48*, D983–D991. [CrossRef]

80. von Mering, C.; Jensen, L.J.; Snel, B.; Hooper, S.D.; Krupp, M.; Foglierini, M.; Jouffre, N.; Huynen, M.A.; Bork, P. STRING: Known and Predicted Protein–Protein Associations, Integrated and Transferred across Organisms. *Nucleic Acids Res.* **2005**, *33*, D433–D437. [CrossRef]

81. Shin, S.; Hudson, R.; Harrison, C.; Craven, M.; Keleş, S. AtSNP Search: A Web Resource for Statistically Evaluating Influence of Human Genetic Variation on Transcription Factor Binding. *Bioinforma. Oxf. Engl.* **2019**, *35*, 2657–2659. [CrossRef] [PubMed]

82. The Gene Ontology Consortium. The Gene Ontology Resource: 20 Years and Still GOing Strong. *Nucleic Acids Res.* **2019**, *47*, D330–D338. [CrossRef] [PubMed]

83. Davis, A.P.; Grondin, C.J.; Johnson, R.J.; Sciaky, D.; Wiegers, J.; Wiegers, T.C.; Mattingly, C.J. Comparative Toxicogenomics Database (CTD): Update 2021. *Nucleic Acids Res.* **2021**, *49*, D1138–D1143. [CrossRef]

International Journal of
Molecular Sciences

Article

Preliminary Study on the Effect of a Night Shift on Blood Pressure and Clock Gene Expression

Barbara Toffoli [1], Federica Tonon [1], Fabiola Giudici [1], Tommaso Ferretti [1], Elena Ghirigato [1], Matilde Contessa [1], Morena Francica [1], Riccardo Candido [1,2], Massimo Puato [3], Andrea Grillo [1,4], Bruno Fabris [1,4] and Stella Bernardi [1,4,*]

[1] Department of Medical Surgical and Health Sciences, University of Trieste, Cattinara Teaching Hospital, Strada di Fiume 447, 34149 Trieste, Italy; btoffoli@units.it (B.T.); effetonon@gmail.com (F.T.); fabiola.giudici@cro.it (F.G.); tomferretti@yahoo.it (T.F.); elenasofia.ghirigato6@gmail.com (E.G.); matilde.contessa@outlook.it (M.C.); morena.francica@asugi.sanita.fvg.it (M.F.); riccardo.candido@units.it (R.C.); andrea.grillo@units.it (A.G.); b.fabris@fmc.units.it (B.F.)
[2] SC Patologie Diabetiche, ASUGI, 34100 Trieste, Italy
[3] SSD Angiologia e Fisiologia Clinica Vascolare Multidisciplinare Cattinara Teaching Hospital, Strada di Fiume 447, 34149 Trieste, Italy; massimo.puato@asugi.sanita.fvg.it
[4] UCO Medicina Clinica, ASUGI Azienda Sanitaria Universitaria Giuliano-Isontina, Cattinara Teaching Hospital, Strada di Fiume 447, 34149 Trieste, Italy
* Correspondence: stella.bernardi@asugi.sanita.fvg.it

Abstract: Night shift work has been found to be associated with a higher risk of cardiovascular and cerebrovascular disease. One of the underlying mechanisms seems to be that shift work promotes hypertension, but results have been variable. This cross-sectional study was carried out in a group of internists with the aim of performing a paired analysis of 24 h blood pressure in the same physicians working a day shift and then a night shift, and a paired analysis of clock gene expression after a night of rest and a night of work. Each participant wore an ambulatory blood pressure monitor (ABPM) twice. The first time was for a 24 h period that included a 12 h day shift (08.00–20.00) and a night of rest. The second time was for a 30 h period that included a day of rest, a night shift (20.00–08.00), and a subsequent period of rest (08.00–14.00). Subjects underwent fasting blood sampling twice: after the night of rest and after the night shift. Night shift work significantly increased night systolic blood pressure (SBP), night diastolic blood pressure (DBP), and heart rate (HR) and decreased their respective nocturnal decline. Clock gene expression increased after the night shift. There was a direct association between night blood pressure and clock gene expression. Night shifts lead to an increase in blood pressure, non-dipping status, and circadian rhythm misalignment. Blood pressure is associated with clock genes and circadian rhythm misalignement.

Keywords: night shift; clock genes; circadian rhythm; blood pressure; ABPM; dipping; physician; internal medicine

Citation: Toffoli, B.; Tonon, F.; Giudici, F.; Ferretti, T.; Ghirigato, E.; Contessa, M.; Francica, M.; Candido, R.; Puato, M.; Grillo, A.; et al. Preliminary Study on the Effect of a Night Shift on Blood Pressure and Clock Gene Expression. *Int. J. Mol. Sci.* **2023**, *24*, 9309. https://doi.org/10.3390/ijms24119309

Academic Editors: Yutang Wang and Dianna Magliano

Received: 13 April 2023
Revised: 19 May 2023
Accepted: 24 May 2023
Published: 26 May 2023

1. Introduction

Shift work has been found to be associated with a higher risk of cardiovascular and cerebrovascular disease [1,2]. The mechanisms underlying this association include the fact that rotating night shifts seem to promote cardiovascular risk factors, such as hypertension, diabetes, and obesity, although results have been variable [3].

The negative effects of night shifts on health seem to be due to the disruption of the normal 24 h circadian rhythm. This rhythm is based on a master clock that resides in the suprachiasmatic nucleus of the hypothalamus, which is entrained by light, and on peripheral clocks that integrate signals coming from the master clock as well as the periphery, such as light and food [4,5]. The clocks regulating circadian rhythm rely on cellular networks of transcription factors (core clock genes) that control circadian variations in cellular gene expression, which in turn regulate most physiological functions over

24 h, such as respiration, metabolism, endocrine and immune function, body temperature, and blood pressure. Blood pressure circadian rhythm, for example, is characterized by a morning surge at wake, a plateau during the day, and a nocturnal decline of 10–20% as compared to the average daytime blood pressure, which is known as dipping [6,7].

A person working at night and sleeping during the day suffers from a state of desynchrony, being unable to undergo complete adaptation because of the many external stimuli promoting a day-oriented schedule, as evidenced by the lack of complete adaptation of the core body temperature, melatonin levels, and cortisol rhythm [8]. Desynchrony or circadian misalignment, which might occur as a result of shift work or significant activity at night, has been associated with an increase in blood pressure, and non-dipping blood pressure status [9], and it can also contribute to organ dysfunction and development of disease [10].

This observational study was carried out in a group of internists with the aim of performing a paired analysis of 24 h blood pressure in the same physicians working a day shift and then a night shift, and a paired analysis of clock gene expression after a night of rest or a night of work.

2. Results

2.1. Population Characteristics

The general characteristics of the 25 internists on shift work are reported in Table 1. The median age was 33 years. There were 15 women (60%) and 10 men (40%), and the average body mass index (BMI) was 22.6. Only one physician was obese. Half of them had a sedentary lifestyle (52%). Only three physicians reported suffering from medical conditions (hypertension, asthma, and Hashimoto's thyroiditis), and five of them were taking medication (olmesartan, on-demand salbutamol inhalation, levothyroxine, and in two cases oral contraceptive pill). Among these physicians, 12 were resident physicians (48%) and 13 were attending physicians (52%). Overall, the average working age was 7 years. During the day shift, physicians took an average of 7720 steps. During the night shift, on average, physicians received 17.5 calls, took care of 8.5 admitted patients, and took 6660 steps.

Table 1. Population characteristics.

Variable		*n* = 25
Age		33 (29–37)
Sex	Female (%)	15 (60%)
	Male (%)	10 (40%)
BMI		22.6 (20.8–24.54)
Physical activity	Yes (%)	12 (48%)
	No (%)	13 (52%)
Medical conditions	Yes (%)	3 (12%)
	No (%)	22 (88%)
Medication	Yes (%)	5 (20%)
	No (%)	20 (80%)
Years of work		7 (2–10)
Working position	Attending physician (%)	13 (52%)
	Resident physician (%)	12 (48%)
Day shift steps		7720 (6504–11,062)
Night shift calls		17.5 (13–20)
Night shift admissions		8.5 (6–10)
Night shift steps		6660 (5785–7822)

Data are expressed as median (1st Qu–3rd-Qu).

2.2. 24 h Ambulatory Blood Pressure Measurements

Figure 1a,b show 24 h blood pressure and Figure 1c,d show heart rates as assessed by two consecutive ambulatory blood pressure (BP) monitors (ABPMs), worn by all 25 internists. The first ABPM (Figure 1a,c; control = cnt) was worn on a day with a day shift at work (8–20) and a night of rest, and the second one (Figure 1b,d, night shift = ns) was worn on a day with a day off from work, a night shift (20–08), and a subsequent period of rest (recovery time).

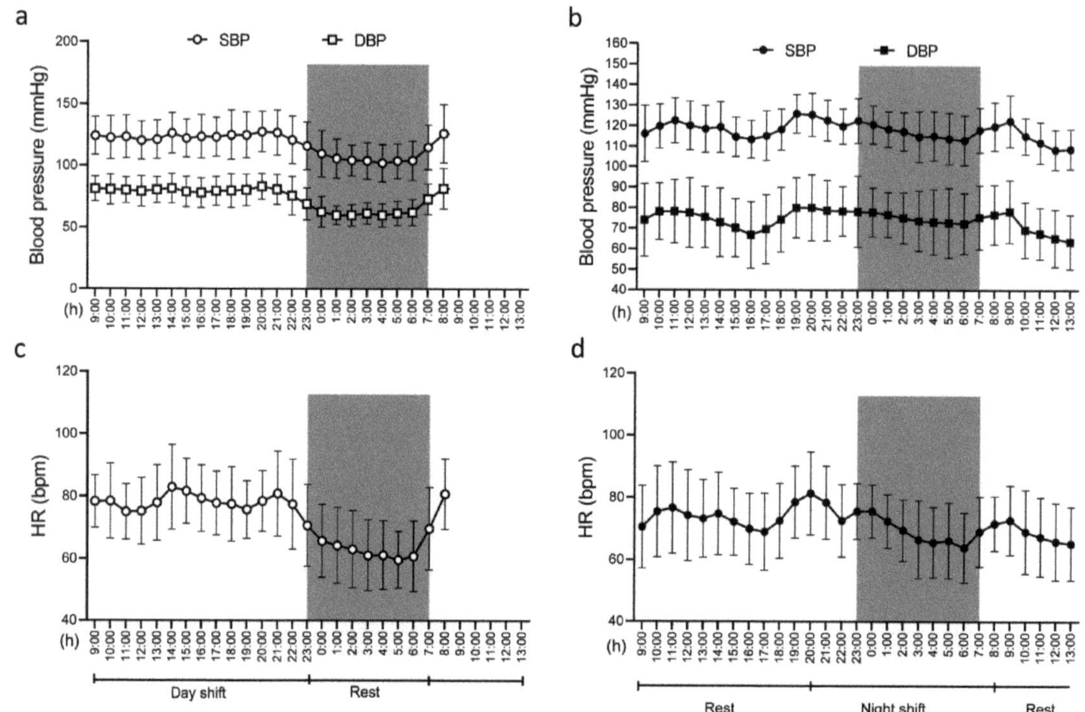

Figure 1. Participants' 24 h blood pressure and heart rate (HR). (**a**) Blood pressure during control (cnt) day; (**b**) blood pressure during night shift (ns) day; (**c**) heart rate during cnt day; (**d**) heart rate during ns day. Gray areas correspond to the sleep hours on cnt day. Data are presented as mean ± SD.

The paired analysis of ABPM results showed that day diastolic blood pressure (DBP) and day heart rate (HR) were higher when physicians were at work (Figure 2b,c). Night systolic blood pressure (SBP), night DBP, and night HR increased during the night shift as compared to the night of rest. DBP decreased during the recovery time, while SBP and HR did not. A repeated-measures ANOVA model was performed to compare SBP, DBP, and their respective dippings during the night of rest, the night shift, and the recovery time after the night shift (Table 2). During the night shift, SBP increased by an average of 10.3 mmHg and DBP increased by an average of 12.1 mmHg as compared to the night of rest. During the recovery time, SBP was reduced by an average of 6.1 mmHg, and DBP was reduced by an average of 9.1 mmHg. In addition, the paired analysis of ABPM results showed that the night shift significantly reduced SBP, DBP, and HR dipping (Figure 2d), corresponding to the nocturnal decline in SBP, DBP, and HR [6]. SBP dipping was 12% (10.5; 14%) on a day with a night of rest, while it was 1.4% (−1.4; 3.7%) on a day with a night of work. DBP dipping was 20.5% (18; 24%) on a day with a night of rest, while it was −0.4% (−2.5; 2.2%) on a day with a night of work. HR dipping was 19% (12; 22%) on a day with a night of rest, while it was 7% (−1.9; 9.6%) on a day with a night of work. In the recovery time after the night shift, dipping remained significantly lower as compared to the dipping during a night of rest (Figure 2d). These data were confirmed by the repeated-measures ANOVA model (Table 2), showing that SBP dipping failed to normalize during the recovery time, as in the recovery time it remained significantly lower as compared to the SBP dipping during the night of rest.

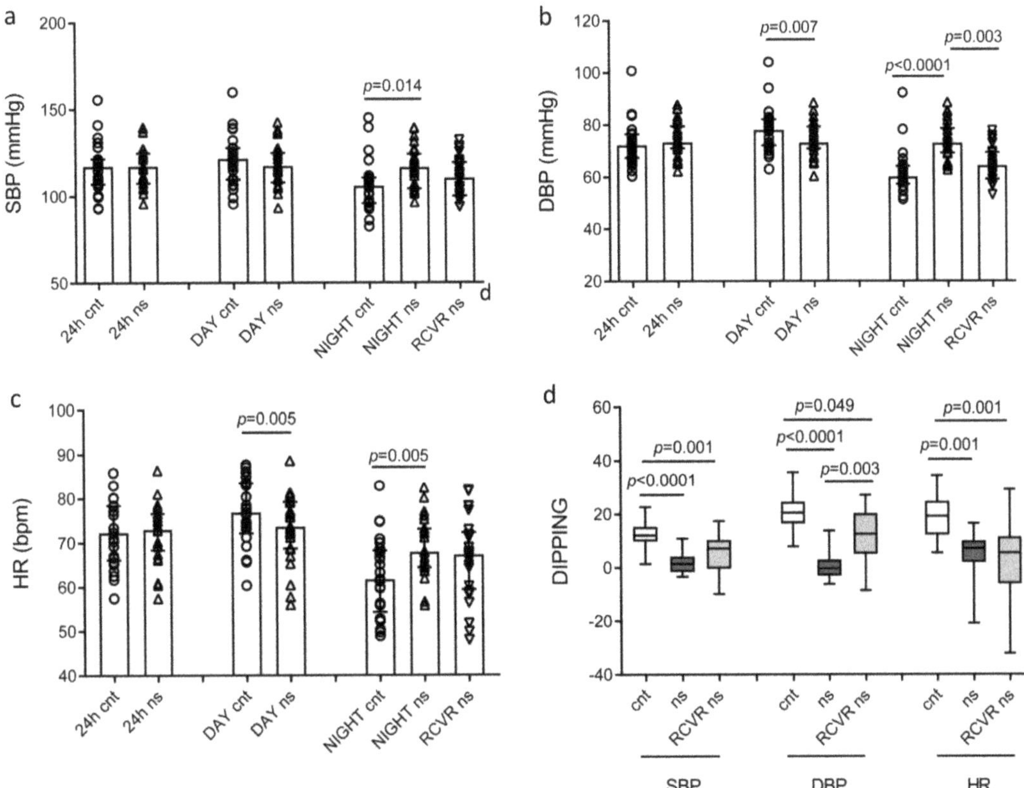

Figure 2. Paired analysis of SBP, DBP, HR, and dippings. (**a**) SBP; (**b**) DBP; (**c**) HR (**d**) dipping. Control (cnt) represents 24 h ABPM worn during the day with a day at work and a night of rest; night shift (ns) represents 24 h ABPM worn during the day with a day off and a night shift; RCVR represents the recovery time after the night shift. Data are presented as median (top of the histogram) with interquartile ranges in (**a–c**) and standard box plots in (**d**).

Table 2. Comparison of night SBP, DBP, and dipping with ANOVA.

	NIGHT cnt (t1)	NIGHT ns (t2)	RCVR ns (t3)	ANOVA	t1 vs. t2	t2 vs. t3	t1 vs. t3
SBP (mmHg)	106.08 ± 14.82	116.38 ± 12.15	110.24 ± 11.66	$F = 10.98$ $p < 0.001$ $\eta_g^2 = 0.1$	$p < 0.001$	$p = 0.006$	$p = 0.28$
DBP (mmHg)	61.85 ± 8.57	73.96 ± 7.12	64.87 ± 6.64	$F = 30.72$ $p < 0.001$ $\eta_g^2 = 0.33$	$p < 0.001$	$p < 0.001$	$p = 0.23$
SBP dipping (%)	12.6 ± 4.82	1.49 ± 3.69	5.01 ± 7.51	$F = 27.56$ $p < 0.001$ $\eta_g^2 = 0.42$	$p < 0.001$	$p = 0.17$	$p < 0.001$
DBP dipping (%)	20.7 ± 6.58	0.47 ± 4.67	11.7 ± 10.6	$F = 42.42$ $p < 0.001$ $\eta_g^2 = 0.55$	$p < 0.001$	$p < 0.001$	$p < 0.001$

"NIGHT cnt" represents the night of rest during the control day (t1); "NIGHT ns" represents the night during the night shift (t2); "RCVR ns" represents the recovery time after the night shift (t3). Data are presented as mean $\pm$ SD.

2.3. Biochemical Parameters and Circadian Rhythm Gene Expression

All subjects underwent blood sampling after fasting twice: after a night of rest and after a night of work. Overall, we found no differences in the levels of glucose, cortisol, melatonin, C-reactive protein (CRP), interleukin (IL)-1β, IL-6, and tumor-necrosis factor (TNF)-α after a night of rest as compared to a night of work (Table 3). On the other hand, after a night shift, we found that there was a significant increase in the expression of most circadian rhythm genes, i.e., *BMAL, CLOCK, PER1, PER2,* and *PER3* (Figure 3a–e), whereas the gene expression of *CRY1, CRY2, IL-6,* and *TNF-α* did not change (Figure 3f–i).

Table 3. Biochemical parameters after a night of rest and after a night of work.

Variable	After Night of Rest	After a Night Shift	*p*-Value
Glucose (mg/dL)	88.5 (80–91)	86 (80–96)	$p = 0.35$
Cortisol (nmol/L)	303 (260.5–362)	331 (269–413)	$p = 0.11$
Melatonin (pg/mL)	108.1 (98–158)	127.8 (99–138)	$p = 0.58$
CRP (mg/L)	0.85 (0.4–1.5)	0.7 (0.6–1.4)	$p = 0.11$
IL-1β (pg/mL)	12.08 (0–69)	12.89 (0–72)	$p = 0.41$
IL-6 (pg/mL)	28.81 (3–117.5)	30.77 (2–156)	$p = 0.15$
TNF-α (pg/mL)	5.13 (0–82.5)	3.20 (0–76)	$p = 1.00$

Data are expressed as median (1st Qu–3rd Qu).

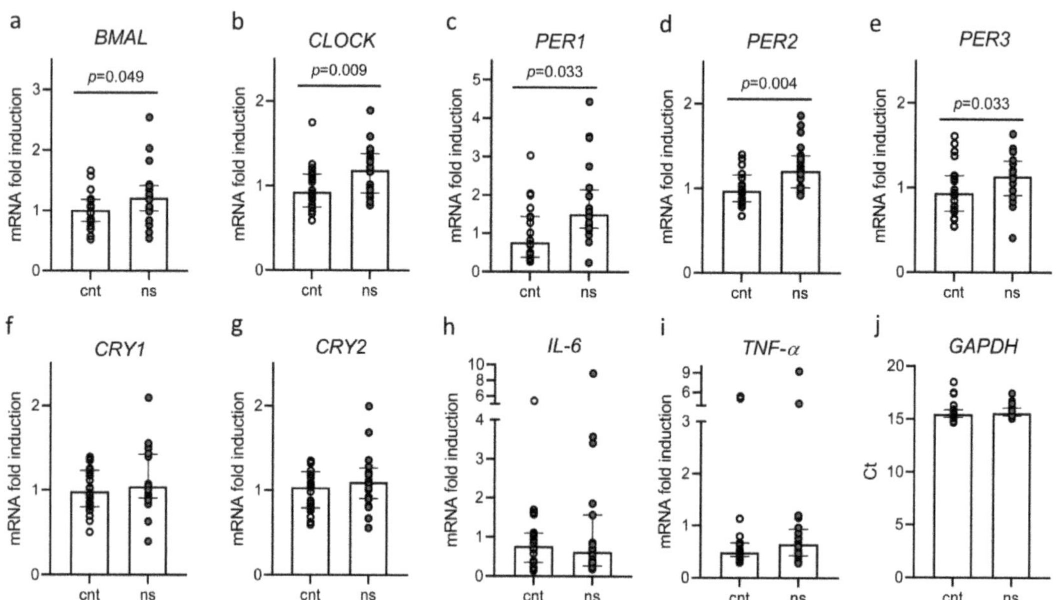

Figure 3. Clock gene expression after a night of rest and after a night shift. Paired analysis of gene expression after a night of rest (cnt) vs. after a night shift (ns). (**a**) *BMAL*; (**b**) *CLOCK*; (**c**) *PER1*; (**d**) *PER2*; (**e**) *PER3*; (**f**) *CRY1*; (**g**) *CRY2*; (**h**) *IL-6*; (**i**) *TNF-α*; (**j**) *GAPDH*. Data are presented as median (top of the histogram) with interquartile ranges. In addition, each individual value is plotted as a point superimposed on the graph.

2.4. Correlation Coefficients and Linear Regression Analysis

Focusing on night blood pressure, night SBP and night DBP were related to sex (Figure 4) and BMI (Table 4), while age and workload, as assessed by the number of calls, admissions, and steps, were not associated with them. On the cnt day, there was a direct correlation between night SBP and *PER3* gene expression on the morning after. On the night shift day, there was a direct correlation between night SBP and night DBP and *PER2* gene expression on the morning after. In addition, *PER2* gene expression on the morning after the night shift was related to the dipping during the recovery period. Otherwise, dippings were not related to age, sex, BMI, or workload.

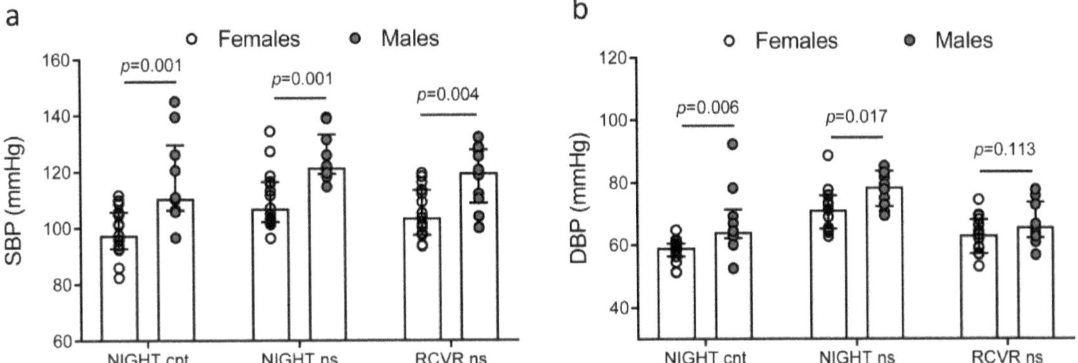

Figure 4. SBP and DBP differences between females and males: (**a**) SBP, systolic blood pressure; (**b**) DBP, diastolic blood pressure. Cnt represents the control day (day with a night of rest), ns represents the night shift day (day with the night shift), RCVR represents the recovery period after the night shift. Data are presented as median (top of the histogram) with interquartile ranges. In addition, each individual value is plotted as a point superimposed on the graph.

Table 4. Correlation coefficients of night SBP and DBP values.

	Cnt Day				Night Shift Day			
Variable	**Night SBP**		**Night DBP**		**Night SBP**		**Night DBP**	
	rho	p	rho	p	rho	p	rho	p
Age	0.01	0.96	0.10	0.65	−0.23	0.28	0.18	0.37
BMI	0.42	0.04 *	0.49	0.01 *	0.40	0.05 *	0.40	0.045 *
BMAL	0.18	0.43	0.09	0.70	−0.22	0.35	−0.09	0.69
CLOCK	0.30	0.20	0.05	0.83	0.36	0.12	0.33	0.15
PER1	−0.02	0.94	−0.09	0.71	−0.17	0.46	−0.05	0.82
PER2	0.23	0.33	0.04	0.87	0.40	0.08 *	0.45	0.04 *
PER3	0.51	0.02 *	0.27	0.25	0.37	0.11	0.35	0.13
CRY1	0.22	0.35	0.09	0.71	0.13	0.58	0.17	0.47
CRY2	0.32	0.17	0.30	0.20	0.23	0.34	0.29	0.22
Night calls	NA	NA	NA	NA	0.15	0.63	0.20	0.54
Night admissions	NA	NA	NA	NA	0.01	0.96	0.07	0.74
Night steps	NA	NA	NA	NA	−0.10	0.66	−0.01	0.96

* Statistically significant variables at a p-value of <0.10 level were selected for multivariate linear regression. NA represents not applicable.

Multivariate linear regression analysis showed the presence of independent associations between night SBP and *PER3* gene expression on the cnt day and the associations between night SBP and DBP and *PER2* gene expression on the night shift day (Table 5).

Table 5. Multivariate linear regression.

Dependent Variable: Night SBP (cnt Day)				
Predictive Variable	**β-Estimate**	**SE**	*p*-**Value**	**Multiple R-Squared**
Sex	9.33	3.96	0.03	
BMI	1.69	0.60	0.01	0.72
PER3	18.84	5.36	0.003	
Dependent Variable: Night SBP (Night Shift Day)				
Predictive Variable	**β-Estimate**	**SE**	*p*-**Value**	**Multiple R-Squared**
Sex	10.52	5.15	0.06	
BMI	0.85	0.77	0.28	0.41
PER2	17.48	8.06	0.045	
Dependent Variable: Night DBP (Night Shift Day)				
Predictive Variable	**β-Estimate**	**SE**	*p*-**Value**	**Multiple R-Squared**
Sex	4.34	3.19	0.19	
BMI	0.50	0.48	0.31	0.48
PER2	11.70	4.99	0.03	

2.5. Random Forest and Linear Regression Analysis

To further explore the relationship between night blood pressure and clock gene expression, because of the large set of covariates and the small sample size, we used the random forest (RF) as a technique to identify predictors of blood pressure, as shown in the variance importance plots (Figures 5 and 6). The main advantage of selecting relevant variables through this algorithmic modeling technique is that it is independent of any assumptions about the relationships among variables and about the distribution of errors.

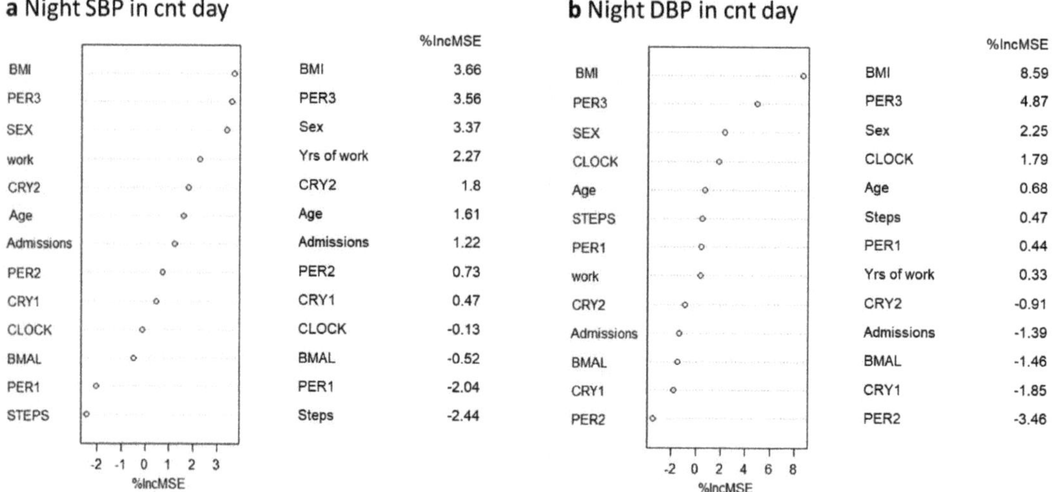

Figure 5. Variance importance plot: %IncMSE represents mean square error (MSE) decrease in accuracy for each feature removal. (**a**) Night SBP, (**b**) Night DBP in a control day.

Consistent with our previous results, the RF showed that the best predictors of night SBP and DBP during the cnt day were BMI, *PER3*, and sex (Figure 5). The best predictors of night SBP during the night shift day were *PER3*, age, and sex, while the best predictors of night DBP during the night shift day were *PER3*, years of work, and sex (Figure 6).

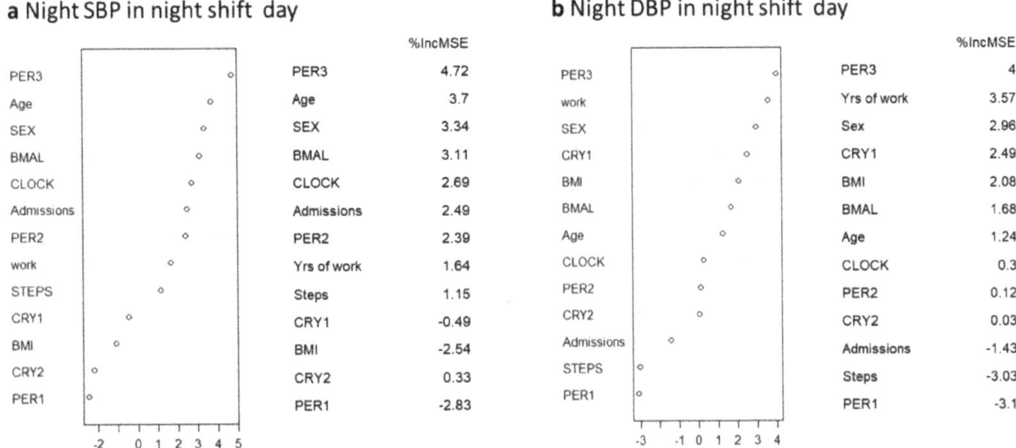

Figure 6. Variance importance plot: %IncMSE represents mean square error (MSE) decrease in accuracy for each feature removal. (**a**) Night SBP, (**b**) Night DBP in a night shift day.

Multivariate linear regression analysis confirmed that sex, BMI, and *PER3* gene expression were independently associated with night SBP on the cnt day. BMI and *PER3* gene expression were independently associated with night DBP on the cnt day. Sex and *PER3* gene expression were independently associated with SBP and DBP on the night shift day (Table 6).

Table 6. Multivariate linear regression after random forest selection of predictive variables.

Dependent Variable: Night SBP (cnt Day)				
Predictive Variable	**β-Estimate**	**SE**	**p-Value**	**Multiple R-Squared**
Sex	−10.36	3.38	0.007	
BMI	1.74	0.52	0.004	0.76
PER3	19.08	4.93	0.001	
Dependent Variable: Night DBP (cnt Day)				
Predictive Variable	**β-Estimate**	**SE**	**p-Value**	**Multiple R-Squared**
Sex	−3.5	2.12	0.12	
BMI	0.99	0.33	0.008	0.61
PER3	6.79	3.10	0.04	
Dependent Variable: Night SBP (Night Shift Day)				
Predictive Variable	**β-Estimate**	**SE**	**p-Value**	**Multiple R-Squared**
Age	−0.11	0.35	0.75	
Sex	−18.37	4.11	<0.001	0.63
PER3	25.34	7.23	0.003	
Dependent Variable: Night DBP (Night Shift Day)				
Predictive Variable	**β-Estimate**	**SE**	**p-Value**	**Multiple R-Squared**
Sex	−8.50	2.81	0.008	
Years of work	−0.23	0.25	0.37	0.51
PER3	12.38	4.99	0.02	

3. Discussion

Here, we show that in a group of internists with a median age of 33 years, the night shift was associated with a significant increase in SBP, DBP, and HR values and a significant reduction in blood pressure and heart rate nocturnal decline (dipping), such that participants had a blood pressure and heart rate non-dipping profile during their night shift. Blood pressure has a circadian rhythm, and, at night, it generally decreases by 10–20%, which is known as dipping. The loss of this blood pressure nocturnal decline is known as non-dipping and it is associated with metabolic and cardiovascular disease [6,11–14]. In our study, night shifts reduced SBP dipping from 12.6% to 1.49% and DBP dipping to 20.7% to 0.47%. These findings are consistent with the work of Yamasaki et al., who found an association between the night shift and non-dipping blood pressure status [9], which could indicate a mechanism whereby rotating night shifts lead to a higher risk of cardiovascular and cerebrovascular disease [1,2,15].

Our data show that during the night shift, SBP increased by an average of 10.3 mmHg and DBP increased by an average of 12.1 mmHg, which is consistent with previous work demonstrating an association between shift work and blood pressure. In general, shift workers seem to be more likely to have hypertension (OR 1.31, 95% CI 1.07–1.6) [16]. In addition, shift workers have a 2.1-fold-increased odds ratio of new antihypertensive medication use as compared to non-shift workers [17], and average sleep duration is inversely associated with blood pressure [18]. The association between shift work and blood pressure has been documented in healthcare workers as well. For example, in a group of physicians working in an emergency department with an age range from 28 to 40 years, DBP was significantly higher during night shift activity, as compared to non-shift awake or sleep [19]. Among female nurses, there was an additive interaction between age and the number of night shifts per month on hypertension prevalence [20]. In particular, the number of night shifts was associated with the prevalence of hypertension in participants in the age range of 36–45 years [20].

Our study also showed that after a night shift, there was a significant increase in the expression of most clock genes, namely *BMAL, CLOCK, PER1, PER2,* and *PER3*, which is consistent with circadian rhythm misalignment. The human 24 h circadian rhythm is controlled by a master clock in the suprachiasmatic nucleus (SNC) and by peripheral clocks that integrate signals coming from the master clock as well as the periphery, such as light and food [4,5]. These inner clocks rely on cellular transcriptional–translational feedback loops whereby BMAL1 heterodimerizes with CLOCK to promote the transcription of two repressor genes, Period (Per homologs *PER1, PER2,* and *PER3*) and Cryptochrome (Cry homologs *CRY1* and *CRY2*). Once they are translated, PER and CRY form a heterodimer that stops the transcription of *BMAL1* and *CLOCK*. Once PER and CRY are degraded, *BMAL1* and *CLOCK* transcription begins again. In normal conditions, PER1, PER2, and PER3 transcripts peak during the biological night, while *BMAL1* transcripts peak during the biological day [21]. It has been shown that sleep deprivation leads to circadian disruption and misalignment with changes in the clock gene expression in human blood cells, such as *PER2* upregulation in a different time period [21]. Additionally, in a mouse model of shift work, two weeks of sleep restriction was associated with changes in core clock gene expression, i.e., an upregulation of *Bmal1* and *Per1* between 0 and 6 am, while overall rhythmicity was preserved [22].

It has been argued that circadian misalignment has an impact on some physiological parameters, such as blood pressure, which is one of the mechanisms whereby shift work can have negative health consequences [3]. In our study, correlation coefficients as well as random forests pointed to an association between circadian rhythm and blood pressure. Although there is a lack of consensus on the appropriate sample size for multiple regression, linear regression analyses confirmed the existence of independent significant associations between night SBP and DBP and clock gene expression, particularly *PER3* expression both on the cnt and night shift day. By contrast, night workload as assessed by steps, the number of calls, and admissions was not related to the blood pressure level, possibly because light exposure (which is experienced during night shifts) may be a sufficient suppressor of circadian rhythm [23,24], which ultimately affects blood pressure. Animal studies have demonstrated an important role for the clock genes—and PER homologs—in the regulation of blood pressure [25]. Deletion of *Per1* in hypertensive male mice resulted in lower blood pressure as compared to wild-type mice, suggesting that loss of *Per1* was protective in the setting of hypertension [26]. Cell culture experiments show that *Per1* is a positive regulator of sodium reabsorption genes [27,28]. Additionally, human studies have shown that *PER1* was upregulated in the kidneys of hypertensive subjects [29]. In addition, it has been shown that clock gene variants explained 7.1% of the variance in SBP, following adjustments for age, sex, BMI, medication, and social characteristics in a non-Hispanic white population [30], and significant associations were found between SBP and *CLOCK*, *PER1*, *PER3*, and *CRY1* [30]. Other studies have shown associations between *CRY1* and *CRY2* polymorphisms and blood pressure in patients with metabolic syndrome [31], or associations between *PER2* and *BMAL1* polymorphisms and the non-dipper phenotype in 372 young hypertensive patients [32]. A genetic association study designed to test the relevance of 19 polymorphisms of *BMAL1* in 1304 individuals showed that two *BMAL1* haplotypes were associated with type 2 diabetes and hypertension [33].

Going back to our study, and its strengths and limitations, its strengths include recruitment of physicians working night shifts in the same setting/hospital and the possibility to perform paired analyses (as the same physicians were tested twice). On the other hand, limitations include the age of physicians (median age was 33), as it is possible that results would differ in older cohorts; the lack of blood sampling and biochemical measurements at night; and the small sample size. The number of physicians recruited did not allow us to assess the impact of other variables on blood pressure in the model for repeated measures, so larger studies are needed to confirm and extend our findings. In particular, we believe that, in future studies, it would be useful to add gene expression and biochemistries at midnight to gain more information on circadian rhythm misalignment and possible hormonal changes during night shifts.

In conclusion, in line with previous work, our study shows that night shift increased blood pressure leading to a non-dipping blood pressure status, which is a cardiovascular disease risk factor. Night shifts increased clock gene expression on the morning after, consistent with circadian misalignment, and there was an independent association between blood pressure during the night shift and *PER2-3* expression. Our data provide further evidence on the negative impact of night work on cardiovascular health as well as on the association between circadian clock misalignment and blood pressure control. These data support interventions aiming to improve sleep and recovery and reduce fatigue, contributing to the sustainable employability of healthcare workers [34].

4. Materials and Methods

4.1. Study Design and Population

This is an observational cross-sectional study, whose aims were (i) to perform a paired analysis of 24 h blood pressure in the same physicians working a day shift and then a night shift; (ii) to perform a paired analysis of clock gene expression after a night of rest or a night of work.

Participants were consecutively recruited among physicians working in the Department of Internal Medicine (Dipartimento di Medicina of Cattinara Teaching Hospital, ASUGI, Trieste, Italy). Inclusion criteria were aged between 25 and 60 years; working night shifts in the same Department of Internal Medicine; consent to take part in the study. Exclusion criteria were history of cardiovascular or endocrine disorders; children with age <2 years; return to work after holidays in the 3 weeks before enrollment; shifts in COVID-19 wards.

After providing the informed consent, each participant wore an ambulatory BP monitor (ABPM) twice. The first time (control, cnt day) was for a 24 h period (from 09.00 to 09.00) that included a 12 h day shift (08.00–20.00) and a night of rest. The second time (night shift, ns day) was for a 30 h period (from 09.00 to 13.00) that included a day off, a night shift (20.00–08.00), and a period of rest (08.00–13.00) the following day. The ABPM was a SpaceLabs Oscillometric Model 90207 (Redmon, WA, USA). The ABPM was applied by a resident physician, who evaluated and calibrated the readings at time of placement and at removal. Before its placement, we measured anthropometric parameters and collected the history of each participant. Participants were also asked to keep a diary of their activities, write down when they slept on the CNT day, and to provide details of their activity during the night shift. In particular, participants were asked to write down the number of steps, the number of calls they received, and admissions they had to take care of during the night shift. In addition, they were asked to fast from 24.00 to 09.00 on both the cnt and ns days. When the ABPM was removed, participants underwent blood sampling after fasting to measure glucose, cortisol, melatonin, CRP, IL-1β, IL-6, and TNF-α and to evaluate clock gene expression.

Dipping was calculated as follows: (mean wake hours BP − mean sleep BP divided by mean wake hours BP) × 100. Non-dipping status was defined as sleep hours BP or HR dipping of less than 10% [35].

This study was conducted in accordance with the Declaration of Helsinki, and the protocol was approved by the Institutional Review Board and Ethics Committee (CEUR 107_2020H).

4.2. Biochemical Parameters and ELISA

Glucose, CRP, and cortisol were measured using an autoanalyzer. Melatonin (E-EL-H2016, Elabscience, Houston, TX, USA), IL-1β (DY201-05, R&D Systems, Minneapolis, MN, USA), IL-6 (DY206-05, R&D Systems), and TNF-α (DY210-05, R&D Systems) were measured using ELISA, according to the manufacturer's instructions. The intra-assay coefficients of variations were 5.28% for melatonin, 3.51% for IL-1β, 5.37% for IL-6, and 3.72% for TNF-α. The inter-assay coefficients of variations were 5.02% for melatonin, 9.91% for IL-1β, 11.40% for IL-6, and 8.74% for TNF-α, as already described [36,37].

4.3. Peripheral Blood Mononuclear Cell Isolation and Gene Expression Analysis

To isolate peripheral blood mononuclear cells (PBMCs), blood samples were collected in EDTA tubes, layered onto the Ficoll-Paque™ Plus (Cytiva Sweden AB, Uppsala, Sweden) solution, and then centrifuged at 2400 rpm for 30 min at room temperature. The mononuclear cell layer that was obtained was used to extract RNA, as already described [38,39].

RNA extraction was performed with the AllPrep DNA/RNA Mini Kit (Qiagen, Hilden, Germany), according to the manufacturer's instructions. A total of 700 ng of RNA was subsequently used to synthesize cDNA with the Superscript First-Strand Synthesis System for RT-PCR (Gibco BRL, Thermo Fisher Scientific, Waltham, MA, USA). The expression of genes related to circadian rhythm (*BMAL, CLOCK, PER1, PER2, PER3, CRY1, CRY2*) and inflammation (*IL-6* and *TNF-α*) was analyzed via RT-qPCR using the SYBR Green system (Thermo Fisher Scientific). Primer sequences used with the SYBR Green system are reported in Table 7. Fluorescence for each cycle was quantitatively analyzed using the StepOnePlus Real-Time PCR System (Applied Biosystems, Thermo Fisher Scientific). The PCR was started with a holding stage at 50 °C for 2 min and 95 °C for 2 min, followed by 40 cycles at 95 °C for 3 s and 60 °C for 30 s. A final dissociation step at 95 °C for 15 s, 60 °C for 1 min, and 95 °C for 15 s completed the program. Gene expression was normalized to *GAPDH* and reported as a ratio compared with the level of expression in controls, which were given an arbitrary value of 1.

Table 7. Primer sequences.

Target (GenBank Accession Number)	Primer Pair	Mature Transcript	Amplicon Size (bp)
BMAL1 (NM_001297719)	(F) 5′-TTACTGTGCTAAGGATGGCTG-3′ (R) 5′-GCCCTGAGAATGAGGTGTTTC-3′	(F) 731–751 (R) 857–837	127
CLOCK (NM_004898)	(F) 5′-CTACATTCACTCAGGACAGGC-3′ (R) 5′-GGCCCATAAGCATAGTACTAGG-3′	(F) 2495–2515 (R) 2614–2593	120
PER1 (NM_002616)	(F) 5′-CCTCCAGTGATAGCAACGG-3′ (R) 5′-ACCAGATGCACATCCTTACAG-3′	(F) 1784–1802 (R) 1874–1854	91
PER2 (NM_022817)	(F) 5′-GCCAGAGTCCAGATACCTTTAG-3′ (R) 5′-TGTGTCCACTTTCGAAGACTG-3′	(F) 505–526 (R) 602–582	98
PER3 (NM_001377275)	(F) 5′-CTGTCTCACTGTGGTTGAAAAG-3′ (R) 5′-TAGGTAACCCAGCAAAGGC-3′	(F) 1063–1084 (R) 1207–1189	145
CRY1 (NM_004075)	(F) 5′-TCCCGTCTGTTTGTGATTCG-3′ (R) 5′-TTAATAGCTGCGTCTCGTTCC-3′	(F) 799–818 (R) 929–909	131
CRY2 (NM_021117)	(F) 5′-TGGATAAGCACTTGGAACGG-3′ (R) 5′-AGAGACAACCAAAGCGCAG-3′	(F) 735–754 (R) 854–836	120
TNF-α (NM_000594)	(F) 5′-ACTTTGGAGTGATCGGCC-3′ (R) 5′-GCTTGAGGGTTTGCTACAAC-3′	(F) 332–349 (R) 470–451	139
IL-6 (NM_000600)	(F) 5′-CCACTCACCTCTTCAGAACG-3′ (R) 5′-CATCTTTGGAAGGTTCAGGTTG-3′	(F) 199–218 (R) 348–327	150
GAPDH (NM_002046)	(F) 5′-CATCCATGACAACTTTGGTATCGT-3′ (R) 5′-CCATCACGCCACAGTTTCC-3′	(F) 565–588 (R) 672–654	108

4.4. Statistical Analyses

Sample size was calculated with openepi.com (https://www.openepi.com/SampleSize/SSMean.htm, accessed on 12 April 2023). To detect a mean difference in blood pressure of 4.6 mmHg [19], with a two-sided significance level of 5% and power of 80%, a paired *t*-test would require 19 subjects. Based on this estimate, we decided to include all the internists working in the department who consented to participate in this study. All statistical analyses were carried out in R system for statistical computing (Version 4.0.2; R development Core Team, 2020, Vienna, Austria). Statistical significance was set at $p < 0.05$. A Shapiro–Wilk test was applied to continuous variables to check for distribution normality. Quantitative variables were reported as median with range (1st Qu–3rd Qu) or mean ± standard deviation, depending on distribution. Categorical variables were reported as absolute frequencies and/or percentages. Continuous variables were compared using a Mann–Whitney test (and Kruskal–Wallis test) or Student's *t*-test (and ANOVA), depending on data distribution and number of groups. A repeated-measures ANOVA model was performed to compare the effect of time on SBP, DBP, SBP dipping, and DBP dipping during the night of rest (t1), the night shift (t2), and the recovery time after the night shift (t3). For every ANOVA model, normality checks were carried out on the dependent variables by group (three different time points), which were approximately normally distributed. If Mauchly's test of sphericity was not met (only for the dipping DBP model) Greenhouse–Geisser sphericity correction was applied. Post hoc analyses were performed with a Bonferroni adjustment. Correlation coefficients were calculated based on data distribution. Associations with a *p*-value below 0.10 were considered for multivariate linear regression to evaluate factors influencing blood pressure. In addition, in order to identify covariates related to blood pressure, we used random forest regression (RF), which is used when there are several covariates and small sample sizes to select a smaller number of relevant predictors. We retained the top 3 predictors based on mean square error (MSE) decrease in accuracy for each feature removal (%IncMSE), and we applied a linear multivariate regression model to estimate the β-regression coefficient of each predictor. Statistical analyses were performed with the randomForest package in R.

Author Contributions: Conceptualization, B.F. and S.B.; methodology, F.G.; formal analysis, B.T. and F.T.; investigation, B.T. and F.T.; data curation, B.T., T.F., E.G., M.C. and M.F.; writing—original draft preparation, B.T. and S.B.; writing—review and editing, R.C., A.G. and M.P.; visualization, B.T.; supervision, S.B.; funding acquisition, B.F. and S.B. All authors have read and agreed to the published version of the manuscript.

Funding: This research received no external funding.

Institutional Review Board Statement: The study was conducted in accordance with the Declaration of Helsinki, and approved by the Institutional Ethics Committee, CEUR-FVG Azienda Regionale di Coordinamento per la Salute (protocol code 107_2020H, 30 December 2022).

Informed Consent Statement: Informed consent was obtained from all subjects involved in the study. Written informed consent was obtained from the participants to publish this paper.

Data Availability Statement: The raw data supporting the conclusions of this article will be made available by the authors upon reasonable request.

Conflicts of Interest: The authors declare no conflict of interest.

References

1. Kawachi, I.; Colditz, G.A.; Stampfer, M.J.; Willett, W.C.; Manson, J.E.; Speizer, F.E.; Hennekens, C.H. Prospective study of shift work and risk of coronary heart disease in women. *Circulation* **1995**, *92*, 3178–3182. [CrossRef] [PubMed]
2. Brown, D.L.; Feskanich, D.; Sanchez, B.N.; Rexrode, K.M.; Schernhammer, E.S.; Lisabeth, L.D. Rotating night shift work and the risk of ischemic stroke. *Am. J. Epidemiol.* **2009**, *169*, 1370–1377. [CrossRef]
3. Hemmer, A.; Mareschal, J.; Dibner, C.; Pralong, J.A.; Dorribo, V.; Perrig, S.; Genton, L.; Pichard, C.; Collet, T.H. The Effects of Shift Work on Cardio-Metabolic Diseases and Eating Patterns. *Nutrients* **2021**, *13*, 4178. [CrossRef]
4. Dunlap, J.C. Molecular bases for circadian clocks. *Cell* **1999**, *96*, 271–290. [CrossRef] [PubMed]
5. Takahashi, J.S.; Hong, H.K.; Ko, C.H.; McDearmon, E.L. The genetics of mammalian circadian order and disorder: Implications for physiology and disease. *Nat. Rev. Genet.* **2008**, *9*, 764–775. [CrossRef] [PubMed]
6. Huart, J.; Persu, A.; Lengele, J.P.; Krzesinski, J.M.; Jouret, F.; Stergiou, G.S. Pathophysiology of the Nondipping Blood Pressure Pattern. *Hypertension* **2023**, *80*, 719–729. [CrossRef] [PubMed]
7. Bilo, G.; Grillo, A.; Guida, V.; Parati, G. Morning blood pressure surge: Pathophysiology, clinical relevance and therapeutic aspects. *Integr. Blood Press. Control* **2018**, *11*, 47–56. [CrossRef]
8. Boivin, D.B.; Tremblay, G.M.; James, F.O. Working on atypical schedules. *Sleep Med.* **2007**, *8*, 578–589. [CrossRef]
9. Yamasaki, F.; Schwartz, J.E.; Gerber, L.M.; Warren, K.; Pickering, T.G. Impact of shift work and race/ethnicity on the diurnal rhythm of blood pressure and catecholamines. *Hypertension* **1998**, *32*, 417–423. [CrossRef]
10. Martino, T.A.; Tata, N.; Belsham, D.D.; Chalmers, J.; Straume, M.; Lee, P.; Pribiag, H.; Khaper, N.; Liu, P.P.; Dawood, F.; et al. Disturbed diurnal rhythm alters gene expression and exacerbates cardiovascular disease with rescue by resynchronization. *Hypertension* **2007**, *49*, 1104–1113. [CrossRef]
11. Ohkubo, T.; Imai, Y.; Tsuji, I.; Nagai, K.; Watanabe, N.; Minami, N.; Kato, J.; Kikuchi, N.; Nishiyama, A.; Aihara, A.; et al. Relation between nocturnal decline in blood pressure and mortality. The Ohasama Study. *Am. J. Hypertens.* **1997**, *10*, 1201–1207. [CrossRef] [PubMed]
12. Chen, J.W.; Jen, S.L.; Lee, W.L.; Hsu, N.W.; Lin, S.J.; Ting, C.T.; Chang, M.S.; Wang, P.H. Differential glucose tolerance in dipper and nondipper essential hypertension: The implications of circadian blood pressure regulation on glucose tolerance in hypertension. *Diabetes Care* **1998**, *21*, 1743–1748. [CrossRef] [PubMed]
13. Davidson, M.B.; Hix, J.K.; Vidt, D.G.; Brotman, D.J. Association of impaired diurnal blood pressure variation with a subsequent decline in glomerular filtration rate. *Arch. Intern. Med.* **2006**, *166*, 846–852. [CrossRef] [PubMed]
14. Salles, G.F.; Reboldi, G.; Fagard, R.H.; Cardoso, C.R.; Pierdomenico, S.D.; Verdecchia, P.; Eguchi, K.; Kario, K.; Hoshide, S.; Polonia, J.; et al. Prognostic Effect of the Nocturnal Blood Pressure Fall in Hypertensive Patients: The Ambulatory Blood Pressure Collaboration in Patients With Hypertension (ABC-H) Meta-Analysis. *Hypertension* **2016**, *67*, 693–700. [CrossRef]
15. Torquati, L.; Mielke, G.I.; Brown, W.J.; Kolbe-Alexander, T. Shift work and the risk of cardiovascular disease. A systematic review and meta-analysis including dose-response relationship. *Scand. J. Work Environ. Health* **2018**, *44*, 229–238. [CrossRef]
16. Manohar, S.; Thongprayoon, C.; Cheungpasitporn, W.; Mao, M.A.; Herrmann, S.M. Associations of rotational shift work and night shift status with hypertension: A systematic review and meta-analysis. *J. Hypertens.* **2017**, *35*, 1929–1937. [CrossRef]
17. Riegel, B.; Daus, M.; Lozano, A.J.; Malone, S.K.; Patterson, F.; Hanlon, A.L. Shift Workers Have Higher Blood Pressure Medicine Use, But Only When They Are Short Sleepers: A Longitudinal UK Biobank Study. *J. Am. Heart Assoc.* **2019**, *8*, e013269. [CrossRef]
18. Matre, D.; Sirnes, P.A.; Goffeng, E.; Skare, O.; Skogstad, M. Sleep Duration, Number of Awakenings and Arterial Stiffness in Industrial Shift Workers: A Five-Week Follow-Up Study. *Int. J. Environ. Res. Public Health* **2022**, *19*, 1964. [CrossRef]
19. Adams, S.L.; Roxe, D.M.; Weiss, J.; Zhang, F.; Rosenthal, J.E. Ambulatory blood pressure and Holter monitoring of emergency physicians before, during, and after a night shift. *Acad. Emerg. Med.* **1998**, *5*, 871–877. [CrossRef]

20. Zhao, B.; Li, J.; Li, Y.; Liu, J.; Feng, D.; Hao, Y.; Zhen, Y.; Hao, X.; Xu, M.; Chen, X.; et al. A cross-sectional study of the interaction between night shift frequency and age on hypertension prevalence among female nurses. *J. Clin. Hypertens.* **2022**, *24*, 598–608. [CrossRef]

21. Moller-Levet, C.S.; Archer, S.N.; Bucca, G.; Laing, E.E.; Slak, A.; Kabiljo, R.; Lo, J.C.; Santhi, N.; von Schantz, M.; Smith, C.P.; et al. Effects of insufficient sleep on circadian rhythmicity and expression amplitude of the human blood transcriptome. *Proc. Natl. Acad. Sci. USA* **2013**, *110*, E1132–E1141. [CrossRef]

22. Barclay, J.L.; Husse, J.; Bode, B.; Naujokat, N.; Meyer-Kovac, J.; Schmid, S.M.; Lehnert, H.; Oster, H. Circadian desynchrony promotes metabolic disruption in a mouse model of shiftwork. *PLoS ONE* **2012**, *7*, e37150. [CrossRef] [PubMed]

23. Depres-Brummer, P.; Levi, F.; Metzger, G.; Touitou, Y. Light-induced suppression of the rat circadian system. *Am. J. Physiol.* **1995**, *268*, R1111–R1116. [CrossRef] [PubMed]

24. Briaud, S.A.; Zhang, B.L.; Sannajust, F. Continuous light exposure and sympathectomy suppress circadian rhythm of blood pressure in rats. *J. Cardiovasc. Pharmacol. Ther.* **2004**, *9*, 97–105. [CrossRef] [PubMed]

25. Douma, L.G.; Gumz, M.L. Circadian clock-mediated regulation of blood pressure. *Free Radic. Biol. Med.* **2018**, *119*, 108–114. [CrossRef]

26. Stow, L.R.; Richards, J.; Cheng, K.Y.; Lynch, I.J.; Jeffers, L.A.; Greenlee, M.M.; Cain, B.D.; Wingo, C.S.; Gumz, M.L. The circadian protein period 1 contributes to blood pressure control and coordinately regulates renal sodium transport genes. *Hypertension* **2012**, *59*, 1151–1156. [CrossRef]

27. Gumz, M.L.; Stow, L.R.; Lynch, I.J.; Greenlee, M.M.; Rudin, A.; Cain, B.D.; Weaver, D.R.; Wingo, C.S. The circadian clock protein Period 1 regulates expression of the renal epithelial sodium channel in mice. *J. Clin. Investig.* **2009**, *119*, 2423–2434. [CrossRef]

28. Gumz, M.L.; Cheng, K.Y.; Lynch, I.J.; Stow, L.R.; Greenlee, M.M.; Cain, B.D.; Wingo, C.S. Regulation of alphaENaC expression by the circadian clock protein Period 1 in mpkCCD(c14) cells. *Biochim. Biophys. Acta* **2010**, *1799*, 622–629. [CrossRef]

29. Marques, F.Z.; Campain, A.E.; Tomaszewski, M.; Zukowska-Szczechowska, E.; Yang, Y.H.; Charchar, F.J.; Morris, B.J. Gene expression profiling reveals renin mRNA overexpression in human hypertensive kidneys and a role for microRNAs. *Hypertension* **2011**, *58*, 1093–1098. [CrossRef]

30. Dashti, H.S.; Aslibekyan, S.; Scheer, F.A.; Smith, C.E.; Lamon-Fava, S.; Jacques, P.; Lai, C.Q.; Tucker, K.L.; Arnett, D.K.; Ordovas, J.M. Clock Genes Explain a Large Proportion of Phenotypic Variance in Systolic Blood Pressure and This Control Is Not Modified by Environmental Temperature. *Am. J. Hypertens.* **2016**, *29*, 132–140. [CrossRef]

31. Kovanen, L.; Donner, K.; Kaunisto, M.; Partonen, T. CRY1, CRY2 and PRKCDBP genetic variants in metabolic syndrome. *Hypertens. Res.* **2015**, *38*, 186–192. [CrossRef] [PubMed]

32. Leu, H.B.; Chung, C.M.; Lin, S.J.; Chiang, K.M.; Yang, H.C.; Ho, H.Y.; Ting, C.T.; Lin, T.H.; Sheu, S.H.; Tsai, W.C.; et al. Association of circadian genes with diurnal blood pressure changes and non-dipper essential hypertension: A genetic association with young-onset hypertension. *Hypertens. Res.* **2015**, *38*, 155–162. [CrossRef] [PubMed]

33. Woon, P.Y.; Kaisaki, P.J.; Braganca, J.; Bihoreau, M.T.; Levy, J.C.; Farrall, M.; Gauguier, D. Aryl hydrocarbon receptor nuclear translocator-like (BMAL1) is associated with susceptibility to hypertension and type 2 diabetes. *Proc. Natl. Acad. Sci. USA* **2007**, *104*, 14412–14417. [CrossRef] [PubMed]

34. van Elk, F.; Robroek, S.J.W.; Smits-de Boer, S.; Kouwenhoven-Pasmooij, T.A.; Burdorf, A.; Oude Hengel, K.M. Study design of PerfectFit@Night, a workplace health promotion program to improve sleep, fatigue, and recovery of night shift workers in the healthcare sector. *BMC Public Health* **2022**, *22*, 779. [CrossRef]

35. Patterson, P.D.; Weiss, L.S.; Weaver, M.D.; Salcido, D.D.; Opitz, S.E.; Okerman, T.S.; Smida, T.T.; Martin, S.E.; Guyette, F.X.; Martin-Gill, C.; et al. Napping on the night shift and its impact on blood pressure and heart rate variability among emergency medical services workers: Study protocol for a randomized crossover trial. *Trials* **2021**, *22*, 212. [CrossRef]

36. Bernardi, S.; Toffoli, B.; Tonon, F.; Francica, M.; Campagnolo, E.; Ferretti, T.; Comar, S.; Giudici, F.; Stenner, E.; Fabris, B. Sex Differences in Proatherogenic Cytokine Levels. *Int. J. Mol. Sci.* **2020**, *21*, 3861. [CrossRef]

37. Tonon, F.; Di Bella, S.; Giudici, F.; Zerbato, V.; Segat, L.; Koncan, R.; Misin, A.; Toffoli, B.; D'Agaro, P.; Luzzati, R.; et al. Discriminatory Value of Adiponectin to Leptin Ratio for COVID-19 Pneumonia. *Int. J. Endocrinol.* **2022**, *2022*, 9908450. [CrossRef]

38. Tonon, F.; Candido, R.; Toffoli, B.; Tommasi, E.; Cortello, T.; Fabris, B.; Bernardi, S. Type 1 diabetes is associated with significant changes of ACE and ACE2 expression in peripheral blood mononuclear cells. *Nutr. Metab. Cardiovasc. Dis.* **2022**, *32*, 1275–1282. [CrossRef]

39. Tonon, F.; Tornese, G.; Giudici, F.; Nicolardi, F.; Toffoli, B.; Barbi, E.; Fabris, B.; Bernardi, S. Children With Short Stature Display Reduced ACE2 Expression in Peripheral Blood Mononuclear Cells. *Front. Endocrinol.* **2022**, *13*, 912064. [CrossRef]

International Journal of
Molecular Sciences

Article

Age-Dependent Changes in the Relationships between Traits Associated with the Pathogenesis of Stress-Sensitive Hypertension in ISIAH Rats

Dmitry Yu. Oshchepkov [1,*], Yulia V. Makovka [1,2], Mikhail P. Ponomarenko [1], Olga E. Redina [1] and Arcady L. Markel [1,2]

[1]   Federal Research Center Institute of Cytology and Genetics, Siberian Branch of Russian Academy of Sciences (SB RAS), Novosibirsk 630090, Russia
[2]   Department of Natural Sciences, Novosibirsk State University, Novosibirsk 630090, Russia
*   Correspondence: diman@bionet.nsc.ru

Abstract: Hypertension is one of the most significant risk factors for many cardiovascular diseases. At different stages of hypertension development, various pathophysiological processes can play a key role in the manifestation of the hypertensive phenotype and of comorbid conditions. Accordingly, it is thought that when diagnosing and choosing a strategy for treating hypertension, it is necessary to take into account age, the stage of disorder development, comorbidities, and effects of emotional–psychosocial factors. Nonetheless, such an approach to choosing a treatment strategy is hampered by incomplete knowledge about details of age-related associations between the numerous features that may contribute to the manifestation of the hypertensive phenotype. Here, we used two groups of male F_2(ISIAHxWAG) hybrids of different ages, obtained by crossing hypertensive ISIAH rats (simulating stress-sensitive arterial hypertension) and normotensive WAG rats. By principal component analysis, the relationships among 21 morphological, physiological, and behavioral traits were examined. It was shown that the development of stress-sensitive hypertension in ISIAH rats is accompanied not only by an age-dependent (FDR < 5%) persistent increase in basal blood pressure but also by a decrease in the response to stress and by an increase in anxiety. The plasma corticosterone concentration at rest and its increase during short-term restraint stress in a group of young rats did not have a straightforward relationship with the other analyzed traits. Nonetheless, in older animals, such associations were found. Thus, the study revealed age-dependent relationships between the key features that determine hypertension manifestation in ISIAH rats. Our results may be useful for designing therapeutic strategies against stress-sensitive hypertension, taking into account the patients' age.

Keywords: principle component analysis; hypertension; stress reactivity; age-related difference; ISIAH rat strain

Citation: Oshchepkov, D.Y.; Makovka, Y.V.; Ponomarenko, M.P.; Redina, O.E.; Markel, A.L. Age-Dependent Changes in the Relationships between Traits Associated with the Pathogenesis of Stress-Sensitive Hypertension in ISIAH Rats. *Int. J. Mol. Sci.* **2023**, *24*, 10984. https://doi.org/10.3390/ijms241310984

Academic Editor: Maria Luisa Balestrieri

Received: 22 May 2023
Revised: 23 June 2023
Accepted: 29 June 2023
Published: 1 July 2023

1. Introduction

Arterial hypertension is a polygenic disorder, the development of which is affected by numerous environmental and endogenous factors. Many risk factors and comorbidities of hypertension are known. The most common are diabetes mellitus, overweight, obesity, and chronic kidney diseases [1–3]. It is believed that some of these pathologies, such as chronic kidney disease, can be either a cause of a hypertensive state or a complication of uncontrolled hypertension [2]. Hypertension is possibly the most powerful modifiable risk factor of heart failure and other cardiovascular complications including myocardial infarction (and cerebrovascular diseases such as stroke). Chronic hypertension drives myocardial remodeling, resulting in hypertensive heart disease [4,5].

Despite major efforts of researchers and physicians to control hypertension, including treatment of hypertension in patients with comorbid conditions, numerous questions about the diagnosis, patient evaluation and monitoring, and the choice of a treatment strategy for hypertension remain highly relevant and not fully resolved [6,7].

By now, it is recognized that it is necessary to take into account different stages of the development of the disorder when diagnosing and choosing a strategy for treating hypertension [6]. It is assumed that at different stages of hypertension progression, various pathophysiological processes may play a key role in the manifestation of the hypertensive phenotype [8]. This notion raises questions about strategies for prescribing antihypertensive drugs depending on age, the disorder stage, and the presence of comorbidities [6,9–11]. It has been emphasized that strategies for the treatment of arterial hypertension in the elderly should take into consideration not only concomitant diseases but also the effects of emotiogenic psychosocial factors [12].

Understanding the importance of the assessment of age-related differences in functional expression of stressor-related negative affect [13,14] is undoubtedly important but is hampered by incomplete knowledge about the details of age-related associations between the numerous features that may contribute to the manifestation of the hypertensive phenotype. Conducting studies on the age-dependent component of stressor-related negative affect in human populations is difficult due to differences in study designs, populations, measurement, in the operational definition, and in the analytic approach, which lead to heterogeneity of findings [13]. Animal strains that model human diseases are often a successful alternative to population studies.

ISIAH (inherited stress-induced arterial hypertension) rats are a proven model of a stress-sensitive form of hypertension and make it possible to investigate the physiological and molecular genetic mechanisms behind the development of a hypertensive state associated with increased stress reactivity under conditions of emotional stress [15–18].

Principal component analysis, which is based on correlation analysis, allows the assessment of the relationships between many traits and the identification of a set of main factors (principal components) that can contribute to the manifestation of certain phenotypic characteristics [19]. Previously, we have used principal component analysis to examine relationships among phenotypic characteristics in a group of 3-month-old male F_2(ISIAHxWAG) hybrids obtained by crossing hypertensive ISIAH rats and normotensive WAG rats [20]. The use of the group of hybrid rats enabled us to obtain a full range of variation between normal and hypertensive phenotypes. In that paper, we analyzed 21 morphological, physiological, and behavioral traits, many of which show interstrain differences and are usually considered key features of the development of the hypertensive phenotype. Behavior was assessed in the open field test, which helps researchers to evaluate basic psychophysiological characteristics: the severity of fear and anxiety reactions, locomotor activity, and levels of exploratory and displacement activities [21]. In that research article, the first principal component was found to be associated with behavioral traits of F_2(ISIAHxWAG) rats in the open field test. The basal blood pressure (BP) level contributed to the second and fourth principal components with the opposite sign. At the same time, the contribution to the second principal component was also made by weight traits: body weight and weights of peripheral organs that are targets of hypertension (kidneys, heart, and adrenal glands). The contribution to the fourth principal component, in addition to the basal level of BP, was made by all traits that are related to the stress-induced state of rats: BP under stress, adrenal weight, plasma corticosterone concentration under stress, and the defecation score in the open field test [20].

The purpose of the present work was to determine the relationships between the same traits in a group of older (six-month-old) males of F_2(ISIAHxWAG) rat hybrids and to identify possible changes in these relationships during maturation of the animals, i.e., with the progression of the disorder.

2. Results

2.1. Age-Dependent Differences in Phenotypic Traits between Two Groups of F_2(ISIAHxWAG) Hybrid Rats of Different Ages

A comparison of the phenotypic traits of F_2(ISIAHxWAG) rats between age groups of 3–4 and 6–7 months (Table 1) uncovered many significant differences. Table 1 shows that as rats grow older, there is a statistically significant increase in the basal level of BP, but the increase in BP under restraint stress in the group of older rats is significantly weaker. Taking into account that the increase in plasma corticosterone concentration in the group of rats aged 6–7 months is also smaller as compared to 3–4-month-old animals, we propose that there is a decrease in stress responsiveness when the rats grow up.

Table 1. Phenotypic traits (mean $\pm$ SE) in the two groups of F_2(ISIAHxWAG) hybrid rats of different ages.

Trait	F_2(ISIAHxWAG) (3–4 Months Old) $n = 101$	F_2(ISIAHxWAG) (6–7 Months Old) $n = 118$
Basal BP, mmHg	158.6 $\pm$ 1.6 ** xxx	164.3 $\pm$ 1.7 xx
BP under stress, mmHg	189.9 $\pm$ 1.8 ***	172.2 $\pm$ 1.8
Increase in BP during stress, mmHg	31.3 $\pm$ 2.0 ***	7.9 $\pm$ 1.9
Plasma corticosterone concentration at rest, µg/100 mL	3.38 $\pm$ 0.35 ** xxx	1.92 $\pm$ 0.17 xxx
Plasma corticosterone concentration during stress, µg/100 mL	31.6 $\pm$ 0.53 ***	25.5 $\pm$ 0.93
Increase in plasma corticosterone concentration under stress, µg/100 mL	28.2 $\pm$ 0.64 ***	23.7 $\pm$ 0.95
Body weight, g	253.9 $\pm$ 3.52 ***	297.4 $\pm$ 3.9
Adrenal weight, mg	36.0 $\pm$ 0.47 ***	39.5 $\pm$ 0.54
Adrenal weight/body weight, mg/100 g body weight	14.4 $\pm$ 0.23 **	13.4 $\pm$ 0.2
Kidney weight, g	1.52 $\pm$ 0.02 ***	1.87 $\pm$ 0.03
Kidney weight/body weight, g/100 g body weight	0.60 $\pm$ 0.006	0.63 $\pm$ 0.005
Heart weight, g	0.95 $\pm$ 0.016 ***	1.09 $\pm$ 0.01
Heart weight/body weight, g/100 g body weight	0.38 $\pm$ 0.0035	0.37 $\pm$ 0.003
Locomotor activity in the first minute of the first trial of the test, crossed squares	51.4 $\pm$ 2.3	47.4 $\pm$ 1.9
Locomotor activity on the periphery of the open-field area, crossed squares	319 $\pm$ 14.9 ***	241.3 $\pm$ 10.1
Grooming at the periphery of the open-field area	3.2 $\pm$ 0.36 ***	5.4 $\pm$ 0.5
Time of grooming at the periphery of the open-field area, s	43.4 $\pm$ 0.52 ***	94.0 $\pm$ 10.2
Rearings at the periphery of the open-field area	19.0 $\pm$ 1.2	19.2 $\pm$ 0.9
Time of rearing at the periphery of the open-field area, s	57.2 $\pm$ 3.8 ***	75.7 $\pm$ 4.0
Defecation	13.0 $\pm$ 0.76	13.2 $\pm$ 0.6
Latency period, s	40.0 $\pm$ 2.1 ***	84.5 $\pm$ 6.2

** $p < 0.01$ and *** $p < 0.001$ for the comparison between F_2(ISIAHxWAG) rats at the age of 3–4 months and F_2(ISIAHxWAG) rats at the age of 6–7 months; xx $p < 0.01$ and xxx $p < 0.001$ as compared to the stress group (Student's *t* test).

Because body weight increases with age and absolute weight values of a target organ are expected to increase, only relative weights of the kidneys, heart, and adrenal glands were used in the further analysis. Such traits as BP under stress and the increase in BP under stress as well as the plasma corticosterone concentration under stress and the increase in plasma corticosterone concentration under stress are interdependent. Therefore, only traits representing stress-induced changes in their values were employed in the subsequent analysis because they reflect stress responsiveness in rats. The increase in corticosterone concentration under the influence of stressors is regarded as one of the key signs of adaptation of the body to the stressful conditions [22]; consequently, the relationships between plasma corticosterone concentration at rest (and its increase under stress) and both morphometric traits and behavioral traits were studied here.

2.2. Analysis of Associations of BP Traits and of Weight Parameters with Plasma Corticosterone Concentration

Figure 1 shows the results of the examination of eight traits (measured in rats of both ages) by principal component analysis. The traits characterizing the weight characteristics of organs are mainly projected onto the horizontal axis (principal component 1); the vector of the absolute body weight and vectors of the relative weights of the adrenal glands, kidneys, and heart have opposite orientations. This finding has a natural explanation: the greater the body weight, the lower the relative weight of the organs. The body weight (a significant negative correlation) and relative weight of the heart and adrenal glands (significant positive correlations) produce the highest loading on the first principal component. Accordingly, the horizontal axis can be described as the axis associated with the weight characteristics of the experimental rats.

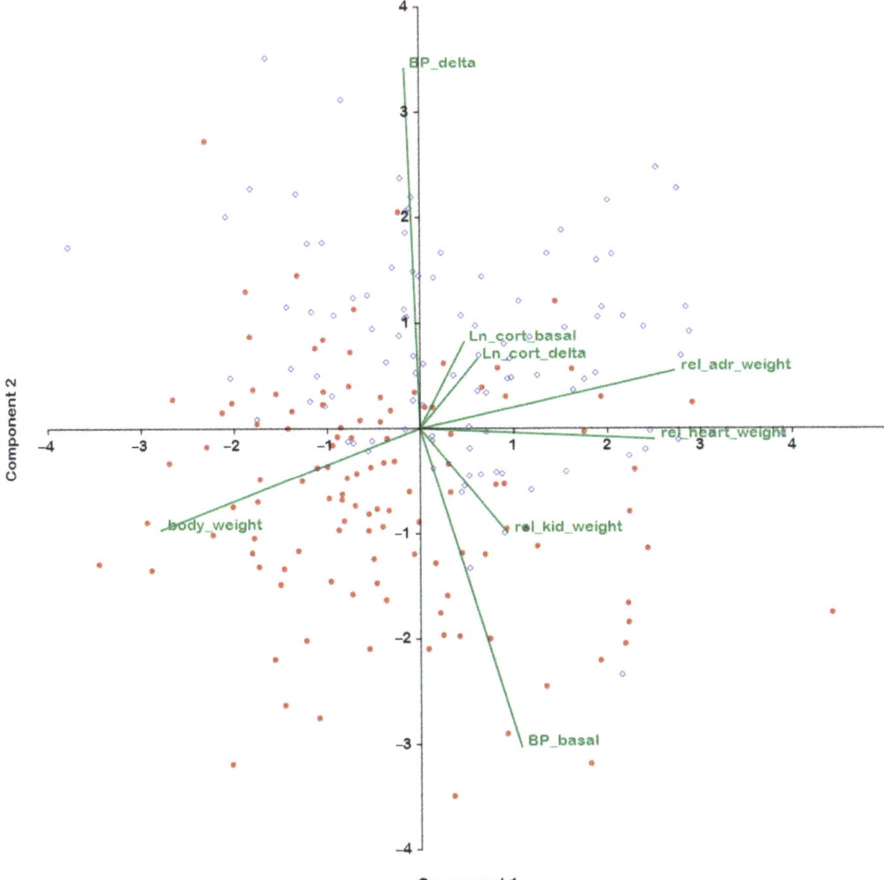

Figure 1. Principal component analysis performed on the combined group of male F_2(ISIAHxWAG) hybrids of different ages. Blue dots: 3–4-month-old rats; red dots: 6–7-month-old rats.

As for the second principal component, the main loadings on it are given by BP values: the basal-BP vector has a significant negative correlation with the second principal component, and the BP increase under stress (ΔBP) vector has the opposite direction and a positive correlation with the second principal component. Thus, the second principal component can be characterized as a factor that determines BP stress reactivity, and an increase in this factor is characteristic of the ISIAH rat strain.

Graphical representation of the contributions of traits to the principal components in the combined group of F_2(ISIAHxWAG) hybrid male rats of different ages (Figure 1) clearly indicated the presence of a significant negative correlation between the second principal component and the age factor. The table of correlations of traits with principal components (Table 2) shows that the second principal component explains 19.32% of the variance of traits associated with age. Overall, the first two principal components account for almost 45% of the variance of the traits presented in Table 2.

Table 2. The table of correlations of traits with principal components in the combined group of male F_2(ISIAHxWAG) hybrid rats of different ages.

Trait	PC 1	PC 2	PC 3	PC 4	PC 5
BP_basal	0.313	**−0.765**	0.029	−0.339	0.184
BP_delta	−0.047	**0.865**	0.178	0.114	0.213
Ln_cort_basal	0.136	0.208	**−0.808**	−0.193	0.269
Ln_cort_delta	0.177	0.166	**0.723**	−0.482	0.274
body_weight	**−0.813**	−0.243	0.216	0.103	0.188
rel_adr_weight	**0.793**	0.139	0.163	−0.058	**−0.428**
rel_kid_weight	0.270	−0.250	0.222	**0.800**	0.116
rel_heart_weight	**0.729**	−0.025	−0.045	0.189	**0.450**
age_indicator	−0.287	**−0.559**	−0.003	0.311	−0.094
% of variance	25.55	19.33	16.65	13.58	8.26

Significant Pearson correlation coefficients (false discovery rate (FDR) < 5%) are boldfaced.

Accordingly, the performed analysis suggests that it is the development of the hypertensive status that is the main age-dependent parameter in which the groups of F_2(ISIAHxWAG) males differ. A positive correlation of age with the basal level of BP and a negative correlation of age with an increase in BP under stress indicate a decline in stress reactivity in the group of male hybrids aged 6–7 months compared with the young group of the animals.

For a more detailed assessment of the influence of the age factor on the variance of the studied parameters, an additional analysis was performed on each age group of rats separately.

The results from comparing the relationships of traits in each age group (Table 3) showed the same contribution of three weight parameters (body weight and relative weights of the heart and adrenal glands) to the first principal component. In both groups, the body weight inversely correlated with the relative weights of the adrenals and heart. The plasma corticosterone concentration in both groups of rats makes the greatest contribution to the second principal component, and both traits of BP make the largest contribution to the third principal component. Nevertheless, we were able to identify some age-dependent differences. The analysis suggested that in the young group of F_2(ISIAHxWAG) hybrid rats, the assessed traits of BP, body weight, and relative weights of the target organs are not associated with the plasma corticosterone concentration and its increase under short-term restraint stress.

In comparison to the young group of rats, in male F_2(ISIAHxWAG) hybrids at the age of 6–7 months, the contribution of the basal BP level to the first principal component is weaker (Table 3). In the second principal component, a contribution of two traits (plasma corticosterone concentration and the relative weight of the kidneys) emerged. In contrast to the young group of rats, in older rats, we in older rats a contribution of the increase in

plasma corticosterone concentration under restraint stress (along with both traits of BP) to the third principal component and to the fourth principal component (along with the relative kidney weight). Therefore, in the adult group of the animals, in contrast to the young group, the trait "the increase in the plasma concentration of corticosterone under restraint stress" manifested an interdependence with both traits of BP and with the relative weight of the kidneys. It should be noted that the contribution of the trait "relative weight of the kidneys" to the second and fourth principal components was opposite.

Table 3. The table of correlations of traits with principal components in groups of F_2(ISIAHxWAG) hybrid males at ages 3–4 and 6–7 months.

3–4 Months Old					
Trait	PC 1	PC 2	PC 3	PC 4	PC 5
BP_basal	**0.606**	−0.227	−**0.540**	−0.006	0.081
BP_delta	−0.416	0.284	**0.713**	−0.143	−0.151
Ln_cort_basal	0.041	−**0.814**	0.313	−0.241	0.087
Ln_cort_delta	0.091	**0.867**	−0.188	0.042	−0.083
body_weight	−**0.869**	−0.026	−0.126	0.135	0.104
rel_adr_weight	**0.709**	0.260	0.212	−**0.516**	−0.147
rel_kid_weight	0.454	0.230	0.426	0.256	**0.701**
rel_heart_weight	**0.526**	−0.191	0.332	**0.603**	−0.440
% of variance	28.64	21.31	16.07	9.91	9.51
6–7 Months Old					
Trait	PC 1	PC 2	PC 3	PC 4	PC 5
BP_basal	0.411	0.044	**0.761**	−0.246	0.117
BP_delta	−0.400	0.370	−**0.643**	0.272	0.243
Ln_cort_basal	0.043	−**0.830**	−0.058	0.095	**0.486**
Ln_cort_delta	0.027	**0.539**	**0.499**	**0.503**	0.379
body_weight	−**0.758**	0.394	0.288	−0.155	−0.030
rel_adr_weight	**0.756**	0.187	−0.154	0.335	−0.290
rel_kid_weight	0.320	**0.466**	−0.351	−**0.618**	0.300
rel_heart_weight	**0.801**	0.200	−0.167	0.024	0.148
% of variance	27.77	19.58	18.78	11.43	8.12

Significant Pearson correlation coefficients (FDR < 5%) are highlighted in bold.

Graphical representation of the data revealed that in the group of 3–4-month-old F_2(ISIAHxWAG) rats, orientations of the vectors of plasma corticosterone concentrations and weight traits are almost orthogonal (Figure 2a). This state of affairs changed significantly by age 6–7 months, when almost all traits began to interact negatively with basal plasma corticosterone levels and positively with corticosterone gains (Figure 2b).

2.3. Analysis of the Associations of Behavior in the Open Field Test with BP and Hormonal Parameters

The analysis of behavior, BP, and hormonal parameters in the combined group of rats of both age ranges showed that six out of seven behavioral characteristics of rats in the open field test make a significant positive contribution to the first principal component, which is responsible for 27% of the total variance of the traits measured in this group (Figure 3, Table 4). A negative contribution to the first principal component was made by the latency period. Neither BP traits nor hormonal characteristics contributed significantly to the first principal component. Also unrelated to the first principal component is the defecation score in the open field test. Thus, the first principal component characterized the behavioral (largely exploratory) activity. The overall level of activity of rats in the open field test did not show a significant association with the age factor, with parameters of hypertensive status, or with glucocorticoid adrenal function.

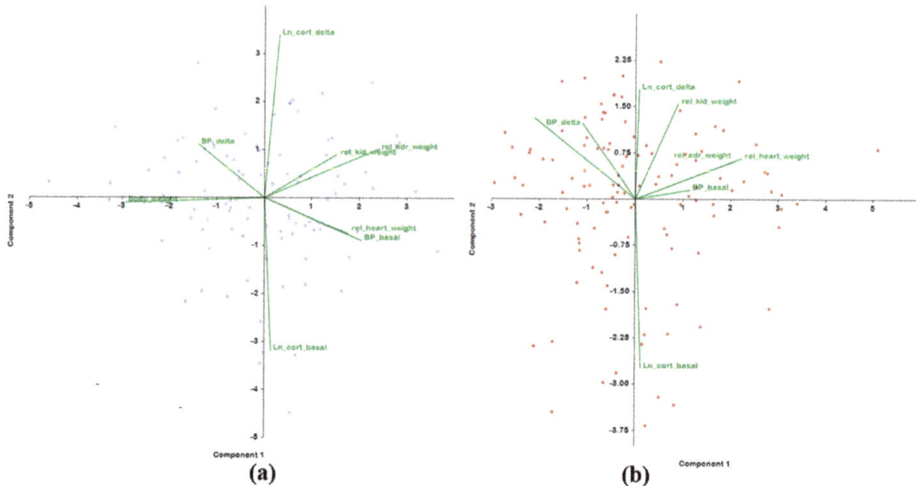

Figure 2. Principal component analysis in groups of F$_2$(ISIAHxWAG) hybrid males of different ages: (**a**) 3–4 months old; (**b**) 6–7 months old.

Table 4. The table of correlations between behavioral traits and principal components in the combined group of male F$_2$(ISIAHxWAG) hybrid rats of different ages.

Trait	PC 1	PC 2	PC 3	PC 4	PC 5
BP_basal	0.131	**0.581**	−0.376	−0.214	**0.539**
BP_delta	−0.207	**−0.680**	0.150	**0.428**	−0.240
Ln_cort_basal	−0.019	−0.089	**0.594**	−0.303	0.316
Ln_cort_delta	−0.155	−0.091	**−0.553**	**0.629**	0.259
LA_1_min	**0.545**	−0.398	0.110	−0.031	0.308
LA_per	**0.649**	−0.470	−0.239	−0.207	−0.003
Ln_G_per	**0.689**	0.426	0.280	**0.439**	−0.047
Ln_R_per	**0.862**	−0.132	−0.197	−0.179	−0.207
Ln_time_G_per	**0.618**	0.467	0.320	**0.468**	−0.040
Ln_time_R_per	**0.797**	0.050	−0.229	−0.192	−0.344
Ln_Def	0.124	−0.045	**0.435**	−0.042	0.127
Ln_Lat	**−0.448**	0.475	−0.020	−0.179	**−0.566**
age_indicator	0.086	**0.577**	0.039	−0.181	−0.280
% of variance	27.11	15.36	11.28	10.62	9.26

Significant Pearson correlation coefficients (FDR < 5%) are boldfaced.

The characteristics of the hypertensive status made the main contribution to the second principal component, which accounted for 15.4% of the total variance, with a positive contribution from basal BP and a negative contribution from the increase in BP under stress. Apparently, the higher the basal BP, the lower the amplitude of the increase in BP under stress. This pattern is possibly due to the tension of the depressor baroreflex with elevated basal pressure. Some characteristics of behavior also correlated with the second principal component: a negative contribution was noted for locomotor activity, and a positive contribution was noted for such parameters as displaced activity (grooming) and the latency period (Table 4). That is, in this case, we are dealing with parameters of behavior that characterize a conflict of several motivations: anxiety, fear, and exploration. "Age indicator" positively and significantly correlated only with the second principal component. Consequently, the second component, associated with the age of rats, can be characterized as a factor of behavioral caution (indecisiveness) and anxiety, which increases with age and with the development of persistent hypertension.

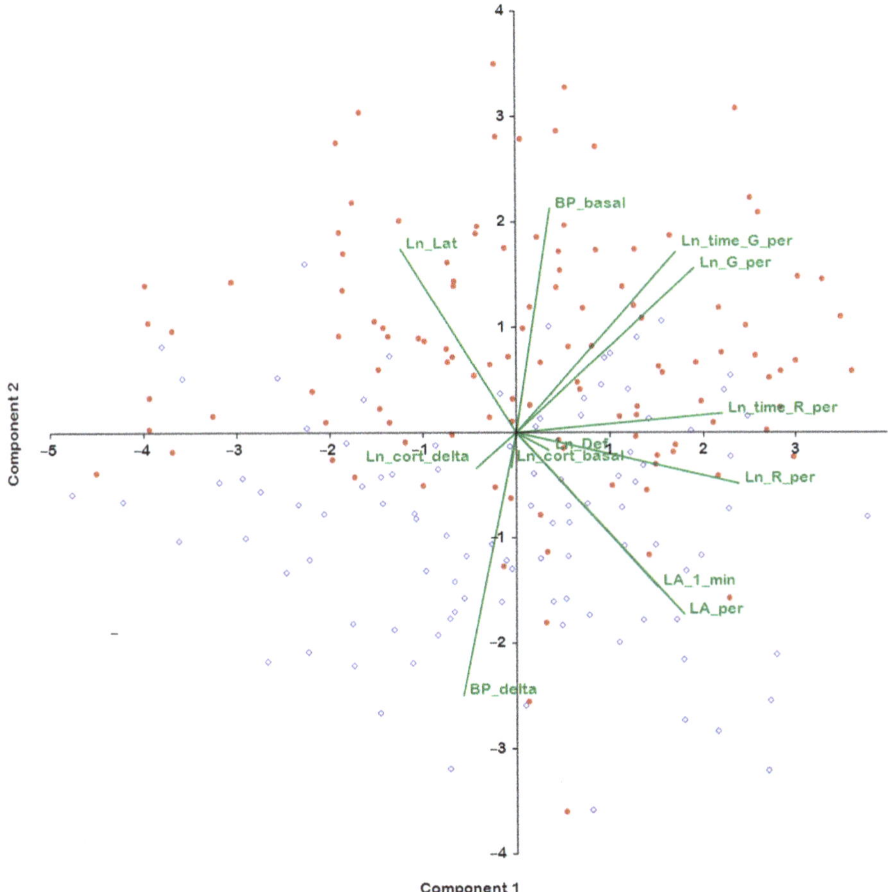

Figure 3. Principal component analysis (of behavioral traits) performed on the combined group of F_2(ISIAHxWAG) hybrid males of the two age ranges. Blue dots: 3–4-month-old rats; red dots: 6–7-month-old rats.

Characteristics of hormonal function—the basal plasma corticosterone level and its increase under stress—were found to make a significant contribution to the third principal component (11.28% of the total variance). By analogy with the basal and stress-induced rise in BP, the vectors of the basal and stress-induced increase in plasma corticosterone levels have mutually opposite directions: positive for the basal level and negative for the increase in plasma corticosterone under stress. Overall, the reciprocity between basal and stress-induced levels of corticosterone can be explained by negative feedback in the hypothalamic–pituitary–adrenal axis, and elevation of the basal level of the hormone should reduce the hormonal response to stress. It is possible that the negative feedback mechanism works more efficiently in young rats, although this hypothesis needs additional evidence. In addition, a significant positive contribution to the third principal component was detected for the defecation score. The association of this parameter in the coordinates of the third principal component with hormonal characteristics gives grounds to regard the defecation score as an indicator of emotionality and to describe the third principal component as a component associated with the magnitude of autonomic hormonal tension or stress.

The fourth principal component was also of interest. It proved to be positively loaded by such traits as increases in BP and plasma corticosterone under stress and by grooming (its frequency and duration) at the periphery of the open-field area. These data allowed us to associate this principal component with anxiety related to stress reactivity (Table 4).

A separate analysis of rats of different ages almost did not alter the structure of the first principal component. In both young and adult rats, the main loading on this component was provided by behavioral traits. The only difference was that in young rats, the negative contribution of the latency period did not reach statistical significance. Therefore, the first principal component in both age groups, just as in the combined group, is a factor of the overall behavioral activity of rats (Table 5). Meanwhile, the set of features associated with subsequent principal components changed rather strongly. Thus, the second principal component in young rats is essentially defined by parameters of the hormonal status: there is a significant positive correlation with the basal level of plasma corticosterone and a negative correlation with the upregulation of corticosterone under stress. On the other hand, reciprocity of the variances of these parameters persisted. In adult rats, the basal level of corticosterone did not contribute to the second principal component, but there remained a significant contribution from the increase in the level of plasma corticosterone under stress. This result confirms the abovementioned hypothesis about the weakening of the negative feedback in the hypothalamic–pituitary–adrenal axis in adult rats compared with young ones.

Furthermore, in adult rats, a significant positive contribution to the second principal component was made by basal BP together with the frequency and total duration of grooming in the open-field area. In young rats, no such correlations were detectable. Obviously, in adult rats, along with an increase in basal BP, the level of behavioral manifestations of anxiety goes up.

The structure of the third principal component is virtually the same between young and adult rats. In both age groups, the third principal component was found to be loaded with BP traits and could be characterized as a principal component associated with stress-dependent enhancement of BP. Moreover, in young and adult rats, the third principal component included grooming parameters, thereby revealing an association of the BP response to stress with behavioral status of the rats.

The fourth principal component indicated that in young rats, anxiety behavior might be somewhat unrelated to BP or plasma corticosterone levels. In adult rats, other traits contributed to the fourth principal component: a positively correlating increase in BP under stress and rearing duration and a negatively relating basal plasma corticosterone level.

The fifth principal component in both groups was shown to be represented by the defecation score in the open field test; this parameter had a very large and significant contribution only to this principal component, thereby enabling us to define the defecation score as a scale of "emotionality". In contrast to the adult group of rats, in young rats, the latency period also made a significant contribution to the fifth principal component. Both traits in young rats did not depend on the level of BP or glucocorticoid function of the adrenal glands. In adult rats, however, there was an inverse relationship between the defecation score and the basal level of corticosterone.

Table 5. The table of correlations of behavioral traits with principal components in groups of F_2(ISIAHxWAG) hybrid males at ages 3–4 and 6–7 months.

3–4 Months Old					
Trait	**PC 1**	**PC 2**	**PC 3**	**PC 4**	**PC 5**
BP_basal	0.139	0.318	**−0.632**	0.439	0.182
BP_delta	−0.071	−0.316	**0.719**	−0.330	0.013
Ln_cort_basal	−0.221	**0.801**	0.153	−0.201	0.101
Ln_cort_delta	0.035	**−0.785**	−0.071	0.397	−0.127
LA_1_min	**0.480**	−0.232	0.059	−0.207	0.128
LA_per	**0.733**	−0.041	−0.139	−0.411	−0.031
Ln_G_per	**0.681**	0.209	0.365	**0.515**	−0.168
Ln_R_per	**0.873**	0.077	−0.117	−0.249	0.011
Ln_time_G_per	**0.580**	0.239	**0.495**	0.512	−0.144
Ln_time_R_per	**0.852**	0.076	−0.092	−0.242	−0.007
Ln_Def	−0.057	0.032	0.388	0.149	**0.790**
Ln_Lat	−0.404	0.323	0.208	−0.106	**−0.502**
% of variance	27.48	14.45	13.02	11.61	8.35
6–7 Months Old					
Trait	**PC 1**	**PC 2**	**PC 3**	**PC 4**	**PC 5**
BP_basal	0.069	**0.676**	**−0.534**	−0.207	−0.081
BP_delta	−0.288	−0.421	**0.548**	**0.474**	−0.058
Ln_cort_basal	0.254	−0.097	0.216	**−0.628**	**−0.465**
Ln_cort_delta	−0.233	**0.591**	−0.063	0.355	0.063
LA_1_min	**0.675**	−0.288	0.025	−0.257	−0.082
LA_per	**0.723**	−0.298	−0.368	0.181	−0.007
Ln_G_per	**0.645**	**0.471**	**0.516**	0.020	0.079
Ln_R_per	**0.837**	−0.108	−0.179	0.340	−0.100
Ln_time_G_per	**0.596**	**0.526**	**0.521**	0.016	0.101
Ln_time_R_per	**0.718**	−0.022	−0.213	**0.469**	−0.099
Ln_Def	0.290	−0.221	−0.048	−0.303	**0.822**
Ln_Lat	**−0.679**	0.065	−0.069	0.176	0.017
% of variance	30.97	14.40	11.62	11.21	7.90

Significant Pearson correlation coefficients (FDR < 5%) are highlighted in bold.

Thus, this analysis allows us to conclude that in the young group of F_2(ISIAHxWAG) hybrid male rats, the behavioral traits assessed in the open field test do not explicitly depend on the plasma corticosterone concentration and on its increase under short-term restraint stress. In rats of the older group, the glucocorticoid function of the adrenal glands can modulate a number of behavioral traits related to emotionality. It is apparent that in contrast to young rats—where only one principal component is loaded with both traits of BP—in rats of the adult group, several principal components that reflect behavior are loaded with BP parameters in different combinations. A graphical representation of the contributions of behavioral traits to the first two principal components in groups of male F_2(ISIAHxWAG) hybrids of different ages is presented in Figure 4.

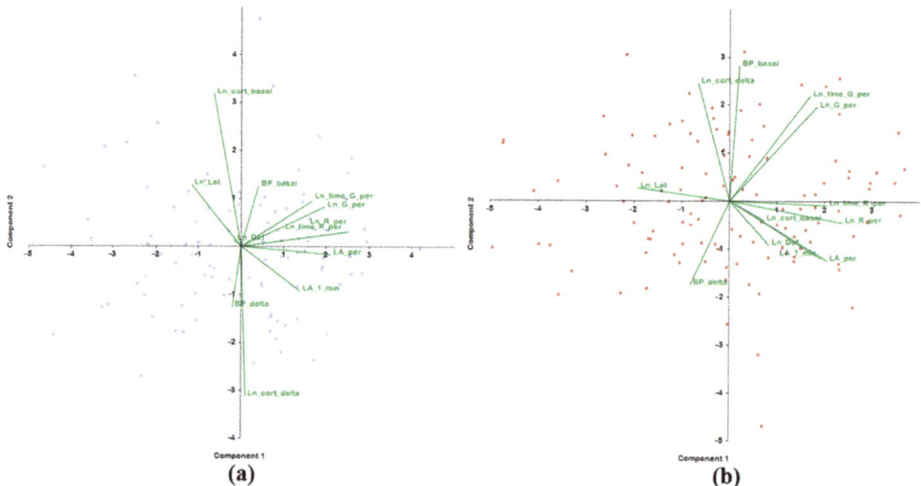

Figure 4. Principal component analysis of behavioral traits in groups of F_2(ISIAHxWAG) hybrid males of different ages: (**a**) 3–4 months old; (**b**) 6–7 months old.

3. Discussion

The development of hypertension is accompanied by alterations to the functioning of numerous physiological systems. In this study, we examined age-dependent changes in the relationships between traits associated with the pathogenesis of stress-sensitive arterial hypertension as well as with the behavioral traits of F_2(ISIAHxWAG) male rats in the open field test. In our work, the use of F_2 hybrids obtained by crossing hypertensive ISIAH rats with normotensive WAG rats can be considered a population model in which the diversity of phenotypes is provided by different combinations of alleles associated with both hypertension and high stress reactivity, as assessed after short-term restraint stress, which can be regarded as emotional stress. The analysis of two groups of rats of different ages made it possible to identify age-dependent phenotypic characteristics associated with the development of stress-sensitive hypertension as well as to evaluate alterations of associations between traits as the animals grew older. In both series of analyses performed on two distinct sets of traits (Tables 2 and 4), a significant (FDR < 5%) relationship with age was found only for the second principal component. In both cases, it was demonstrated that the development of a hypertensive status is the main age-dependent parameter in which the groups of F_2(ISIAHxWAG) males differ. This result indicates the adequacy of the chosen model for identifying associations related specifically to the hypertensive phenotype. The observed increase in BP with age is well known and points to aggravation of the development of hypertension. The vascular resistance to blood flow is enhanced due to many factors changing over time [23]. Our experiment showed that the presence of a positive relationship between age and basal BP is accompanied by a negative relationship between age and the increase in BP under stress. These results suggest that there was a significant decline of stress reactivity in the group of male F_2 (ISIAHxWAG) hybrids at the age of 6–7 months compared with the group of rats aged 3–4 months. These data are in good agreement with research on the age-dependent differences in stress responsiveness among spontaneously hypertensive rats (SHRs). The hypothalamic–pituitary–adrenocortical response to acute stress is markedly enhanced in SHRs during early development of hypertension and does not differ in older SHRs from that in normotensive Wistar–Kyoto (WKY) and Wistar rats [24]. A decrease in stress reactivity of the cardiovascular function with ageing has been associated with a decline in β-adrenergic responsiveness with age [25]. Additionally, an impaired ability to recover from exposure to stressful stimuli in the elderly may be a consequence of chronically elevated glucocorticoid levels, which are characteristic of many age-related disorders, including hypertension [26]. The results of our experiment are in good agreement with these notions and indicate in both age groups of rats the presence of a negative correlation between the concentration of corticosterone in the blood plasma at rest and its increase under restraint stress.

Cortisol (corticosterone in rats) is a major stress-related glucocorticoid hormone. Its activation under stress contributes to the restoration of homeostasis and exerts effects on cardiovascular function, metabolism, muscle function, behavior, and the immune system [22]. Elevation of the cortisol level during stress helps to restore the functioning of various organ systems. Cortisol concentration in bodily fluids is considered a reliable indicator of stress. Cortisol reactivity is regarded as one of possible mechanisms through which psychosocial stress may influence the risk of hypertension [27].

According to our principal component analysis, in the young group of animals in our experiment, the traits of plasma corticosterone were not directly related to any of the traits being analyzed (Tables 3 and 5). These findings are consistent with a report that in infants, the cortisol level, its reactivity to stress, and behavior are unrelated [28].

By contrast, in our group of rats aged 6–7 months, straightforward relationships were detected between traits of plasma corticosterone concentration and other assessed traits. Moreover, the increase in corticosterone concentration contributed to the third principal component, in addition to the positive contribution of the basal level of BP and the negative contribution of the increase in BP under stress (Table 3).

There is an opinion that low reactivity of BP to stress indicates the presence of an existing disease or disorder. The blunted cardiovascular and cortisol reactions to acute psychological stress have been linked to depression, obesity, behavioral disorders, and poorer cognitive ability, as reviewed in refs. [14,29]. It has been hypothesized that the detection of a blunted stress response may be informative for identifying patients who require closer medical attention [29]. One of the reasons for the blunted stress reactivity is genetic polymorphisms [30,31]. Because the ISIAH rat strain has been created by selection for a sharp increase in BP under short-term restraint (emotional) stress, we can theorize that genetic features of ISIAH rats may contribute to a decline in stress reactivity during the progression of the disorder in question. This supposition is supported by a report that many genes whose transcription differs between ISIAH and normotensive control rats are associated with the stress response [32–36]. This observation suggests that ISIAH rats—even at rest—are in a state of chronic functional stress.

We have previously shown that in adult ISIAH rats (4–5 months old), just as in normotensive WAG rats, significant elevation of the plasma corticosterone concentration takes place 15 min after the onset of restraint stress. During the next 15 min, however, the level of corticosterone in normotensive rats does not increase, while in hypertensive ISIAH rats, it continues to rise sharply and becomes significantly higher than that in control rats [17]. These differences signify an exaggerated stress response in hypertensive ISIAH rats compared to normotensive control WAG rats.

This phenomenon may be not only be characteristic of ISIAH and WAG rats, as evidenced by results of studies on the activation of immediate early genes described below. A number of studies on the hypothalamus of hypertensive and normotensive rats show differences in kinetics of transcription of the *Fos* gene, which is a marker of neuronal activation under stress. In normotensive Sprague–Dawley rats exposed to restraint stress, the peak of *Fos* gene activation in the paraventricular nucleus of the hypothalamus is observed 15 min after the onset of stress, and after 1 h, it returns to the control level [37]. In the research on *Fos* gene activation in the hypothalamus under the action of short-term restraint stress in hypertensive rats, somewhat different results were obtained. The peak of transient upregulation of *Fos* gene transcription in the hypothalamus of borderline hypertensive male rats occurred after 30 min of immobilization stress and returned to baseline only after 3 h [38]. Another report, on *Fos* gene activation in the paraventricular nucleus of the hypothalamus of male hypertensive SHRSP (spontaneously hypertensive, stroke-prone) rats under restraint stress, has also shown a peak of *Fos* gene activity at

30 min after the onset of stress [39]. As the hypertensive SHRSP rats mature and the pathophysiological state worsens, the *Fos* gene is activated under short-term (30 min) stress to a greater extent than in normotensive control animals [40]. Accordingly, those researchers proposed that the prolonged activation of the *Fos* gene may be a consequence of the pathophysiological state of rats associated with features of the manifestation of hypertensive status.

In the pathophysiological state of the body, a prolonged stress response, including prolonged elevation of BP following stress, is linked with the inability to shut off stress responses essential for adaptation, homeostasis, and survival, even under conditions in which these responses are no longer required (e.g., when a stressor is removed or a stressful situation is over) [14,41]. The impaired recovery from stress (magnitude and/or duration of a stress response) is connected with the development or worsening of a disease [14]. Based on the foregoing, it can be hypothesized that the processes of hypertension progression in ISIAH rats and the increase in the duration of recovery from stress may be related.

In our experiment, we also revealed age-related correlations with other traits associated with the manifestation of hypertensive status. In contrast to the young group of F_2(ISIAHxWAG) hybrid rats, in which all the assessed parameters do not explicitly depend on the concentration of plasma corticosterone and its increase under short-term restraint stress, in adult rats, both plasma corticosterone traits (along with other features) contribute to several principal components (Tables 3 and 5). These results support the notion that during aging, there are alterations to the activity of various endocrine systems, including changes in hormonal secretion and modulation of feedback sensitivity [42]. Our findings are in good agreement with the known concept that there are several stages in the development of hypertension [8] as well as with the reports that the outcome of pharmacotherapy of hypertension may depend on age, i.e., on the stage of disorder development [43]. The fact that a phenotype controlled by many genes can be influenced by different genetic loci at different ages has been demonstrated both in research on the molecular genetic mechanisms of the development of the pathology in ISIAH rats [44–46] and in articles on rats that are models of other types of hypertension [47–49] as well as in a study on anthropometric parameters (height, weight, body mass index, and systolic BP) in humans [50].

One of the advantages of the approach to data analysis presented here is that it reveals a dual nature of features by revealing a set of features involved in direct and inverse relationships. This information may be useful for interpreting genetic mapping data on these traits and for identifying candidate genes at common genetic loci involved in the control over these traits.

Our analysis suggests that the development of a hypertensive status in ISIAH rats is accompanied by an increase in anxiety (Table 4). Furthermore (Table 5), one can see that in the young group of rats of F_2(ISIAHxWAG) hybrids, the basal value of BP and the BP increase under short-term restraint stress contribute only to one (second) principal component. By contrast, in adult rats, the BP traits (along with other traits) contribute to several principal components; these other traits include basal levels and an increase in the corticosterone concentration, thereby pointing to the emergence of a relationship between behavior and hormonal and hypertensive status in rats with age. This result is in good agreement with conclusions from an analysis of population data suggesting that the elderly with hypertension are more likely to have depression and anxiety [51]. These mental problems are associated with a lower quality of life, a lower rate of treatment compliance, and higher mortality among elderly individuals [51]. The relevance of this problem is also emphasized by the fact that hypertension is also considered a risk factor for dementia [52].

Although some results of our study and the conclusions drawn are consistent with the data obtained earlier both in studies on other strains of hypertensive rats and in human population studies, we report some age-dependent associations for the first time. Our work confirms the complex nature of associations between key hypertension-related traits and identifies combinations of positive and negative relationships between some of them. Our findings suggest that ISIAH rats are an adequate model for identifying molecular genetic mechanisms linked with the development of anxiety during aging and the exacerbation of the pathogenesis of the stress-sensitive form of hypertension.

As with any experimental work, our study has several limitations. Certainly, one limitation is that we analyzed only males. In addition, it should be mentioned that the patterns observed on model animals cannot be considered an exact replica of the processes in the human body; nevertheless, no one doubts that the molecular and physiological mechanisms and their regulation are largely similar between humans and rodents.

Our paper does not contain real clinical evidence but does offer avenues and clear guidelines for further research because we determined the parameters that should be examined in human population studies on the nature of the stress-sensitive form of hypertension and on age-dependent features of its manifestation.

4. Materials and Methods

4.1. Animals

We used two groups of F_2(ISIAHxWAG) hybrid males at ages 3–4 ($n = 103$) and 6–7 months ($n = 126$). The hybrids were obtained by crossing ISIAH/Icgn rats, which are a breeding model for inherited stress-induced hypertension, with normotensive WAG/GSto-Icgn rats (Wistar Albino Glaxo). The rats were kept under standard conditions in a vivarium at the Institute of Cytology and Genetics SB RAS (Novosibirsk, Russia), with free access to feed and water. Experiments were conducted in accordance with the International Standards for the Care and Use of Laboratory Animals and were approved by the Bioethics Committee of the Institute of Cytology and Genetics for the work with experimental animals (protocol No. 69 of 20 January 2021).

4.2. Measurements of BP and Other Traits

Interstrain differences in the evaluated traits of ISIAH and WAG rats and of their F_1(ISIAHxWAG) hybrids and F_2(ISIAHxWAG) hybrids at the age of 3–4 months are given in ref. [20]. A graphical abstract summarizing the time course of the experimental data collection is presented in Figure 5.

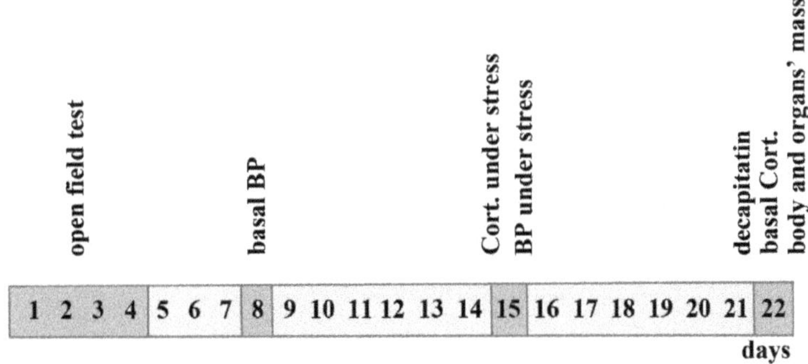

Figure 5. An outline of experimental data collection. BP: blood pressure; Cort.: corticosterone. Days 1–4: open field test; day 8: basal BP measurement; day 15: BP measurement under stress and blood collection for corticosterone quantification under stress; day 22: decapitation, blood collection for basal corticosterone quantification, and body and organ weight measurement.

BP measurements in rats were performed by the tail-cuff method (described in detail elsewhere [17]). Briefly, basal values of BP were measured in anesthetized rats to exclude the stress associated with the tail-cuff procedure. To measure the stress-induced BP and plasma corticosterone response, rats were placed in a small wire-mesh cylinder cage for 0.5 h. BP after stress was measured without anesthesia. Blood was collected from the tail vein immediately after restraint. Then, the rats were returned to their home cages.

One week later, the animals were rapidly decapitated. Samples of plasma were assayed for the basal corticosterone level. The target organs of hypertension (heart, kidneys, and adrenals) were excised and weighed. Plasma corticosterone levels were assayed by the competitive protein-binding radioassay technique [53]. Procedures for measuring BP and blood sampling for measuring the concentration of corticosterone in plasma were carried out at the same time of day, namely from 13:00 to 15:00 h local time.

4.3. Parameters of Rat Behavior

The parameters of rat behavior analyzed in the open field test were selected on the basis of previous studies because these parameters were the most informative [19]. Behavioral characteristics were evaluated in an open-field arena with automatic registration as described earlier [46]. The test lasted 6 min. The following behavioral characteristics of rats were analyzed: (1) locomotor activity in the first minute of the first trial of the open field test (number of crossed squares); (2) locomotor activity on the periphery of the open-field area (the sum of crossed squares for four trials); (3) grooming on the periphery of the open-field area (number of grooming acts in four trials); (4) grooming time, s (sum over four trials); (5) standing up on the hind legs (rearing) on the periphery of the open-field area (number of rearings in four trials); (6) rearing time, s (sum over four trials); (7) defecation (number of boluses in four trials); (8) latency period, s (time from placement on the platform to the start of locomotion, total for four trials). Rats were tested on 4 consecutive days (four trials). The sum of values from the four trials gave an integral assessment of behavior, which largely eliminated the influence of random factors on such a labile indicator as behavior. The locomotor activity in the first minute of the first trial was analyzed separately from the general locomotor activity because the locomotor activity of the rat in the first minute mainly corresponds to the reaction of avoiding an unfamiliar environment, in contrast to the locomotor activity at the end of the first and subsequent trials, which contributes mostly to the principal component associated with exploratory behavior [19].

4.4. Statistical Analysis

Statistical analysis was performed using PAST software version 4.03 (PAST 4.03) [54]. Several traits were log-transformed by taking the natural logarithm to eliminate the skewness and kurtosis. Results of our previous study have shown that behavioral traits of rats in the open field test and morphometric traits contribute to different principal components [20]. Taking this into account, in the present work, the analyses of behavioral and morphometric traits were carried out separately. Principal component analysis was performed in two versions: (1) separately for 3–4- and 6−7-month-old rats; (2) jointly for the two age groups, where the parameter "age indicator" served as an independent categorical (dummy) variable in the calculation of correlations with principal components. The calculation of the significance of the Pearson correlation coefficient (r) was conducted via the Benjamini–Hochberg correction for multiple comparisons. The data were considered significant at FDR < 5%.

5. Conclusions

BP and an increase in BP under stress are the traits that we consider the main endpoints in this paper. Nonetheless, they are formed by combined action of many endogenous and exogenous factors. In turn, high BP may have an impact on the same factors (a feedback loop). Elevation of BP and an increase in the concentration of plasma corticosterone during stress are traits that characterize the response to stress. Weakening of the response to stress may be due to self-regulation of stress response systems: an increase in basal function automatically reduces the response to an additional stimulus, thereby leading to a state of chronic stress that is characterized by a persistent increase in basal BP and a rise of basal plasma corticosterone concentration. In our work, the use of the ISIAH rat strain, which simulates the stress-sensitive form of arterial hypertension, enabled us to analyze the traits of emotionality that rats exhibit in the open field test. Our data clearly show that

the development of hypertensive status in F_2(ISIAHxWAG) rats is accompanied by the enhancement of anxiety. The study revealed age-dependent relationships between key features and their changes with age and worsening of the hypertensive state. In particular, it was demonstrated that the concentration of plasma corticosterone at rest and its increase under short-term restraint stress in a young group of rats does not have a straightforward relationship with other assessed traits. By contrast, in older animals, such associations were found. Because emotion-related brain activity has been reported to be an important factor for the development of hypertension in humans [55], we hope that our analysis of relationships between the investigated traits (which are associated with a genetically determined hypertensive phenotype and high stress reactivity under conditions of short-term restraint (emotional) stress) will be useful for the design of strategies for the treatment of hypertension, taking into account the age of the patients.

Author Contributions: Conceptualization, O.E.R., D.Y.O. and A.L.M.; data curation, Y.V.M. and D.Y.O.; formal analysis, Y.V.M., D.Y.O. and M.P.P.; funding acquisition, O.E.R.; investigation, D.Y.O., M.P.P., A.L.M. and O.E.R.; methodology, D.Y.O., O.E.R. and A.L.M.; supervision, O.E.R. and A.L.M.; writing—original draft, D.Y.O., O.E.R. and A.L.M.; writing—review and editing, Y.V.M., D.Y.O., M.P.P., O.E.R. and A.L.M. All authors have read and agreed to the published version of the manuscript.

Funding: The study was supported by the Russian Science Foundation (grant No. 22−14-00082).

Institutional Review Board Statement: All procedures were conducted in compliance with European Communities Council Directive 210/63/EU of 22 September 2010. The study protocol was approved by the Bioethical Council of the federal research center Institute of Cytology and Genetics SB RAS (Novosibirsk, Russia), protocol No. 69 of 20 January 2021.

Informed Consent Statement: Not applicable.

Data Availability Statement: Not applicable.

Acknowledgments: The authors are thankful to the multi-access center Bioinformatics for the use of computational resources (Institute of Cytology and Genetics, SB RAS, Novosibirsk, Russia) as supported by Russian government project FWNR-2022-0020. The authors are also grateful to the Vivarium for Conventional Animals (Institute of Cytology and Genetics, SB RAS, Novosibirsk, Russia) for the breeding, regular care, and maintenance of the experimental rats, as supported by Russian government project FWNR-2022-0019. The authors are grateful to Nikolai A. Shevchuk for help with correcting the English version of the manuscript.

Conflicts of Interest: The authors declare no conflict of interest.

References

1. Bozkurt, B.; Aguilar, D.; Deswal, A.; Dunbar, S.B.; Francis, G.S.; Horwich, T.; Jessup, M.; Kosiborod, M.; Pritchett, A.M.; Ramasubbu, K.; et al. Contributory Risk and Management of Comorbidities of Hypertension, Obesity, Diabetes Mellitus, Hyperlipidemia, and Metabolic Syndrome in Chronic Heart Failure: A Scientific Statement from the American Heart Association. *Circulation* **2016**, *134*, e535–e578. [CrossRef] [PubMed]
2. Hamrahian, S.M.; Falkner, B. Hypertension in Chronic Kidney Disease. *Adv. Exp. Med. Biol.* **2017**, *956*, 307–325. [CrossRef] [PubMed]
3. Teck, J. Diabetes-Associated Comorbidities. *Prim. Care* **2022**, *49*, 275–286. [CrossRef] [PubMed]
4. Di Palo, K.E.; Barone, N.J. Hypertension and Heart Failure: Prevention, Targets, and Treatment. *Heart Fail Clin.* **2020**, *16*, 99–106. [CrossRef]
5. Omote, K.; Verbrugge, F.H.; Borlaug, B.A. Heart Failure with Preserved Ejection Fraction: Mechanisms and Treatment Strategies. *Annu. Rev. Med.* **2022**, *73*, 321–337. [CrossRef]
6. Flack, J.M.; Adekola, B. Blood pressure and the new ACC/AHA hypertension guidelines. *Trends Cardiovasc. Med.* **2020**, *30*, 160–164. [CrossRef]
7. Ding, P.; Pan, Y.; Wang, Q.; Xu, R. Prediction and evaluation of combination pharmacotherapy using natural language processing, machine learning and patient electronic health records. *J. Biomed. Inform.* **2022**, *133*, 104164. [CrossRef]
8. Johnson, R.J.; Rodriguez-Iturbe, B.; Kang, D.H.; Feig, D.I.; Herrera-Acosta, J. A unifying pathway for essential hypertension. *Am. J. Hypertens.* **2005**, *18*, 431–440. [CrossRef]
9. Feldstein, C.; Julius, S. Establishing targets for hypertension control in patients with comorbidities. *Curr. Hypertens. Rep.* **2010**, *12*, 465–473. [CrossRef]

10. Magvanjav, O.; Cooper-DeHoff, R.M.; McDonough, C.W.; Gong, Y.; Segal, M.S.; Hogan, W.R.; Johnson, J.A. Antihypertensive therapy prescribing patterns and correlates of blood pressure control among hypertensive patients with chronic kidney disease. *J. Clin. Hypertens.* **2019**, *21*, 91–101. [CrossRef]

11. Lee, J.H.; Kim, K.I.; Cho, M.C. Current status and therapeutic considerations of hypertension in the elderly. *Korean J. Intern. Med.* **2019**, *34*, 687–695. [CrossRef]

12. Oliveros, E.; Patel, H.; Kyung, S.; Fugar, S.; Goldberg, A.; Madan, N.; Williams, K.A. Hypertension in older adults: Assessment, management, and challenges. *Clin. Cardiol.* **2020**, *43*, 99–107. [CrossRef]

13. Stawski, R.S.; Scott, S.B.; Zawadzki, M.J.; Sliwinski, M.J.; Marcusson-Clavertz, D.; Kim, J.; Lanza, S.T.; Green, P.A.; Almeida, D.M.; Smyth, J.M. Age differences in everyday stressor-related negative affect: A coordinated analysis. *Psychol. Aging.* **2019**, *34*, 91–105. [CrossRef]

14. Whittaker, A.C.; Ginty, A.; Hughes, B.M.; Steptoe, A.; Lovallo, W.R. Cardiovascular Stress Reactivity and Health: Recent Questions and Future Directions. *Psychosom. Med.* **2021**, *83*, 756–766. [CrossRef]

15. Markel, A.L. Development of a new strain of rats with inherited stress-induced arterial hypertension. In *Genetic Hypertension*; Sassard, J., Libbey, J., Eurotext, Eds.; Colloque INSERM: Paris, France, 1992; Volume 218, pp. 405–407.

16. Markel, A.L.; Maslova, L.N.; Shishkina, G.T.; Bulygina, V.V.; Machanova, N.A.; Jacobson, G.S. Developmental influences on blood pressure regulation in ISIAH rats. In *Development of the Hypertensive Phenotype: Basic and Clinical Studies*; McCarty, R., Blizard, D.A., Chevalier, R.L., Birkenhager, W.H., Reid, J.L., Eds.; Handbook of hypertension; Elsevier: Amsterdam, The Netherlands, 1999; Volume 19, pp. 493–526.

17. Markel, A.L.; Redina, O.E.; Gilinsky, M.A.; Dymshits, G.M.; Kalashnikova, E.V.; Khvorostova, Y.V.; Fedoseeva, L.A.; Jacobson, G.S. Neuroendocrine profiling in inherited stress-induced arterial hypertension rat strain with stress-sensitive arterial hypertension. *J. Endocrinol.* **2007**, *195*, 439–450. [CrossRef]

18. Redina, O.E.; Markel, A.L. Stress, Genes, and Hypertension. Contribution of the ISIAH Rat Strain Study. *Curr. Hypertens. Rep.* **2018**, *20*, 66. [CrossRef]

19. Markel, A.L.; Galaktionov Yu, K.; Efimov, V.M. Factor analysis of rat behavior in an open field test. *Neurosci. Behav. Physiol.* **1989**, *19*, 279–286. [CrossRef]

20. Efimov, V.M.; Kovaleva, V.Y.; Markel, A.L. A new approach to the study of genetic variability of complex characters. *Heredity* **2005**, *94*, 101–107. [CrossRef]

21. Sturman, O.; Germain, P.L.; Bohacek, J. Exploratory rearing: A context- and stress-sensitive behavior recorded in the open-field test. *Stress* **2018**, *21*, 443–452. [CrossRef]

22. Johnson, E.O.; Kamilaris, T.C.; Chrousos, G.P.; Gold, P.W. Mechanisms of stress: A dynamic overview of hormonal and behavioral homeostasis. *Neurosci. Biobehav. Rev.* **1992**, *16*, 115–130. [CrossRef] [PubMed]

23. Marin, J. Age-related changes in vascular responses: A review. *Mech. Ageing. Dev.* **1995**, *79*, 71–114. [CrossRef] [PubMed]

24. Hausler, A.; Girard, J.; Baumann, J.B.; Ruch, W.; Otten, U.H. Stress-induced secretion of ACTH and corticosterone during development of spontaneous hypertension in rats. *Clin. Exp. Hypertens A* **1983**, *5*, 11–19. [CrossRef] [PubMed]

25. O'Malley, K.; Docherty, J.R.; Kelly, J.G. Adrenoceptor status and cardiovascular function in ageing. *J. Hypertens Suppl.* **1988**, *6*, S59–S62.

26. Yiallouris, A.; Tsioutis, C.; Agapidaki, E.; Zafeiri, M.; Agouridis, A.P.; Ntourakis, D.; Johnson, E.O. Adrenal Aging and Its Implications on Stress Responsiveness in Humans. *Front Endocrinol.* **2019**, *10*, 54. [CrossRef]

27. Hamer, M.; Steptoe, A. Cortisol responses to mental stress and incident hypertension in healthy men and women. *J. Clin. Endocrinol. Metab.* **2012**, *97*, E29–E34. [CrossRef]

28. Ramsay, D.; Lewis, M. Reactivity and regulation in cortisol and behavioral responses to stress. *Child. Dev.* **2003**, *74*, 456–464. [CrossRef]

29. Carroll, D.; Ginty, A.T.; Whittaker, A.C.; Lovallo, W.R.; de Rooij, S.R. The behavioural, cognitive, and neural corollaries of blunted cardiovascular and cortisol reactions to acute psychological stress. *Neurosci. Biobehav. Rev.* **2017**, *77*, 74–86. [CrossRef]

30. Lovallo, W.R.; Cohoon, A.J.; Sorocco, K.H.; Vincent, A.S.; Acheson, A.; Hodgkinson, C.A.; Goldman, D. Early-Life Adversity and Blunted Stress Reactivity as Predictors of Alcohol and Drug use in Persons with COMT (rs4680) Val158Met Genotypes. *Alcohol. Clin. Exp. Res.* **2019**, *43*, 1519–1527. [CrossRef]

31. Oshchepkov, D.; Chadaeva, I.; Kozhemyakina, R.; Zolotareva, K.; Khandaev, B.; Sharypova, E.; Ponomarenko, P.; Bogomolov, A.; Klimova, N.V.; Shikhevich, S.; et al. Stress Reactivity, Susceptibility to Hypertension, and Differential Expression of Genes in Hypertensive Compared to Normotensive Patients. *Int. J. Mol. Sci.* **2022**, *23*, 2835. [CrossRef]

32. Fedoseeva, L.A.; Klimov, L.O.; Ershov, N.I.; Alexandrovich, Y.V.; Efimov, V.M.; Markel, A.L.; Redina, O.E. Molecular determinants of the adrenal gland functioning related to stress-sensitive hypertension in ISIAH rats. *BMC Genomics* **2016**, *17* (Suppl. 14), 989. [CrossRef]

33. Fedoseeva, L.A.; Klimov, L.O.; Ershov, N.I.; Efimov, V.M.; Markel, A.L.; Orlov, Y.L.; Redina, O.E. The differences in brain stem transcriptional profiling in hypertensive ISIAH and normotensive WAG rats. *BMC Genom.* **2019**, *20*, 297. [CrossRef] [PubMed]

34. Fedoseeva, L.A.; Ryazanova, M.A.; Ershov, N.I.; Markel, A.L.; Redina, O.E. Comparative transcriptional profiling of renal cortex in rats with inherited stress-induced arterial hypertension and normotensive Wistar Albino Glaxo rats. *BMC Genet.* **2016**, *17* (Suppl. 1), 12. [CrossRef]

35. Klimov, L.O.; Ershov, N.I.; Efimov, V.M.; Markel, A.L.; Redina, O.E. Genome-wide transcriptome analysis of hypothalamus in rats with inherited stress-induced arterial hypertension. *BMC Genet.* **2016**, *17* (Suppl. 1), 13. [CrossRef]

36. Ryazanova, M.A.; Fedoseeva, L.A.; Ershov, N.I.; Efimov, V.M.; Markel, A.L.; Redina, O.E. The gene-expression profile of renal medulla in ISIAH rats with inherited stress-induced arterial hypertension. *BMC Genet.* **2016**, *17*, 155–171. [CrossRef]

37. Girotti, M.; Pace, T.W.; Gaylord, R.I.; Rubin, B.A.; Herman, J.P.; Spencer, R.L. Habituation to repeated restraint stress is associated with lack of stress-induced c-fos expression in primary sensory processing areas of the rat brain. *Neuroscience* **2006**, *138*, 1067–1081. [CrossRef]

38. Mansi, J.A.; Rivest, S.; Drolet, G. Effect of immobilization stress on transcriptional activity of inducible immediate-early genes, corticotropin-releasing factor, its type 1 receptor, and enkephalin in the hypothalamus of borderline hypertensive rats. *J. Neurochem.* **1998**, *70*, 1556–1566. [CrossRef]

39. Imaki, T.; Shibasaki, T.; Chikada, N.; Harada, S.; Naruse, M.; Demura, H. Different expression of immediate-early genes in the rat paraventricular nucleus induced by stress: Relation to corticotropin-releasing factor gene transcription. *Endocr. J.* **1996**, *43*, 629–638. [CrossRef]

40. Imaki, T.; Naruse, M.; Harada, S.; Chikada, N.; Nakajima, K.; Yoshimoto, T.; Demura, H. Stress-induced changes of gene expression in the paraventricular nucleus are enhanced in spontaneously hypertensive rats. *J. Neuroendocrinol.* **1998**, *10*, 635–643. [CrossRef]

41. Brosschot, J.F. Ever at the ready for events that never happen. *Eur. J. Psychotraumatol.* **2017**, *8*, 1309934. [CrossRef]

42. van den Beld, A.W.; Kaufman, J.M.; Zillikens, M.C.; Lamberts, S.W.J.; Egan, J.M.; van der Lely, A.J. The physiology of endocrine systems with ageing. *Lancet Diabetes Endocrinol.* **2018**, *6*, 647–658. [CrossRef]

43. Toczek, M.; Baranowska-Kuczko, M.; Grzeda, E.; Pedzinska-Betiuk, A.; Weresa, J.; Malinowska, B. Age-specific influences of chronic administration of the fatty acid amide hydrolase inhibitor URB597 on cardiovascular parameters and organ hypertrophy in DOCA-salt hypertensive rats. *Pharmacol. Rep.* **2016**, *68*, 363–369. [CrossRef] [PubMed]

44. Redina, O.E.; Smolenskaya, S.E.; Maslova, L.N.; Markel, A.L. Genetic Control of the Corticosterone Level at Rest and Under Emotional Stress in ISIAH Rats with Inherited Stress-Induced Arterial Hypertension. *Clin. Exp. Hypertens.* **2010**, *32*, 364–371. [CrossRef] [PubMed]

45. Redina, O.E.; Smolenskaya, S.E.; Maslova, L.N.; Markel, A.L. The genetic control of blood pressure and body composition in rats with stress-sensitive hypertension. *Clin. Exp. Hypertens.* **2013**, *35*, 484–495. [CrossRef]

46. Redina, O.E.; Smolenskaya, S.E.; Markel, A.L. Genetic control of the behavior of ISIAH rats in the open field test. *Russ. J. Genet.* **2022**, *58*, 791–803. [CrossRef]

47. Samani, N.J.; Gauguier, D.; Vincent, M.; Kaiser, M.A.; Bihoreau, M.T.; Lodwick, D.; Wallis, R.; Parent, V.; Kimber, P.; Rattray, F.; et al. Analysis of quantitative trait loci for blood pressure on rat chromosomes 2 and 13. Age-related differences in effect. *Hypertension* **1996**, *28*, 1118–1122. [CrossRef]

48. Kovacs, P.; van den Brandt, J.; Kloting, I. Effects of quantitative trait loci for lipid phenotypes in the rat are influenced by age. *Clin. Exp. Pharmacol. Physiol.* **1998**, *25*, 1004–1007. [CrossRef]

49. Garrett, M.R.; Dene, H.; Rapp, J.P. Time-course genetic analysis of albuminuria in Dahl salt-sensitive rats on low-salt diet. *J. Am. Soc. Nephrol.* **2003**, *14*, 1175–1187. [CrossRef]

50. Beck, S.R.; Brown, W.M.; Williams, A.H.; Pierce, J.; Rich, S.S.; Langefeld, C.D. Age-stratified QTL genome scan analyses for anthropometric measures. *BMC Genet.* **2003**, *4*, S31. [CrossRef]

51. Turana, Y.; Tengkawan, J.; Chia, Y.C.; Shin, J.; Chen, C.H.; Park, S.; Tsoi, K.; Buranakitjaroen, P.; Soenarta, A.A.; Siddique, S.; et al. Mental health problems and hypertension in the elderly: Review from the HOPE Asia Network. *J. Clin. Hypertens.* **2021**, *23*, 504–512. [CrossRef]

52. Liang, X.; Shan, Y.; Ding, D.; Zhao, Q.; Guo, Q.; Zheng, L.; Deng, W.; Luo, J.; Tse, L.A.; Hong, Z. Hypertension and High Blood Pressure Are Associated with Dementia Among Chinese Dwelling Elderly: The Shanghai Aging Study. *Front Neurol.* **2018**, *9*, 664. [CrossRef]

53. Tinnikov, A.A.; Bazhan, N.M. Determination of glucocorticoids in blood and adrenal gland cultures by competitive protein binding without preliminary extraction. *Lab Delo* **1984**, *12*, 709–713.

54. Hammer, O.; Harper, D.A.T.; Ryan, P.D. Paleontological statistics software package for education and data analysis. *Palaeontol. Electron.* **2001**, *4*, 9–18.

55. Schaare, H.L.; Blochl, M.; Kumral, D.; Uhlig, M.; Lemcke, L.; Valk, S.L.; Villringer, A. Associations between mental health, blood pressure and the development of hypertension. *Nat. Commun.* **2023**, *14*, 1953. [CrossRef]

International Journal of
Molecular Sciences

Article

Prognostic Significance of Activated Monocytes in Patients with ST-Elevation Myocardial Infarction

Mohamed Abo-Aly [1,2], Elica Shokri [1], Lakshman Chelvarajan [1], Wadea M. Tarhuni [3], Himi Tripathi [1,4] and Ahmed Abdel-Latif [1,4,*]

[1] Gill Heart and Vascular Institute, University of Kentucky, Lexington, KY 40536, USA
[2] Cardiovascular Division, Department of Medicine, Perelman School of Medicine, University of Pennsylvania, Philadelphia, PA 19104, USA
[3] Canadian Cardiac Research Center, Department of Internal Medicine, Division of Cardiology, University of Saskatchewan, Saskatoon, SK S7N 5A2, Canada
[4] Cardiovascular Division, Department of Medicine, University of Michigan, Ann Arbor, MI 48109, USA
* Correspondence: aalatif@umich.edu

Abstract: Circulating monocytes have different subsets, including classical (CD14++CD16−), intermediate (CD14++CD16+), and nonclassical (CD14+CD16++), which play different roles in cardiovascular physiology and disease progression. The predictive value of each subset for adverse clinical outcomes in patients with coronary artery disease is not fully understood. We sought to evaluate the prognostic efficacy of each monocyte subset in patients with ST-elevation myocardial infarction (STEMI). We recruited 100 patients with STEMI who underwent primary percutaneous coronary intervention (PCI). Blood samples were collected at the time of presentation to the hospital (within 6 h from onset of symptoms, baseline (BL)) and then at 3, 6, 12, and 24 h after presentation. Monocytes were defined as CD45+/HLA-DR+ and then subdivided based on the expression of CD14, CD16, CCR2, CD11b, and CD42. The primary endpoint was a composite of all-cause death, hospitalization for heart failure, stent thrombosis, in-stent restenosis, and recurrent myocardial infarction. Univariate and multivariate Cox proportional hazards models, including baseline comorbidities, were performed. The mean age of our cohort was 58.9 years and 25% of our patients were females. Patients with high levels (above the median) of CD14+CD16++ monocytes showed an increased risk for the primary endpoint in comparison to patients with low levels; adjusted hazard ratio (aHR) for CD14+/CD16++ cells was 4.3 (95% confidence interval (95% CI) 1.2–14.8, $p = 0.02$), for CD14+/CD16++/CCR2+ cells was 3.82 (95% CI 1.06–13.7, $p = 0.04$), for CD14+/CD16++/CD42b+ cells was 3.37 (95% CI 1.07–10.6, $p = 0.03$), for CD14+/CD16++/CD11b+ was 5.17 (95% CI 1.4–18.0, $p = 0.009$), and for CD14+ HLA-DR+ was 7.5 (95% CI 2.0–28.5, $p = 0.002$). CD14++CD16−, CD14++CD16+, and their CD11b+, CCR2+, and CD42b+ aggregates were not significantly predictive for our composite endpoint. Our study shows that CD14+ CD16++ monocytes and their subsets expressing CCR2, CD42, and CD11b could be important predictors of clinical outcomes in patients with STEMI. Further studies with a larger sample size and different coronary artery disease phenotypes are needed to verify the findings.

Keywords: myocardial infarction; human; inflammation; clinical outcomes

Citation: Abo-Aly, M.; Shokri, E.; Chelvarajan, L.; Tarhuni, W.M.; Tripathi, H.; Abdel-Latif, A. Prognostic Significance of Activated Monocytes in Patients with ST-Elevation Myocardial Infarction. *Int. J. Mol. Sci.* **2023**, 24, 11342. https://doi.org/10.3390/ijms241411342

Academic Editors: Yutang Wang and Dianna Magliano

Received: 26 June 2023
Revised: 4 July 2023
Accepted: 9 July 2023
Published: 12 July 2023

1. Introduction

Cardiovascular diseases are the leading cause of mortality and morbidity worldwide [1]. ST-segment elevation myocardial infarction (STEMI) has the worst prognosis among cardiovascular diseases. STEMI is caused by the rupture of atherosclerotic plaque in the coronary arteries. The incidence of STEMI comprises around 40% of myocardial infarction (MI) presentations [2]. Despite significant advances in timely revascularization and medical therapy, 1-year mortality of STEMI can be as high as 20% [3]. Although the mortality rate for STEMI has significantly improved in the last decade, the risk of adverse clinical outcomes widely differs among various subgroups of STEMI patients [4].

Therefore, risk stratification of individual patients with STEMI is of paramount importance for individualized management strategies and allocating health care resources. Acute myocardial infarction (AMI) triggers a systemic and local inflammatory response, which initiates the mobilization and recruitment of a wide variety of inflammatory cells, including monocytes [5,6]. Monocytes have different subsets; classical (CD14++CD16−), intermediate (CD14++CD16+), and nonclassical (CD14+CD16++), which play different roles in cardiovascular physiology and cardiovascular disease progression [7]. CD14++CD16− monocytes play a significant role in triggering the inflammatory response after myocardial injury in STEMI and contribute to atherosclerosis [8]. Although previous clinical studies have shown that CD14++CD16− can predict adverse clinical outcomes in patients with coronary artery disease [8], other studies have shown that CD14++CD16− monocytes do not change after ischemic events, such as stroke [9]. CD14+CD16++ cells can have pro-inflammatory functions by secreting tumor necrosis alpha (TNF-α) and interleukin-1 (IL-1), which enhance the immune cell recruitment [10,11]. Moreover, they can have athero-protective properties by enhancing efferocytosis [7]. Similarly, CD14++CD16+ monocytes have mixed functions and can be both pro-inflammatory and anti-inflammatory [9,12–14]. In this study, we provide a comprehensive temporal analysis of the dynamic changes in circulating innate immune cells, including monocytes and inflammatory markers, and their prognostic values in patients with STEMI.

2. Results

2.1. Baseline Characteristics

This prospective study included 100 patients with STEMI enrolled between August 2017 and July 2020. The baseline characteristics of the cohort are shown in Table 1. The study population was predominantly male (75%), and the mean age was 58.9 years (interquartile range-IQR 1.03). The median BMI of the study population was 28.8 (IQR 0.62), and the study population was predominantly white (92%). The right coronary artery was the culprit artery in 50% of patients, followed by the left anterior descending artery (31%) and the left circumflex artery (15%). In this study, 31% of our patients had DM, 67% had hypertension, 35% had hyperlipidemia, and 55% were current smokers. The mean follow-up of enrolled patients was 1.3 years.

Table 1. Baseline characteristics of study population.

Demographic and Clinical Variables	N (%), Mean (SD), or Median (IQR)
Age	58.9 (1.03)
Sex, female	25 (25%)
BMI	28.8 (0.62)
Diabetes mellitus	31 (31%)
Hypertension	67 (67%)
Hyperlipidaemia	35 (35%)
Current smoker	55 (55%)
Congestive heart failure	1 (1%)
Baseline LVEF	46.07 (13.8)
Previous myocardial infarction	14 (14%)
Previous coronary revascularization	24 (24%)
Previous coronary bypass surgery	1 (1%)
Previous stroke	4 (4%)
Peripheral vascular disease	2 (2%)
Chronic kidney disease	1 (1%)
Door to balloon time in minutes *	33.5 (28.4)
<u>Race</u>	
White	92 (92%)

Table 1. *Cont.*

Demographic and Clinical Variables	N (%), Mean (SD), or Median (IQR)
African American	5 (5%)
Culprit artery	
Left anterior descending	31 (31%)
Left circumflex	15 (15%)
Right coronary artery	50 (50%)
Medications at discharge	
Statins	97 (97%)
Angiotensin converting enzyme	63 (63%)
Angiotensin receptor blocker	10 (10%)
Beta blocker	93 (93%)
Aspirin	96 (96%)
P2Y12 inhibitors	100 (100%)
Ticagrelor	85 (85%)
Clopidogrel	11 (11%)
Prasugrel	4 (4%)

BMI, body mass index; LVEF, left ventricular ejection fraction. * Door to balloon time was available in only 41 patients.

2.2. Mobilization of Inflammatory Cells

We assessed the number of circulating inflammatory cells in an attempt to characterize their dynamic mobilization after STEMI in humans. We characterized monocyte populations based on the expression of CD14 and CD16 into classical monocytes (CD14++CD16−), non-classical monocytes (CD14+CD16++), and intermediate monocytes (CD14++CD16+). CD14++CD16− monocytes peaked at 12 h and returned to nadir levels at 24 h after injury (Figure 1). Similarly, CD14+CD16++ monocytes peaked at 6 h after STEMI (Figure 2). However, CD14++CD16+ monocytes reached their peak 24 h after injury (Figure 3).

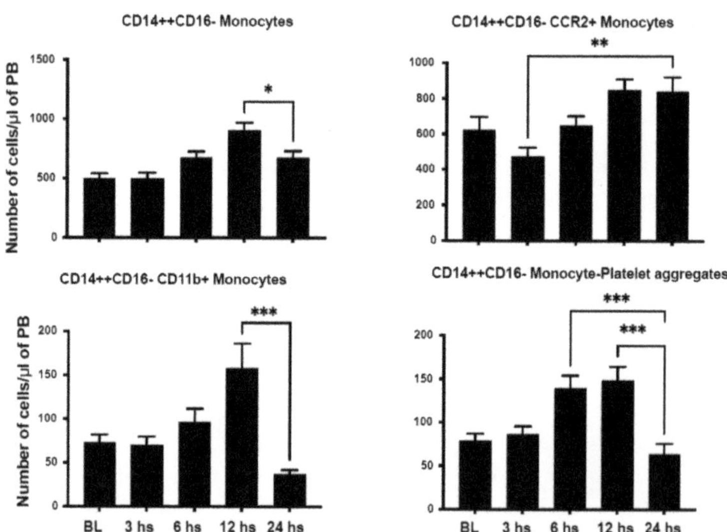

Figure 1. Classical monocytes expressing CD14++/CD16− are mobilized after myocardial infarction in humans. Bar graphs showing the number of circulating CD14++/CD16− monocytes in the peripheral blood after STEMI. The data demonstrate a peak of classical monocytes in the peripheral blood at 6 h after STEMI. The subset of classical monocytes expressing the activation markers CCR2, CD11b, and CD42 (monocyte–platelet aggregates) also peaked within 12–24 h after STEMI (data shown as mean $\pm$ SEM, * $p < 0.05$, ** $p < 0.01$, *** $p < 0.001$ compared to 24 h value).

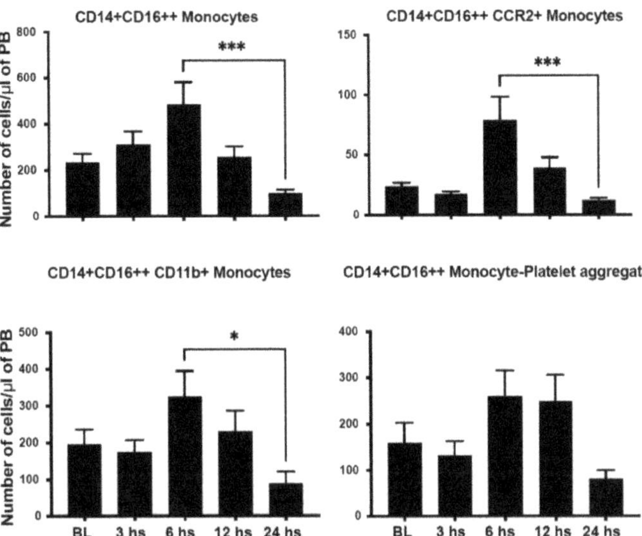

Figure 2. Non-classical monocytes expressing CD14+/CD16++ are mobilized after myocardial infarction in humans. Bar graphs showing the number of circulating CD14+/CD16++ monocytes in the peripheral blood after STEMI. The data demonstrate a peak of non-classical monocytes in the peripheral blood at 6 h after STEMI. The subset of non-classical monocytes expressing the activation markers CCR2, CD11b, and CD42 (monocyte–platelet aggregates) also peaked within 6–12 h after STEMI (data shown as mean $\pm$ SEM, * $p < 0.05$, *** $p < 0.001$ compared to 24 h value).

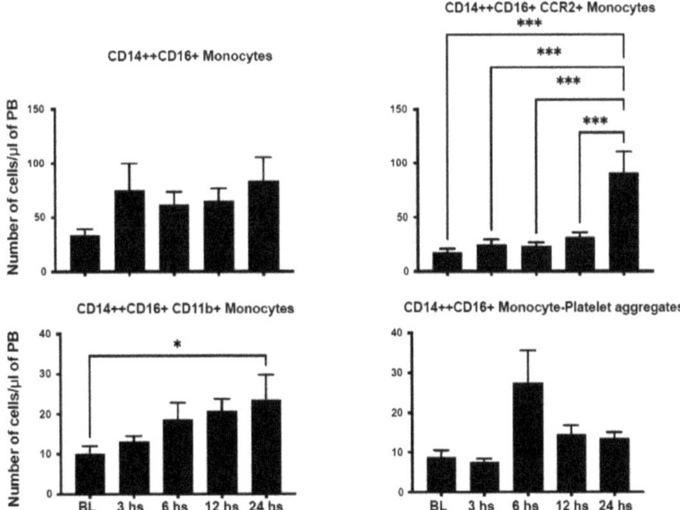

Figure 3. Intermediate monocytes expressing CD14++/CD16+ are mobilized after myocardial infarction in humans. Bar graphs showing the number of circulating CD14++/CD16+ monocytes in the peripheral blood after STEMI. The data demonstrate a bimodal peak of intermediate monocytes in the peripheral blood at 3 and 24 h after STEMI. The subset of intermediate monocytes expressing the activation markers CCR2 and CD11b also peaked within 24 h after STEMI. However, intermediate monocytes expressing CD42 (monocyte–platelet aggregates) peaked at 6 h after STEMI (data shown as mean $\pm$ SEM, * $p < 0.05$, *** $p < 0.001$ compared to 24 h value).

C-C Motif Chemokine Receptor 2 (CCR2) plays a critical role in immune cell chemotaxis to monocyte chemoattractant protein-1 (MCP-1), a chemokine that specifically mediates monocyte chemotaxis. MCP-1 is upregulated in tissue injury such as myocardial infarction and directs the mobilization of monocytes and their influx into the injured myocardium. We noticed a peak in CD14+CD16−/CCR2+ at 24 h after injury (Figure 1). However, we observed a dip in CD14+CD16++/CCR2+ counts at 3 h after injury, which then peaked at 6 h after injury (Figure 2). Similarly, CD14++CD16+/CCR2+ increased progressively to its maximum level at 24 h after STEMI (Figure 3). The expression of CD11b is a marker of immune cell activation/inflammation. CD14+CD16− monocytes expressing CD11b peaked at 12 h after injury and returned to nadir levels by 24 h after STEMI (Figure 2). The expression of CD11b on the surface of CD14+CD16++ peaked at 6 h (Figure 2) but at 24 h on CD14++CD16+ monocytes (Figure 3). Platelet–leukocyte aggregates are indicative of monocyte activation and have been linked to adverse clinical events. Our data indicate that platelet–monocyte aggregates peaked early after STEMI among most monocyte subpopulations. This is consistent with the early peak inflammatory response as represented by the cytokine data.

2.3. Changes in Plasma Cytokines after STEMI in Humans

Chemokines and cytokines play an important role in the communication between injured tissue and hematopoietic cells and their progenitors in the bone marrow and spleen to regulate immune cell production and their mobilization to areas of need. We have recently shown that PB levels of markers of neutrophil activation such as myeloperoxidase (MPO) and S100A8/A9 are upregulated early after STEMI [15,16]. We assessed the plasma levels of classical chemokines/cytokines involved in myelopoiesis and monocytosis, such as IL-1β, IL-6, GM-CSF, and RANTES. Plasma levels of IL-1β, GM-CSF, and RANTES peaked early after STEMI, while the level of IL-6 peaked within 6–12 h after acute injury (Figure 4).

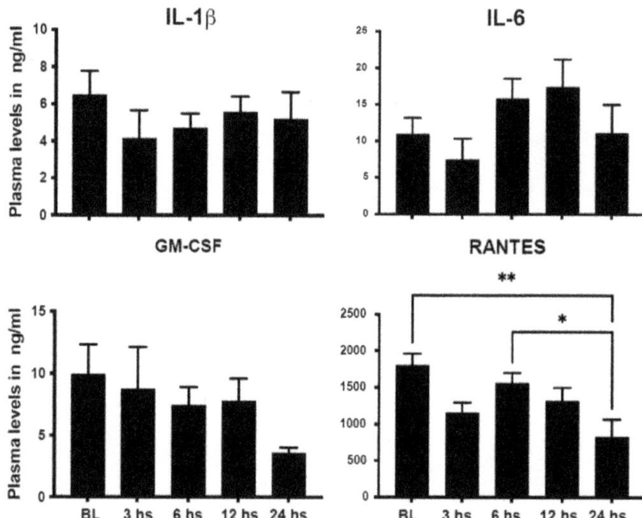

Figure 4. Plasma levels of pro-inflammatory cytokines show dynamic changes after STEMI. Bar graphs showing the plasma levels of the pro-inflammatory cytokines: interleukin-1 beta, interleukin-6, granulocyte-macrophage colony-stimulating factor, and RANTES (Regulated upon Activation, Normal T Cell Expressed and Presumably Secreted) cytokines. These data show dynamic changes in the plasma level with an early peak that precedes the mobilization of activated monocytes (data shown as mean ± SEM, * $p < 0.05$, ** $p < 0.01$, compared to 24 h value).

2.4. Correlation between Circulating Monocytes and Clinical Outcomes

We assessed the relationship between the number of circulating inflammatory cells, inflammatory cytokines, and clinical outcomes in our patient cohort. Patients were followed up for a median of 470 days. Overall, we observed 17 major adverse cardiac events during the follow-up period. Most of these clinical events were in the form of unplanned recurrent revascularization. Patients were divided based on the median number of circulating inflammatory cells at their peak into high level (above median) and low level (below median). In the univariate cox proportional hazards model, among all baseline characteristics, only age (HR = 1.06, 95% CI 1.01–1.11, p = 0.01), congestive heart failure (HR = 10.22, 95% CI 1.29–80.68, p = 0.02), and stroke (HR = 9.20, 95% CI 1.94–43.43, p = 0.005) significantly predicted the composite endpoint (Supplemental Table S1). In the univariate cox proportional hazards model, only total monocytes (CD14+/HLA-DR+) (3.80, 95% CI 1.19–12.13, p = 0.02), CD14+CD16++ monocytes (3.48, 95% CI 1.08–11.26, p = 0.03), and CD14+CD16++ monocytes expressing CD11b (4.99, 95% CI 1.46–7.06, p = 0.01) significantly predicted the clinical composite endpoint (Figure 5). Other monocyte populations such as CD14+CD16++/CD42b+ monocytes (2.86, 95% CI 0.95–8.55, p = 0.05) and CD14+CD16++/CCR2+ monocytes (2.73, 95% CI 0.91–8.11, p = 0.07) showed borderline significance towards predicting our composite endpoint (Supplemental Table S2). A close observation of the survival curves shows an early separation of the curves for all outcomes measured.

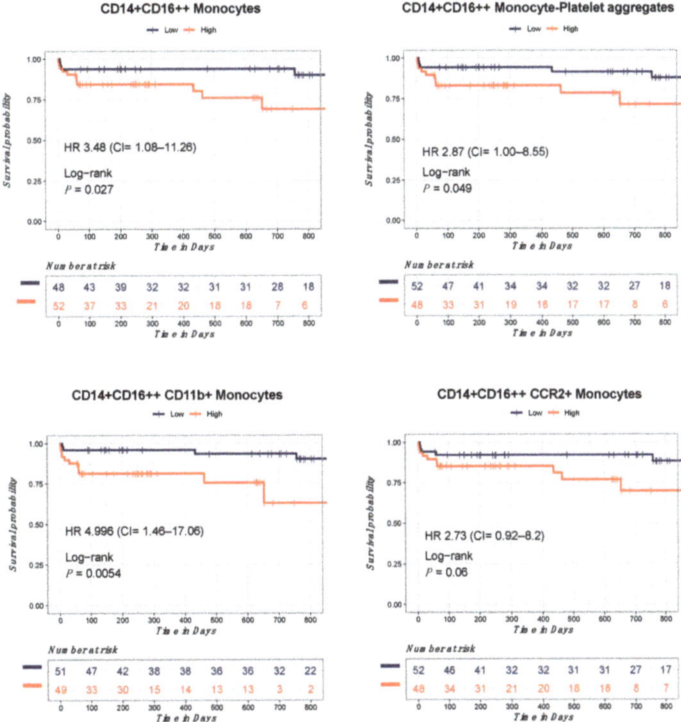

Figure 5. Elevated numbers of CD14+/CD16++ monocytes correlate with long-term adverse clinical events in STEMI patients. Survival curves for the probability of developing all-cause mortality, heart failure hospitalization, recurrent myocardial infarction, and stent thrombosis. These curves show a strong correlation between elevated numbers of circulating non-classical monocytes and their subsets expressing CD11b, CCR2, and their aggregates with platelets, and clinical events during long-term follow-up in STEMI patients.

In the multivariate cox proportional hazards model, after adjusting for age, stroke, and congestive heart failure, CD14+CD16++ monocytes (adjusted HR (aHR) = 4.30, 95% CI 1.25–14.81, p = 0.02), CD14+CD16++/CCR2+ monocytes (aHR = 3.82, 95% CI = 1.06–13.76, p = 0.04), CD14+CD16++/CD42b+ monocytes (aHR = 3.37, 95% CI = 1.07–10.64, p = 0.03), and CD14+CD16++/CD11b+ monocytes (aHR = 5.17, 95% CI = 1.48–18.06, p = 0.009) remained significantly predictive of the clinical composite endpoint (Table 2).

Table 2. Multivariate cox proportional hazards model for a composite endpoint of all-cause death, myocardial infarction, recurrent hospitalization for heart failure, stent thrombosis, or in-stent restenosis.

Models	Hazardous Ratio	95% Confidence Interval	p Value
Model 1			
CD14+CD16++ monocytes	4.30	1.25–14.81	0.02
Age	1.06	1.01–1.11	0.007
Stroke	14.03	2.61–75.31	0.002
Congestive heart failure	18.41	1.92–176.40	0.01
Model 2			
CD14+CD16++/CCR2+	3.82	1.06–13.76	0.04
Age	1.05	1.01–1.11	0.01
Stroke	17.89	2.77–115.24	0.002
Congestive heart failure	16.52	1.69–161.28	0.01
Model 3			
CD14+CD16++/CD42b+	3.37	1.07–10.64	0.03
Age	1.06	1.01–1.11	0.007
Stroke	12.02	2.29–63.12	0.003
Congestive heart failure	18.04	1.87–174.13	0.01
Model 4			
CD14+CD16++/CD11b+	5.17	1.48–18.06	0.009
Age	1.05	1.01–1.10	0.01
Stroke	14.58	2.76–76.94	0.001
Congestive heart failure	15.64	1.62–150.95	0.01
Model 5			
CD14+ HLA-DR+	7.57	2.00–28.57	0.002
Age	1.09	1.03–1.16	0.001
Stroke	14.86	2.68–82.30	0.001
Congestive heart failure	23.59	2.38–233.17	0.006

3. Discussion

Myocardial infarction and the resulting heart failure are leading causes of morbidity and mortality worldwide. Several risk prediction models are available for risk stratification after STEMI, including the TIMI risk score and GRACE risk model. Recent studies as well as clinical guidelines have demonstrated the value of risk stratification for personalized medical care in the setting of myocardial infarction [17–19]. However, risk stratification models that incorporate biological variables, such as the immune response and markers of inflammation, are scarce. Changes in the epidemiological characteristics of MI and the availability of new biomarkers warrant an assessment of the performance of these scores in contemporary practice. We propose the addition of inflammatory and immune parameters to clinical predictive models to enhance their accuracy. Post-myocardial infarction inflammation is a major risk factor for adverse cardiac remodeling, heart failure, and major adverse cardiac events in short- and long-term follow-up. In this study, we systematically characterize innate immune cell mobilization and inflammatory cytokines in patients presenting with STEMI. Our data demonstrate dynamic changes in the number of circulating monocytes and their subsets as well as plasma cytokine levels. Peak numbers of monocytes followed cytokine levels suggesting a well-orchestrated immune response. The peak number of circulating monocytes and their subsets significantly predicted clinical outcomes in STEMI patients. Our results provide justification for incorporating inflammatory immune parameters in clinical risk stratification models.

Myocardial injury triggers a series of signaling events to communicate with peripheral blood cells (PBCs) and bone marrow (BM) through processes that are just now being elucidated [20]. Monocytosis after AMI is a poor prognostic indicator, in part because monocytes may contribute to infarct expansion and impair cardiac remodeling, thereby promoting the progression to HF [21,22]. In the first wave of PBC response to AMI, classical (CD14++CD16−) monocytes and neutrophils aid in the clean-up after tissue damage through phagocytosis and the release of proteolytic enzymes. However, this initial injury response may actually confer long-term harm because the reduction in the initial recruitment of monocytes can reduce infarct size and prevent cardiac remodeling after AMI [23]. In addition to effects on the myocardium, monocytosis following AMI accelerates experimental atherosclerosis in animal models, thus initiating a vicious cycle; indeed, this type of cycle may contribute to recurrent coronary events in humans [24]. Our comprehensive study provides additional information regarding the prognostic value of specific monocyte subsets after STEMI. While the CD14++/CD16− cells have been traditionally described as the pro-inflammatory subset associated with adverse clinical events, our study suggests that other monocyte populations such as CD14+/CD16++ monocytes and their subsets were associated with adverse clinical events [8,13,21]. This can be explained by our specific monocyte markers that further defined certain populations of activated monocytes that express CD11b, CCR2, and the platelet marker CD42, which delineates platelet–monocyte complexes, a strong predictor of clinical events in coronary artery disease patients. We also examined the dynamics of circulating monocyte populations while other studies focused on one timepoint. These differences could explain the discrepancy between our results and others and provide additional markers for clinical risk stratifying in STEMI patients. Overall, we noticed that most of the monocyte populations tested peak within 12–24 h after STEMI and therefore, these timepoints represent the optimal timepoint in the clinical setting.

Circulating monocyte infiltration into the heart after myocardial infarction is largely considered a maladaptive response to sterile injury leading to scar expansion and cardiac dysfunction. Circulating monocytes are a heterogeneous population of immune cells that play a role in inflammation and tissue repair. In patients with STEMI, monocyte subsets have been associated with different aspects of prognosis [8,25–29]. Elevated monocyte levels upon hospital admission have been associated with adverse clinical outcomes in STEMI patients, such as recurrent infarction, heart failure, and increased mortality [28,30]. Furthermore, specific monocyte subsets have demonstrated differential prognostic implications in the context of STEMI. The pro-inflammatory CD14++CD16+ monocyte subset, in particular, has been linked to higher rates of cardiovascular events and poor clinical outcomes [13]. Additionally, a heightened CD14++CD16− monocyte count has been shown to independently predict future cardiovascular events [8]. These findings underscore the potential of circulating monocyte levels as a valuable prognostic tool in STEMI patients. In recent studies, monocyte–platelet aggregates have been found to correlate with poor outcomes in patients with acute myocardial infarction, suggesting a possible synergistic role between monocytes and platelets in driving the inflammatory response post-infarction [31,32]. This interaction is hypothesized to contribute to the initiation and propagation of inflammation and thrombosis, further aggravating the inflammatory response in the myocardium. Consequently, monitoring circulating monocyte levels and their subtypes could improve risk stratification and help tailor personalized therapeutic strategies for STEMI patients, potentially reducing the burden of adverse outcomes in this population. Our findings corroborate the available literature and provide additional insights into the monocyte subsets and their dynamic changes after STEMI and in relation to circulating inflammatory cytokines. Our study suggests that CD14+/CD16++ monocytes that express activation markers such as CD11b and CCR2 peak late after STEMI and are associated with adverse clinical events in STEMI patients.

Our study highlights the prognostic value of circulating immune cells and inflammatory markers in STEMI patients. However, our study has multiple limitations. Given our cohort's relatively small number, it is impossible to directly compare the prognostic value of circulatory monocytes with other models verified in large datasets, such as the TIMI and GRACE scores. Hence, future large cohort studies powered to examine these prognostic scores, either independently or in combination, are warranted. We conducted a stringent statistical analysis to account for the confounding factors on measured outcomes; however, there may be unmeasured or uncontrolled confounding factors that could influence the relationship between monocyte subsets and clinical outcomes in patients with STEMI that have not been assessed in our cohort. The study enrolled patients with revascularized STEMI; hence, the results may not be generalizable to other patient populations such as those with other forms of myocardial injury such as non-ST elevation myocardial infarction or patients undergoing elective percutaneous coronary interventions. Along the same lines, patients with non-revascularized STEMI, a rare occurrence in contemporary clinical practice, may have different patterns of circulating immune cells and inflammation, and accordingly, different outcomes.

In conclusion, our study demonstrates that circulating activated monocytes are associated with adverse clinical outcomes in patients with STEMI. This data provide evidence that incorporating biological markers of inflammation can enhance risk stratification models for STEMI patients and guide physicians to allocate more aggressive therapies to patients who need them the most. Future large, randomized studies utilizing biological parameters such as monocyte subsets are needed for risk stratification, the development of novel therapies, and long-term coronary artery disease monitoring.

4. Materials and Methods

4.1. Patient Enrollment

The study population consists of 100 patients with acute STEMI enrolled at the University of Kentucky hospitals between August 2017 and July 2020. STEMI was diagnosed based on EKG findings of new (or increased) and persistent ST-segment elevation in at least two contiguous leads of ≥ 1 mm in all leads other than leads V2–V3 where the following cut-off points apply: ≥ 2.5 mm in men <40 years old ≥ 2 mm in men >40 years old. Samples were collected at the time of presentation to the hospital (0 h; within 6 h from onset of symptoms; baseline (BL)) and then at 3, 6, 12, and 24 h after presentation. Five matched controls with similar comorbidities, but no active myocardial ischemia/infarction, were included in the analysis. Supposing momentarily that there is a single timepoint with a significance level of 5%, we estimate that a sample size of 67 patients will provide 80% or better power to detect an expected 15% change in circulating monocyte counts after STEMI based on published reports. The study protocol complies with the Declaration of Helsinki and was approved by the University of Kentucky's Institutional Review Board and Ethics Committees [33]. All patients provided written informed consent at enrollment. Our primary outcome was a clinical composite endpoint of all-cause death (death from any cause as assessed by examining the hospital medical records and national social security death index), hospitalization for heart failure (hospital admissions with heart failure as the primary discharge diagnosis), stent thrombosis (recurrent myocardial infraction with confirmation of probable or definite stent thrombosis in the discharge summary), in-stent restenosis (confirmed stent restenosis on a coronary angiogram), and recurrent myocardial infarction (defined as rehospitalization of myocardial infarction that is diagnosed based on the STEMI EKG criteria detailed above or a significant rise and fall in cardiac biomarkers which is more than 5-fold the upper level of normal in the setting of symptoms and clinical criteria of myocardial injury). The clinical outcomes were assessed by a member of the research team (M.A-A.) and adjudicated by the senior author (A.AL.)

4.2. Flow Cytometry

For human peripheral blood inflammatory monocytes' quantification, peripheral blood (PB) samples were stained against CD14 PE (Biolegend, San Diego, CA, USA), CD16 FITC (Biolegend, San Diego, CA, USA), HLA DR APC/Cy7 (Biolegend, San Diego, CA, USA), CD 42b PE/Cy7 (Biolegend, San Diego, CA, USA), CCR2 PerCP/Cy5.5 (Biolegend, San Diego, CA, USA), and CD11b APC (Biolegend, San Diego, CA, USA). Monocytes were defined as CD45+/HLA-DR+ and then subdivided based on the expression of CD14 and CD16 into classical (CD14++/CD16−), intermediate (CD14++/CD16+), and non-classical (CD14+/CD16++) (Supplemental Figure S1). Monocyte subpopulations were then classified based on their expression of CCR2, CD11b, and CD42. All samples were stained after lysis of the red blood cells and staining buffers included FC blocking to reduce nonspecific staining. Samples were acquired using an LSR II (Becton Dickinson, Mountainview, CA, USA) system and analyzed using FlowJo (version 7) software to generate dot plots and analyze the data.

4.3. Luminex Assay

At the pre-defined timepoints, plasma was collected using the PB collection protocol detailed earlier. Inflammatory biomarkers, such as interleukin-1 beta (IL-1β), IL-6, granulocyte-macrophage colony-stimulating factor (GM-CSF), and Chemokine ligand 5, known as Regulated upon Activation, Normal T Cell Expressed and Presumably Secreted (RANTES), were quantified using the Milliplex cytokine magnetic kit (MILLIPLEX MAP for Luminex xMap Technology, Millipore, Burlington, MA, USA) according to the manufacturer's protocol.

4.4. Statistical Analysis

Baseline demographic characteristics were presented as mean with standard error (SE), median with interquartile range, or frequencies with percentages as appropriate. Patients were divided based on the median number of monocyte subpopulations. Patients who were above the median were considered the high-level group, and patients who were below the median were considered the low-level group. A univariate Cox proportional hazards model was performed to estimate the hazard ratio of our composite clinical endpoint in the high- compared to low-level groups for these parameters. Another univariate Cox proportional hazards model, including age, sex, body mass index, diabetes mellitus, hypertension, congestive heart failure, troponin-I, previous MI, stroke, chronic kidney disease, and peripheral vascular disease, was performed to estimate the hazards ratio of the same composite endpoint. Inflammatory cells that were significantly predictable to the composite endpoint in the univariate Cox proportional hazards model were entered in a multivariate Cox proportional hazards model with the baseline characteristic variables that were significantly predictable to the composite endpoint in the univariate Cox proportional hazards model. The dynamics of inflammatory cells over time were compared using repeated measure ANOVA or the Friedman test as appropriate. Because the hospital changed the troponin assay in the middle of the study, the numerical values of troponin-1 ranged from single digits to four digits. To overcome this irregularity in the data, we divided troponin-1 in each assay separately into quartiles, and each quartile received the same code in both assays. Hence, troponin-1 was analyzed as an ordinal variable rather than a numerical variable in our study. All statistical analyses were performed using R studio, version 1.4.1106 (R Foundation for Statistical Computing) and IBM SPSS Statistics, version 28.0, Armonk, NY, USA: IBM Corp.

Supplementary Materials: The following supporting information can be downloaded at: https://www.mdpi.com/article/10.3390/ijms241411342/s1.

Author Contributions: Conceptualization, A.A.-L., M.A.-A. and L.C.; methodology, A.A.-L. and M.A.-A.; software, M.A.-A.; validation, A.A.-L., W.M.T., L.C., H.T. and M.A.-A.; formal analysis, M.A.-A.; investigation, L.C., H.T. and M.A.-A.; resources, A.A.-L.; data curation, L.C., H.T. and M.A.-A.; writing—original draft preparation, A.A.-L.; writing—review and editing, A.A.-L., M.A.-A., W.M.T. and E.S.; visualization, A.A.-L. and M.A.-A.; supervision, A.A.-L.; project administration, M.A.-A.; funding acquisition, A.A.-L. All authors have read and agreed to the published version of the manuscript.

Funding: This research was funded by NIH Grant R01 HL124266.

Institutional Review Board Statement: The study protocol complies with the Declaration of Helsinki and was approved by the University of Kentucky's Institutional Review Board and Ethics Committees; approval # 46103.

Informed Consent Statement: Informed consent was obtained from all subjects involved in the study.

Data Availability Statement: All data included in this paper are available upon reasonable request to the corresponding author.

Conflicts of Interest: The authors declare no conflict of interest.

References

1. Thom, T.; Haase, N.; Rosamond, W.; Howard, V.J.; Rumsfeld, J.; Manolio, T.; Zheng, Z.J.; Flegal, K.; O'Donnell, C.; Kittner, S.; et al. Heart disease and stroke statistics–2006 update: A report from the American Heart Association Statistics Committee and Stroke Statistics Subcommittee. *Circulation* **2006**, *113*, e85–e151. [PubMed]
2. Peterson, E.D.; Roe, M.T.; Chen, A.Y.; Fonarow, G.C.; Lytle, B.L.; Cannon, C.P.; Rumsfeld, J.S. The NCDR ACTION Registry-GWTG: Transforming contemporary acute myocardial infarction clinical care. *Heart* **2010**, *96*, 1798–1802. [CrossRef] [PubMed]
3. Park, J.; Choi, K.H.; Lee, J.M.; Kim, H.K.; Hwang, D.; Rhee, T.M.; Kim, J.; Park, T.K.; Yang, J.H.; Song, Y.B.; et al. Prognostic Implications of Door-to-Balloon Time and Onset-to-Door Time on Mortality in Patients with ST -Segment-Elevation Myocardial Infarction Treated with Primary Percutaneous Coronary Intervention. *J. Am. Heart Assoc.* **2019**, *8*, e012188. [CrossRef] [PubMed]
4. Sanchis-Gomar, F.; Perez-Quilis, C.; Leischik, R.; Lucia, A. Epidemiology of coronary heart disease and acute coronary syndrome. *Ann. Transl. Med.* **2016**, *4*, 256. [CrossRef] [PubMed]
5. Gabay, C.; Kushner, I. Acute-phase proteins and other systemic responses to inflammation. *N. Engl. J. Med.* **1999**, *340*, 448–454. [CrossRef]
6. Buffon, A.; Biasucci, L.M.; Liuzzo, G.; D'Onofrio, G.; Crea, F.; Maseri, A. Widespread coronary inflammation in unstable angina. *N. Engl. J. Med.* **2002**, *347*, 5–12. [CrossRef]
7. Tahir, S.; Steffens, S. Nonclassical monocytes in cardiovascular physiology and disease. *Am. J. Physiol. Cell Physiol.* **2021**, *320*, C761–C770. [CrossRef]
8. Berg, K.E.; Ljungcrantz, I.; Andersson, L.; Bryngelsson, C.; Hedblad, B.; Fredrikson, G.N.; Nilsson, J.; Björkbacka, H. Elevated CD14++CD16− monocytes predict cardiovascular events. *Circ. Cardiovasc. Genet.* **2012**, *5*, 122–131. [CrossRef]
9. Urra, X.; Villamor, N.; Amaro, S.; Gomez-Choco, M.; Obach, V.; Oleaga, L.; Planas, A.M.; Chamorro, A. Monocyte subtypes predict clinical course and prognosis in human stroke. *J. Cereb. Blood Flow. Metab.* **2009**, *29*, 994–1002. [CrossRef]
10. Auffray, C.; Fogg, D.; Garfa, M.; Elain, G.; Join-Lambert, O.; Kayal, S.; Sarnacki, S.; Cumano, A.; Lauvau, G.; Geissmann, F. Monitoring of blood vessels and tissues by a population of monocytes with patrolling behavior. *Science* **2007**, *317*, 666–670. [CrossRef]
11. Hamers, A.A.J.; Dinh, H.Q.; Thomas, G.D.; Marcovecchio, P.; Blatchley, A.; Nakao, C.S.; Kim, C.; McSkimming, C.; Taylor, A.M.; Nguyen, A.T.; et al. Human Monocyte Heterogeneity as Revealed by High-Dimensional Mass Cytometry. *Arter. Thromb. Vasc. Biol.* **2019**, *39*, 25–36. [CrossRef]
12. Rossol, M.; Kraus, S.; Pierer, M.; Baerwald, C.; Wagner, U. The CD14(bright) CD16+ monocyte subset is expanded in rheumatoid arthritis and promotes expansion of the Th17 cell population. *Arthritis Rheum.* **2012**, *64*, 671–677. [CrossRef]
13. Rogacev, K.S.; Cremers, B.; Zawada, A.M.; Seiler, S.; Binder, N.; Ege, P.; Grosse-Dunker, G.; Heisel, I.; Hornof, F.; Jeken, J.; et al. CD14++CD16+ monocytes independently predict cardiovascular events: A cohort study of 951 patients referred for elective coronary angiography. *J. Am. Coll. Cardiol.* **2012**, *60*, 1512–1520. [CrossRef]
14. Azeredo, E.L.; Neves-Souza, P.C.; Alvarenga, A.R.; Reis, S.R.; Torrentes-Carvalho, A.; Zagne, S.M.; Nogueira, R.M.; Oliveira-Pinto, L.M.; Kubelka, C.F. Differential regulation of toll-like receptor-2, toll-like receptor-4, CD16 and human leucocyte antigen-DR on peripheral blood monocytes during mild and severe dengue fever. *Immunology* **2010**, *130*, 202–216. [CrossRef]
15. Nagareddy, P.R.; Sreejit, G.; Abo-Aly, M.; Jaggers, R.M.; Chelvarajan, L.; Johnson, J.; Pernes, G.; Athmanathan, B.; Abdel-Latif, A.; Murphy, A.J. NETosis Is Required for S100A8/A9-Induced Granulopoiesis after Myocardial Infarction. *Arter. Thromb. Vasc. Biol.* **2020**, *40*, 2805–2807. [CrossRef]

16. Sreejit, G.; Abdel-Latif, A.; Athmanathan, B.; Annabathula, R.; Dhyani, A.; Noothi, S.K.; Quaife-Ryan, G.A.; Al-Sharea, A.; Pernes, G.; Dragoljevic, D.; et al. Neutrophil-Derived S100A8/A9 Amplify Granulopoiesis after Myocardial Infarction. *Circulation* **2020**, *141*, 1080–1094. [CrossRef]

17. Coelho-Lima, J.; Georgiopoulos, G.; Ahmed, J.; Adil, S.E.R.; Gaskin, D.; Bakogiannis, C.; Sopova, K.; Ahmed, F.; Ahmed, H.; Spray, L.; et al. Prognostic value of admission high-sensitivity troponin in patients with ST-elevation myocardial infarction. *Heart* **2021**, *107*, 1881–1888. [CrossRef]

18. Ibanez, B.; James, S.; Agewall, S.; Antunes, M.J.; Bucciarelli-Ducci, C.; Bueno, H.; Caforio, A.L.P.; Crea, F.; Goudevenos, J.A.; Halvorsen, S.; et al. 2017 ESC Guidelines for the management of acute myocardial infarction in patients presenting with ST-segment elevation. *Rev. Esp. Cardiol.* **2017**, *70*, 1082. [CrossRef]

19. McLeod, P.; Adamson, P.D.; Coffey, S. Do we need early risk stratification after ST-elevation myocardial infarction? *Heart* **2021**, *107*, 1852–1853. [CrossRef]

20. Tripathi, H.; Al-Darraji, A.; Abo-Aly, M.; Peng, H.; Shokri, E.; Chelvarajan, L.; Donahue, R.R.; Levitan, B.M.; Gao, E.; Hernandez, G.; et al. Autotaxin inhibition reduces cardiac inflammation and mitigates adverse cardiac remodeling after myocardial infarction. *J. Mol. Cell Cardiol.* **2020**, *149*, 95–114. [CrossRef]

21. Maekawa, Y.; Anzai, T.; Yoshikawa, T.; Asakura, Y.; Takahashi, T.; Ishikawa, S.; Mitamura, H.; Ogawa, S. Prognostic significance of peripheral monocytosis after reperfused acute myocardial infarction: A possible role for left ventricular remodeling. *J. Am. Coll. Cardiol.* **2002**, *39*, 241–246. [CrossRef] [PubMed]

22. Panizzi, P.; Swirski, F.K.; Figueiredo, J.L.; Waterman, P.; Sosnovik, D.E.; Aikawa, E.; Libby, P.; Pittet, M.; Weissleder, R.; Nahrendorf, M. Impaired infarct healing in atherosclerotic mice with Ly-6C(hi) monocytosis. *J. Am. Coll. Cardiol.* **2010**, *55*, 1629–1638. [CrossRef] [PubMed]

23. Zouggari, Y.; Ait-Oufella, H.; Bonnin, P.; Simon, T.; Sage, A.P.; Guerin, C.; Vilar, J.; Caligiuri, G.; Tsiantoulas, D.; Laurans, L.; et al. B lymphocytes trigger monocyte mobilization and impair heart function after acute myocardial infarction. *Nat. Med.* **2013**, *19*, 1273–1280. [CrossRef] [PubMed]

24. Dutta, P.; Courties, G.; Wei, Y.; Leuschner, F.; Gorbatov, R.; Robbins, C.S.; Iwamoto, Y.; Thompson, B.; Carlson, A.L.; Heidt, T.; et al. Myocardial infarction accelerates atherosclerosis. *Nature* **2012**, *487*, 325–329. [CrossRef]

25. Askari, N.; Lipps, C.; Voss, S.; Staubach, N.; Grun, D.; Klingenberg, R.; von Jeinsen, B.; Wolter, J.S.; Kriechbaum, S.; Dorr, O.; et al. Circulating Monocyte Subsets Are Associated with Extent of Myocardial Injury but not with Type of Myocardial Infarction. *Front. Cardiovasc. Med.* **2021**, *8*, 741890. [CrossRef]

26. Dutta, P.; Nahrendorf, M. Monocytes in myocardial infarction. *Arter. Thromb. Vasc. Biol.* **2015**, *35*, 1066–1070. [CrossRef]

27. Mangold, A.; Hofbauer, T.M.; Ondracek, A.S.; Artner, T.; Scherz, T.; Speidl, W.S.; Krychtiuk, K.A.; Sadushi-Kolici, R.; Jakowitsch, J.; Lang, I.M. Neutrophil extracellular traps and monocyte subsets at the culprit lesion site of myocardial infarction patients. *Sci. Rep.* **2019**, *9*, 16304. [CrossRef]

28. Tapp, L.D.; Shantsila, E.; Wrigley, B.J.; Pamukcu, B.; Lip, G.Y. The CD14++CD16+ monocyte subset and monocyte-platelet interactions in patients with ST-elevation myocardial infarction. *J. Thromb. Haemost.* **2012**, *10*, 1231–1241. [CrossRef]

29. Zhou, X.; Liu, X.L.; Ji, W.J.; Liu, J.X.; Guo, Z.Z.; Ren, D.; Ma, Y.Q.; Zeng, S.; Xu, Z.W.; Li, H.X.; et al. The Kinetics of Circulating Monocyte Subsets and Monocyte-Platelet Aggregates in the Acute Phase of ST-Elevation Myocardial Infarction: Associations with 2-Year Cardiovascular Events. *Medicine* **2016**, *95*, e3466. [CrossRef]

30. Krychtiuk, K.A.; Lenz, M.; Richter, B.; Hohensinner, P.J.; Kastl, S.P.; Mangold, A.; Huber, K.; Hengstenberg, C.; Wojta, J.; Heinz, G.; et al. Monocyte subsets predict mortality after cardiac arrest. *J. Leukoc. Biol.* **2021**, *109*, 1139–1146. [CrossRef]

31. Sarma, J.; Laan, C.A.; Alam, S.; Jha, A.; Fox, K.A.; Dransfield, I. Increased platelet binding to circulating monocytes in acute coronary syndromes. *Circulation* **2002**, *105*, 2166–2171. [CrossRef]

32. Shantsila, E.; Lip, G.Y. The role of monocytes in thrombotic disorders. Insights from tissue factor, monocyte-platelet aggregates and novel mechanisms. *Thromb. Haemost.* **2009**, *102*, 916–924.

33. World Medical Association. World Medical Association Declaration of Helsinki: Ethical principles for medical research involving human subjects. *Jama* **2013**, *310*, 2191–2194. [CrossRef]

International Journal of
Molecular Sciences

Article

Augmentation of Cathepsin Isoforms in Diabetic db/db Mouse Kidneys Is Associated with an Increase in Renal MARCKS Expression and Proteolysis

Mohammed F. Gholam [1,2], Niharika Bala [1], Yunus E. Dogan [1] and Abdel A. Alli [1,3,*]

[1] Department of Physiology and Aging, University of Florida College of Medicine, Gainesville, FL 32610, USA
[2] Department of Basic Medical Sciences, College of Medicine, King Saud bin Abdulaziz University for Health Sciences, Jeddah 22384, Saudi Arabia
[3] Department of Medicine Division of Nephrology, Hypertension and Renal Transplantation, University of Florida College of Medicine, Gainesville, FL 32610, USA
* Correspondence: aalli@ufl.edu; Tel.: +1-352-273-7877

Abstract: The expression of the myristoylated alanine-rich C-kinase substrate (MARCKS) family of proteins in the kidneys plays an important role in the regulation of the renal epithelial sodium channel (ENaC) and hence overall blood pressure regulation. The function of MARCKS is regulated by post-translational modifications including myristoylation, phosphorylation, and proteolysis. Proteases known to cleave both ENaC and MARCKS have been shown to contribute to the development of high blood pressure, or hypertension. Here, we investigated protein expression and proteolysis of MARCKS, protein expression of multiple protein kinase C (PKC) isoforms, and protein expression and activity of several different proteases in the kidneys of diabetic db/db mice compared to wild-type littermate mice. In addition, MARCKS protein expression was assessed in cultured mouse cortical collecting duct (mpkCCD) cells treated with normal glucose and high glucose concentrations. Western blot and densitometric analysis showed less abundance of the unprocessed form of MARCKS and increased expression of a proteolytically cleaved form of MARCKS in the kidneys of diabetic db/db mice compared to wild-type mice. The protein expression levels of PKC delta and PKC epsilon were increased, while cathepsin B, cathepsin S, and cathepsin D were augmented in diabetic db/db kidneys compared to those of wild-type mice. An increase in the cleaved form of MARCKS was observed in mpkCCD cells cultured in high glucose compared to normal glucose concentrations. Taken together, these results suggest that high glucose may contribute to an increase in the proteolysis of renal MARCKS, while the upregulation of the cathepsin proteolytic pathway positively correlates with increased proteolysis of MARCKS in diabetic kidneys, where PKC expression is augmented.

Keywords: MARCKS; PKC; cathepsin; proteolysis; kidney

Citation: Gholam, M.F.; Bala, N.; Dogan, Y.E.; Alli, A.A. Augmentation of Cathepsin Isoforms in Diabetic db/db Mouse Kidneys Is Associated with an Increase in Renal MARCKS Expression and Proteolysis. *Int. J. Mol. Sci.* **2023**, *24*, 12484. https://doi.org/10.3390/ijms241512484

Academic Editors: Yutang Wang and Dianna Magliano

Received: 31 March 2023
Revised: 29 July 2023
Accepted: 31 July 2023
Published: 5 August 2023

1. Introduction

Several studies have shown protein kinase C (PKC) activation in kidney cells cultured in high glucose [1,2] and in kidney tissue of diabetic animal models [3,4]. PKC activation in response to hyperglycemic conditions leads to production of the profibrotic cytokine transforming growth factor β (TGF-β1) and expansion of the extracellular matrix, which contributes to pathophysiological features of diabetic nephropathy [5]. The activation of PKC in diabetic kidneys plays an important role in pathophysiology, since PKC phosphorylates and regulates a myriad of proteins including those that regulate the actin cytoskeleton.

The myristoylated alanine-rich C-kinase substrate (MARCKS) family of proteins includes MARCKS and MARCKS-like protein 1 (MLP1); these proteins are among the most prominent substrates of PKC. In the kidneys, MARCKS and MLP1 have been shown to play an important role in maintaining the integrity of the actin cytoskeleton and regulating the function of epithelial transport mechanisms at the luminal plasma membrane. MARCKS [6–8] and MLP1 [9,10] both interact with epithelial sodium channels (ENaC) in renal epithelial cells. These proteins increase the density of ENaC at the membrane and its open probability by sequestering and increasing the local concentration of phosphatidylinositol 4,5-bisphosphate (PIP2) in close proximity to ENaC [6]. PIP2 stabilizes ENaC in an open confirmation at the membrane [10,11]. Phospholipase C (PLC) hydrolyses PIP2 and reduces its concentration at the membrane [12]. The interaction between MARCKS and ENaC in the kidneys is regulated by post-translational modifications, proteases, and protease inhibitors [7]. A previous study by our group showed a strong association between MARCKS and ENaC in diabetic db/db kidneys, which was attenuated by human alpha-1 antitrypsin treatment [13].

Although the regulation of MARCKS expression and function by PKC has been studied in various cells [14,15], its regulation in diabetic kidneys is not well understood. In order to address these knowledge gaps, the goals of this study were to (1) investigate the proteolysis and expression of MARCKS in diabetic db/db kidneys, (2) investigate changes in the expression and activity of multiple proteases, including different isoforms of cathepsins in diabetic kidneys, (3) investigate changes in PLC protein expression in diabetic kidneys, and (4) investigate whether acute high glucose treatment of ENaC expressing renal epithelial cells itself induces proteolysis of MARCKS.

2. Results

2.1. Increased Proteolysis of MARCKS Proteins in Diabetic db/db Kidneys

First, to confirm these animals were diabetics, we measured their blood glucose concentrations as shown in Table 1. The mean blood glucose concentration for the wild-type mice was 115.14 ± 14.84 mg/dL and for the db/db mice was 536.14 ± 12.7 mg/dL. Furthermore, urinary albumen and creatinine were comparable between the groups as shown in Table 1.

Table 1. Characteristics between groups of mice.

Group	Wild-Type	db/db	*p*-Value
Urinary albumin (mg/dL)	0.0588 ± 0.0114	0.0669 ± 0.0253	0.77
Urinary creatinine (mg/dL)	9.187 ± 2.206	8.327 ± 1.771	0.76
Body weight (gm)	23.484 ± 0.513	34.624 ± 2.653	0.0053
Blood glucose (mg/dL)	115.143 ± 14.838	536.143 ± 12.695	8.581×10^{-11}

The subcellular localization of MARCKS is known to be regulated by phosphorylation and proteolysis. Therefore, we investigated changes in the cleaved form of MARCKS in the kidneys of diabetic db/db mice and wild-type littermate mice. There was more MARCKS proteolysis, indicated by an increase in the 37 kDa band in diabetic db/db kidneys compared to those of wild-type mice, as shown in a Western blot that was probed for various forms of MARCKS using a recombinant antibody against the MARCKS protein (Figure 1). Additionally, there was less abundance of a 75–80 kDa form of MARCKS in the diabetic db/db kidneys, while the 100 kDa immunoreactive band did not show an appreciable difference between the two groups (Figure 1).

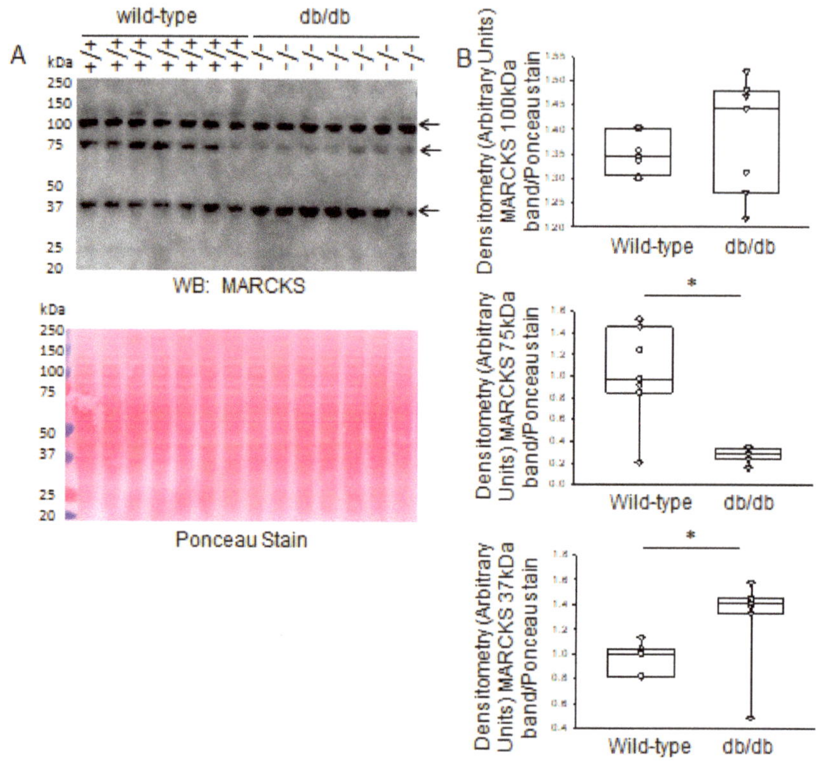

Figure 1. Western blot and densitometric analysis of myristoylated alanine-rich C-kinase substrate (MARCKS) protein expression in the kidneys of healthy wild-type and diabetic db/db mice. (**A**) Western blot for MARCKS protein. Arrows indicate immunoreactive bands corresponding to the uncleaved and cleaved forms of MARCKS protein. Ponceau stain was used to assess lane loading. (**B**) Densitometric analysis of the immunoreactive bands in panel A indicated by an arrow. *N* = 7 mice in each group. * represents a *p*-value of <0.05.

2.2. Increased PKC Isoform Expression in Diabetic db/db Kidneys

PKC activation in diabetic kidneys is known. For example, it has been reported that high glucose causes activation of PKC-alpha and PKC-epsilon isoforms in whole kidney lysates of a streptozotocin-induced rat model of type 1 diabetes [16]. Here, we investigated whether PKC isoforms are increased in the kidney cortex of db/db mice representing a model of type 2 diabetes. Protein expression of PKC-alpha in the kidneys did not show an appreciable difference between diabetic db/db mice and wild-type littermate mice (Figure 2).

Next, we investigated changes in the protein expression of PKC-delta in the diabetic db/db kidneys compared to those of wild-type littermate mice. Western blot and densitometric analysis showed a greater amount of PKC-delta protein being expressed in the diabetic db/db kidneys compared to the healthy wild-type kidneys (Figure 3).

PKC epsilon is also expressed in the mouse kidneys. Similar to PKC delta, Western blot and densitometric analysis showed greater levels of PKC epsilon protein being expressed in the kidneys of diabetic db/db mice compared to those of healthy wild-type control mice (Figure 4).

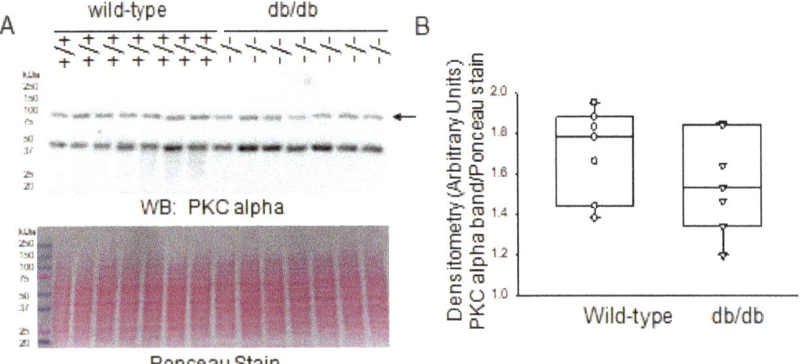

Figure 2. Western blot and densitometric analysis of PKC alpha protein expression in the kidneys of healthy wild-type and diabetic db/db mice. (**A**) Western blot for PKC alpha protein. Arrow indicates the immunoreactive band corresponding to PKC alpha protein. Ponceau stain was used to assess lane loading. (**B**) Densitometric analysis of the immunoreactive band in panel A indicated by an arrow. *N* = 7 mice in each group.

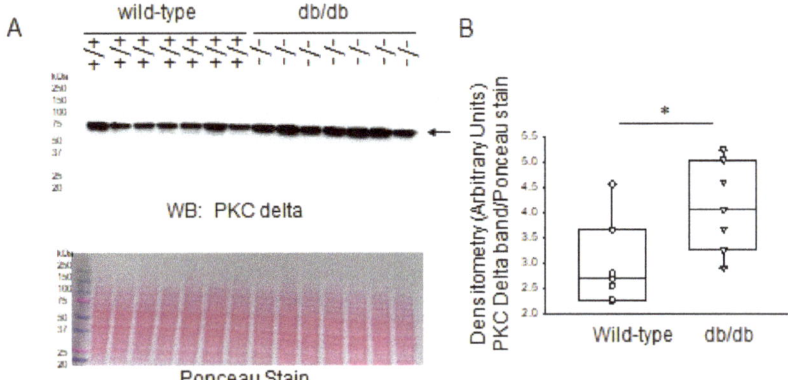

Figure 3. Western blot and densitometric analysis of PKC delta protein expression in the kidneys of healthy wild-type and diabetic db/db mice. (**A**) Western blot for PKC delta protein. Arrow indicates the immunoreactive band corresponding to PKC delta protein. Ponceau stain was used to assess lane loading. (**B**) Densitometric analysis of the immunoreactive band in panel A indicated by an arrow. *N* = 7 mice in each group. * represents a *p*-value of <0.05.

2.3. Expression Levels of Proteases in Diabetic db/db Kidneys Compared to Kidneys of Wild-Type Control Mice

In addition to being regulated by PKC phosphorylation, the function of MARCKS is regulated by proteolysis. Our group previously investigated the role of MARCKS and ENaC proteolysis in the kidneys [8]. Calpain-1 and calpain-2 protein expression and activity in cardiac mitochondria is known to be increased in diabetes leading to myocardial injury and dysfunction [17]. The expression and activities of other proteases, including cathepsins, in diabetic kidneys compared to healthy kidneys have not been fully investigated. Here, we focused on investigating changes in expression and activity of specific proteases within the kidneys of diabetic and healthy mice that are known to cleave MARCKS and/or ENaC. First, we investigated changes in kidney kallikrein 1 (KLK1) protease expression between the two groups. Western blot and densitometric analysis did not show an appreciable change in KLK1 between the two groups (Figure 5).

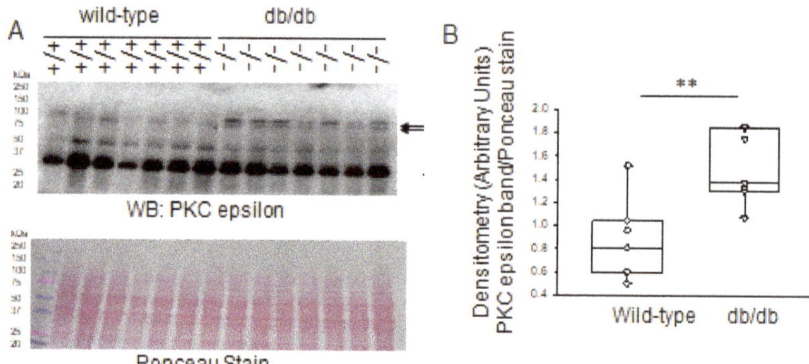

Figure 4. Western blot and densitometric analysis of PKC epsilon protein expression in the kidneys of healthy wild-type and diabetic db/db mice. (**A**) Western blot for PKC epsilon protein. Arrow indicates the immunoreactive band corresponding to PKC epsilon protein. Ponceau stain was used to assess lane loading. (**B**) Densitometric analysis of the immunoreactive band in panel A indicated by an arrow. $N = 7$ mice in each group. ** represents a *p*-value of <0.01.

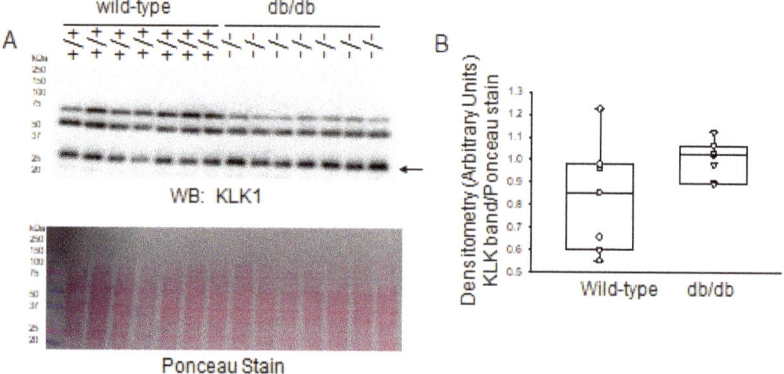

Figure 5. Western blot and densitometric analysis of Kallikrein 1 (KLK1) protein expression in the kidneys of healthy wild-type and diabetic db/db mice. (**A**) Western blot for KLK1 protein. Arrow indicates the immunoreactive band corresponding to kallikrein 1 protein. Ponceau stain was used to assess lane loading. (**B**) Densitometric analysis of the immunoreactive band in panel A indicated by an arrow. $N = 7$ mice in each group.

Next, we investigated changes in the protein expression of furin between the two groups. There were significantly lower levels of furin activity in the diabetic db/db kidneys compared to the kidneys of wild-type mice (Figure 6).

It is well established that MARCKS is cleaved by cathepsins. However, protein expression and activity of various members of the cathepsin family has not been investigated in diabetic db/db kidneys compared to those of healthy wild-type mice. Western blot and densitometric analysis showed greater basal levels of cathepsin B protein expression in the kidneys of diabetic db/db mice compared to the control group (Figure 7A,B). Consistent with the changes in cathepsin B protein expression, there was greater cathepsin B activity in the diabetic db/db kidneys compared to those of wild-type littermate mice (Figure 7C).

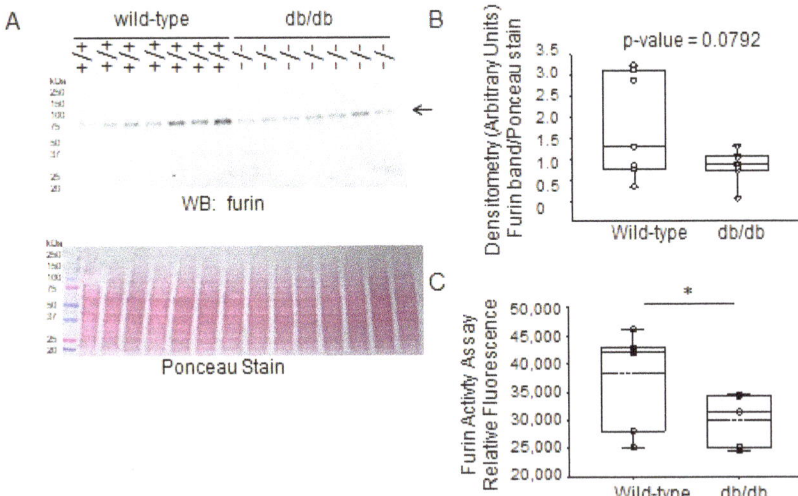

Figure 6. Western blot and densitometric analysis of furin protein expression in the kidneys of healthy wild-type and diabetic db/db mice. (**A**) Western blot for furin protein. Arrow indicates the immunoreactive band corresponding to furin protein. Ponceau stain was used to assess lane loading. (**B**) Densitometric analysis of the immunoreactive bands in panel A indicated by an arrow. (**C**) Furin activity (shown as relative fluorescence) in kidney lysates from diabetic db/db mice compared to wild-type mice. $N = 7$ mice in each group. * represents a *p*-value of <0.05.

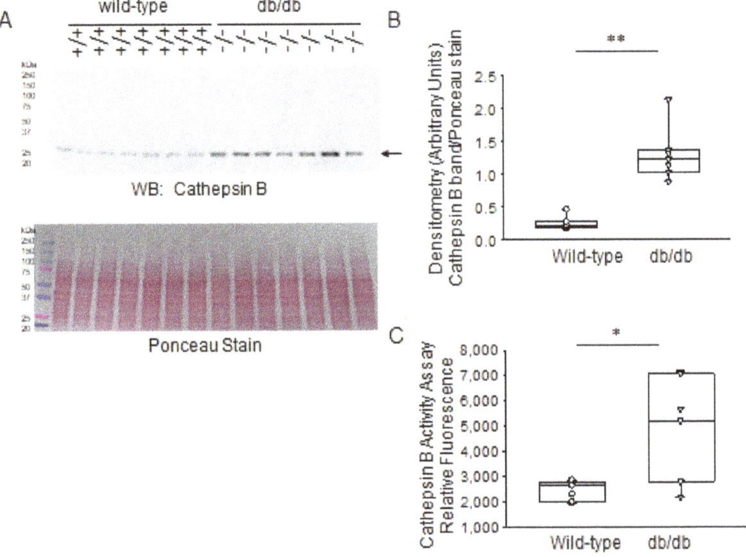

Figure 7. Western blot and densitometric analysis of cathepsin B protein expression in the kidneys of healthy wild-type and diabetic db/db mice. (**A**) Western blot for cathepsin B protein. Arrow indicates the immunoreactive band corresponding to cathepsin B protein. Ponceau stain was used to assess lane loading. (**B**) Densitometric analysis of the immunoreactive band in panel A indicated by an arrow. (**C**) Cathepsin B activity (shown as relative fluorescence) in kidney lysates from diabetic db/db mice compared to wild-type mice. $N = 7$ mice in each group. * represents a *p*-value of <0.05. ** represents a *p*-value of <0.01.

Cathepsin D is another abundantly expressed member of the cathepsin family that is expressed in the kidneys. Similar to cathepsin B, Western blot and densitometric analysis showed greater basal levels of cathepsin D protein expression in the diabetic db/db kidneys compared to those of healthy wild-type littermate mice (Figure 8).

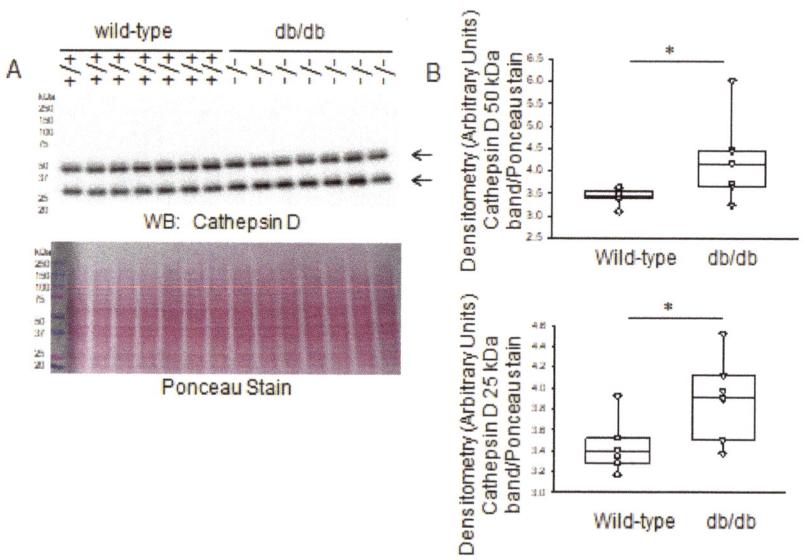

Figure 8. Western blot and densitometric analysis of cathepsin D protein expression in kidneys of healthy wild-type and diabetic db/db mice. (**A**) Western blot for cathepsin D protein. Arrow indicates the immunoreactive band corresponding to cathepsin D protein. Ponceau stain was used to assess lane loading. (**B**) Densitometric analysis of the immunoreactive bands in panel A indicated by an arrow. N = 7 mice in each group. * represents a p-value of <0.05.

Next, we examined the expression of another prominent cathepsin family member in the mouse kidneys. Similar to cathepsin B and cathepsin D, Western blot and densitometric analysis showed greater basal levels of cathepsin S protein expression in the diabetic db/db kidneys compared to the kidneys of healthy wild-type mice (Figure 9A,B). The activity levels of cathepsin S was shown to be comparable between the two groups (Figure 9C).

Next, we compared the expression of prostasin in the kidneys of diabetic db/db mice compared to healthy wild-type mice. As shown in Figure 10, the levels of prostasin were comparable between the two groups.

2.4. Expression Levels of Phospholipases in Diabetic db/db Kidneys Compared to Kidneys of Wild-Type Control Mice

In addition to PKC-mediated phosphorylation of serine residues within the effector domain of MARCKS and proteolysis of the protein by various proteases, the interaction between basic amino acids within the effector domain of MARCKS and anionic phospholipid phosphates (e.g., PIP2) within the plasma membrane plays a role in MARCKS being associated with the inner leaflet of the lipid bilayer. Therefore, we investigated changes in the expression of two different members of the phospholipase C family of proteins that are responsible for hydrolyzing PIP2 and regulating their availability at the membrane. As shown in Figure 11, basal protein expression levels of phospholipase C beta 3 were lower in the diabetic db/db kidneys compared to those of healthy wild-type littermate mice.

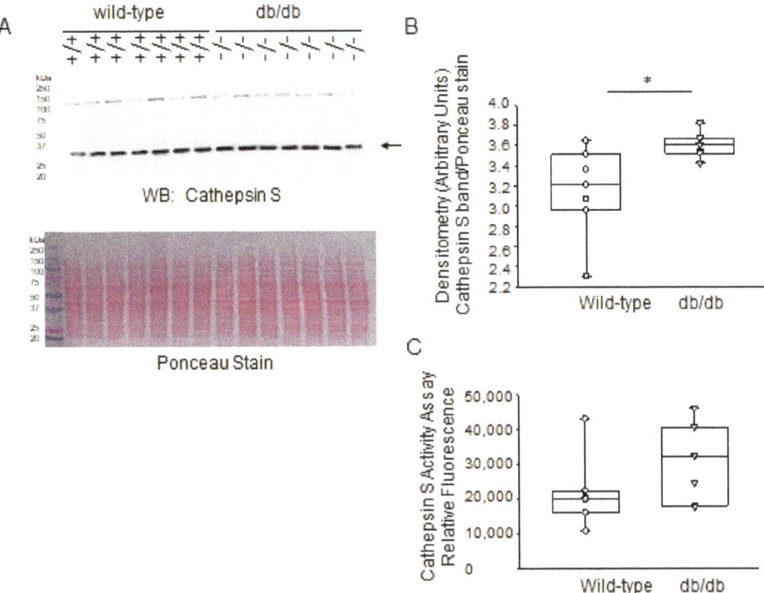

Figure 9. Western blot and densitometric analysis of cathepsin S protein expression in the kidneys of healthy wild-type and diabetic db/db mice. (**A**) Western blot for cathepsin S protein. Arrow indicates the immunoreactive band corresponding to cathepsin S protein. Ponceau stain was used to assess lane loading. (**B**) Densitometric analysis of the immunoreactive band in panel A indicated by an arrow. (**C**) Cathepsin S activity (shown as relative fluorescence) in kidney lysates from diabetic db/db mice compared to wild-type mice. $N = 7$ mice in each group. * represents a *p*-value of <0.05.

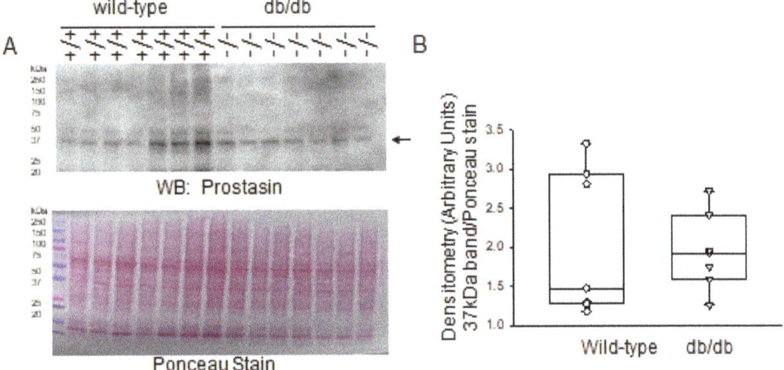

Figure 10. Western blot and densitometric analysis of prostasin protein expression in the kidneys of healthy wild-type and diabetic db/db mice. (**A**) Western blot for Prostasin protein. Arrow indicates the immunoreactive band corresponding to Prostasin protein. Ponceau stain was used to assess lane loading. (**B**) Densitometric analysis of the immunoreactive band in panel A indicated by an arrow. $N = 7$ mice in each group.

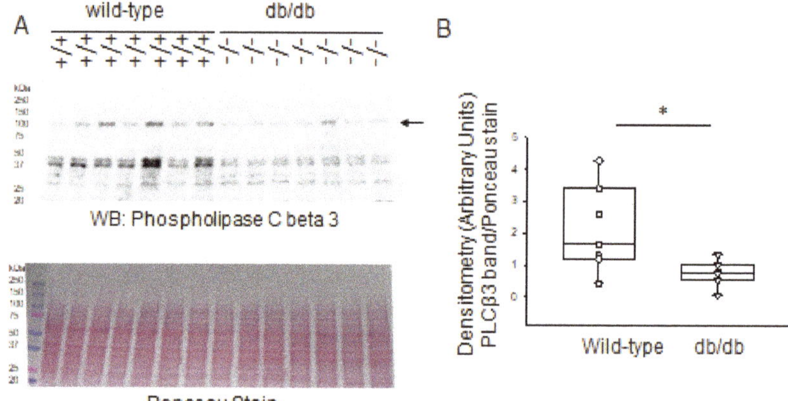

Figure 11. Western blot and densitometric analysis of phospholipase C beta 3 protein expression in the kidneys of healthy wild-type and diabetic db/db mice. (**A**) Western blot for phospholipase C beta 3 protein. Arrow indicates the immunoreactive band corresponding to phospholipase C beta 3 protein. Ponceau stain used to assess lane loading. (**B**) Densitometric analysis of the immunoreactive band in panel A indicated by an arrow. $N = 7$ mice in each group. * represents a p-value of <0.05.

Next, we investigated protein expression of phospholipase C gamma 1 in the two groups. Converse to phospholipase beta 3, basal protein expression levels of phospholipase C gamma 1 were greater in the kidneys of diabetic db/db mice compared to wild-type mice (Figure 12).

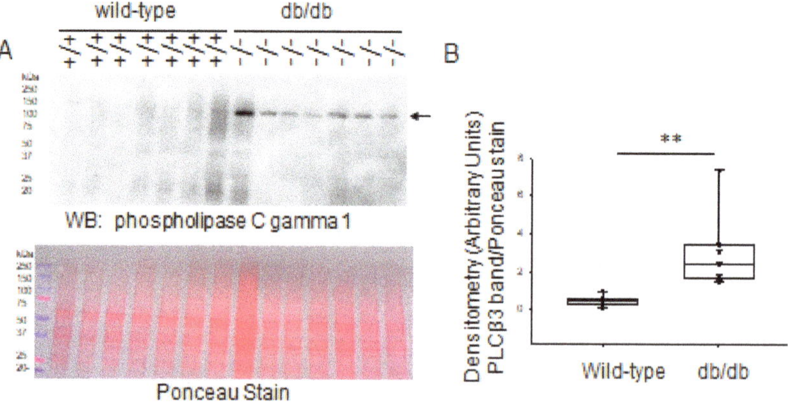

Figure 12. Western blot and densitometric analysis of phospholipase C gamma 1 protein expression in the kidneys of healthy wild-type and diabetic db/db mice. (**A**) Western blot for phospholipase C gamma 1 protein. Arrow indicates the immunoreactive band corresponding to phospholipase C gamma 1 protein. Ponceau stain used to assess lane loading. (**B**) Densitometric analysis of the immunoreactive band in panel A indicated by an arrow. $N = 7$ mice in each group. ** represents a p-value of <0.01.

2.5. Expression and Proteolysis of MARCKS Protein in Mouse Cortical Collecting Duct Cells Cultured in Normal Glucose Compared to High Glucose Conditions

To investigate whether the increase in MARCKS expression and proteolysis could be due to high glucose, we cultured mpkCCD cells in normal and high concentrations of glucose before harvesting the cells for protein and then probing for MARCKS protein expression by Western blot and assessing the immunoreactive bands by densitometric analysis. As shown in Figure 13, MARCKS protein was augmented in mpkCCD cells cultured in high glucose conditions compared to normal glucose conditions (Figure 13). Another group of cells was treated with mannitol as an osmotic control. The expression and proteolysis of MARCKS protein were comparable in cells treated with normal glucose and mannitol (Figure 13).

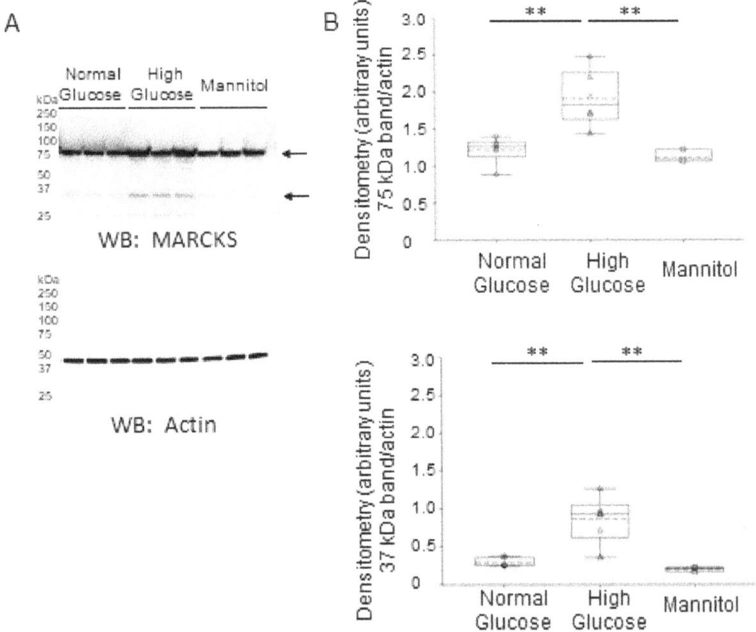

Figure 13. Western blot and densitometric analysis of MARCKS protein expression in mpkCCD cells. (**A**) Western blot for MARCKS protein in mpkCCD cells treated with normal glucose (5.5 mM), high glucose conditions (25 mM), or mannitol (25 mM). Western blot for actin was used to assess lane loading. Arrows indicate the immunoreactive bands for the uncleaved and cleaved forms of MARCKS protein. (**B**) Densitometric analysis of the immunoreactive bands in panel A normalized to actin. $N = 6$ independent experiments for the normal glucose and high glucose groups and $N = 3$ independent experiments for the mannitol control group. The top arrow indicates the uncleaved form (75 kDa) of MARCKS protein and the bottom arrow indicates the cleaved form (37 kDa) of MARCKS protein. ** represents a *p*-value of <0.01.

To corroborate the changes in protein expression of MARCKS and specific proteases, kinases, and phospholipases observed by Western blot and densitometric analysis, we performed immunohistochemistry. As shown in Figure 14, MARCKS protein expression was elevated in the diabetic db/db kidneys compared to the healthy wild-type kidneys. Similarly, PLC gamma 1, PKC delta, PKC epsilon, and cathepsin B, S, and D isoforms were elevated in the diabetic db/db kidneys compared to the control group (Figure 14).

To further demonstrate that cathepsin B regulates MARCKS protein in the mouse collecting duct, we transfected mpkCCD cells with non-targeting (NT) control siRNA or cathepsin B specific siRNA. Western blot and densitometric analysis showed MAR-CKS protein was reduced in cells with an siRNA-mediated knockdown of cathepsin B. (Figure 15).

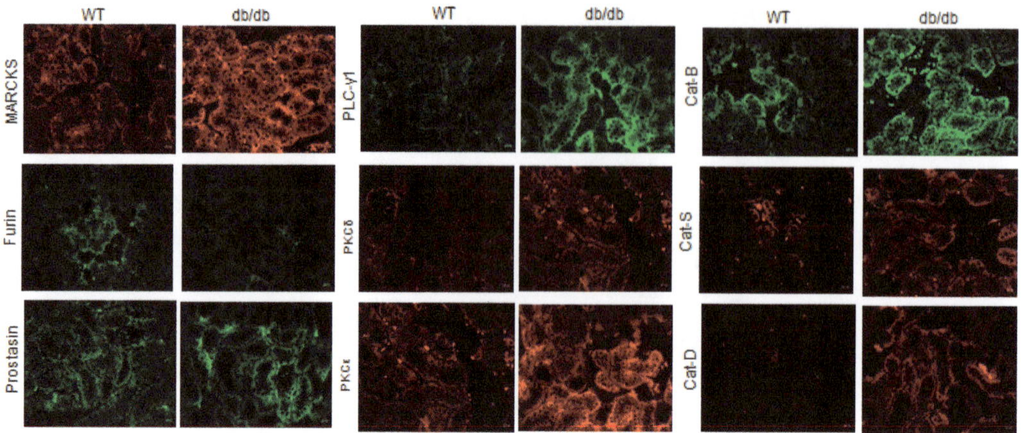

Figure 14. Immunohistochemistry of MARCKS, furin, prostasin, PKC and cathepsin isoforms in the diabetic and healthy mouse kidneys. N = 4 mice in each group. The scale bar represents 200 μm for each image. Cat B represents cathepsin B, Cat S represents cathepsin S, and Cat D represents cathepsin D.

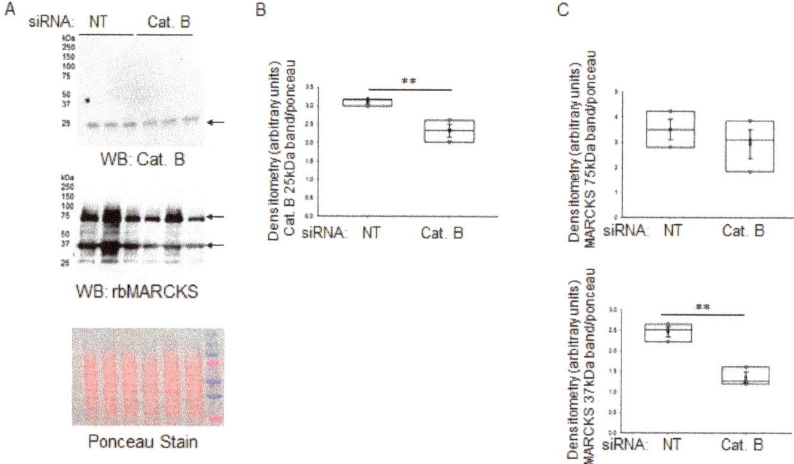

Figure 15. Representative Western blot of cathepsin B and MARCKS proteins after siRNA-mediated knockdown in mpkCCD cells. (**A**) Representative Western blot showing protein expression of cathepsin B (Cat. B) protein (top) and MARCKS protein (middle) from mpkCCD cells transfected with non-targeting (NT) siRNA or cathepsin B siRNA. Arrows indicate the immunoreactive bands for cathepsin B protein and MARCKS protein. Ponceau staining (bottom) was used to assess lane loading. (**B**) Densitometric analysis of the immunoreactive cathepsin B band in panel A normalized to Ponceau staining. (**C**) Densitometric analysis of the MARCKS bands in panel A normalized to Ponceau staining. N = 3 representing 3 independent siRNA experiments. ** represents a *p*-value of <0.01.

3. Discussion

We previously investigated renal MARCKS protein expression and proteolysis in salt-loaded hypertensive diabetic db/db mice [13]. To our knowledge this is the first study showing greater basal expression and activity of various proteases, and proteolysis of MARCKS protein, in diabetic db/db kidneys compared to the kidneys of healthy wild-type littermate mice. Here we show that a 75–80 kDa immunoreactive band corresponding to the uncleaved form of MARCKS is decreased, while a 37 kDa cleaved form of MARCKS is increased in the diabetic db/db kidneys compared to those of wild-type mice. We also show that the protein expression and activity of cathepsin B, cathepsin D, and cathepsin S are augmented in the diabetic db/db kidneys compared to those of healthy wild-type mice. These results suggest that hallmarks associated with diabetes may play a role in the upregulation of specific members of the cathepsin family of proteases and proteolysis of renal MARCKS. To investigate whether the increase in MARCKS proteolysis could at least in part be due to high glucose conditions, as in the hyperglycemic environment of the diabetic db/db kidneys, we placed mouse cortical collecting duct cells in either normal or high glucose and then measured MARCKS proteolysis by Western blotting. These experiments suggested that the cleavage of renal MARCKS may be due to high glucose conditions.

Since MARCKS is a prominent substrate of PKC, we also investigated changes in protein expression of various PKC isoforms in this study. PKC is a family of serine/threonine protein kinases that are crucial in regulating many biological processes, such as cell division, growth, and apoptosis, as well as cellular responses to environmental stressors. PKC is made up of a family of at least 12 isoforms that are divided into three groups based on how they respond to calcium and phospholipids [18]. Each PKC isoform may have distinct activities, but they have not yet been fully characterized due to the lack of specific inhibitors for each isoform and its distinct tissue distribution and subcellular localization [18]. PKC alpha, PKC delta, and PKC epsilon were reported to be activated in rat glomeruli 2 weeks after streptozotocin treatment [19]. Here, we investigated basal protein expression levels of PKC alpha, delta, and epsilon in the diabetic db/db kidneys compared to those of wild-type mice. Western blot and densitometric analyses showed that PKC delta and PKC epsilon, but not PKC alpha, were elevated in the diabetic db/db kidneys compared to the control group. The increase in protein expression of various PKC isoforms in the kidneys of diabetic db/db mice that was observed in this study is consistent with other published studies.

This is also the first study to show differential protein expression of PLC isoforms in diabetic and healthy kidneys. PIP2 plays an essential role in stabilizing ENaC at the luminal membrane through the interaction with MARCKS protein. PIP2 levels in the kidneys would presumably be elevated when PLC expression and activity are low since this enzyme hydrolyses PIP2 to yield the second messengers DAG and IP3. Although DAGs activate PKC, it is generated from either various intracellular lipid species or synthesized during de novo lipid biosynthesis of triacylglycerols and phospholipids and during catabolism of triacylglycerols stored in the endoplasmic reticulum or cytoplasmic associated lipid droplets [20]. The results presented here for the differential expression of PLCβ3 and PLCγ1 protein expression in the diabetic db/db kidneys compared to the healthy wild-type kidneys show that there is a need to further investigate the role of PLC isoforms in diabetic kidneys to better understand the interplay between PIP2 availability at the membrane and subcellular localization of MARCKS protein in diabetic kidneys.

Although this study presents novel findings on the regulation of MARCK protein in diabetic kidneys in a putative mechanism involving increased protein expression and activity of multiple cathepsins and differential protein expression of PLC isoforms, there are some limitations. First, we did not investigate whether the increased proteolysis of MARCKS by cathepsins prevents PKC mediated phosphorylation in diabetic kidneys. Our previous study suggests calpain-2 mediated cleavage of the carboxy terminal tail of MARCKS prevents its phosphorylation by PKC and translocation from the membrane [8]. Another limitation is that we did not investigate whether increased MARCKS proteolysis in diabetic db/db kidneys increases the risk of developing diabetic nephropathy. In this study we used young adult 8-week-old diabetic db/db mice, and diabetic nephropathy is typically studied several weeks later in these mice [21–23]. Finally, this study did not investigate how altered proteolysis of MARCKS affects the dynamics and organization of the actin cytoskeleton in the diabetic db/db kidney.

Collectively, these results suggest that MARCKS cleavage and association with the plasma membrane is augmented in diabetic kidneys, presumably by increased protein expression and activity of multiple cathepsin family members and decreased PLC beta 3 protein expression (Figure 16). Although high glucose appears to increase renal MARCKS proteolysis in vitro, the subcellular localization of renal MARCKS and its function at the plasma membrane warrant additional investigation.

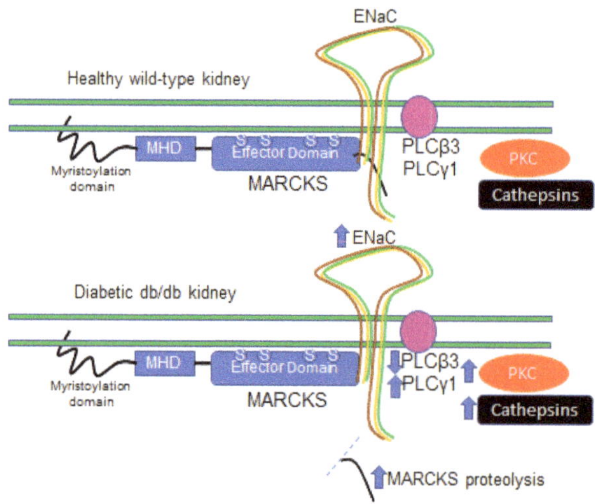

Figure 16. Proposed model for the regulation of MARCKS in diabetic db/db kidneys. The expression of protein kinase C (PKC) delta and epsilon isoforms is augmented in diabetic kidneys. Additionally, the expression and activity of cathepsin B, D and S isoforms are increased in diabetic kidneys. MARCKS proteolysis is greater in diabetic kidneys compared to the kidneys of healthy wild-type littermate mice. There are lower basal levels of phospholipase C β3 (PLCβ3) protein expression and greater basal levels of PLCγ1 protein expression in the kidneys of diabetic db/db mice compared to wild-type littermate mice. The proteolysis of MARCKS increases the density of the protein at the apical plasma membrane, presumably, by preventing PKC from accessing and phosphorylating serine residues within the effector domain of the protein. At the apical plasma membrane, MARCKS plays an essential role in maintaining the organization of the actin cytoskeleton and stabilizing ENaC in an open confirmation.

4. Materials and Methods

4.1. Animal Studies

Seven 8-week-old male db/db mice (BKS.Cg-Dock7 m +/+ Leprdb 116/J; Stock No: 000642) and seven age-matched wild-type littermate male mice were purchased from the Jackson Laboratory (Bar Harbor, ME, USA). The mice were individually placed in metabolic cages for 24 h urine collection for one week. These animal studies were performed under an approved University of Florida Institutional Animal Care and Use Committees protocol and National Institutes of Health "Guide for the Care and Use of Laboratory Animals" guidelines.

4.2. Tail Bleeds for Measuring Blood Glucose

Blood glucose was measured in each group of mice using a digital glucometer (CVS Health, Woonsocket, SD, USA).

4.3. Measurement of Urinary Albumin

Albumin concentration was measured using an albumin assay kit (Proteintech; Rosemont, IL, USA) (Table 2) according to the manufacturer's instructions.

Table 2. Sources of assays used in this study. Assay kits were purchased from Abcam (Waltham, MA, USA), BPS Bioscience (San Diego, CA, USA), and Proteintech (Rosemont, IL, USA).

Assay	Manufacturer	Catalog Number
Cathepsin B	Abcam	ab65300
Cathepsin S	Abcam	ab65307
Furin	BPS Bioscience	78040
Albumin	Proteintech	KE00076
Creatinine	Abcam	ab204537

4.4. Measurement of Urinary Creatinine

Creatinine concentration was measured using a creatinine assay kit (Abcam; Waltham, MA, USA) (Table 2) according to the manufacturer's instructions.

4.5. Cathepsin Activity Assays

Cathepsin B and S activities from soluble fraction kidney cortex lysates were determined by performing cathepsin activity assays (Table 2) while following the manufactures instructions.

4.6. Furin Activity Assay

Furin activity in soluble fraction kidney cortex lysates was determined after performing a furin activity assay (Table 2) while following the manufactures instructions.

4.7. Tissue Homogenization and BCA Assay

Fifty milligrams of kidney cortex tissue were homogenized in 500 µL of tissue protein extraction reagent (TPER; Thermo Fisher Scientific, Waltham, MA, USA) using an Omni TH homogenizer (Warrenton, VA, USA) after being washed in $1\times$ PBS (prepared from a $10\times$ PBS solution. The tissue lysates were centrifuged at 13,000 rpm for 10 min at room temperature in a micromax benchtop centrifuge (Thermo IEC) before 450 µL of the supernatant was subject to ultracentrifugation at 34,000 rpm for 30 min at 4 °C using an optima L-90K ultracentrifuge (Beckman Coulter; Schaumburg, IL, USA) and an SW55 rotor (Beckman Coulter; Schaumburg, IL, USA). The pellets were resuspended in 200 µL TPER before being sonicated twice for 3 s at a time while on ice. A bicinchoninic acid protein assay (BCA) (Thermo Fisher Scientific) was used to determine the protein concentrations.

4.8. Western Blot and Densitometric Analysis

Total protein in the amount of 50µg per sample was separated on 4–20% tris HCl polyacrylamide gels at 200 V for 50 min at room temperature using a Criterion electrophoresis machine (BioRad, Hercules, CA, USA). The separated proteins were transferred onto nitrocellulose membranes (Thermo Fisher Scientific) in Towbin buffer (25 mM tris, 192 mM glycine, 20% methanol (*vol/vol*) using a Criterion transfer apparatus (BioRad). The membranes were blocked in 5% nonfat dry milk (prepared in $1\times$ tris-buffered saline (TBS)) (Bio-Rad) (*w/v*) for 1 h at room temperature. Next, the blots were incubated in a 1:1000 dilution of primary antibody (Table 3) while rocking overnight at 4 °C. Afterwards, the membranes were rinsed three times for five minutes in $1\times$ TBS, and then incubated with horseradish peroxidase-conjugated goat anti-rabbit secondary antibody (BioRad) (diluted 1:3000 in blocking solution) for 1 h at room temperature. The membranes were developed on an imager (Bio-Rad) after being washed four times with $1\times$ TBS for 4 min and then incubated with ECL reagent (BioRad) for 7 min at room temperature. Image J software (IJ 1.46r) was used for the densitometric analysis.

Table 3. Sources of antibodies used in this study. Antibodies were purchased from Abcam (Waltham, MA, USA), Cell Signaling Technologies (Danvers, MA, USA), ThermoFisher Scientific Invitrogen, (Waltham, MA, USA), Boster Biological Technology (Pleasanton, CA, USA), and Santa Cruz Biotechnologies (Dallas, TX, USA).

Antibody	Application	Manufacturer	Catalog Number
MARCKS	WB	Abcam	ab72459
Cathepsin B	WB, IHC	Cell Signaling Technologies	31718
Cathepsin D	WB	Cell Signaling Technologies	69854
Cathepsin S	WB	Abcam	ab232740
Prostasin	WB, IHC	ThermoFisher Scientific Invitrogen	PA5-27977
Furin	WB	Cell Signaling Technologies	64709
PKC alpha	WB	Cell Signaling Technologies	2056
PKC delta	WB	Cell Signaling Technologies	9616
PKC epsilon	WB	Cell Signaling Technologies	2683
KLK1	WB	Boster Biological Technology	PA1709
PLCβ3	WB	Cell Signaling Technologies	14247
PLCγ1	WB, IHC	Cell Signaling Technologies	5690
MARCKS	IHC	Santa Cruz Biotechnologies	sc-100777
Furin	IHC	Santa Cruz Biotechnologies	sc-133142
Cathepsin S	IHC	Santa Cruz Biotechnologies	sc-271619
Cathepsin D	IHC	Santa Cruz Biotechnologies	sc-377299
PKC delta	IHC	Santa Cruz Biotechnologies	sc-8402
PKC epsilon	IHC	Santa Cruz Biotechnologies	sc-1681

4.9. Immunohistochemistry

The left kidney from each mouse was fixed in formalin for 24 h, rinsed with $1\times$ phosphate buffered saline (PBS) (137 mM NaCl, 2.7 mM KCl, 10 mM Na_2HPO_4, 1.8 mM KH_2PO_4), placed in 70% ethanol, and then cut into 4-micrometer sections for immunohistochemistry. Paraffin-embedded kidney tissue sections were subjected to two exchanges of xylene (Fisher Scientific, Pittsburgh, PA, USA), two exchanges of 100% ethanol, one exchange of 95% ethanol, one exchange of 70% ethanol, one exchange of 50% ethanol, and finally one exchange of type-1 water for intervals of three minutes each. Next, the slides were boiled in citrate buffer (Vector Labs, Inc., Burlingame, CA, USA) for 20 min and then washed in type-1 water for three minutes. The slides were washed with $1\times$ PBS for a period of five minutes and then blocked with normal horse serum at a concentration of 2.5% (Vector laboratories, Inc.) for 20 min in a humidified chamber. Afterwards, 200 uL of primary antibody (Table 3) was applied to each tissue section and incubated for 60 min in a humidified environment. The slides were washed three times with $1\times$ PBS for a total of two minutes each time. After adding one drop of VectaFluor Duet Reagent (Vector labs, Inc.), the tissue was placed in a humidified chamber and allowed to incubate for 30 min at room temperature. The tissue was then washed three times for two minutes each with $1\times$ PBS. One drop of Vectashield anti-fade mounting solution (Vector Labs, Inc.) was added and a 22 × 22-1 glass coverslip (Fisher Scientific) was applied before the slides were imaged on an Olympus BX41 microscope equipped with a $40\times$ objective.

4.10. Culture and Treatment of mpkCCD Cells

Mouse mpkCCD cells were maintained in a 1:1 mixture of DMEM and Ham's F-12 medium (GIBCO; Grand Island, NY, USA) supplemented with 20 mM HEPES, 2 mM l-glutamine, 50 nM dexamethasone, 1 nM triiodothyronine, $1\times$ penicillin-streptomycin, and 2% fetal bovine serum (Corning-Mediatech; Woodland, CA, USA). The cells were maintained in a humidified incubation at 5% CO_2 and at 37 °C. The amount of glucose present in the complete growth media was 5.5 mM, which was considered a normal glucose level. A separate batch of mpkCCD cell media was supplemented with additional glucose (25 mM final concentration), which was considered a high glucose level.

4.11. siRNA Transfection

Non-targeting control siRNA or cathepsin B (Ctsb) siRNA siGENOME SMARTpool was purchased from Horizon Discovery Biosciences (Waterbeach, Cambridge, UK), and mpkCCD cells were seeded in 6 well plates and transected with DharmaFECT reagent (Horizon Discovery Biosciences) when 70 percent confluent according to the manufacturer's instructions.

4.12. Statistical Analysis

The data presented here are shown as mean values $\pm$ SEM. A Student's *t*-test was performed to determine whether there was a statistically significant difference between the two groups. A one-way ANOVA was used to compare more than two groups. SigmaPlot software (Jandel Scientific, Corte Madera, CA, USA) was used to plot the data. A p value of <0.05 between the groups was considered to be statistically significant.

Author Contributions: M.F.G., N.B., Y.E.D. and A.A.A. performed experiments and analyzed data. A.A.A. was involved in the supervision of this study, provided resources, and acquired funding for this project. M.F.G., N.B., Y.E.D. and A.A.A. were involved in the writing and editing of this manuscript. All authors have read and agreed to the published version of the manuscript.

Funding: This work was supported by a National Institute of Diabetes and Digestive and Kidney Diseases Grant R01 DK123078-01A1 (to A.A.A.).

Institutional Review Board Statement: All animal studies were performed under an approved Institutional Animal Care and Use Committee protocol (IACUC Study #202011157; approved 12 November 2020) at the University of Florida.

Informed Consent Statement: Not applicable.

Data Availability Statement: The individual data points from each experiment are plotted and shown within the figures of this manuscript.

Acknowledgments: We thank the University of Florida College of Medicine for providing resources and equipment to help carry out this project.

Conflicts of Interest: The authors declare no conflict of interest.

References

1. Craven, P.A.; Studer, R.K.; Negrete, H.; DeRubertis, F.R. Protein kinase C in diabetic nephropathy. *J. Diabetes Complicat.* **1995**, *9*, 241–245. [CrossRef] [PubMed]
2. Derubertis, F.R.; Craven, P.A. Activation of protein kinase C in glomerular cells in diabetes. Mechanisms and potential links to the pathogenesis of diabetic glomerulopathy. *Diabetes* **1994**, *43*, 1–8. [CrossRef] [PubMed]
3. Ha, H.; Yu, M.R.; Choi, Y.J.; Lee, H.B. Activation of protein kinase c-δ and c-ε by oxidative stress in early diabetic rat kidney. *Am. J. Kidney Dis.* **2001**, *38* (Suppl. 1), S204–S207. [CrossRef] [PubMed]
4. Murphy, M.; McGinty, A.; Godson, C. Protein kinases C: Potential targets for intervention in diabetic nephropathy. *Curr. Opin. Nephrol. Hypertens* **1998**, *7*, 563–570. [CrossRef]
5. Meier, M.; Menne, J.; Haller, H. Targeting the protein kinase C family in the diabetic kidney: Lessons from analysis of mutant mice. *Diabetologia* **2009**, *52*, 765–775. [CrossRef]
6. Alli, A.A.; Bao, H.-F.; Alli, A.A.; Aldrugh, Y.; Song, J.Z.; Ma, H.-P.; Yu, L.; Al-Khalili, O.; Eaton, D.C. Phosphatidylinositol phosphate-dependent regulation of Xenopus ENaC by MARCKS protein. *Am. J. Physiol.-Ren. Physiol.* **2012**, *303*, F800–F811. [CrossRef]
7. Alli, A.A.; Bao, H.-F.; Liu, B.-C.; Yu, L.; Aldrugh, S.; Montgomery, D.S.; Ma, H.-P.; Eaton, D.C. Calmodulin and CaMKII modulate ENaC activity by regulating the association of MARCKS and the cytoskeleton with the apical membrane. *Am. J. Physiol. Physiol.* **2015**, *309*, F456–F463. [CrossRef]
8. Montgomery, D.S.; Yu, L.; Ghazi, Z.M.; Thai, T.L.; Al-Khalili, O.; Ma, H.-P.; Eaton, D.C.; Alli, A.A. ENaC activity is regulated by calpain-2 proteolysis of MARCKS proteins. *Am. J. Physiol. Physiol.* **2017**, *313*, C42–C53. [CrossRef]
9. Song, C.; Yue, Q.; Moseley, A.; Al-Khalili, O.; Wynne, B.M.; Ma, H.; Wang, L.; Eaton, D.C. Myristoylated alanine-rich C kinase substrate-like protein-1 regulates epithelial sodium channel activity in renal distal convoluted tubule cells. *Am. J. Physiol. Physiol.* **2020**, *319*, C589–C604. [CrossRef]
10. Yue, Q.; Al-Khalili, O.; Moseley, A.; Yoshigi, M.; Wynne, B.M.; Ma, H.; Eaton, D.C. PIP2 Interacts Electrostatically with MARCKS-like Protein-1 and ENaC in Renal Epithelial Cells. *Biology* **2022**, *11*, 1694. [CrossRef]

11. Zhai, Y.-J.; Wu, M.-M.; Linck, V.A.; Zou, L.; Yue, Q.; Wei, S.-P.; Song, C.; Zhang, S.; Williams, C.R.; Song, B.-L.; et al. Intracellular cholesterol stimulates ENaC by interacting with phosphatidylinositol-4,5-bisphosphate and mediates cyclosporine A-induced hypertension. Biochim. et Biophys. *Acta BBA -Mol. Basis Dis.* **2019**, *1865*, 1915–1924. [CrossRef] [PubMed]

12. Tuna, K.M.; Liu, B.-C.; Yue, Q.; Ghazi, Z.M.; Ma, H.-P.; Eaton, D.C.; Alli, A.A. Mal protein stabilizes luminal membrane PLC-β3 and negatively regulates ENaC in mouse cortical collecting duct cells. *Am. J. Physiol. Physiol.* **2019**, *317*, F986–F995. [CrossRef] [PubMed]

13. Lugo, C.I.; Liu, L.P.; Bala, N.; Morales, A.G.; Gholam, M.F.; Abchee, J.C.; Elmoujahid, N.; Elshikha, A.S.; Avdiaj, R.; Searcy, L.A.; et al. Human Alpha-1 Antitrypsin Attenuates ENaC and MARCKS and Lowers Blood Pressure in Hypertensive Diabetic db/db Mice. *Biomolecules* **2022**, *13*, 66. [CrossRef]

14. Arnold, T.P.; Standaert, M.L.; Hernandez, H.; Watson, J.; Mischak, H.; Kazanietz, M.G.; Zhao, L.; Cooper, D.R.; Farese, R.V. Effects of insulin and phorbol esters on MARCKS (myristoylated alanine-rich C-kinase substrate) phosphorylation (and other parameters of protein kinase C activation) in rat adipocytes, rat soleus muscle and BC3H-1 myocytes. *Biochem. J.* **1993**, *295*, 155–164. [CrossRef]

15. Chappell, D.S.; Patel, N.A.; Jiang, K.; Li, P.; Watson, J.E.; Byers, D.M.; Cooper, D.R. Functional involvement of protein kinase C-βII and its substrate, myristoylated alanine-rich C-kinase substrate (MARCKS), in insulin-stimulated glucose transport in L6 rat skeletal muscle cells. *Diabetologia* **2009**, *52*, 901–911. [CrossRef] [PubMed]

16. Kang, N.; Alexander, G.; Park, J.K.; Maasch, C.; Buchwalow, I.; Luft, F.C.; Haller, H. Differential expression of protein kinase C isoforms in streptozotocin-induced diabetic rats. *Kidney Int.* **1999**, *56*, 1737–1750. [CrossRef] [PubMed]

17. Zheng, D.; Cao, T.; Zhang, L.-L.; Fan, G.-C.; Qiu, J.; Peng, T.-Q. Targeted inhibition of calpain in mitochondria alleviates oxidative stress-induced myocardial injury. *Acta Pharmacol. Sin.* **2021**, *42*, 909–920. [CrossRef] [PubMed]

18. Huwiler, A.; Schulze-Lohoff, E.; Fabbro, D.; Pfeilschifter, J. Immunocharacterization of protein kinase C isoenzymes in rat kidney glomeruli, and cultured glomerular epithelial and mesangial cells. *Nephron Exp. Nephrol.* **1993**, *1*, 19–25.

19. Babazono, T.; Kapor-Drezgic, J.; Dlugosz, J.A.; Whiteside, C. Altered expression and subcellular localization of diacylglycerol-sensitive protein kinase C isoforms in diabetic rat glomerular cells. *Diabetes* **1998**, *47*, 668–676. [CrossRef]

20. Kolczynska, K.; Loza-Valdes, A.; Hawro, I.; Sumara, G. Diacylglycerol-evoked activation of PKC and PKD isoforms in regulation of glucose and lipid metabolism: A review. *Lipids Health Dis.* **2020**, *19*, 113. [CrossRef]

21. Park, C.; Zhang, Y.; Zhang, X.; Wu, J.; Chen, L.; Cha, D.; Su, D.; Hwang, M.-T.; Fan, X.; Davis, L.; et al. PPARα agonist fenofibrate improves diabetic nephropathy in db/db mice. *Kidney Int.* **2006**, *69*, 1511–1517. [CrossRef] [PubMed]

22. Seo, J.-W.; Kim, Y.G.; Lee, S.H.; Lee, A.; Kim, D.-J.; Jeong, K.-H.; Lee, K.H.; Hwang, S.J.; Woo, J.S.; Lim, S.J.; et al. Mycophenolate Mofetil Ameliorates Diabetic Nephropathy in db/db Mice. *BioMed Res. Int.* **2015**, *2015*, 301627. [CrossRef] [PubMed]

23. Arai, H.; Tamura, Y.; Murayama, T.; Minami, M.; Yokode, M. Differential effect of statins on diabetic nephropathy in db/db mice. *Int. J. Mol. Med.* **2011**, *28*, 683–687. [CrossRef] [PubMed]

International Journal of
Molecular Sciences

Article

Tumor Growth Ameliorates Cardiac Dysfunction and Suppresses Fibrosis in a Mouse Model for Duchenne Muscular Dystrophy

Laris Achlaug, Lama Awwad, Irina Langier Goncalves, Tomer Goldenberg and Ami Aronheim *,†

Department of Cell Biology and Cancer Science, Ruth and Bruce Rappaport Faculty of Medicine, Technion—Israel Institute of Technology, P.O. Box 9649, Haifa 31096, Israel; laris1010.ab@gmail.com (L.A.); lamaaw@campus.technion.ac.il (L.A.); irinalan@campus.technion.ac.il (I.L.G.); tomergo@campus.technion.ac.il (T.G.)
* Correspondence: aronheim@technion.ac.il; Tel.: +972-77-8875454
† Current address: 7 Efron St. Bat-Galim, P.O. Box 9649, Haifa 31096, Israel.

Abstract: The interplay between heart failure and cancer represents a double-edged sword. Whereas cardiac remodeling promotes cancer progression, tumor growth suppresses cardiac hypertrophy and reduces fibrosis deposition. Whether these two opposing interactions are connected awaits to be determined. In addition, it is not known whether cancer affects solely the heart, or if other organs are affected as well. To explore the dual interaction between heart failure and cancer, we studied the human genetic disease Duchenne Muscular Dystrophy (DMD) using the MDX mouse model. We analyzed fibrosis and cardiac function as well as molecular parameters by multiple methods in the heart, diaphragm, lungs, skeletal muscles, and tumors derived from MDX and control mice. Surprisingly, cardiac dysfunction in MDX mice failed to promote murine cancer cell growth. In contrast, tumor-bearing MDX mice displayed reduced fibrosis in the heart and skeletal and diaphragm muscles, resulting in improved cardiac function. The latter is at least partially mediated via M2 macrophage recruitment to the heart and diaphragm muscles. Collectively, our data support the notion that the effect of heart failure on tumor promotion is independent of the improved cardiac function in tumor-bearing mice. Reduced fibrosis in tumor-bearing MDX mice stems from the suppression of new fibrosis synthesis and the removal of existing fibrosis. These findings offer potential therapeutic strategies for DMD patients, fibrotic diseases, and cardiac dysfunction.

Keywords: Duchenne Muscular Dystrophy; fibrosis; cardiac remodeling; cardiac dysfunction; tumor; macrophage recruitment

Citation: Achlaug, L.; Awwad, L.; Langier Goncalves, I.; Goldenberg, T.; Aronheim, A. Tumor Growth Ameliorates Cardiac Dysfunction and Suppresses Fibrosis in a Mouse Model for Duchenne Muscular Dystrophy. *Int. J. Mol. Sci.* **2023**, *24*, 12595. https://doi.org/10.3390/ijms241612595

Academic Editors: Dianna Magliano and Yutang Wang

Received: 10 July 2023
Revised: 31 July 2023
Accepted: 6 August 2023
Published: 9 August 2023

1. Introduction

Fibrosis, the thickening or scarring of a tissue, is part of the wound-healing process [1,2]. Pathophysiological fibrosis is observed in several diseases, including ones involving the liver [3], heart [4,5], skeletal muscles [6], kidneys [7], and lungs [8]. Collectively, fibrotic diseases account for up to half of deaths in the developed world and are clearly an unmet clinical need [9]. We have previously shown that heart failure in tumor-bearing mice results in an altered gene expression program in the heart, tumor, and other organs [10,11]. These changes lead to the secretion of multiple factors that promote tumor growth and metastasis spread [10–12]. Yet, we have also found that tumor growth suppresses cardiac remodeling, reduces cardiac hypertrophy, and inhibits de novo synthesis of fibrosis in mice with cardiac dysfunction [13]. Whether the interplay between heart failure and tumor growth is connected to cardiac dysfunction amelioration in tumor-bearing mice is yet to be determined. It is also unknown whether the tumor-dependent effects on cardiac function are specific to the failing heart or represent a systemic effect. In order to investigate these inquiries within the context of a human disease with clinical relevance, we utilized the MDX mouse model for Duchenne

Muscular Dystrophy (DMD). DMD is a muscular disorder that impacts approximately one in 3500 males, resulting from an X-linked loss-of-function mutation in the dystrophin gene. Individuals affected by DMD experience muscle fibrosis in skeletal, cardiac, and diaphragm muscles, as well as lung complications [14,15]. Additionally, DMD patients commonly exhibit cardiomyopathy and muscle pathologies. By subcutaneously implanting cancer cells into C57Bl/10 (control) and MDX mice, we were able to monitor tumor growth, fibrosis, and cardiac contractile function. Although MDX mice displayed clear cardiac dysfunction, no potentiation of tumor growth was observed. Nevertheless, tumor growth did, however, significantly suppress de novo synthesis of fibrosis deposition and significantly reduce existing fibrosis in the heart and skeletal and diaphragm muscles, resulting in improved cardiac contractile function. This occurred, at least in part, via macrophage M2 recruitment to the heart. These cancer paradigms may provide novel therapeutic strategies towards the treatment of human fibrotic diseases.

2. Results

2.1. Tumor Growth Significantly Improved the Contractile Function of MDX Heart and Diaphragm Muscles

To examine how cardiac dysfunction affects tumor growth in the DMD mouse model, breast cancer PyMT cells were implanted into the flanks of six-month-old MDX male mice. As the MDX mouse strain is under a C57Bl/10 background, C57Bl/10 mice served as a disease-free control. Tumor growth was monitored over time, and cardiac function was assessed using echocardiography prior to sacrifice, after which fractional shortening (FS) was calculated (Figure 1A). MDX male mice displayed significantly lower contractile function than C57Bl/10 mice, with the former reaching a FS of 20% compared to 25% in the latter (Figure 1B and Supplemental Table S1). Nevertheless, tumor volume was similar in the two cohorts and throughout the experiment (Figure 1C), and the tumor weight was similar at the endpoint (Figure 1D). Interestingly, cardiac dysfunction was not accompanied by heart hypertrophy, as ventricular weight to body weight ratio (VW/BW) was similar in both C57Bl/10 and MDX mice (Figure 1E). Similar results were obtained when MDX and C57Bl/10 male mice were implanted with Lewis lung carcinoma cells (LLC) (Supplemental Figure S1A–D, Supplemental Table S2). These findings suggest that the cardiac dysfunction observed in MDX mice occurs independent of cardiac hypertrophy and is not sufficient to promote tumor growth in this model. In addition, though the FS in C57Bl/10 tumor-bearing mice is unchanged as compared with C57Bl/10 non-tumor-bearing, in the tumor-bearing MDX mice (PyMT and LLC), cardiac function was significantly improved (Figures 1B and S1D).

Reduced cardiac contractile function in the MDX mouse model is typically associated with fibrosis in the heart. Therefore, we examined the extent of fibrosis in heart sections derived from C57Bl/10 (B10), PyMT tumor-bearing, and non-tumor-bearing MDX mice by staining them with Masson's Trichrome. Indeed, tumor-bearing MDX mice displayed much lower fibrosis staining compared to aged-matched MDX mice with no tumor (Figure 2A). Diaphragms of tumor-bearing MDX mice were also stained with Masson's Trichome in order to evaluate the extent of fibrosis. Similar to the heart, diaphragms of tumor-bearing MDX mice displayed much lower fibrosis staining compared to aged-matched MDX mice with no tumor (Figure 2B). These reductions were accompanied by reduced levels of fibrosis hallmark gene markers, as evaluated by qRT-PCR using mRNA derived from the heart, diaphragm muscles, and skeletal muscles (Figure 2C–E). Interestingly, tumors derived from MDX mice exhibited elevated transcription levels of fibrosis hallmark gene markers compared to tumors of a similar size derived from C57Bl/10 mice (Figure 2F). These results suggest that tumors display increased fibrosis gene programming in MDX mice compared to C57Bl/10 mice, and the heart, diaphragm muscles, and skeletal muscles exhibit reduced fibrosis.

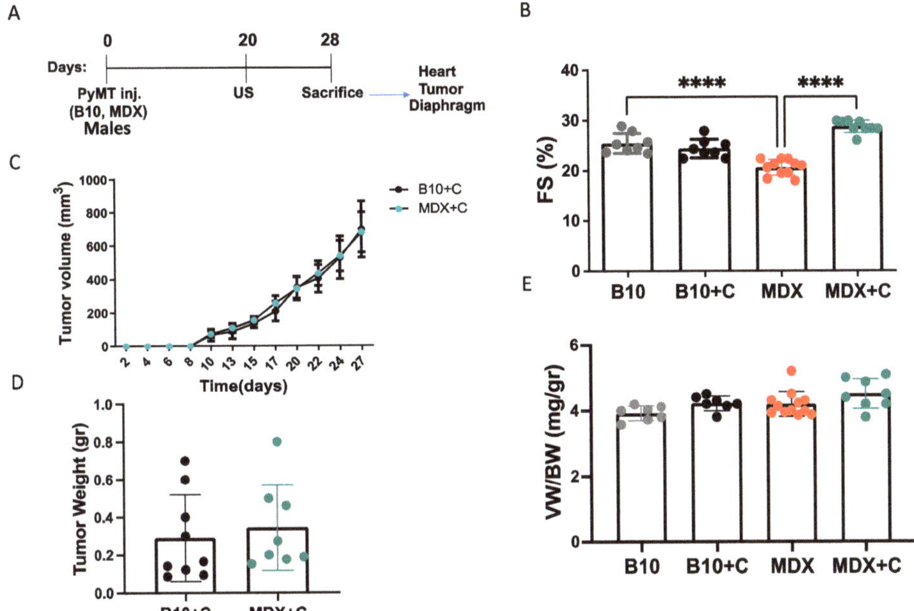

Figure 1. Cardiac dysfunction fails to promote tumor growth, but tumor growth is accompanied by improved cardiac contractile function. (**A**) Schematic representation of the experimental timeline. C57Bl/10 and MDX male mice (6 months old) were injected in the flanks with PyMT cells (10^6 cells per mouse) or left untreated (control). Echocardiography (US) was performed 8 days prior to sacrifice. (**B**) The measured fractional shortening (FS) in C57Bl/10 (B10), MDX, and tumor-bearing mouse groups (B10+C and MDX+C). FS was assessed by echocardiography and calculated using the formula: FS (%) = [(LVDd- LVDs)/LVDd]. (**C**) Tumor volume (width^2 × length × 0.5) was monitored over time in tumor-bearing C57Bl/10 and MDX mice (**D**) Tumor weight at the endpoint in C57Bl/10 and MDX mice cohorts. (**E**) Ventricular weight to body weight ratio (VW/BW). Data are presented as mean ± SE. One-way ANOVA followed by Tukey post-test (**B,E**); two-way ANOVA with Bonferroni repeated measure (**C**) or Student's *t*-test (**D**). **** $p < 0.0001$. Each dot represents one mouse.

Similar experiments were repeated in an orthotopic cancer model. Namely, PyMT cells were implanted into the mammary fat pad of MDX and C57Bl/10 female mice (Supplemental Figure S2A). Consistent with the MDX male mice, tumor-bearing MDX female mice exhibited improved cardiac contractile function, as observed by elevated FS (Supplemental Figure S2B, Supplemental Table S3) and a reduction in fibrosis hallmark gene markers in heart and diaphragm muscles following tumor implantation, as assessed by qRT-PCR (Supplemental Figure S2C,D). In addition, the lung weight to body weight ratio was lower in MDX female mice as compared with C57Bl/10 female mice. The LW/BW ratio was elevated in tumor-bearing female MDX mice (Supplemental Figure S3A). This increase in lung weight in tumor-bearing mice was accompanied by a reduction in fibrosis hallmark gene markers compared to the MDX female mouse cohort (Supplemental Figure S3B).

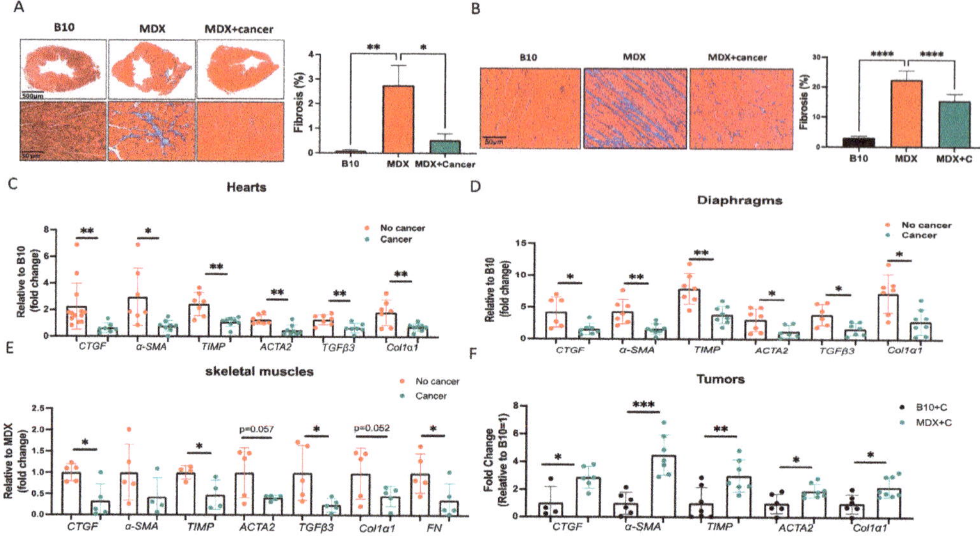

Figure 2. Tumor growth suppresses fibrosis hallmark gene markers transcription in MDX mice. (**A**,**B**) Left panel: Representative image of heart (**A**) and diaphragm (**B**) sections of B10, (left), MDX mouse (middle) and tumor-bearing MDX mouse (right) stained for Masson's trichrome. Scale bar: 500 μm (top) and 50 μm (bottom). Right panel: Percent of interstitial fibrosis, quantified using ImageJ software, based on at least five fields from each individual mouse in each cohort (n = 5). (**C–E**) Transcription levels of fibrosis hallmark gene markers CTGF, αSMA, TIMP, ACTA2, TGFβ3, Col1α1, and FN (in skeletal muscles) in the heart (**C**), diaphragm (**D**), and skeletal muscles (**E**) of non-tumor-bearing and tumor-bearing MDX mice, measured using qRT-PCR normalized to Hsp90 (heart) mb2m (diaphragms and skeletal muscles). (**F**) Transcription levels of fibrosis hallmark gene markers CTGF, αSMA, TIMP, ACTA2, and Col1α1 in the tumors of C57Bl/10 and MDX mice, measured using qRT-PCR normalized to β-actin. Data are presented as the relative expression compared to C57/Bl/10 mice (determined as 1) in the hearts, diaphragms, and tumors and compared to MDX (determined as 1) in skeletal muscles. Results are presented as mean ± SE; one-way repeated measures ANOVA followed by Tukey posttests (**A–D**) or Student's *t*-test (**E,F**). * $p < 0.05$; ** $p < 0.01$; *** $p < 0.001$, **** $p < 0.0001$. Each dot represents one mouse.

2.2. Improved Contractile Function and Reduction in Fibrosis in Tumor-Bearing Mice Is Mediated by M2 Macrophage Recruitment to the Heart

A previous study suggested that innate immune cells are involved in tumor-dependent cardiac dysfunction amelioration [13]. Therefore, we sought to examine their role in global fibrosis reduction in the MDX mice model. Towards this end, we used qRT-PCR analysis of mRNA derived from PyMT tumors, hearts, diaphragms, and spleens of the male mice cohorts. We first compared the level of F4/80, a gene marker representing the total macrophage population, in the tumors derived from C57Bl/10 and MDX mice (Figure 3A). In general, we observed reduced F4/80 expression levels in tumors derived from MDX mice as compared with C57Bl/10 tumors. We then compared F4/80 expression levels in tissues derived from MDX mice in the absence and presence of tumors. F4/80 expression levels were higher in the heart and the diaphragm muscles of PyMT tumor-bearing MDX mice compared to non-tumor-bearing mice (Figure 3B,C). In contrast, F4/80 expression levels were lower in the spleens of tumor-bearing MDX mice (Figure 3D).

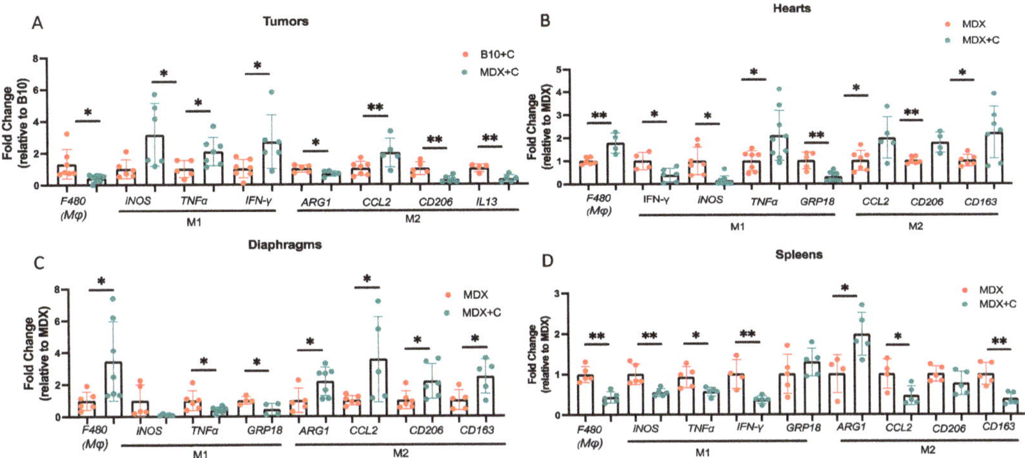

Figure 3. Tumor growth in MDX mice induces the expression of M2-polarizing hallmark gene markers. (**A–D**) qRT-PCR measuring transcription levels of macrophage hallmark gene markers in the tumors as compared with C57Bl/10 (**A**) in the hearts (**B**), diaphragm muscles (**C**), and spleens (**D**). Gene expression compares non-tumor-bearing and PyMT-tumor-bearing MDX male mice (same experimental cohort as in Figure 1). Measurements were obtained using qRT-PCR, normalized to housekeeping gene Hsp90. Results are presented as mean ± SE, Student's *t*-test. * $p < 0.05$; ** $p < 0.01$. Each dot represents one mouse.

Macrophages are largely divided into two main functionally distinct forms, pro-inflammatory (M1) and anti-inflammatory (M2) [16,17]. To distinguish between the two macrophage populations, we examined the expression levels of specific M1 (iNOS, TNFα, IFN-γ) and M2 macrophage gene markers (ARG-1, CD206, CCL2, and IL-13) [16,17]. In tumors derived from MDX mice, we observed higher levels of M1 gene markers, whereas there were lower levels in the M2 gene markers as compared with tumors derived from C57Bl/10 (Figure 3A). We then compared the levels of M1-M2 hallmark gene markers in various tissues derived from tumor-bearing and non-tumor-bearing MDX mice. In the heart and diaphragm muscles of tumor-bearing MDX mice, we observed higher levels of M2 gene markers and lower levels of M1 gene markers (Figure 3B,C). In contrast, we observed a significant reduction in M1 gene markers in the spleens of tumor-bearing mice, whereas inconsistent changes were observed for the M2 markers (higher levels of Arg1 and lower levels of CCL2 and CD163) (Figure 3D). These results suggest that M1-to-M2 polarization occurs in the hearts and diaphragms of tumor-bearing MDX mice compared with MDX mice. This polarization is only partially observed in the spleens of tumor-bearing MDX mice.

Macrophage recruitment was further supported by fluorescence-activated cell sorting (FACS) of hearts of non-tumor-bearing vs. tumor-bearing MDX mice. In the hearts, a significant elevation in the macrophage population (F480+, CD64+) was observed in tumor-bearing MDX mice compared with non-tumor-bearing mice (Figure 4A,B), even though M2 macrophage polarization occurred already in naive MDX mice, and no further elevation in M2 macrophage population occurred in tumor-bearing MDX mice as shown by RT-PCR using M1/M2 markers (Figure 4D–F, Supplemental Figure S4A,B). Nonetheless, the percent (%) of macrophages in the hearts of tumor-bearing mice was significantly elevated compared with the hearts of naïve MDX mice. Collectively, these results support the systemic alteration in the macrophage cell population towards M2 polarization, which is prominent in the heart.

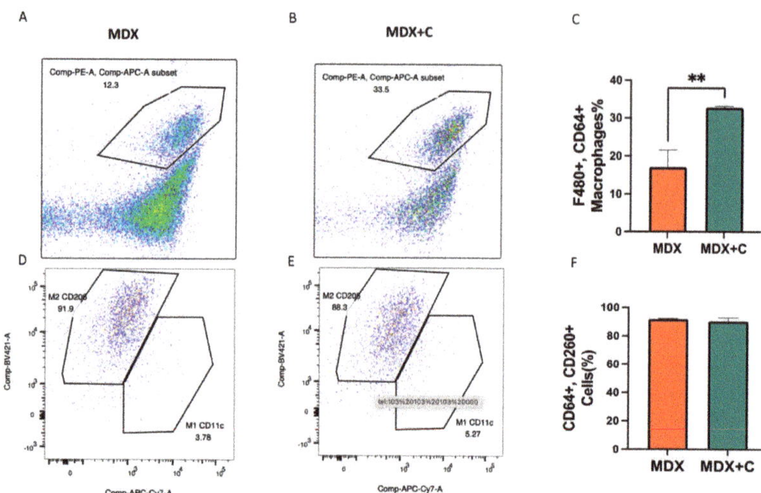

Figure 4. Tumor growth in MDX mice induces significant macrophage recruitment. (**A,B**) FACS analysis of total macrophage population in the hearts of naïve (n = 3) (**A**) and tumor-bearing (n = 5) MDX mice (**B**). These data are shown as the macrophages (%) F480+, CD64+ in (**C**). Additional analysis of M2 macrophage population CD64+CD206 in the hearts (**D,E**) and the quantification of % macrophages in each group (**F**). Results are presented as mean ± SEM, Student's *t*-test. ** *p* < 0.01.

Macrophage recruitment was further supported by fluorescence-activated cell sorting (FACS) of hearts of non-tumor-bearing vs. tumor-bearing MDX mice. In the hearts, a significant elevation in the macrophage population (F480+, CD64+) was observed in tumor-bearing MDX mice compared with non-tumor-bearing mice (Figure 4A,B), even though M2 macrophage recruitment occurred already in naive MDX mice, and the total macrophage population was significantly increased. Collectively, these results support the systemic alteration in the macrophage cell population towards M2 recruitment, which is prominent in the heart.

We next examined two master regulatory factors associated with M1-M2 macrophages' switch, G-CSF [18] and IL-13 [19] in the serum derived from MDX mice and tumor-bearing MDX mice. Significantly higher levels of both G-CSF and IL-13 were observed in the serum of tumor-bearing MDX mice as compared to MDX mice (Figure 5). This observation implies that these secreted factors could potentially play a pivotal role in promoting M2 macrophage recruitment to the heart.

Collectively, our findings indicate that the MDX mice display reduced cardiac contractile function and massive fibrosis. Surprisingly, tumor-bearing MDX mice feature M2 macrophage recruitment to the heart. M2 macrophages are associated with significant muscle repair functions, as demonstrated by reduced matrix deposition and a reduction in the presence of fibrosis, which leads to the amelioration of cardiac contractile function and an overall beneficial phenotype. A schematic summary of the main manuscript findings and conclusions is provided in Figure 6.

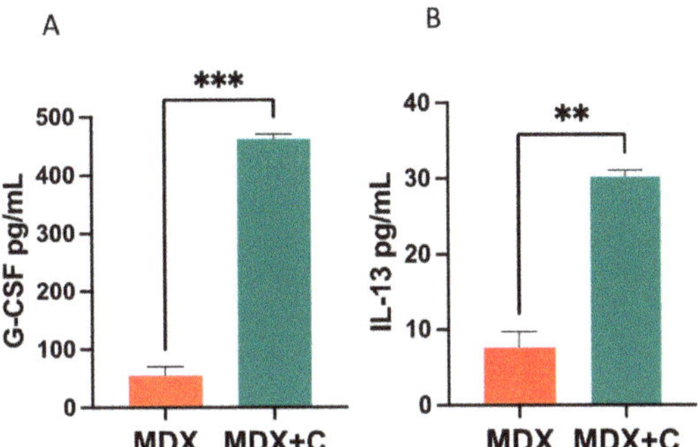

Figure 5. M2 polarizing cytokines G-CSF and IL-13 are elevated in the serum of PyMT-bearing compared with non-tumor-bearing MDX mice. Serum levels as in (**A**), obtained by ELISA for G-CSF and (**B**) IL-13. Pooled blood serum of MDX (n = 5 each) and PyMT-bearing MDX mice (n = 5) was used. Results are presented as mean ± SE, Student's *t*-test. ** $p < 0.01$, *** $p < 0.001$.

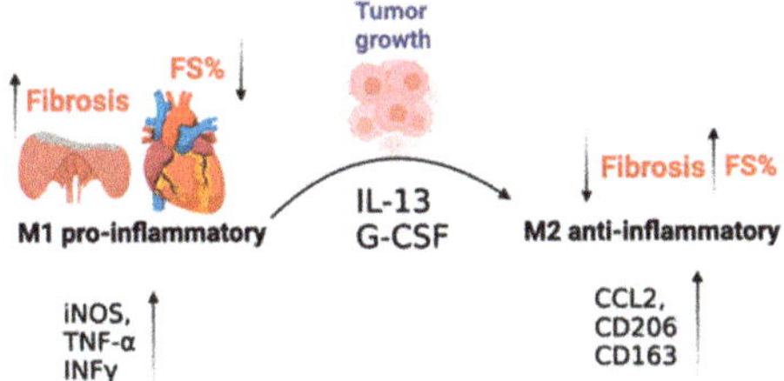

Figure 6. Graphical abstract describing the manuscript's main findings. Tumor growth reduces fibrosis in the heart and diaphragm muscles and ameliorates cardiac contractile function. This occurs at least partially via M2 macrophage recruitment.

3. Discussion

Fibrosis is an unmet clinical need in various diseases involving skeletal and cardiac muscles, lungs, liver, and kidneys [9]. One such disease is DMD, an X-linked genetic disease that involves cardiac, diaphragm, and skeletal muscle fibrosis [14]. Whereas previous studies have demonstrated that heart failure [10,11,20,21] promotes cancer progression, we found no promotion of cancer growth in the MDX mouse model for DMD, despite the fact that MDX mice demonstrate apparent cardiac dysfunction (FS 20%). Yet the tumor promotion phenotype was identified in multiple other mice models, such as those reflecting myocardial infarction (MI) [20–22], pressure overload (TAC) [10], and heart hypertrophy (ATF3 transgene) [11]. In addition, a chronic hypertension mouse model with heart hypertrophy lacking cardiac dysfunction resulted in enhanced tumor growth [12]. The MDX mouse model displayed reduced cardiac contractile function in the absence of cardiac hypertrophy. Therefore, we can conclude that the tumor promotion phenotype is dependent on cardiac hypertrophy and not a direct result of cardiac dysfunction.

To study the effect of tumor growth on cardiac function and fibrosis in the MDX mouse model, we analyzed the heart contractile function of MDX mice in the presence and absence of tumors. Cardiac contractile function was improved in tumor-bearing MDX mice and was comparable to the levels found in naïve C57Bl/10 mice. Naïve C57Bl/10 mice displayed reduced basal cardiac contractile function as compared to naïve C57Bl/6 tumor growth. Nevertheless, in tumor-bearing C57Bl/10, no improvement of cardiac contractile function was observed as compared with non-tumor-bearing C57Bl/10 mice.

The transcription of fibrosis hallmark gene markers in the lungs, heart, and skeletal and diaphragm muscles of MDX mice was suppressed significantly, the fibrosis staining revealed a considerable reduction in fibrosis both in the heart and diaphragm. This finding strongly indicates that tumor growth not only inhibited the development of new fibrosis hallmark gene markers but also promoted the dissolution of existing fibrotic scars. Moreover, the observation of decreased expression of fibrosis hallmark gene markers in the skeletal and diaphragm muscles and lungs suggests that the suppression of fibrosis due to tumor growth is not limited to specific tissues but is rather a widespread phenomenon. These results may have broader implications and relevance for other fibrotic disease models, suggesting potential systemic effects of tumor-induced fibrosis suppression.

Our findings indicate that these beneficial tumor effects on fibrosis are mediated via macrophage recruitment. In the tumor, the M1 macrophage population is increased, whereas in the heart and diaphragm muscles, there was an increase mostly in the total M2 macrophage population. This was demonstrated by qRT-PCR, FACS analysis, and ELISA using M2-polarizing secreted factors in the serum. Specifically, in the heart and diaphragm muscles, the local synthesis of the M2-polarizing cytokine CCL2 was elevated. Notably, CCL2 has been shown to play an important role in the prevention of LV dysfunction and remodeling after MI [23]. Similarly, M2 macrophages are thought to accelerate cardiac and tissue repair processes [17,24,25]. In contrast, in the tumor, the M1-polarizing cytokines iNOS, TNF-α, and INFγ are prominent [17]. Importantly, M2 macrophages' contribution to reducing fibrosis in various organs strongly indicates their therapeutic side effect under the scenario of a growing tumor. The noted reduction in the spleen in M1 hallmark gene markers and the elevation in M2 gene markers may indicate that spleen-derived macrophages initiate the polarization of classically activated (M1) to alternatively activated (M2) macrophages. The final macrophage polarization occurs upon their recruitment to the fibrotic site, such as in the lungs, heart, and diaphragm muscles [26]. The recruitment of the F480$^+$/CD64$^+$ cell population in the heart in tumor-bearing MDX mice was further supported by FACS analysis. Furthermore, the M2-polarizing factors G-CSF and IL-13 were shown by ELISA to be elevated in tumor-bearing MDX mice as compared with non-tumor-bearing MDX mice. A previous study showed that IL-13 induces M2 polarization, leading to improved cardiac function and reduced heart injury in a viral myocarditis mouse model [19]. G-CSF was shown to induce macrophage polarization and mobilization in previous studies [27]. Furthermore, G-CSF has been shown to decrease inflammatory processes and to act positively on the process of peripheral nerve regeneration during the course of muscular dystrophy [18]. It was shown that already in four-week-old MDX mice, the M2 macrophage population dominates skeletal muscles [28]. Nonetheless, the M2 population supports fibrosis, whereas the increase in the M2 population in tumor-bearing MDX mice appears to be associated with repair and improvement in cardiac function.

In summary, in the MDX mouse model for DMD, cardiac dysfunction in the absence of heart hypertrophy fails to promote tumor growth. On the other hand, tumor growth in MDX mice reduces fibrosis in the lungs, heart, and skeletal and diaphragm muscles. The tumor paradigms identified here could serve as novel therapeutic strategies for the treatment of the devastating DMD disease. They may also have a beneficial systemic outcome for other fibrotic diseases.

Limitations: The study is limited to concluding the effect of cancer growth on cardiac fibrosis and function in MDX mice. In DMD human patients, fibrosis treatment is limited, and cancer is not considered an optional treatment. The identification of factors that are responsible for the beneficial effects here may be used as a potential therapeutic treatment.

Clinical perspectives: The manuscript describes, for the first time, a beneficial effect of cancer on cardiac dysfunction and fibrosis in a clinically relevant mouse model. Tumor growth ameliorates cardiac dysfunction and reduces overall fibrosis. Fibrosis diseases account for more than half deaths worldwide and represent an unmet need. Harnessing tumor paradigms may serve as a novel strategy to improve cardiac function and fibrosis diseases.

4. Materials and Methods

4.1. Animals

The C57BL/10ScSn-Dmd (MDX) and C57BL/10J (B10) mice were acquired from the Jackson Laboratory. They were bred and raised from birth at the Pre-Clinical Research Authority, located at the Ruth and Bruce Rappaport Faculty of Medicine, Technion. The mice were kept in mating cages with regular access to food and water. At three weeks of age, they were weaned into individual cages.

4.2. Cell Culture

Polyoma Middle T (PyMT) murine breast carcinoma cells were derived from primary tumor-bearing transgenic mice expressing PyMT under the control of the murine mammary tumor virus promoter [29]. The PyMT cells were kindly provided by Prof. Tsonwin Hai (Ohio State University, Columbus, OH, USA). The LLC cancer cell line was purchased from the American Type Culture Collection ATCC. Both cell lines were tested by IDEXX BioAnalytics and found to be free of *Mycoplasma* and viral contaminations. Cells were cultured in DMEM containing 10% FBS, 1% streptomycin and penicillin, 1% l-glutamine, and 1% sodium pyruvate at 37 °C in a humidified atmosphere containing 5% CO_2. Cancer cell implantation was conducted at maximal passage number five.

4.3. Cancer Cell Implantation

PyMT cancer cells and LLC cancer cells were injected into the mammary fat pad of female mice (10^6 cells per mouse) and into the back flanks of mice. The tumor size was measured using a caliper, and the tumor volume was calculated using the formula: $Width^2 \times Length \times 0.5$. The humane endpoint is considered to be when a tumor reaches a size of 1500 mm^3, according to the Institutional Animal Care and Use Committee.

4.4. Echocardiography

Mice were anesthetized with 1% isoflurane and kept on a 37 °C heated plate throughout the procedure. Echocardiography was performed with a Vevo3100 micro-ultrasound imaging system (VisualSonics, Fujifilm, Tokyo, Japan) equipped with 13–38 MHz (MS 400) and 22–55 MHz (MS550D) linear array transducers. Cardiac size, shape, and function were analyzed using conventional two-dimensional imaging and M-mode recordings. Maximal left ventricular end-diastolic (LVDd) and end-systolic (LVDs) dimensions were measured in short-axis M-mode images. Fractional shortening (FS) was calculated with the following formula: $FS\% = [(LVDd - LVDs)/LVIDd] \times 100$. FS values are based on the average of at least three measurements for each mouse.

4.5. RNA Extraction

RNA was extracted from lungs, hearts, diaphragms, and tumors using an Aurum total RNA fatty or fibrous tissue kit (no. 732–6830, Bio-Rad, Hercules, CA, USA), according to the manufacturer's instructions. cDNA was synthesized from 1000 ng purified mRNA with an iScript cDNA Synthesis Kit (no. 170–8891, Bio-Rad), according to the manufacturer's instructions.

4.6. Quantitative Real-Time PCR

Quantitative real-time polymerase chain reaction (qRT-PCR) was performed with a QuantStudio3 (Thermofisher Scientific, 5823 Newton drive, Carlsbad, CA 92008, USA). Serial dilutions of a standard sample were included for each gene to generate a standard curve. Values were normalized to Hsp90, β-actin, and mb2M expression levels for the heart, diaphragm, tumor tissue, and lungs, respectively. All the oligonucleotide sequences that were used are listed in Supplementary Table S4.

4.7. Fibrosis Staining

Heart tissues and diaphragms were fixed in 4% formaldehyde overnight, embedded in paraffin, serially sectioned at 10 μm intervals, and then mounted on slides. Masson trichrome staining was performed according to the standard protocol. Images were acquired using a 3DHistech Pannoramic 250 Flash III (3DHISTECH Ltd, H-1141 Budapest, Öv u. 3., Hungary). Each section was fully scanned. The percent of interstitial fibrosis was determined as the ratio of the fibrosis area to the total area of the section using Image Pro Plus software. Each dot represents the mean of the values taken from at least five fields, derived from a single mouse.

4.8. Heart Single-Cell Suspension and Flow Cytometry

Heart single-cell suspension and flow cytometry was prepared as previously described [30]. Briefly, hearts were perfused, extracted, finely minced, and then incubated with digestive enzymes at 37 °C on a rocking shaker at 50 rpm for 45–60 min. Samples were homogenized with a 40 μm cell strainer. Red blood cells were lysed using ammonium-chloride-potassium (ACK) lysis buffer. Next, samples were centrifuged at $400 \times g$ for 5 min at 4 °C, and the pellet was then suspended with FACS buffer. Cells were immune-stained with the following anti-mouse antibodies: CD45-Alexa Fluor[®®] 700 (BioLegend, 103128, San Diego, CA, USA), CD11b-PerCP (BioLegend, 101228, CA, USA), F480-PE (BioLegend, 123110, CA, USA), CD206-BV421 (BioLegend, 141717, CA, USA), CD64-APC (BioLegend, 161006, CA, USA), Ly-6G- Brilliant Violet 510™ (BioLegend, 127633, CA, USA), Ly-6C-PE/Cyanine7 (BioLegend, 128018, CA, USA), Propidium iodide (Sigma, 25535-16-4, Darmstadt, Germany). Cells were incubated (30 min, 4 °C) with the antibody mixture in a staining buffer (PBS containing 1% bovine serum albumin and 0.05% sodium azide) and then washed twice with a staining buffer. Cells were acquired using a LSRFortessa flow cytometer (BD Biosciences, Franklin Lakes, NJ, USA). The data were analyzed using FlowJo V.10 software (FlowJo, Ashland, OR, USA).

4.9. ELISA

ELISA quantification of candidate secreted factors' protein levels in the blood was performed with a granulocyte-colony stimulating factor (G-CSF) mouse ELISA Kit (MCS00, R&D Systems, Minneapolis, MN, USA) and a Mouse IL-13 ELISA Kit (M1300CB, R&D systems), according to the manufacturer's instructions.

4.10. Statistical Analysis

The data are presented as mean ± standard error (SE). All mice were included in each statistical analysis, unless they were euthanized for humane reasons before reaching the experimental endpoint. During data collection, the experimental groups were blinded to the researchers. The mice for each group were selected randomly. Each experimental group consisted of at least n = 5 mice.

To determine the statistical significance of tumor volume, a two-way repeated-measures ANOVA followed by the Bonferroni posttest was used. For comparisons between several means, a one-way ANOVA followed by the Tukey posttest was performed. For comparisons between two means, either a two-tailed Student's *t*-test or Mann–Whitney U test was utilized.

All statistical analyses were conducted using GraphPad Prism 10 software. A significance level of $p < 0.05$ was considered statistically significant.

Supplementary Materials: The following supporting information can be downloaded at: https://www.mdpi.com/article/10.3390/ijms241612595/s1.

Author Contributions: Conceptualization: A.A. and L.A. (Laris Achlaug). Methodology: A.A., L.A. (Laris Achlaug), L.A. (Lama Awwad), I.L.G. and T.G. Investigation: A.A. and L.A. (Laris Achlaug). Visualization: A.A. and L.A. (Laris Achlaug). Funding acquisition: A.A. Project administration: A.A. Supervision: A.A. Writing—original draft: A.A. Writing—review and editing: A.A., L.A. (Laris Achlaug), L.A. (Lama Awwad) and I.L.G. All authors have read and agreed to the published version of the manuscript.

Funding: This work was partially supported by the Ministry of Health grant # 3-17783 (to A. Aronheim) and by the Swiss Technion Society.

Institutional Review Board Statement: All experimental protocols were approved by the Institutional Committee for Animal Care and Use at the Technion—Israel Institute of Technology, Faculty of Medicine, Haifa, Israel. Approval number IL-010-02-21. All study procedures comply with the guidelines of the NIH Guide for the Care and Use of Laboratory Animals.

Informed Consent Statement: Not applicable.

Data Availability Statement: All the obtained data used to support the findings of this study are available from the corresponding author upon reasonable request.

Acknowledgments: The authors wish to thank Sharon Aviram and all the members of the Aronheim lab, Regina Katsman, for taking care of the animal colony.

Conflicts of Interest: The authors declare that they have no competing interests.

References

1. Darby, I.A.; Laverdet, B.; Bonte, F.; Desmouliere, A. Fibroblasts and myofibroblasts in wound healing. *Clin. Cosmet. Investig. Dermatol.* **2014**, *7*, 301–311. [CrossRef] [PubMed]
2. Wynn, T.A. Cellular and molecular mechanisms of fibrosis. *J. Pathol.* **2008**, *214*, 199–210. [CrossRef]
3. Yanguas, S.C.; Cogliati, B.; Willebrords, J.; Maes, M.; Colle, I.; van den Bossche, B.; de Oliveira, C.; Andraus, W.; Alves, V.A.F.; Leclercq, I.; et al. Experimental models of liver fibrosis. *Arch. Toxicol.* **2016**, *90*, 1025–1048. [CrossRef] [PubMed]
4. Frangogiannis, N.G. Cardiac fibrosis. *Cardiovasc. Res.* **2021**, *117*, 1450–1488. [CrossRef] [PubMed]
5. Diez, J.; de Boer, R.A. Management of cardiac fibrosis is the largest unmet medical need in heart failure. *Cardiovasc. Res.* **2022**, *118*, e20–e22. [CrossRef]
6. Bulfield, G.; Siller, W.G.; Wight, P.A.; Moore, K.J. X chromosome-linked muscular dystrophy (mdx) in the mouse. *Proc. Natl. Acad. Sci. USA* **1984**, *81*, 1189–1192. [CrossRef]
7. Humphreys, B.D. Mechanisms of Renal Fibrosis. *Annu. Rev. Physiol.* **2018**, *80*, 309–326. [CrossRef]
8. Wijsenbeek, M.; Cottin, V. Spectrum of Fibrotic Lung Diseases. *N. Engl. J. Med.* **2020**, *383*, 958–968. [CrossRef]
9. Murtha, L.A.; Schuliga, M.J.; Mabotuwana, N.S.; Hardy, S.A.; Waters, D.W.; Burgess, J.K.; Knight, D.A.; Boyle, A.J. The Processes and Mechanisms of Cardiac and Pulmonary Fibrosis. *Front. Physiol.* **2017**, *8*, 777. [CrossRef]
10. Avraham, S.; Abu-Sharki, S.; Shofti, R.; Haas, T.; Korin, B.; Kalfon, R.; Friedman, T.; Shiran, A.; Saliba, W.; Shaked, Y.; et al. Early Cardiac Remodeling Promotes Tumor Growth and Metastasis. *Circulation* **2020**, *142*, 670–683. [CrossRef]
11. Awwad, L.; Aronheim, A. Cardiac Dysfunction Promotes Cancer Progression via Multiple Secreted Factors. *Cancer Res.* **2022**, *82*, 1753–1761. [CrossRef] [PubMed]
12. Awwad, L.; Goldenberg, T.; Langier-Goncalves, I.; Aronheim, A. Cardiac Remodeling in the Absence of Cardiac Contractile Dysfunction Is Sufficient to Promote Cancer Progression. *Cells* **2022**, *11*, 1108. [CrossRef]
13. Awwad, L.; Shofti, R.; Haas, T.; Aronheim, A. Tumor Growth Ameliorates Cardiac Dysfunction. *Cells* **2023**, *12*, 1853. [CrossRef] [PubMed]
14. Duan, D.; Goemans, N.; Takeda, S.; Mercuri, E.; Aartsma-Rus, A. Duchenne muscular dystrophy. *Nat. Rev. Dis. Primers* **2021**, *7*, 13. [CrossRef] [PubMed]
15. Verhaart, I.E.C.; Aartsma-Rus, A. Therapeutic developments for Duchenne muscular dystrophy. *Nat. Rev. Neurol.* **2019**, *15*, 373–386. [CrossRef]
16. Orecchioni, M.; Ghosheh, Y.; Pramod, A.B.; Ley, K. Macrophage Polarization: Different Gene Signatures in M1(LPS+) vs. Classically and M2(LPS−) vs. Alternatively Activated Macrophages. *Front. Immunol.* **2019**, *10*, 1084. [CrossRef]
17. Kim, Y.; Nurakhayev, S.; Nurkesh, A.; Zharkinbekov, Z.; Saparov, A. Macrophage Polarization in Cardiac Tissue Repair Following Myocardial Infarction. *Int. J. Mol. Sci.* **2021**, *22*, 2715. [CrossRef]
18. Simoes, G.F.; Benitez, S.U.; Oliveira, A.L. Granulocyte colony-stimulating factor (G-CSF) positive effects on muscle fiber degeneration and gait recovery after nerve lesion in MDX mice. *Brain Behav.* **2014**, *4*, 738–753. [CrossRef]
19. Yang, H.; Chen, Y.; Gao, C. Interleukin-13 reduces cardiac injury and prevents heart dysfunction in viral myocarditis via enhanced M2 macrophage polarization. *Oncotarget* **2017**, *8*, 99495–99503. [CrossRef]
20. Meijers, W.C.; Maglione, M.; Bakker, S.J.L.; Oberhuber, R.; Kieneker, L.M.; de Jong, S.; Haubner, B.J.; Nagengast, W.B.; Lyon, A.R.; van der Vegt, B.; et al. Heart Failure Stimulates Tumor Growth by Circulating Factors. *Circulation* **2018**, *138*, 678–691. [CrossRef]

21. Koelwyn, G.J.; Newman, A.A.C.; Afonso, M.S.; van Solingen, C.; Corr, E.M.; Brown, E.J.; Albers, K.B.; Yamaguchi, N.; Narke, D.; Schlegel, M.; et al. Myocardial infarction accelerates breast cancer via innate immune reprogramming. *Nat. Med.* **2020**, *26*, 1452–1458. [CrossRef]

22. Koelwyn, G.J.; Aboumsallem, J.P.; Moore, K.J.; de Boer, R.A. Reverse cardio-oncology: Exploring the effects of cardiovascular disease on cancer pathogenesis. *J. Mol. Cell Cardiol.* **2022**, *163*, 1–8. [CrossRef] [PubMed]

23. Morimoto, H.; Takahashi, M.; Izawa, A.; Ise, H.; Hongo, M.; Kolattukudy, P.E.; Ikeda, U. Cardiac overexpression of monocyte chemoattractant protein-1 in transgenic mice prevents cardiac dysfunction and remodeling after myocardial infarction. *Circ. Res.* **2006**, *99*, 891–899. [CrossRef]

24. Ley, K. M1 Means Kill; M2 Means Heal. *J. Immunol.* **2017**, *199*, 2191–2193. [CrossRef]

25. Sager, H.B.; Kessler, T.; Schunkert, H. Monocytes and macrophages in cardiac injury and repair. *J. Thorac. Dis.* **2017**, *9*, S30–S35. [CrossRef] [PubMed]

26. Mulder, R.; Banete, A.; Basta, S. Spleen-derived macrophages are readily polarized into classically activated (M1) or alternatively activated (M2) states. *Immunobiology* **2014**, *219*, 737–745. [CrossRef] [PubMed]

27. Wen, Q.; Kong, Y.; Zhao, H.Y.; Zhang, Y.Y.; Han, T.T.; Wang, Y.; Xu, L.P.; Zhang, X.H.; Huang, X.J. G-CSF-induced macrophage polarization and mobilization may prevent acute graft-versus-host disease after allogeneic hematopoietic stem cell transplantation. *Bone Marrow Transplant.* **2019**, *54*, 1419–1433. [CrossRef] [PubMed]

28. Villalta, S.A.; Nguyen, H.X.; Deng, B.; Gotoh, T.; Tidball, J.G. Shifts in macrophage phenotypes and macrophage competition for arginine metabolism affect the severity of muscle pathology in muscular dystrophy. *Hum. Mol. Genet.* **2009**, *18*, 482–496. [CrossRef]

29. Wolford, C.C.; McConoughey, S.J.; Jalgaonkar, S.P.; Leon, M.; Merchant, A.S.; Dominick, J.L.; Yin, X.; Chang, Y.; Zmuda, E.J.; O'Toole, S.A.; et al. Transcription factor ATF3 links host adaptive response to breast cancer metastasis. *J. Clin. Investig.* **2013**, *123*, 2893–2906. [CrossRef]

30. Aronoff, L.; Epelman, S.; Clemente-Casares, X. Isolation and Identification of Extravascular Immune Cells of the Heart. *J. Vis. Exp.* **2018**, *138*, 58114.

International Journal of
Molecular Sciences

Article

Effect of Hydralazine on Angiotensin II-Induced Abdominal Aortic Aneurysm in Apolipoprotein E-Deficient Mice

Yutang Wang [1,*,†], Owen Sargisson [1,†], Dinh Tam Nguyen [1], Ketura Parker [1], Stephan J. R. Pyke [1], Ahmed Alramahi [1], Liam Thihlum [1], Yan Fang [1], Morgan E. Wallace [1], Stuart P. Berzins [1], Ernesto Oqueli [2,3], Dianna J. Magliano [4,*] and Jonathan Golledge [5,6]

[1] Discipline of Life Science, Institute of Innovation, Science and Sustainability, Federation University Australia, Ballarat, VIC 3353, Australia; owensargisson@students.federation.edu.au (O.S.); ngutam372@gmail.com (D.T.N.); m.wallace@federation.edu.au (M.E.W.); s.berzins@federation.edu.au (S.P.B.)
[2] Cardiology Department, Grampians Health Ballarat, Ballarat, VIC 3350, Australia; ernesto.oqueliflores@bhs.org.au
[3] School of Medicine, Faculty of Health, Deakin University, Geelong, VIC 3220, Australia
[4] Diabetes and Population Health, Baker Heart and Diabetes Institute, Melbourne, VIC 3004, Australia
[5] Queensland Research Centre for Peripheral Vascular Disease, College of Medicine and Dentistry, James Cook University, Townsville, QLD 4811, Australia; jonathan.golledge@jcu.edu.au
[6] Department of Vascular and Endovascular Surgery, The Townsville University Hospital, Townsville, QLD 4811, Australia
* Correspondence: yutang.wang@federation.edu.au (Y.W.); dianna.magliano@baker.edu.au (D.J.M.)
† These authors contributed equally to this work.

Citation: Wang, Y.; Sargisson, O.; Nguyen, D.T.; Parker, K.; Pyke, S.J.R.; Alramahi, A.; Thihlum, L.; Fang, Y.; Wallace, M.E.; Berzins, S.P.; et al. Effect of Hydralazine on Angiotensin II-Induced Abdominal Aortic Aneurysm in Apolipoprotein E-Deficient Mice. *Int. J. Mol. Sci.* **2023**, *24*, 15955. https://doi.org/10.3390/ijms242115955

Academic Editor: José Luis Martin-Ventura

Received: 21 July 2023
Revised: 31 October 2023
Accepted: 1 November 2023
Published: 3 November 2023

Abstract: The rupture of an abdominal aortic aneurysm (AAA) causes about 200,000 deaths worldwide each year. However, there are currently no effective drug therapies to prevent AAA formation or, when present, to decrease progression and rupture, highlighting an urgent need for more research in this field. Increased vascular inflammation and enhanced apoptosis of vascular smooth muscle cells (VSMCs) are implicated in AAA formation. Here, we investigated whether hydralazine, which has anti-inflammatory and anti-apoptotic properties, inhibited AAA formation and pathological hallmarks. In cultured VSMCs, hydralazine (100 μM) inhibited the increase in inflammatory gene expression and apoptosis induced by acrolein and hydrogen peroxide, two oxidants that may play a role in AAA pathogenesis. The anti-apoptotic effect of hydralazine was associated with a decrease in caspase 8 gene expression. In a mouse model of AAA induced by subcutaneous angiotensin II infusion (1 μg/kg body weight/min) for 28 days in apolipoprotein E-deficient mice, hydralazine treatment (24 mg/kg/day) significantly decreased AAA incidence from 80% to 20% and suprarenal aortic diameter by 32% from 2.26 mm to 1.53 mm. Hydralazine treatment also significantly increased the survival rate from 60% to 100%. In conclusion, hydralazine inhibited AAA formation and rupture in a mouse model, which was associated with its anti-inflammatory and anti-apoptotic properties.

Keywords: abdominal aortic aneurysm; hydralazine; inflammation; atherosclerosis

1. Introduction

Abdominal aortic aneurysm (AAA) rupture is estimated to be responsible for approximately 200,000 deaths worldwide each year [1]. Open surgical and endovascular aortic repair are the only current treatments for AAA, but they are associated with safety and durability problems [2]. The perioperative mortality rate of open surgical AAA repair is approximately 5% [3,4], and for endovascular AAA repair, it is about 2% [2]. Up to 20% of patients need re-intervention after endovascular AAA repair [2]. AAA repair is only recommended for women with an AAA diameter of ≥50 mm and men with an AAA diameter of ≥55 mm, as randomized controlled trials suggest repair of small aneurysms does not reduce mortality [2,5]. Patients with small AAAs are managed conservatively using repeated imaging to monitor the size of the AAA. Up to 70% of patients undergoing imaging surveillance eventually require AAA repair [2,5]. Medications are needed to limit AAA growth, but previously tested drugs have been ineffective in randomized clinical trials [6,7].

Aortic inflammation and loss of vascular smooth muscle cells (VSMCs) are two hallmark features of AAA pathology [2]. Hydralazine, an antihypertensive medication, has been shown to inhibit inflammation [8,9] in various disease models (spinal cord injury [10], kidney injury [11], and sepsis [12]), and also limit apoptosis of cardiomyocytes [13]. Whether hydralazine inhibits the apoptosis of VSMCs is unknown. In this study, we aimed to investigate whether hydralazine inhibits apoptosis of VSMCs in vitro and whether it inhibits AAA formation in vivo.

To investigate whether hydralazine inhibits the apoptosis of VSMCs in vitro, we used two apoptosis inducers, i.e., acrolein and hydrogen peroxide (H_2O_2). The oxidant acrolein is considered one of the most toxic and harmful components of cigarette smoke [14,15], the latter being a major risk factor for AAA formation [2]. The oxidant H_2O_2 is thought to play an important role in AAA pathogenesis [16].

To investigate whether hydralazine inhibits AAA formation in vivo, we used a mouse model in which AAA was induced by a subcutaneous infusion of angiotensin II. We compared the AAA incidence, size, and survival between two groups of mice treated with or without hydralazine. As this angiotensin II infusion-induced AAA model is also a model of atherosclerosis [17] and cardiac hypertrophy [18,19], we simultaneously investigated whether hydralazine affected these two conditions.

2. Results

2.1. Hydralazine Decreased Inflammatory Cytokine Expression in Cultured VSMCs

We used two separate oxidants (acrolein and H_2O_2) to induce cytokine expression in VSMCs. Incubation of the cells with either of these two compounds increased the expression of inflammatory cytokines (Figure 1). Co-incubation of the cells with hydralazine (100 μM) decreased gene expression of interleukin-1 (IL-1), IL-6, tumor necrosis factor-alpha (TNFα), and interferon-gamma (IFNγ, Figure 1). Enzyme-linked immunosorbent assay (ELISA) results confirmed that H_2O_2 increased IL-6 protein levels, and hydralazine (100 μM) mitigated such an increase (Figure S1).

2.2. Hydralazine Decreased Acrolein- and H_2O_2-Induced Cell Death in Cultured VSMCs

Acrolein and H_2O_2 are two commonly used apoptosis inducers [20,21] and may play a role in AAA pathogenesis [14–16]. Incubation of VSMCs with acrolein and H_2O_2 increased caspase 8 gene expression and did not affect caspases 3 and 9 (Figure 2), suggesting that these two compounds activated the extrinsic apoptotic pathway. Co-incubation of the cells with hydralazine (100 μM) mitigated the increase in caspase 8 gene expression, suggesting that hydralazine may inhibit apoptosis induced by acrolein and H_2O_2. Indeed, hydralazine attenuated cell death induced by acrolein and H_2O_2 (Figure 3). Flow cytometry results confirmed that hydralazine (100 μM) decreased the percentage of cells undergoing apoptosis induced by H_2O_2, as assessed by annexin V staining (Figure 4).

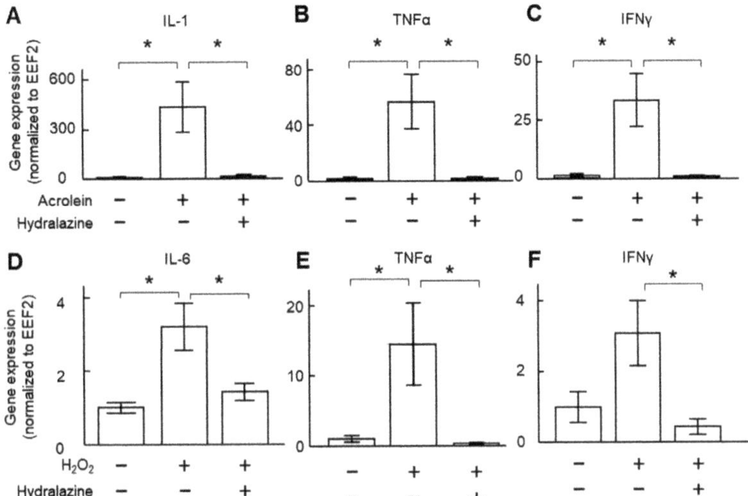

Figure 1. Effect of hydralazine on inflammatory gene expression. (**A–C**) VSMCs were incubated with acrolein (2 μg/mL) in the presence (+) or absence (−) of hydralazine (100 μM) for 24 h. mRNA was then extracted, and gene expression of IL-1 (**A**), TNFα (**B**), and IFNγ (**C**) was analyzed using qPCR. (**D–F**), VSMCs were treated with hydrogen peroxide (H_2O_2, 800 μM) in the presence or absence of hydralazine (100 μM) for 24 h. mRNA was then extracted, and gene expression of IL-6 (**D**), TNFα (**E**), and IFNγ (**F**) was analyzed using qPCR. *n*= 5–6. Error bars = standard error. Data were analyzed using a Kruskal–Wallis one-way ANOVA. * $p < 0.05$. EEF2, eukaryotic translation elongation factor 2; IFNγ, interferon-gamma; IL, interleukin; TNFα, tumor necrosis factor-alpha; VSMCs, vascular smooth muscle cells.

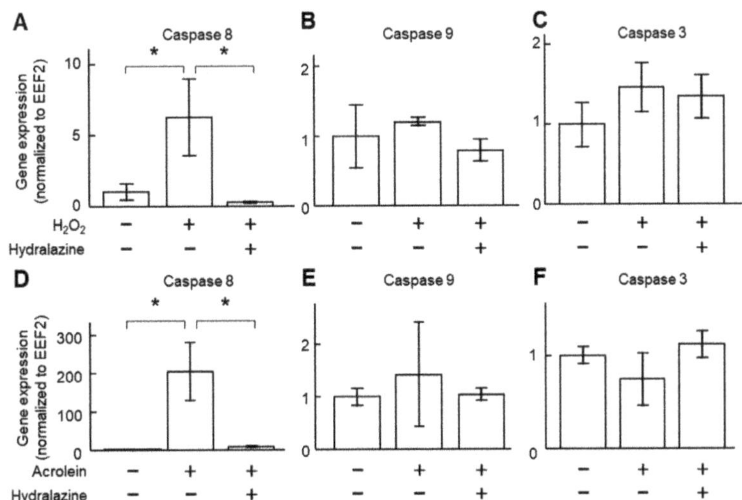

Figure 2. Effect of hydralazine on caspase gene expression. VSMCs were treated with acrolein (2 μg/mL, (**A–C**)) or hydrogen peroxide (800 μM, (**D–F**)) in the presence (+) or absence (−) of hydralazine (100 μM) for 24 h. mRNA was then extracted, and gene expression of caspase 8 (**A,D**), caspase 9 (**B,E**), and caspase 3 (**C,F**) was analyzed using qPCR. *n* = 5–6. Error bars = standard error. Data were analyzed using a Kruskal–Wallis one-way ANOVA. * $p < 0.05$. EEF2, eukaryotic translation elongation factor 2; H_2O_2, hydrogen peroxide; VSMCs, vascular smooth muscle cells.

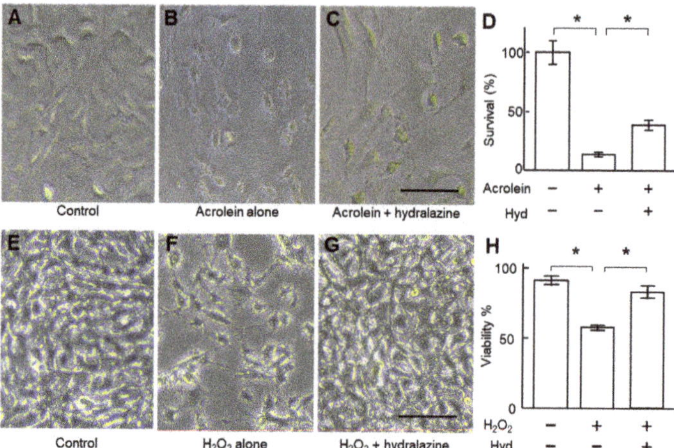

Figure 3. Effect of hydralazine on VSMC survival. (**A–D**), VSMCs were incubated with acrolein (4 µg/mL) in the presence (+) or absence (−) of hydralazine (100 µM) for 24 h. Representative cell images of control cells (**A**) and cells incubated with acrolein in the absence (**B**) or presence of hydralazine (**C**). Scale bar = 50 µm. (**D**) The survival of cells, assessed by the MTS assay; $n = 8$. (**E–H**), VSMCs were incubated with hydrogen peroxide (H_2O_2) in the presence (+) or absence (−) of hydralazine (100 µM) for 24 h. Representative cell images of control cells (**E**) and cells treated with H_2O_2 (800 µM) in the absence (**F**) or presence of hydralazine (**G**). (**H**) The cell viability, assessed by the trypan blue assay; $n = 4$. Error bars = standard error. Data were analyzed using a Kruskal–Wallis one-way ANOVA. * $p < 0.05$. VSMCs, vascular smooth muscle cells.

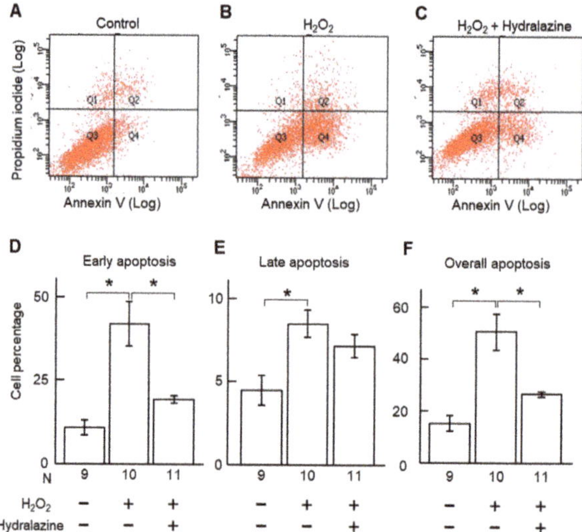

Figure 4. Effect of hydralazine on VSMC apoptosis induced by H_2O_2. (**A–C**), Flow cytometry graphs of control VSMC cells (**A**), and cells treated with H_2O_2 (100 µM, **B**) or co-treated with H_2O_2 (100 µM) and hydralazine (100 µM, **C**). Cells were harvested and then stained with propidium iodide and annexin V. (**D–F**), Percentage of apoptotic cells. Cells in the early apoptotic stage (**A**) were defined as annexin V-high and propidium iodide-low. Cells in the late apoptotic stage (**E**) were defined as annexin V-high and propidium iodide-high. Overall apoptotic cells (**F**) were defined as annexin V-high. Error bars = standard error. Data were analyzed using a Kruskal–Wallis one-way ANOVA. * $p < 0.05$. H_2O_2, hydrogen peroxide; VSMCs, vascular smooth muscle cells.

2.3. Hydralazine Treatment Protected Apolipoprotein E-Deficient (ApoE$^{-/-}$) Mice against AAA

Angiotensin II infusion increased blood pressure in ApoE$^{-/-}$ mice, and hydralazine treatment attenuated this increase (Figure 5A,B). Angiotensin II or hydralazine did not affect heart rate (Figure 5C).

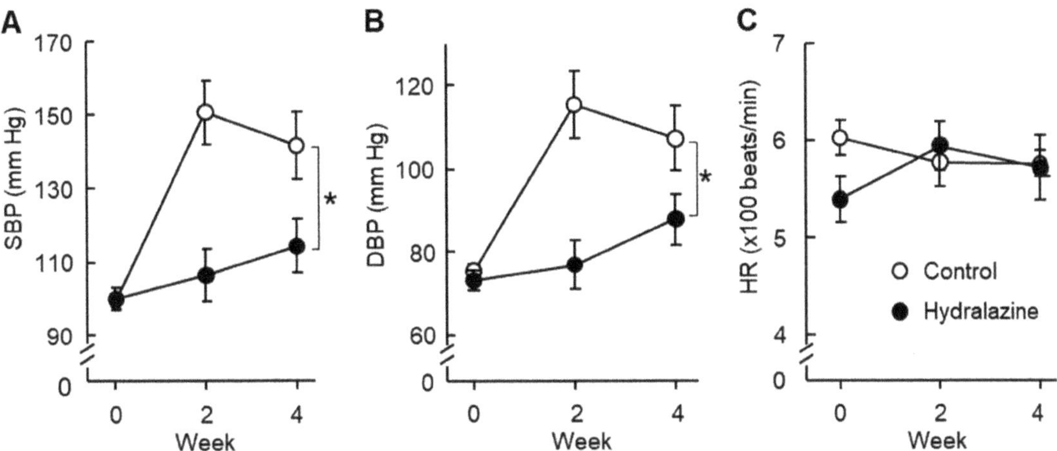

Figure 5. Effect of hydralazine treatment on blood pressure and heart rate. The basal (Week 0) systolic blood pressure (SBP, (**A**)), diastolic blood pressure (DBP, (**B**)), and heart rate (HR, (**C**)) of the mice were measured by the tail-cuff method. Then, the mice in the hydralazine-treated group were treated with hydralazine (24 mg/kg/day, filled circles), and the mice in the control group received vehicle (water, open circles) until the end of the experiment. 3 days after the initiation of the hydralazine treatment, angiotensin II (1 μg/kg body weight/min) was subcutaneously administered to all the mice for 28 days. Blood pressure and heart rate were measured again at 2 and 4 weeks after angiotensin II administration. Error bars = standard error. * $p < 0.05$. Hydralazine group, $n = 10$. Control groups: Week 0, $n = 10$; week 2, $n = 6$; and week 4, $n = 6$.

Angiotensin II infusion induced AAA formation in ApoE$^{-/-}$ mice (Figures 6A,B and S2). Hydralazine treatment decreased the incidence of AAA from 80% to 20% ($p < 0.05$, Figure 6C) and suprarenal aortic diameter by 32%, from 2.26 mm to 1.53 mm ($p < 0.05$, Figure 6D). When the four animals that died of aortic rupture were excluded, similar results were obtained (Figures 6E and S3). Hydralazine treatment also decreased aortic arch diameter, thoracic aorta diameter, and mean maximum aortic diameter ($p < 0.05$, Figures S3 and 6F). In addition, hydralazine treatment protected the mice against aortic rupture and enhanced the survival rate from 60% to 100% ($p < 0.05$, Figure 6G). Consistent with the in vitro results, in vivo data showed that hydralazine treatment decreased apoptosis in mouse suprarenal aortas (Figure S4).

2.4. Hydralazine Treatment Protected Mice against Atherosclerosis and Cardiac Hypertrophy

As this AAA model is also a model of atherosclerosis [17] and cardiac hypertrophy [18,19], we also investigated whether hydralazine affected those two conditions in this animal model. The results showed that hydralazine treatment significantly reduced atherosclerosis at the aortic arch and right common carotid artery (Figure 7), heart weight, and cardiomyocyte width (Figure S5).

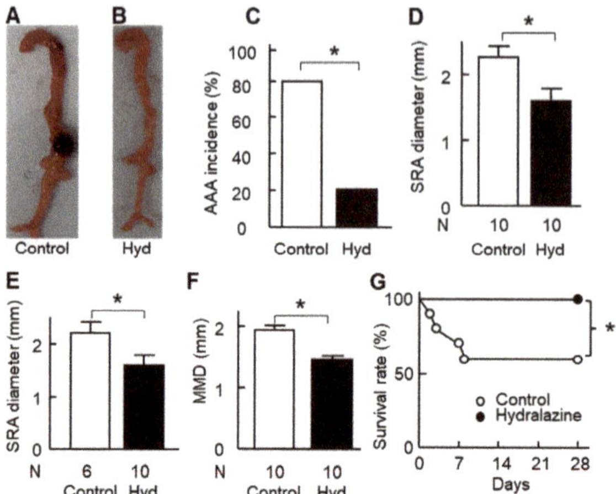

Figure 6. Effect of hydralazine treatment on abdominal aortic aneurysm (AAA). Two groups of mice were used: the control group and the hydralazine-treated group (Hyd). Mice in the latter group were treated with hydralazine (24 mg/kg/day) and mice in the former group were treated with vehicle (water) until the end of the experiment. 3 days after the initiation of the hydralazine treatment, angiotensin II (1 µg/kg body weight/min) was subcutaneously administered to all the mice for 28 days to induce AAA. (**A**,**B**), Representative images of aortas of mice in the absence (**A**) or presence of hydralazine (**B**). (**C**), AAA incidence of the mice. (**D**,**E**), Diameter of the suprarenal aorta (SRA) of the mice. (**D**) panel included all the mice (surviving mice plus dead mice due to aortic rupture) and the (**E**) panel included surviving mice only. (**F**), Mean maximum aortic diameter (MMD) of the mice which was the mean of the maximum diameter of the following four aortic segments: aortic arch, thoracic aorta, suprarenal aorta, and infrarenal aorta. (**G**), Kaplan–Meier survival curve. Error bars = standard error. * $p < 0.05$.

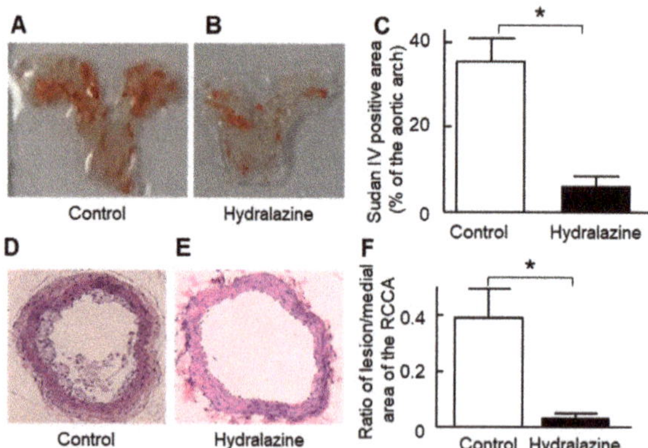

Figure 7. Effect of hydralazine treatment on atherosclerosis. Mice, treated with or without hydralazine (24 mg/kg/day), were sacrificed at the end of the 28-day subcutaneous infusion of angiotensin II (1 µg/kg body/min) for atherosclerosis assessment. (**A**,**B**) Representative images of Sudan IV staining of the aortic arch in the control (**A**) and hydralazine-treated mice (**B**). (**C**) The ratio of the Sudan IV positive area over the entire arch surface area. (**D**,**E**) The representative image of H&E staining of the right common carotid artery (RCCA) in the control (**D**) and hydralazine-treated mice (**E**). (**F**), The ratio of the lesion area over the medial area. $n = 6$ in the control group and $n = 10$ in the hydralazine group. Error bars = standard error. Image magnification = 10×. Data were analyzed using the Mann–Whitney U test. * $p < 0.05$. H&E: hematoxylin and eosin.

3. Discussion

This study found that incubation of VSMCs with hydralazine in vitro inhibited the expression of pro-inflammatory genes and reduced apoptosis. Hydralazine also inhibited AAA formation and rupture in ApoE$^{-/-}$ mice. To our knowledge, this is the first report suggesting that hydralazine decreases AAA formation. Our results are consistent with previous reports on cerebral aneurysms. In one study, hydralazine administration decreased the incidence of advanced-stage cerebral aneurysms from 50% to 9% in a rat model [22]. In another study, hydralazine protected against cerebral aneurysmal rupture in a mouse model [23].

In contrast, our results are different from those reported by Cassis and colleagues [24], who found that hydralazine did not affect angiotensin II-induced AAA formation in ApoE$^{-/-}$ mice. The reason for this disparity is unclear, but may relate to differences in study designs, reporting, and methods [24]. In keeping with many previous reports, approximately 40% of mice had an aortic rupture in the current study [25,26]. In the investigation by Cassis and colleagues, aortic rupture was not reported, and it is possible such mice were excluded. Drinking water containing hydralazine was freshly prepared three times a week in the current study, and the preparation of hydralazine-containing drinking water was not reported in the prior study [24]. Personal communication with the authors confirmed that they had freshly prepared hydralazine every three days, which was slightly less frequent than in the current study. Whether this slight difference in hydralazine preparation contributed to the discrepancy observed is unknown. The systolic blood pressure of the hydralazine-treated mice in the previous study was 135 mm Hg [24], compared to 114 mm Hg in the current study, suggesting that hydralazine was having a more powerful effect in our investigation, where there was more frequent replenishment of hydralazine in drinking water. This suggests that differences in hydralazine preparation might have contributed to the disparity in findings.

Aortic inflammation and VSMC apoptosis are key features of AAA pathology [2]. Inhibition of inflammation, e.g., by blocking IL-6 [27] and C-X-C motif chemokine receptor 2 (CXCR2) [28], inhibits AAA formation in mouse models. In addition, interventions that inhibit VSMC apoptosis, such as 2-hydroxypropyl-β-cyclodextrin [29] and mitochondrial fission inhibitors [30], have been shown to attenuate AAA development in mouse models. Our results showed that hydralazine inhibited expression of some pro-inflammatory genes and apoptosis of VSMCs, suggesting that the anti-AAA effect of hydralazine may be mediated, at least in part, by its anti-inflammatory and anti-apoptotic properties.

Whether lowering blood pressure contributed to the anti-AAA effect of hydralazine is unknown. It has been shown that hypertension is an important risk factor for AAA in humans [31]. Increasing blood pressure alone does not result in AAA formation in animal studies [24], and antihypertensive drugs such as beta-blockers do not inhibit AAA progression in humans [32]. Therefore, lowering blood pressure may not be the key mechanism underlying the anti-AAA effect of hydralazine reported in this study.

Hydralazine did not increase the heart rate of mice with hypertension induced by angiotensin II in the present study. This is consistent with previous reports that hydralazine did not affect heart rate in mice with hypertension induced by angiotensin II [33] or in genetically hypertensive mice [34]. However, it has been reported that hydralazine increases heart rate in normotensive mice [35,36]. The reason underlying this inconsistency is unclear. One might speculate that hydralazine decreases the blood pressure of normotensive mice to a sub-normal range, and this may compromise blood supply to vital organs. Consequently, mice may increase their heart rate to increase cardiac output and maintain sufficient blood supply to vital organs. However, in hypertensive mice, hydralazine may decrease blood pressure to a normal range; in this case, blood supply may not be compromised, and thus the mice may not respond with an increase in heart rate.

We also observed that hydralazine has an anti-atherosclerotic effect in ApoE$^{-/-}$ mice. Our results are consistent with some previous reports. Hydralazine has been shown to decrease atherosclerosis induced by deoxycorticosterone acetate salt and high-fat diet administration [37]. In addition, hydralazine decreased atherosclerosis in ApoE$^{-/-}$ mice fed a normal diet [38]. The anti-atherosclerotic effect of hydralazine may be due to its blood pressure-lowering and anti-inflammatory properties, as hypertension and inflammation are key treatment targets for limiting atherosclerosis-associated cardiovascular events [39–41]. Nonetheless, some reports have suggested that hydralazine does not decrease atherosclerosis. For example, one study showed that hydralazine did not affect atherosclerosis in the aortic root of ApoE$^{-/-}$ mice fed a high-fat diet [42], although the authors of this study did not investigate whether hydralazine affects atherosclerosis in other vascular territories. Other investigators [35,43] have reported that hydralazine does not significantly reduce the severity of aortic atherosclerosis in comparison to controls.

Angiotensin II infusion also induces cardiac hypertrophy in mice [18,19]. The current study suggests that hydralazine protects mice against cardiac hypertrophy, as indicated by lighter heart weights and smaller widths of cardiomyocytes. Our results are consistent with the report from Vial and colleagues [44], who showed that hydralazine protected against cardiac hypertrophy in Wistar rats treated with deoxycorticosterone acetate salt. This protective effect of hydralazine may result from its blood pressure-lowering and anti-inflammatory properties. Hypertension is a major risk factor for cardiac hypertrophy [45], and lowering blood pressure with antihypertensive drugs reduces cardiac hypertrophy in humans [46]. It is known that cardiac inflammation is increased in cardiac hypertrophy, and inhibition of inflammation could decrease cardiomyocyte enlargement [47].

In conclusion, our study showed that hydralazine inhibited AAA formation in ApoE$^{-/-}$ mice infused with angiotensin II, which may be mediated by its anti-inflammatory and anti-apoptotic effects. A limitation of this study is its sample size. Whether hydralazine inhibits AAA in other animal models or in humans needs to be investigated in the future. In addition, in this angiotensin II infusion mouse model, hydralazine treatment protected the mice from developing atherosclerosis and cardiac hypertrophy. The underlying mechanisms by which hydralazine exerts its effects need to be further investigated in the future.

4. Methods

4.1. Animals

Twenty male ApoE$^{-/-}$ mice were purchased from the Animal Resources Centre, Perth, Australia. All experiments were conducted in a temperature-controlled animal house (21 ± 1 °C) under a 12:12 h light–dark cycle, and mice were given a normal diet and water ad libitum.

4.2. Experimental Protocol

The mice (14 weeks old) were randomized into a control group and a hydralazine-treated group ($n = 10$ per group). The mice in the former group received plain drinking water, whereas those in the latter group received hydralazine via drinking water (24 mg/kg/day) throughout the experiment [48]. The hydralazine-containing drinking water was freshly prepared three times a week (Mondays, Wednesdays, and Fridays). Three days after the initiation of the hydralazine treatment, angiotensin II was administered subcutaneously to all the mice via a micro-osmotic pump (Model 2004, ALZET, DURECT Corporation, Cupertino, CA, USA) at a rate of 1 μg/kg body weight/min for 28 days to induce AAA [26]. Blood pressure and heart rate were measured in conscious mice at 0 weeks (before hydralazine treatment), 2 weeks, and 4 weeks after angiotensin II infusion using the tail-cuff method [49].

4.3. Morphometry Measurement of Aortic Diameter

Morphometry measurements were performed on aortas harvested from the surviving mice following euthanasia at the end of the experiment or at the autopsy of the mice who died of aortic rupture during the experiment, as previously described [26]. In brief, the harvested aortas were digitally photographed together with a ruler. Maximum external diameters of the aortic arch, thoracic aorta, suprarenal aorta, and infrarenal aorta were determined from the images using computer-aided analysis (Adobe Photoshop, version 24.6, Adobe Systems Incorporated, San Jose, CA, USA). The mean maximum aortic diameter was calculated as the mean of the maximum diameter of the aortic arch, thoracic aorta, suprarenal aorta, and infrarenal aorta. This morphometrical method has been shown to be highly reproducible [26].

4.4. Quantification of the Atherosclerotic Lesion Area

Atherosclerosis in the aortic arch was quantified by en face Sudan IV staining of the aortic arch as described previously [17] and by morphometry analysis of the right common carotid artery after hematoxylin and eosin (H&E) staining as described previously [17].

4.5. Cell Width of Cardiomyocytes

Five hearts from control mice and four hearts from hydralazine-treated mice were randomly selected. The left ventricles of the mice were formalin-fixed and paraffin-embedded before sectioning (5 μm thick). The sections were stained with H&E, and images were captured using a Nikon Eclipse 80i microscope (Nikon, Tokyo, Japan) with a CCD (charge-coupled device) camera at ×20 magnification. Twenty cardiomyocytes were randomly selected from each image, and their widths were measured 10 times using Adobe Photoshop. The mean width of cardiomyocytes was then calculated.

4.6. Cell Culture

VSMCs were isolated from the mouse aorta [50] and cultured as previously described [17]. In brief, VSMCs were cultured at 37 °C in Dulbecco's modified Eagle medium (DMEM) supplemented with 10% *v/v* fetal bovine serum, 100 U/mL penicillin, and 100 μg/mL streptomycin in a standard cell culture incubator.

4.7. MTS Assay

VSMCs were cultured in 96-well plates (2×10^4 cells/well) for 24 h before the addition of acrolein (4 μg/mL) with or without hydralazine (100 μM) [51,52] for another 24 h. Cell numbers were then assessed using an MTS assay kit (Abcam, Cambridge, UK) as previously described [50].

4.8. Trypan Blue Assay

The trypan blue assay was conducted as previously described [53]. In brief, 2 mL of VSMCs (5×10^5 cells/mL) were placed in wells of 6-well plates and incubated in an incubator at 37 °C for 24 h. Then, the cells were incubated with H_2O_2 (800 μM) with or without hydralazine (100 μM). After another 24 h, the cells were trypsinized, stained with trypan blue, and the viability of the cells was determined using the Countess Automated Cell Counter (Invitrogen, Waltham, MA, USA).

4.9. Gene Expression Analysis

Effect of hydralazine on inflammatory and apoptotic gene expression in VSMCs treated with acrolein: VSMCs (1×10^6 cells per well) were cultured in 3 wells of 6-well plates; Well 1 served as a control; Well 2 was incubated with acrolein alone (2 μg/mL); and Well 3 was co-incubated with acrolein (2 μg/mL) and hydralazine (100 μM). RNA was then extracted using the TRI reagent (Merck).

Effect of hydralazine on inflammatory and apoptotic gene expression in VSMCs treated with H_2O_2: the cells were treated as described in the previous paragraph, except that the acrolein was replaced with H_2O_2 (800 μM).

The extracted RNA was reverse transcribed to cDNA using the High-Capacity Reverse Transcription Kit (Life Technologies, Carlsbad, CA, USA). Gene expression was assessed by quantitative PCR using SYBR reagents (Bioline Global Pty Ltd, Gregory Hills, Australia). Primer sets were outlined in Table S1. The cycling conditions were 40 cycles of 95 °C for 15 s, 58 °C for 20 s, and 72 °C for 20 s. Relative gene expression was assessed using the $2^{-\Delta\Delta Ct}$ method [54]. Eukaryotic translation elongation factor 2 (EEF2) served as a housekeeping gene for normalizing gene expression [55].

4.10. Enzyme-Linked Immunosorbent Assay (ELISA)

VSMCs (5×10^5 cells/well) were placed in wells of 6-well plates and incubated in an incubator at 37 °C. After 24 h, the DMEM supplemented with 10% fetal bovine serum in the wells was replaced with 2 mL of fresh DMEM supplemented with 1% fetal bovine serum. The cells were then incubated with H_2O_2 (200 µM) with or without hydralazine (100 µM). After another 24 h, the culture medium was collected, and interleukin 6 protein levels in the culture medium were quantified using an enzyme-linked immunosorbent assay kit (Sigma-Aldrich Pty Ltd, Bayswater, Australia) according to the manufacturer's instructions.

4.11. Flow Cytometry Assay

VSMCs (5×10^5 cells/well) were placed in wells of 6-well plates and were incubated with H_2O_2 (200 µM) with or without hydralazine (100 µM). After 24 h, the nonadherent cells in the culture medium were collected, and the adherent cells were trypsinized. Total cells (both adherent and nonadherent) were then used to determine apoptosis using BD Pharmingen reagents: Annexin V– BV421, Propidium Iodide (PI), and Annexin binding buffer, according to the manufacturer's instructions [56]. Briefly, cells were washed and resuspended in binding buffer at a concentration of 2×10^6 cells per ml, before being filtered and stained with annexin V and propidium iodide. Cell numbers were counted using the Becton-Dickson LSRFortessa X-20 flow cytometer (Franklin Lakes, NJ, USA) and analyzed using the BD FACSDiva software (v9.0). Apoptotic states were defined as previously described [57]: early apoptotic cells were defined as annexin V-high and propidium iodide-low; late apoptotic cells were defined as annexin V-high and propidium iodide-high; and overall apoptotic cells were defined as annexin V-high.

4.12. Terminal Deoxynucleotidyl Transferase Mediated dUTP Nick-End Labeling (TUNEL) Assay

Cell apoptosis was assessed by the TUNEL kit (Abcam) according to the manufacturer's instructions as previously described [58]. In brief, 5 mice from each group were randomly chosen, and 5 mm thick paraffin-embedded suprarenal aortic sections were dewaxed and treated with proteinase K. After endogenous peroxidase was blocked with 3% H_2O_2, apoptotic cells were labeled with TdT Enzyme and detected using a 3,30-diaminobenzidine (DAB) substrate. The sections were then counterstained with methyl green. Four high magnification fields ($\times$40) were randomly chosen from each sample, and TUNEL-positive cells were counted. The mean of the positive cells was calculated to represent cell apoptosis for each sample.

4.13. Statistical Analyses

The difference between two groups was analyzed using the Mann–Whitney U-test [59], and the difference among three groups was analyzed using a Kruskal–Wallis one-way ANOVA. The difference in blood pressure and heart rate between two groups (with or without hydralazine) was analyzed using multiple linear regression: dependent variable = blood pressure or heart rate, and independent variables = groups (with or without hydralazine) and time. Kaplan–Meier survival curves were analyzed using the log-rank (Mantel-Cox) test. The difference in AAA incidence was analyzed using Fisher's exact test [59,60]. All tests were two-sided, and a p value of <0.05 was regarded as statistically significant. All analyses were performed using SPSS version 27.0 (IBM SPSS Statistics for Windows, Armonk, NY, USA, IBM Corporation).

Supplementary Materials: The following supporting information can be downloaded at: https://www.mdpi.com/article/10.3390/ijms242115955/s1.

Author Contributions: Conceptualization, Y.W. and J.G.; investigation, Y.W., O.S., D.T.N., K.P. and S.J.R.P., A.A., L.T., Y.F., M.E.W. and S.P.B.; writing—original draft preparation Y.W., O.S., E.O. and D.J.M.; writing—review and editing, Y.W., O.S., D.T.N., K.P., S.J.R.P., A.A., L.T., Y.F., M.E.W., S.P.B., E.O., D.J.M. and J.G.; funding acquisition, Y.W. All authors have read and agreed to the published version of the manuscript.

Funding: This research was funded by the National Health and Medical Research Council of Australia, grant number 1062671. Owen Sargisson was supported by a Fee-Offset Scholarship through Federation University Australia.

Institutional Review Board Statement: The study was conducted in accordance with the Declaration of Helsinki and approved by the Institutional Ethics Committee of Federation University Australia (protocol code, 14-019; date of approval, 21 October 2014).

Informed Consent Statement: Not applicable.

Data Availability Statement: Not applicable.

Conflicts of Interest: The authors declare no conflict of interest.

References

1. Sampson, U.K.; Norman, P.E.; Fowkes, F.G.; Aboyans, V.; Yanna, S.; Harrell, F.E., Jr.; Forouzanfar, M.H.; Naghavi, M.; Denenberg, J.O.; McDermott, M.M.; et al. Global and regional burden of aortic dissection and aneurysms: Mortality trends in 21 world regions, 1990 to 2010. *Global Heart* **2014**, *9*, 171–180.e110. [CrossRef]
2. Golledge, J.; Krishna, S.M.; Wang, Y. Mouse models for abdominal aortic aneurysm. *Br. J. Pharmacol.* **2022**, *179*, 792–810. [CrossRef] [PubMed]
3. Landry, G.J.; Liem, T.K.; Abraham, C.Z.; Jung, E.; Moneta, G.L. Predictors of perioperative morbidity and mortality in open abdominal aortic aneurysm repair. *Am. J. Surg.* **2019**, *217*, 943–947. [CrossRef]
4. Stather, P.W.; Sidloff, D.; Dattani, N.; Choke, E.; Bown, M.J.; Sayers, R.D. Systematic review and meta-analysis of the early and late outcomes of open and endovascular repair of abdominal aortic aneurysm. *Br. J. Surg.* **2013**, *100*, 863–872. [CrossRef]
5. Brewster, D.C.; Cronenwett, J.L.; Hallett, J.W., Jr.; Johnston, K.W.; Krupski, W.C.; Matsumura, J.S. Guidelines for the treatment of abdominal aortic aneurysms. Report of a subcommittee of the Joint Council of the American Association for Vascular Surgery and Society for Vascular Surgery. *J. Vasc. Surg.* **2003**, *37*, 1106–1117. [CrossRef]
6. Su, Z.; Guo, J.; Gu, Y. Pharmacotherapy in Clinical Trials for Abdominal Aortic Aneurysms: A Systematic Review and Meta-Analysis. *Clin. Appl. Thromb. Hemost.* **2022**, *28*, 10760296221120423. [CrossRef] [PubMed]
7. Weaver, L.M.; Loftin, C.D.; Zhan, C.G. Development of pharmacotherapies for abdominal aortic aneurysms. *Biomed. Pharmacother.* **2022**, *153*, 113340. [CrossRef]
8. Kesavan, S.K.; Bhat, S.; Golegaonkar, S.B.; Jagadeeshaprasad, M.G.; Deshmukh, A.B.; Patil, H.S.; Bhosale, S.D.; Shaikh, M.L.; Thulasiram, H.V.; Boppana, R.; et al. Proteome wide reduction in AGE modification in streptozotocin induced diabetic mice by hydralazine mediated transglycation. *Sci. Rep.* **2013**, *3*, 2941. [CrossRef] [PubMed]
9. Ikeda, F.; Azuma, K.; Ogihara, T.; Toyofuku, Y.; Otsuka, A.; Mita, T.; Hirose, T.; Tanaka, Y.; Kawamori, R.; Watada, H. Angiotensin II type 1 receptor blocker reduces monocyte adhesion to endothelial cells in spontaneously hypertensive rats. *Endocr. J.* **2007**, *54*, 605–612. [CrossRef]
10. Quan, X.; Ma, T.; Guo, K.; Wang, H.; Yu, C.Y.; Qi, C.C.; Song, B.Q. Hydralazine Promotes Central Nervous System Recovery after Spinal Cord Injury by Suppressing Oxidative Stress and Inflammation through Macrophage Regulation. *Curr. Med. Sci.* **2023**, *43*, 749–758. [CrossRef]
11. Chiang, C.H.; Chen, C.; Fang, S.Y.; Lin, S.C.; Chen, J.W.; Chang, T.T. Xanthine oxidase/NADPH oxidase inhibition by hydralazine attenuates acute kidney injury and prevents the transition of acute kidney injury to chronic kidney disease. *Life Sci.* **2023**, *327*, 121863. [CrossRef] [PubMed]
12. Santos, D.M.D.; Da Silva, E.A.P.; Oliveira, J.Y.S.; Marinho, Y.Y.M.; Santana, I.R.; Heimfarth, L.; Pereira, E.W.M.; Júnior, L.J.Q.; Assreuy, J.; Menezes, I.A.C.; et al. The Therapeutic Value of Hydralazine in Reducing Inflammatory Response, Oxidative Stress, and Mortality in Animal Sepsis: Involvement of the PI3K/AKT Pathway. *Shock* **2021**, *56*, 782–792. [CrossRef]
13. Li, C.; Su, Z.; Ge, L.; Chen, Y.; Chen, X.; Li, Y. Cardioprotection of hydralazine against myocardial ischemia/reperfusion injury in rats. *Eur. J. Pharmacol.* **2020**, *869*, 172850. [CrossRef]
14. Jia, L.; Liu, Z.; Sun, L.; Miller, S.S.; Ames, B.N.; Cotman, C.W.; Liu, J. Acrolein, a toxicant in cigarette smoke, causes oxidative damage and mitochondrial dysfunction in RPE cells: Protection by (R)-alpha-lipoic acid. *Investig. Ophthalmol. Vis. Sci.* **2007**, *48*, 339–348. [CrossRef] [PubMed]
15. Hikisz, P.; Jacenik, D. The Tobacco Smoke Component, Acrolein, as a Major Culprit in Lung Diseases and Respiratory Cancers: Molecular Mechanisms of Acrolein Cytotoxic Activity. *Cells* **2023**, *12*, 879. [CrossRef]

16. Clark, J.B. Role of Hydrogen Peroxide in the Development of Abdominal Aortic Aneurysms. Master's Thesis, Emory University, Atlanta, GA, USA, 2011. Available online: https://etd.library.emory.edu/concern/etds/cn69m4331?locale=en (accessed on 16 March 2023).

17. Wang, Y.; Nguyen, D.T.; Anesi, J.; Alramahi, A.; Witting, P.K.; Chai, Z.; Khan, A.W.; Kelly, J.; Denton, K.M.; Golledge, J. Moxonidine Increases Uptake of Oxidised Low-Density Lipoprotein in Cultured Vascular Smooth Muscle Cells and Inhibits Atherosclerosis in Apolipoprotein E-Deficient Mice. *Int. J. Mol. Sci.* **2023**, *24*, 3857. [CrossRef]

18. Tsuruda, T.; Sekita-Hatakeyama, Y.; Hao, Y.; Sakamoto, S.; Kurogi, S.; Nakamura, M.; Udagawa, N.; Funamoto, T.; Sekimoto, T.; Hatakeyama, K.; et al. Angiotensin II Stimulation of Cardiac Hypertrophy and Functional Decompensation in Osteoprotegerin-Deficient Mice. *Hypertension* **2016**, *67*, 848–856. [CrossRef] [PubMed]

19. Matsumoto, E.; Sasaki, S.; Kinoshita, H.; Kito, T.; Ohta, H.; Konishi, M.; Kuwahara, K.; Nakao, K.; Itoh, N. Angiotensin II-induced cardiac hypertrophy and fibrosis are promoted in mice lacking Fgf16. *Genes Cells* **2013**, *18*, 544–553. [CrossRef]

20. Shao, D.; Gao, Z.; Zhao, Y.; Fan, M.; Zhao, X.; Wei, Q.; Pan, M.; Ma, B. Sulforaphane Suppresses H_2O_2-Induced Oxidative Stress and Apoptosis via the Activation of AMPK/NFE2L2 Signaling Pathway in Goat Mammary Epithelial Cells. *Int. J. Mol. Sci.* **2023**, *24*, 1070. [CrossRef]

21. Wang, H.T.; Chen, T.Y.; Weng, C.W.; Yang, C.H.; Tang, M.S. Acrolein preferentially damages nucleolus eliciting ribosomal stress and apoptosis in human cancer cells. *Oncotarget* **2016**, *7*, 80450–80464. [CrossRef]

22. Kimura, N.; Shimizu, H.; Eldawoody, H.; Nakayama, T.; Saito, A.; Tominaga, T.; Takahashi, A. Effect of olmesartan and pravastatin on experimental cerebral aneurysms in rats. *Brain Res.* **2010**, *1322*, 144–152. [CrossRef]

23. Tada, Y.; Wada, K.; Shimada, K.; Makino, H.; Liang, E.I.; Murakami, S.; Kudo, M.; Kitazato, K.T.; Nagahiro, S.; Hashimoto, T. Roles of Hypertension in the Rupture of Intracranial Aneurysms. *Stroke* **2014**, *45*, 579–586. [CrossRef]

24. Cassis, L.A.; Gupte, M.; Thayer, S.; Zhang, X.; Charnigo, R.; Howatt, D.A.; Rateri, D.L.; Daugherty, A. ANG II infusion promotes abdominal aortic aneurysms independent of increased blood pressure in hypercholesterolemic mice. *Am. J. Physiol. Heart Circ. Physiol.* **2009**, *296*, H1660–H1665. [CrossRef] [PubMed]

25. Ghoshal, S.; Loftin, C.D. Cyclooxygenase-2 inhibition attenuates abdominal aortic aneurysm progression in hyperlipidemic mice. *PLoS ONE* **2012**, *7*, e44369. [CrossRef] [PubMed]

26. Krishna, S.M.; Li, J.; Wang, Y.; Moran, C.S.; Trollope, A.; Huynh, P.; Jose, R.; Biros, E.; Ma, J.; Golledge, J. Kallistatin limits abdominal aortic aneurysm by attenuating generation of reactive oxygen species and apoptosis. *Sci. Rep.* **2021**, *11*, 17451. [CrossRef] [PubMed]

27. Paige, E.; Clément, M.; Lareyre, F.; Sweeting, M.; Raffort, J.; Grenier, C.; Finigan, A.; Harrison, J.; Peters, J.E.; Sun, B.B.; et al. Interleukin-6 Receptor Signaling and Abdominal Aortic Aneurysm Growth Rates. *Circ. Genom. Precis. Med.* **2019**, *12*, e002413. [CrossRef] [PubMed]

28. Nie, H.; Wang, H.X.; Tian, C.; Ren, H.L.; Li, F.D.; Wang, C.Y.; Li, H.H.; Zheng, Y.H. Chemokine (C-X-C motif) receptor 2 blockade by SB265610 inhibited angiotensin II-induced abdominal aortic aneurysm in Apo E(-/-) mice. *Heart Vessel.* **2019**, *34*, 875–882. [CrossRef] [PubMed]

29. Lu, H.; Sun, J.; Liang, W.; Chang, Z.; Rom, O.; Zhao, Y.; Zhao, G.; Xiong, W.; Wang, H.; Zhu, T.; et al. Cyclodextrin Prevents Abdominal Aortic Aneurysm via Activation of Vascular Smooth Muscle Cell Transcription Factor EB. *Circulation* **2020**, *142*, 483–498. [CrossRef]

30. Cooper, H.A.; Cicalese, S.; Preston, K.J.; Kawai, T.; Okuno, K.; Choi, E.T.; Kasahara, S.; Uchida, H.A.; Otaka, N.; Scalia, R.; et al. Targeting mitochondrial fission as a potential therapeutic for abdominal aortic aneurysm. *Cardiovasc. Res.* **2021**, *117*, 971–982. [CrossRef] [PubMed]

31. Kobeissi, E.; Hibino, M.; Pan, H.; Aune, D. Blood pressure, hypertension and the risk of abdominal aortic aneurysms: A systematic review and meta-analysis of cohort studies. *Eur. J. Epidemiol.* **2019**, *34*, 547–555. [CrossRef] [PubMed]

32. Siordia, J.A. Beta-Blockers and Abdominal Aortic Aneurysm Growth: A Systematic Review and Meta-Analysis. *Curr. Cardiol. Rev.* **2021**, *17*, e230421187502. [CrossRef] [PubMed]

33. Inanaga, K.; Ichiki, T.; Matsuura, H.; Miyazaki, R.; Hashimoto, T.; Takeda, K.; Sunagawa, K. Resveratrol attenuates angiotensin II-induced interleukin-6 expression and perivascular fibrosis. *Hypertens. Res.* **2009**, *32*, 466–471. [CrossRef]

34. Kai, T.; Kino, H.; Ishikawa, K. Role of the renin-angiotensin system in cardiac hypertrophy and renal glomerular sclerosis in transgenic hypertensive mice carrying both human renin and angiotensinogen genes. *Hypertens. Res.* **1998**, *21*, 39–46. [CrossRef]

35. Custodis, F.; Baumhäkel, M.; Schlimmer, N.; List, F.; Gensch, C.; Böhm, M.; Laufs, U. Heart rate reduction by ivabradine reduces oxidative stress, improves endothelial function, and prevents atherosclerosis in apolipoprotein E-deficient mice. *Circulation* **2008**, *117*, 2377–2387. [CrossRef] [PubMed]

36. Kim, S.M.; Chen, L.; Mizel, D.; Huang, Y.G.; Briggs, J.P.; Schnermann, J. Low plasma renin and reduced renin secretory responses to acute stimuli in conscious COX-2-deficient mice. *Am. J. Physiol. Renal Physiol.* **2007**, *292*, F415–F422. [CrossRef] [PubMed]

37. Weiss, D.; Taylor, W.R. Deoxycorticosterone acetate salt hypertension in apolipoprotein E-/- mice results in accelerated atherosclerosis: The role of angiotensin II. *Hypertension* **2008**, *51*, 218–224. [CrossRef]

38. Galvani, S.; Coatrieux, C.; Elbaz, M.; Grazide, M.H.; Thiers, J.C.; Parini, A.; Uchida, K.; Kamar, N.; Rostaing, L.; Baltas, M.; et al. Carbonyl scavenger and antiatherogenic effects of hydrazine derivatives. *Free Radic. Biol. Med.* **2008**, *45*, 1457–1467. [CrossRef]

39. Alexander, R.W. Hypertension and the Pathogenesis of Atherosclerosis. *Hypertension* **1995**, *25*, 155–161. [CrossRef]

40. Kosmas, C.E.; Silverio, D.; Sourlas, A.; Montan, P.D.; Guzman, E.; Garcia, M.J. Anti-inflammatory therapy for cardiovascular disease. *Ann. Transl. Med.* **2019**, *7*, 147. [CrossRef] [PubMed]
41. Arnett Donna, K.; Blumenthal Roger, S.; Albert Michelle, A.; Buroker Andrew, B.; Goldberger Zachary, D.; Hahn Ellen, J.; Himmelfarb Cheryl, D.; Khera, A.; Lloyd-Jones, D.; McEvoy, J.W.; et al. 2019 ACC/AHA Guideline on the Primary Prevention of Cardiovascular Disease: Executive Summary. *J. Am. Coll. Cardiol.* **2019**, *74*, 1376–1414. [CrossRef] [PubMed]
42. Wu, H.; Cheng, X.W.; Hu, L.; Hao, C.N.; Hayashi, M.; Takeshita, K.; Hamrah, M.S.; Shi, G.P.; Kuzuya, M.; Murohara, T. Renin inhibition reduces atherosclerotic plaque neovessel formation and regresses advanced atherosclerotic plaques. *Atherosclerosis* **2014**, *237*, 739–747. [CrossRef]
43. Noda, K.; Hosoya, M.; Nakajima, S.; Ohashi, J.; Fukumoto, Y.; Shimokawa, H. Anti-atherogenic effects of the combination therapy with olmesartan and azelnidipine in diabetic apolipoprotein E-deficient mice. *Tohoku J. Exp. Med.* **2012**, *228*, 305–315. [CrossRef]
44. Vial, J.H.; Yong, A.C.; Boyd, G.W. Structural change in the rat hindlimb during deoxycorticosterone acetate hypertension; its reversibility and prevention. *J. Hypertens.* **1989**, *7*, 143–150. [CrossRef]
45. Aronow, W.S. Hypertension and left ventricular hypertrophy. *Ann. Transl. Med.* **2017**, *5*, 310. [CrossRef]
46. Dahlöf, B.; Pennert, K.; Hansson, L. Reversal of left ventricular hypertrophy in hypertensive patients. A metaanalysis of 109 treatment studies. *Am. J. Hypertens.* **1992**, *5*, 95–110. [CrossRef] [PubMed]
47. Smeets, P.J.; Teunissen, B.E.; Planavila, A.; de Vogel-van den Bosch, H.; Willemsen, P.H.; van der Vusse, G.J.; van Bilsen, M. Inflammatory pathways are activated during cardiomyocyte hypertrophy and attenuated by peroxisome proliferator-activated receptors PPARalpha and PPARdelta. *J. Biol. Chem.* **2008**, *283*, 29109–29118. [CrossRef] [PubMed]
48. Loufrani, L.; Henrion, D. Vasodilator treatment with hydralazine increases blood flow in mdx mice resistance arteries without vascular wall remodelling or endothelium function improvement. *J. Hypertens.* **2005**, *23*, 1855–1860. [CrossRef] [PubMed]
49. Wang, Y.; Krishna, S.M.; Moxon, J.; Dinh, T.N.; Jose, R.J.; Yu, H.; Golledge, J. Influence of apolipoprotein E, age and aortic site on calcium phosphate induced abdominal aortic aneurysm in mice. *Atherosclerosis* **2014**, *235*, 204–212. [CrossRef] [PubMed]
50. Wang, Y.; Nguyen, D.T.; Yang, G.; Anesi, J.; Chai, Z.; Charchar, F.; Golledge, J. An Improved 3-(4,5-Dimethylthiazol-2-yl)-5-(3-Carboxymethoxyphenyl)-2-(4-Sulfophenyl)-2H-Tetrazolium Proliferation Assay to Overcome the Interference of Hydralazine. *Assay. Drug Dev. Technol.* **2020**, *18*, 379–384. [CrossRef]
51. Haefliger, I.; Pedrini, M.; Anderson, D.R. Relaxing effect of CEDO 8956 and hydralazine HCl in cultured smooth muscle cells versus pericytes: A preliminary study. *Klin. Monbl. Augenheilkd.* **2002**, *219*, 277–280. [CrossRef]
52. Burcham, P.C.; Raso, A.; Kaminskas, L.M. Chaperone heat shock protein 90 mobilization and hydralazine cytoprotection against acrolein-induced carbonyl stress. *Mol. Pharmacol.* **2012**, *82*, 876–886. [CrossRef]
53. Wang, Y.; Nguyen, D.T.; Yang, G.; Anesi, J.; Kelly, J.; Chai, Z.; Ahmady, F.; Charchar, F.; Golledge, J. A Modified MTS Proliferation Assay for Suspended Cells to Avoid the Interference by Hydralazine and β-Mercaptoethanol. *Assay. Drug Dev. Technol.* **2021**, *19*, 184–190. [CrossRef]
54. Livak, K.J.; Schmittgen, T.D. Analysis of relative gene expression data using real-time quantitative PCR and the 2− ΔΔCT method. *Methods* **2001**, *25*, 402–408. [CrossRef] [PubMed]
55. Wang, Y.; Dinh, T.N.; Nield, A.; Krishna, S.M.; Denton, K.; Golledge, J. Renal Denervation Promotes Atherosclerosis in Hypertensive Apolipoprotein E-Deficient Mice Infused with Angiotensin II. *Front. Physiol.* **2017**, *8*, 215. [CrossRef] [PubMed]
56. Lakshmanan, I.; Batra, S.K. Protocol for Apoptosis Assay by Flow Cytometry Using Annexin V Staining Method. *Bio-Protocol* **2013**, *3*, e374. [CrossRef]
57. Kuystermans, D.; Avesh, M.; Al-Rubeai, M. Online flow cytometry for monitoring apoptosis in mammalian cell cultures as an application for process analytical technology. *Cytotechnology* **2016**, *68*, 399–408. [CrossRef]
58. Liu, Y.; Nadeem, A.; Sebastian, S.; Olsson, M.A.; Wai, S.N.; Styring, E.; Engellau, J.; Isaksson, H.; Tägil, M.; Lidgren, L.; et al. Bone mineral: A trojan horse for bone cancers. Efficient mitochondria targeted delivery and tumor eradication with nano hydroxyapatite containing doxorubicin. *Mater. Today Bio* **2022**, *14*, 100227. [CrossRef] [PubMed]
59. Wang, Y. Definition, prevalence, and risk factors of low sex hormone-binding globulin in US adults. *J. Clin. Endocrinol. Metab.* **2021**, *106*, e3946–e3956. [CrossRef]
60. Qian, T.; Sun, H.; Xu, Q.; Hou, X.; Hu, W.; Zhang, G.; Drummond, G.R.; Sobey, C.G.; Charchar, F.J.; Golledge, J.; et al. Hyperuricemia is independently associated with hypertension in men under 60 years in a general Chinese population. *J. Hum. Hypertens.* **2021**, *35*, 1020–1028. [CrossRef] [PubMed]

MDPI
St. Alban-Anlage 66
4052 Basel
Switzerland
www.mdpi.com

International Journal of Molecular Sciences Editorial Office
E-mail: ijms@mdpi.com
www.mdpi.com/journal/ijms

www.ingramcontent.com/pod-product-compliance
Lightning Source LLC
LaVergne TN
LVHW071529170726
843515LV00008B/2093